A History of the Radiological Sciences

▶ ▶ ▶ ▶ ▶ ▶ ▶ ▶ ▶ ▶ ▶ ▶ ▶ ▶ ▶

Radiation Oncology

Executive Editor
Raymond A. Gagliardi, M.D.

Editor
J. Frank Wilson, M.D.

Series Advisor
Nancy Knight

Copyright 1996 Radiology Centennial, Inc.
1891 Preston White Drive, Reston, VA 20191.
No part of this book may be reproduced without the written consent of the publisher.
Printed in the U.S.A.

CONTENTS

I.   1   **The Early Years of Radiation Therapy**
Nancy Knight
J. Frank Wilson, M.D.

II.   21   **Clinical Practice**
James D. Cox, M.D.

III.   43   **Developments in Technology**
Martin S. Weinhous, Ph.D.
Luther W. Brady, M.D.

IV.   89   **Medical Physics and Radiation Oncology**
R. J. Shalek, Ph.D., J.D.

V.   129   **Clinical and Radiobiological Research**
Eric J. Hall, D.Sc.

VI.   165   **Training and Education**
Nancy Knight

VII.	185	**Brachytherapy**
		Basil S. Hilaris, M.D.
VIII.	201	**Intersociety, Government, and Economic Relations**
		Carl R. Bogardus, Jr., M.D.
IX.	231	**Women in Radiation Oncology and Radiation Physics**
		Kate Knepper
		Sarah S. Donaldson, M.D.
X.	263	**African American Radiation Oncologists**
		Carl M. Mansfield, M.D., Sc.D.
XI.	277	**Radiomedical Fraud and Popular Perceptions of Radiation**
		Roger M. Macklis, M.D.
XII.	293	**The Future of Radiation Oncology**
		Herman D. Suit, M.D., D. Phil.
	313	**Index**

Preface

When roentgenology emerged as a recognized medical discipline from the confusion following the discovery of the X ray, its first application was as a diagnostic tool. But, as history relates, only a very short time later the possibilities of radiation as an instrument of therapy were realized. For many years the general radiologist was expected to be an expert on both diagnosis and therapeutic applications, as well as possess a superior knowledge of physics. From our vantage point it is almost impossible to believe there was a time when the body of knowledge in the field was so limited that each practitioner was expected to command it all. Moreover the amount and quality of work accomplished by these early physicians and their staffs are remarkable.

As both technical and scientific knowledge grew by quantum leaps, the use of ionizing radiation was only in its fifth decade when the specialty of radiation therapy began to emerge, with the general radiologist remaining only in small communities. Gradually the larger specialty separated, and radiation oncology was established as a separate practice entity. Dr. J. Frank Wilson and his coauthors tell here the story of the evolution of a medical specialty, from its beginnings as an empirical leap of faith to its present status as the major player in the treatment of malignant disease.

Raymond A. Gagliardi, M.D.
Executive Editor

Foreword

J. Frank Wilson, M.D.

▶ ▶ ▶ ▶ ▶ ▶

At our distant remove, it is almost impossible to imagine the difficulties faced by those who first investigated the healing powers of Wilhelm Conrad Röntgen's new rays. But, even after a hundred years, it is possible to feel a measure of their wonder and to share their enthusiasm as they aimed the X ray at aches and pains, skin infections, and more deadly diseases. In fact, Röntgen's revelation immediately unleashed an insatiable thirst throughout society for medical benefits from further investigation of the powers of the new ray, a search which continues unabated today.

As we read about the transition from radiation experimenter to radiation therapist to radiation oncologist during the last one hundred years, what emerges is a profound respect for the extraordinary courage and dedication of those who were willing to strike out into the unknown. We see the attempts to fashion a coherent field of practice out of disparate techniques and methods, where results could not be compared and follow-ups were virtually unknown. We agonize as these early practitioners only slowly come to terms with the central paradox of ionizing radiation: its ability to harm as well as to heal. And we understand what drew these innovative men and women from different professions and backgrounds into this novel endeavor: the promise that this new technology could heal diseases previously considered hopeless. The personal commitment and fortitude demonstrated by our intrepid forebears in overcoming the obstacles and frustrations encountered in the path towards their goals were extraordinary. These noble examples should fortify our own efforts to prevail as a profession as we confront modern challenges and vicissitudes.

It has been a singular and rewarding honor to assemble this book of essays, accurately detailing the origins and developments of our specialty. Sage and neophyte alike will appreciate that much of the lore and legend of American radiation oncology, previously preserved mostly through verbal tradition, has also been successfully captured. Altogether, the volume stands without precedent and is a comprehensive source of the provenance upon which modern radiation oncology is based.

Our authors were given complete freedom in their areas of focus and in the way they approached their topics. They took to their task with a relish and sense of excitement, as is self-evident in each of their masterly contributions. Each author comes from a different viewpoint, yielding a balanced, if not always seamless, view of a hard field which has, itself, been shaped by a century of accretive and sometimes hard-won consensus. Doubtless, many worthy names and events have been omitted from this volume, a fact which calls for continued efforts at the collection and writing of the history of radiation oncology. A well organized repository of properly archived documentary sources will be essential. The history of radium and natural radition, touched on here by Dr. Hilaris and others, should be expanded in a separate volume to mark the centennial of its discovery.

In preparation for the writing of this volume the editors surveyed both primary and secondary literature from one hundred years of radiation ther-

apy. We found few women and minorities in these histories, in no small measure because of both overt and more subtle exclusionary mechanisms at work historically within organized medicine. We hope that through the chapters by Drs. Mansfield and Donaldson some of the "lost" names in the history of the field will emerge into the light of future documentation, and that these will prove of special interest to women and minorities entering the field more recently.

Special recognition and great thanks must go to Nancy Knight, whose guiding expertise, knowledge, literary skills and editorial assistance so generously facilitated this project and left an obvious mark on its end product. Without her participation its completion would have been much less certain. We also extend our sincere gratitude to Victoria Giannotti and Constance Marx for their assistance with research and the preparation of manuscripts, and to the publications staff of the American Roentgen Ray Society for their work in getting the final manuscript in press.

Finally, this book is dedicated to the guiding spirit of radiation oncology history, Juan A. del Regato. His own remarkable life in the field as researcher, clinician, organizer, teacher, author, and historian nonpareil serves as a beacon to us all. In introducing his book, *Radiological Oncologists,* he astutely opined, "A history of radiation oncology is a difficult undertaking, for radiotherapy is a richly braided fabric of interwoven threads—physics, radiobiology, tumor pathology, and clinical medicine. Each component had its own rhythm and progression, not always historically synchronous with the others. Moreover, time amalgamates the original with the erroneous and the irrelevant. In order to be a faithful and reliable courier of scientific excellence, the historian must endeavor to seek out the gold from a mixed bag of ordinary change. In addition, a history of our specialty implies an account of its rise within the ranks of organized medicine to a position of recognized authority and academic respectability." We hope and trust that he has found our efforts worthy.

On behalf of all who have contributed to this Centennial project, it is a genuine pleasure to present this unique historical chronicle of American radiation oncology to its intended audience.

CONTRIBUTORS

Carl R. Bogardus, Jr., M.D., Medical Director, Department of Radiation Oncology, Cancer Treatment Center of Oklahoma, Midwest City, Oklahoma

Luther T. Brady, M.D., Professor and Chair, Department of Radiation Oncology, Hahnemann University, Philadelphia, Pennsylvania

James D. Cox, M.D., Professor and Head, Department of Radiation Oncology, University of Texas M.D. Anderson Cancer Center, Houston, Texas

Sarah S. Donaldson, M.D., Professor, Department of Radiation Oncology, Stanford University, Stanford, California

Eric J. Hall, Ph.D., Director, Center for Radiological Research, Columbia University, New York, New York

Basil S. Hilaris, M.D., Professor and Chair, Department of Radiation Medicine, New York Medical College, New York, New York

Kate Knepper, graduate student, History Department, Stanford University, Stanford, California

Nancy Knight, Consultant, Center for the American History of Radiology, Reston, Virginia

Roger M. Macklis, M.D., Chair, Department of Radiation Oncology, Cleveland Clinic Foundation, Cleveland, Ohio

Carl M. Mansfield, M.D., Sc. D., Associate Director, Division of Cancer Treatment, National Institutes of Health, Bethesda, Maryland

Juan A. del Regato, M.D., Professor of Radiology, University of South Florida College of Medicine, Tampa, Florida

Robert J. Shalek, Ph.D., J.D., Professor of Biophysics Emeritus, University of Texas M.D. Anderson Cancer Center, Houston, Texas

Herman D. Suit, M.D., D. Phil., Chair, Department of Radiation Oncology, Harvard Medical School at Massachusetts General Hospital, Boston, Massachusetts

Martin S. Weinhous, Ph.D., Chief of Medical Physics, Cleveland Clinic Foundation, Cleveland, Ohio

J. Frank Wilson, M.D., Professor and Chair, Department of Radiation Oncology, Medical College of Wisconsin, Milwaukee, Wisconsin

Early radiation therapy was empirical, unprotected, and offered amazing results, as in this 1899 treatment for epithelioma of the nose. (Courtesy of the Center for the American History of Radiology, Reston, Va.)

CHAPTER ONE

THE EARLY YEARS OF RADIATION THERAPY

Nancy Knight and J. Frank Wilson, M.D.

▶ ▶ ▶ ▶ ▶ ▶ ▶ ▶ ▶ ▶ ▶ ▶ ▶ ▶ ▶

On Friday, 8 November 1895, a German physicist and university rector worked late in his laboratory.[1] Like many busy academics he was forced to fit his research into time not already allotted to administrative and teaching tasks. His own interest lay in investigating the phenomena of fluorecent emanations from electrically charged sealed glass tubes. In his darkened workroom he shielded a Hittorf-Crookes tube with cardboard in preparation for a series of experiments. While testing the soundness of the shielding he noticed, to his surprise, the glow of a barium platinocyanide-coated screen on a bench about a meter away. He turned his electrical source on and off; the screen glowed on and off as well. This could not, he knew, be the cathode rays familiar to physicists, as they could not have an effect at such a distance. He investigated further.

At some point on that day or the ones that immediately followed, he moved his hand near the glowing screen.[2] He was startled at what appeared to be the shadows of moving bones. After weeks of careful experimentation and the reassuring verification of a photograph of the living bones of his wife's hand, Wilhelm Conrad Röntgen made his announcement. He had discovered "Eine neue art von strahlen," a new kind of ray (Figs. 1.1 and 1.2).[3] From a small physics laboratory in Würzburg came a discovery that would revolutionize medicine. The X ray would change forever the physician's orientation to the interior of the living body. It would introduce into medicine the first costly, ever-changing, and absolutely necessary machine. And the X ray would touch the lives of millions of sufferers in ways that Röntgen could not have imagined in 1895.

X-RAY MANIA

Within weeks news of the magic ray that could see through living human flesh had spread to the United States. "Hidden Solids Revealed!" proclaimed the *New York Times* science writer on 16 January 1896.[4] The detailed explanation of the simple apparatus needed to generate X rays led hundreds of electricians, photographers, physicists, and physicians in setting up their own experimental "Roentgen outfits." Anything that could be photographed with a regular

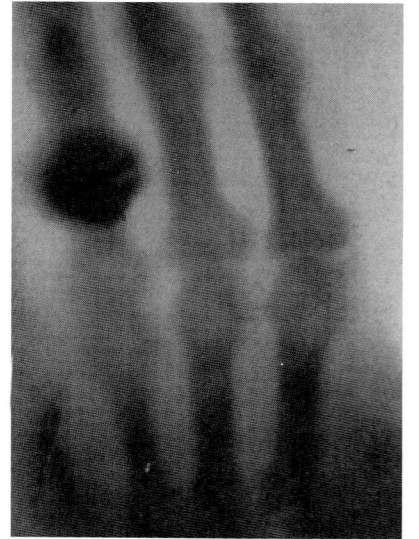

camera was soon the subject of radiography. "Startling Results" announced in the press by Yale physics professors accompanied radiographs of nutmeats inside uncracked walnuts.[5] Radiographs of feet in boots and ringed hands adorned popular and technical journals. Poems, cartoons, patent medicines, and songs employed the X ray as a metaphor for piercing vision and discernment.[6]

Much has been written about the American public's early fascination with X rays and later radium.[7] X-ray proof lead underwear was marketed, bone portrait studios opened, and Edison hoped to see an X-ray unit (of his manufacture, of course) in every home (Fig. 1.3). Underlying the many humorous and bizarre examples from a public smitten with the new technology was the notion of infinite possibility inherent in the "new light." The United States in the 1890s seemed a place of limitless opportunity. New inventions crowded the marketplace, industry promised to revolutionize the quality and delivery of goods, and great prosperity seemed just within the reach of anyone with a novel idea. A question formed in the minds of many of the professionals and tradespersons experimenting with X rays: If this magic ray could see through solid objects, what else might it do?

It was exactly this curiosity and the entrepreneurial spirit of the time that gave some medical writers reservations about Röntgen's discovery. In February 1896 the editors of the *Journal of the American Medical Association (J.A.M.A.)* cautioned:

> There will doubtless be an extensive advertisement of cathode ray baths, X-ray treatments, etc., but it is to be hoped that any active exploitations of these will, until the matter is more elucidated by accurate scientific researches, be confined to the irregulars who have no standing in the regular medical profession.[8]

MANY PRACTITIONERS

Readers of *J.A.M.A.* understood who were the "regulars" and the "irregulars" in the American medical landscape of the

Fig. 1.1 Wilhelm Conrad Röntgen (1845–1923) (Courtesy of the Center for the American History of Radiology, Reston, Va.)

Fig. 1.2 The radiograph of Bertha Röntgen's hand was widely reproduced in Europe and the United States. (Courtesy of the Center for the American History of Radiology, Reston, Va.)

Fig. 1.3 So great was the fascination with the new discovery that special studios opened where persons with and without physical problems could sit for their "roentgen portraits." (Courtesy of the Center for the American History of Radiology, Reston, Va.)

Fig. 1.4 Glass and mica plate static generators in use for over a decade by electrotherapists would prove immediately suitable for experimentation with the newly discovered X rays. (Courtesy of the Center for the American History of Radiology, Reston, Va.)

Fig. 1.5 Electrotherapists covered a broad spectrum, from well-trained physicians incorporating mild electricity into accepted treatment methods to those who advocated more exotic treatments, like this electric foot bath. (Courtesy of the Center for the American History of Radiology, Reston, Va.)

1890s, but this was an understanding that sometimes varied with the perspective of the individual physician. Medical practice in the United States was far from a monolithic and unified profession with agreed-upon rules of practice. In the east physicians trained in teaching hospitals with European-influenced programs were considered the cream of medicine. But they constituted only a small percentage of practicing physicians, and an even smaller number of this elite would become directly involved with the new rays. A different class of practitioners, trained at what would later be called "B" medical schools and through extended preceptorships, made up the bulk of American practitioners. Most of these doctors had general practices, many with hospital and clinic affiliations. Membership in medical organizations was open to all doctors with a "legitimate" medical degree, and even this was not scrutinized too closely for quality.

Hovering about the fringe of this corps of physicians who liked to call themselves the "regulars" was a growing number of practitioners who adhered to exotic and sometimes quite colorful theories about the body and healing. Electrotherapists, hydrotherapists, mesmerists, and osteopaths were lumped together with frank charlatans and quacks in the minds of the "regulars." The public, however, flocked to the new therapies and seemed to welcome with open arms each new "ism" (as the new treatments were mocked in mainstream medical journals). It may be worthwhile here to note the heroic methods still in vogue in much of "regular" medicine when compared to the relatively soothing and noninvasive manipulations of many of these peripheral practitioners.

Electrotherapy in particular was gaining in popularity and legitimacy in the 1890s. Several noted "regulars" had joined the ranks of the electrotherapists, while electrotherapy equipment had begun to turn up in even the most conservative of eastern urban hospitals (Fig. 1.4).[9] Treatment consisted in the application of small amounts of static electricity or current to the afflicted part, though electrical "baths" and other gear were also popular (Fig. 1.5). Electricity was applied to a bewildering range of diseases and dysfunctions:

body and head aches, dermatological problems, impotence, malaise, over-excitability, hair loss, memory loss, tuberculosis, and that catch-all of nineteenth-century women's ailments, neurasthenia.[10]

Despite an influential national organization, the American Electrotherapeutic Association, and a growing acceptance among both the public and referring physicians, the electrotherapists struggled for professional recognition.[11] Many had entered medicine "through the back door" as electricians or technicians who acquired first the necessary apparatus and second (often by mail) medical degrees. Electrotherapy, with its vague relief of often even more vague symptoms offered little of the therapeutic decisiveness favored by the "regulars."

The announcement that a miraculous new ray had been found—and that it could be produced out of the same machines the electrotherapists had been using for decades—brought both hope and the "regulars" to the doorstep of electrotherapists across the United States. Physicians in hospitals wanted to borrow their machines and wanted advice in operating them to secure diagnostic X rays. Several electricians and electrotherapists hung out new shingles as "X-Ray Laboratories."[12] The door to full acceptance in legitimate medical practice had been opened by their expertise in the new technology. Electrotherapists and their "regular" colleagues forged ahead to test the powers of the X ray.

Post hoc rationales for innovative treatments often save medical workers from looking foolhardy. Many accounts of early X ray treatment begin with notations of redness and erythema after diagnostic work, followed by the reasoned conclusion that "perhaps there might be some beneficial effect on diseased tissues."[13] In truth, the application of the rays to disease was a natural extension for this body of practitioners who were accustomed to applying electricity for a range of maladies. Even the same apparatus was used (Fig. 1.6). The earliest therapeutic applications of the rays were almost entirely empirical and without rationale or reason. After the fact, when amazing results were noted, electrother-

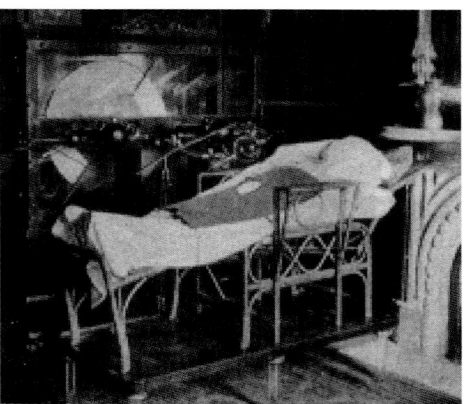

Fig. 1.6 Static machines were easily converted to radiation therapy devices with the addition of a tube and a little rewiring. Here the patient was being treated for cancer of the pelvis, 1899. (Courtesy of the Center for the American History of Radiology, Reston, Va.)

apists and other physicians struggled to make their adventure into the unknown sound scientific. One author, after candidly admitting that his first efforts at radiation therapy had been "purely speculative and empirical" went on to gamely catalog a list of factors he recalled as his rationale—a jumble of thoughts on sunlight, bacteria, ozone, chemicals, and ultraviolet light.[14] More direct was the author who recalled that "We forged ahead with what glimmer the new light gave us, shining it on whatever diseases came to hand, with the hopes that here and there we might uncover a new and fruitful use."[15]

MANY APPLICATIONS

The therapeutic applications to which the rays were immediately turned were direct reflections of the medical and health concerns of the time. The notion that bacteria were instrumental in the cause and transmission of some diseases had only recently been accepted. Tuberculosis and its related involvements were a primary concern of the age, both as a public health issue and in medical research. In addition, reported incidences of cancer were up dramatically at the turn of the century, causing many physicians to look closely for both the cause and cure for malignancies.[28] Finally, some practitioners noticed an unexpected and beneficial side-effect of the rays in disease: a striking diminution in pain.

Benign conditions

In February 1896 Thomas Edison wrote "What can be easier than to turn

Émil Grubbé
(1875–1960)

(Courtesy of the Center for the American History of Radiology, Reston, Va.)

On the day the first notices appeared in American papers an electrical engineer and chemical assayer in Chicago was pleased to find that he already possessed the necessary equipment to generate the new rays. An active experimenter with Crookes tubes, Émil Grubbé (1875–1960) had in his assayer's office the tubes, induction coils, electrical generators, and fluorescent chemicals needed to duplicate Röntgen's work.[16] Chicago newspapers, like those in other cities, carried detailed directions for setting up the equipment and making radiographs on plain photographic plates.[17] Grubbé set to work immediately, investigating a series of innovations in tubes and generators.

Within two weeks he had developed a troublesome and suppurating erythema on his left hand. He had been using the hand several times daily to test the "penetrating powers" of his new tubes. On 27 January 1896 Grubbé was seen at the Hahnemann Medical College in Chicago. Three professors, J.E. Gilman, A.C. Halphide, and R. Ludlam, offered advice on the burn-like lesion. According to Grubbé, Gilman then stated:

> ...any physical agent capable of doing so much damage to normal cells and tissues might offer possiblities, if used as a therapeutic agent, in the treatment of pathologic conditions in which pronounced irritative, blistering, or even destructive effects might be desirable.[18]

Drs. Halphide and Ludlam concurred and each agreed to send a patient to Grubbé for trial treatment.

Grubbé later reported that on 28 January 1896 he treated a Mrs. Rose Lee, a fifty-five-year-old patient who suffered with an open inoperable recurrent carcinoma of the left breast. With lead sheets around the breast as shielding, a Crookes tube was suspended about three inches above the site. The treatment lasted one hour—followed by similar treatments over the next seventeen days. The day after Mrs. Lee's treatment was initiated, Grubbé saw a Mr. Carr, who was eighty years old with extensive ulcerous lupus vulgaris on his face and neck. He, too, had one-hour exposures daily until mid-February. "And so," noted Grubbé in his overwrought style nearly fifty years later, "without the blaring of trumpets or the beating of drums, a new therapeutic agent had arrived."[19]

This strange quiet has been precisely the problem for other physicians and historians who have looked at Grubbé's achievements. Although the silence was broken in the 1930s by his own hornblowing and drumbeating, there is little evidence to back up Grubbé's claims and still more reason to doubt that his account was entirely accurate.

First, he did not publish his version of these activities for a number of years.[20] He would defend his uncharacteristic reserve by stating that in 1896 he was not yet a physician, a distinction he achieved with a somewhat dubious medical degree from Hahnemann in 1898. He claimed to have treated many patients between 1896 and the opening of his own X-ray laboratory in 1898, but the referring physicians took credit for his work.[21] This, he asserted, accounted for the unwillingness of any of these physicians to step forward to affirm his earliest therapeutic efforts.

Second, there is the troubling question of the two original patients. Grubbé produced handwritten referral slips on Mrs. Lee and Mr. Carr in the 1930s as proof of events as he described them. Later submitted to FBI analysis, the paper and ink were verified as consistent with those in use at the turn of the century.[22] What was not questioned was the sudden appearance of these crucial documents or the very real oddity of formal referrals written in quite similar styles by physicians sending patients to the factory room of a chemical assayer.

And what became of these patients? Grubbé, in his several accounts of these patients, gave no clue as to the immediate effects of the treatments. He reported that each patient died within a month of the original treatment, Mrs. Lee of systemic carcinoma and Mr. Carr from a skull fracture after falling off one of Chicago's elevated sidewalks.[23] Exhaustive searches of the warehoused certificates of death for Cook County have located neither a Rose Lee nor an A. Carr for all of 1896 and 1897.[24] On the chance that Grubbé might have changed their names in respect for patient anonymity, a second search was conducted attempting to match similar ages and causes of death. Again, no matches were found.

Finally there is the matter of Grubbé himself. Victim to his many unshielded experiments with the rays, he suffered over one hundred surgical procedures and amputations over his long life. Many acquaintances ascribed his relentless bitterness and contentiousness to his disfigurement and his often-thwarted quest for recognition as "the father of radiotherapy." A more disagreeable character can hardly be found in radiology's history. Grubbé's personal papers and memoirs, on file at the Center for the American History of Radiology, reveal that he was, throughout his life, a difficult and often mean-spirited man. At his death in 1960 he left his extensive library and modest fortune to the University of Chicago with one stipulation: that a biography be written recounting his life and achievements. Paul Hodges, longtime head of the radiology department, reluctantly agreed to take on the task. The more he learned about the departed Grubbé the less he liked him, and the biography is a balanced but pejorative assessment of Grubbé's life work.[25] Hodges's advice to an historian in later years was, "If you're going to be fool enough to leave your money to have your biography written, then try to lead an exemplary life. Failing that, for God's sake remember to tell your lawyer to stipulate that it be a positive biography."[26]

There is no definitive answer about Grubbé's priority in the therapeutic use of the roentgen rays. It is true that he was one of the first to experiment with the rays in the United States and to experience their negative effects, and among the first to set up and successfully run both an X-ray clinic and a radiological school. He later treated successfully many patients with benign and cancerous lesions. But he took credit for, among other things, the introduction of "a new therapeutic era...of radiation therapy," and, by extension, "treatment with thorium, mesothorium, radium, alpha rays, beta rays, gamma rays, radon, and all the isotopic chemicals which can now be made from the cyclatron [sic], betatron, or other atomic fission devices."[27] His bombast may have made Grubbé a particularly unappealing pioneer to his medical colleagues and blinded them to the validity of some, if not all, of his claims to early and innovative work in the field.

the rays on the lungs of persons afflicted with consumption?"[29] Two weeks later James Burry reported in *J.A.M.A.* on irradiation of tuberculosis bacilli for two hours in an attempt to assess the "germicidal effects."[30] Numerous researchers wrote of successful *in vitro* applications of the rays over the next few years.[31] By 1898 reports in the literature covered the beneficial effects of X rays on pulmonary tuberculosis in both guinea pigs and human subjects.[32]

One physician (and ex-electrotherapist), C. H. Brauer of David City, Nebraska, detailed his X-ray treatment of four patients with tuberculosis of the lungs. "For one week I gave daily sittings of eight minutes duration placing a Crooke's tube ten inches from the body and using such penetration that I could see finger bones with the fluoroscope at four feet."[33] Such descriptions were meant to be helpful to those who wished to duplicate the author's results, which included diminished coughs, returned appetite, weight gain, and, in one case, "a complete disappearance" of the disease.

Physicians who sought to cure pulmonary tuberculosis with the rays must have been encouraged by the immediate and salutary effects of treatment in lupus vulgaris and associated disfigurements. The first case reported in the United States was by Philip Mill Jones of San Francisco in 1900, but was preceded in 1898 with reports by Freund and Schiff in Germany.[34] In his 1901 textbook Francis Williams summarized more than a dozen cases of lupus vulgaris treated by the X ray with spectacular results.[35] William Pusey and Eugene Caldwell described a number of lupus cases in detail, as did other early textbook authors (Figs. 1.7 and 1.8).[36] These patients were clearly successes, many relieved in a matter of weeks of scars they had borne for years.

Fig. 1.7 I.R. Kelly of Oakland, California, is seen treating a lupus lesion on the nose in 1900. Note the complete absence of protection for Dr. Kelly and his assistant. (Courtesy of the Center for the American History of Radiology, Reston, Va.)

Most anthologists of treatments listed lupus vulgaris first in the heading of "tubercular problems," going on to summarize curative attempts in lupus erythematosus, tubercular ulcers, tubercular glands, tuberculosis of the bones and joints, tuberculosis of the conjunctiva, leprosy, actinomycosis, rhinoscleroma, blastomycosis, and a number of other disease entities not seen today as related.[37] In many cases the immediate relief occasioned by the drying effects of the rays may have been confused with a cure. In others, genuine and permanent improvement was noted.

These successes led to immediate applications to any and all dermatological or surface lesions. By 1902 numerous cases had been reported of favorable X-ray treatment of hypertrichosis, acne, psoriasis, alopecia areata, favea, tinea tonsurans, even excessive facial perspiration. Such laundry lists of diseases and symptoms cannot convey the broadness and optimism with which the X ray was applied to disease. And since the rays had proven effective in shrinking swelling in dermatological conditions, some physicians reasoned that internal swellings might be equally susceptible. The X ray was used to treat gout, goiters (Fig. 1.9), and in 1905 was first applied to thymic enlargement in infants.[38]

Leopold Freund, who had pioneered radiation therapy in Europe with the irra-

Fig. 1.8 Excellent cosmetic results were seen in early treatments of lupus vulgaris. (Courtesy of the Center for the American History of Radiology, Reston, Va.)

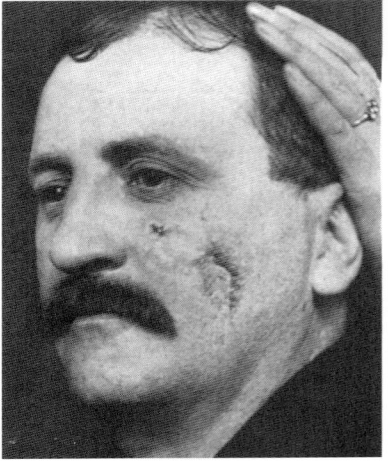
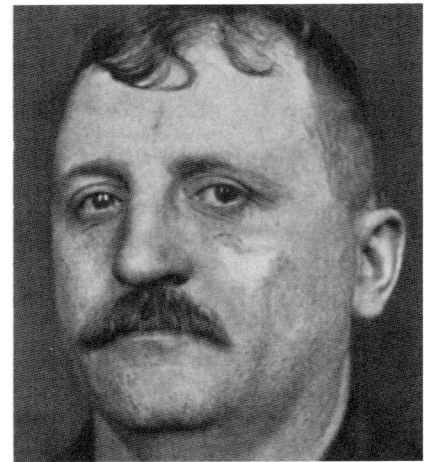

Fig. 1.9 In this early photograph of X-ray treatment for goiter the patient wears a pointed metal shield to protect the chin. (Courtesy of the Center for the American History of Radiology, Reston, Va.)

dition of a disfiguring hairy nevus in 1896, advocated the use of the rays as a medical depilatory.[39] In France and Italy X-ray beauty clinics were opened for the cosmetic removal of unwanted hair. In the United States these treatments were administered discreetly in doctors' offices (Fig. 1.10).

If some of the early applications of the rays seem farfetched, it is important to remember that there was no more compelling reason to use them on lupus than for hair removal; no system of medical thinking that made X rays more promising in gout than in cancer. So many beneficial results were observed, and these were so unpredictable and varied, that many X-ray workers felt called to find new and potentially rewarding applications. This explains the eagerness of Mihran Kassabian, who in 1904 obtained permission from the Insane Department of the Philadelphia Hospital to irradiate a number of "epileptic" patients (Fig. 1.11). Looking at the pictures of these patients as they underwent months of irradiation and hair loss, we can see other possible diagnoses. But like many practitioners who experienced mixed results, Kassabian remained upbeat about the experiment, noting that the patients' seizures had diminished in number and severity and that their hair had grown back with renewed vigor and shine.[40]

What emerged from these numerous uses for the X ray in benign conditions was not a well-codified system of therapeutics. Instead, the medical literature and textbooks carried exhaustive descriptions of individual cases, outcomes, technique, and idiosyncratic observations. The practitioner was free to choose, with little guidance, from the smorgasbord of treatment sites and methods. There was no consensus on the most favorable approaches.

By 1910 most physicians who used radiation therapy in their practices agreed that the rays had not lived up to their early promise in the treatment of pulmonary tuberculosis. The gradual realization that there were dangers in unlimited applications of radiation led to conservatism among some practitioners. But great successes had been noted in a range of diseases and symptoms, and radiation

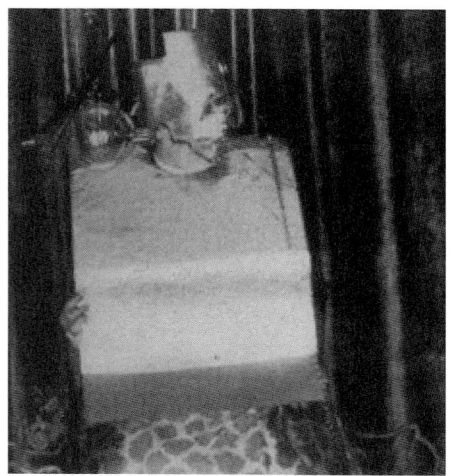

Fig. 1.10 Samuel Monell's 1899 set-up for "removing whiskers from women's faces." The extensive shielding was unusual for contemporary treatments, but may reflect Monell's interest in franchising such a system with X-ray depilatory booths in hair salons. (Courtesy of the Center for the American History of Radiology, Reston, Va.)

Fig. 1.11 "Epileptic" inmates from the Insane Department of the Philadelphia Hospital, irradiated over a series of months in 1904 by Mihran Kassabian. (Courtesy of the Center for the American History of Radiology, Reston, Va.)

would remain the treatment of choice for many of these benign conditions well into the middle of the twentieth century.

The X Rays in Cancer

The effects of the rays in benign skin disease led to similar applications in surface cancers. But cancer had been uppermost in the minds of many observers since the announcement of Röntgen's discovery. In 1896 Nikola Tesla, the electrical inventor, wondered if it might be possible to "load" X rays with cancer-fighting drugs or chemicals and project them into the body.[41] Articles questioned whether cancer, the "modern disease," might be treated with the new rays.

The works of Tage Sjögren and Thor Stenbeck in 1898 and 1899 are often cited as the first treatments of skin cancer, with a successfully treated case of basal cell carcinoma and one of squamous cell carcinoma.[42] At the same time this work was in progress, a pathologist and medical student in Washington, D. C., had begun the treatment of a patient diagnosed with epithelioma of the skin. Wallace Merrill and Walter Johnson published their results with this and other patients in 1900, noting:

> We are firmly convinced that, by means of the proper application of X rays under conditions of no practical discomfort to the patient, we can bring about the painless removal of the slow-growing epithelioma.[43]

It is noteworthy that both the Swedes and Merrill and Johnson combined these initial treatments with surgery.

The X rays offered hope previously unavailable, and soon applications had been made to a range of surface lesions. By 1901 and 1902 numerous reports of promising treatment of "cutaneous carcinoma" appeared in medical journals.[44] Sequeira, at the London Hospital, reported on more tha one hundred cases he had treated by 1903. He noted a number of recurrences after treatment, but stated optimistically that "as a matter of practical import the recurrences are usually easily removed by fresh applications of the rays."[45] Many practitioners seem to have had difficulty in distinguishing the relief of symptoms from a total extirpation of the disease. For some this was perhaps the result of lack of experience with cancer. Surgeons knew the tenacity of the disease, even when it appeared to have been excised.

William Pusey described the number of cases of skin cancers treated by the rays by 1904 as "a veritable torrent of literature."[46] Kassabian listed numerous authors and their results in 1907, including his own minority opinion that surgery as an adjunct to radiation was unnecessary and could potentially diminish the healing effects of the rays.[47] He conceded that there was little agreement on methods or results:

> Some prefer the soft and others the hard tube. Views also vary as to the duration of the seances and their frequency. It is asserted by some that a slight dermatitis is always to be aimed at, in order to obtain the proper action. The great variety of cases encountered will allow of no special technic; the peculiarities of the epithelioma themselves will frequently dictate the method to be pursued.[48]

Using these many approaches, X rays had by 1907 been used to treat carcinoma of the orbit and eyelid, as well as "subdural nodular masses."[49] Pusey presented a chart of sixty-nine patients treated for various forms of epithelioma, including history, record of treatment, outcome, and follow up. Most were listed as "well" after periods of eighteen to twenty-four months (Fig. 1.12).[50]

Émil Grubbé stated that he had treated a woman for cancer of the breast in 1896. In 1897 Gocht in Germany reported two cases of inoperable breast cancer treat-

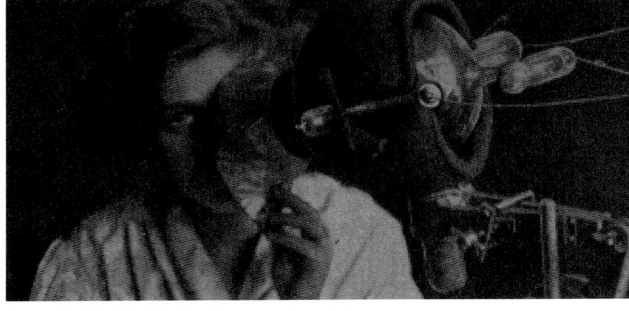

▶ Fig. 1.12 Skin cancer treatment, around 1902. Note that the shielding is held in the patient's exposed hand. (Courtesy of the Center for the American History of Radiology, Reston, Va.)

ed without beneficial results.[51] George Hopkins in Philadelphia in 1901 treated two cases, one of which was deemed a cure.[52] As with cutaneous involvements, treatments were usually given every two to four days until positive results were noted or "to the limit of the patient's tolerance." Kassabian detailed six cases, reporting that he had irradiated more cancers of the breast with varying outcomes, but "would report only on those cases which gave good results."[53]

Pusey praised the cosmetic healing in patients who had appeared first with open cancers following mastectomies. Of one patient he said, "It seems hardly credible that this scar represents the site of the previous tumor mass" (Fig. 1.13).[54] He also recorded the first treatment of a male patient with breast cancer.[55] In summarizing his patients up to 1904, Pusey noted that eighteen of thirty-one breast cancer patients had died. Most had been referred as inoperable and with open, suppurating involvements. He accurately observed that the treatment prolonged life, added comfort, and provided relief from painful dressings. In a few cases it seemed to have effected a complete cure.[56]

With improved and specialized treatment tubes, the rays were aimed at cancers of the larynx and esophagus and at cervical cancer. In 1902 Clarence E. Skinner began treating a patient for a large ovarian tumor. Over fifteen months she received 136 treatments and was pronounced cured by Skinner.[57] Seven years later she was reported to have had no recurrence of the large mass, although she was bothered by a persistent and worsening X-ray burn over the exposed area.[58]

Treatments of similar deep-seated cancers were reported with increasing frequency but inconsistent results. The equipment of the period was simply inadequate to heal effectively deep within the body without damaging healthy tissues. Many practitioners reported excellent results in the short term, but others were skeptical. Kassabian was direct:

> I believe that in many instances, where brilliant results were achieved in irradiation of sarcomas, in all probability there was a mistake in diagnosis and a less malignant affection was present or else the operators were a little too enthusiastic when making their reports.[59]

Most deserving of enthusiasm in the treatment of deep-seated disease were the initial attempts to improve the condition of patients with leukemia and pseudoleukemia (at the time synonymous with Hodgkin's disease). The latter had been considered uniformly fatal to children and adults who presented with the characteristic symptoms. In 1901 Pusey was the first to treat patients with Hodgkin's and reported two cases "symptomatically cured" at the meeting of the Chicago Medical Society in February 1902.[60] His first patient, a four-year-old with extensive lymphatic involvement, had been discharged from the hospital as incurable (Fig. 1.14). Initial radiation therapy reduced the symptoms, which returned in six months. Again, the X rays diminished the swelling and improved the general clinical picture, but a few months later the child died "of inspiration pneumonia." Pusey considered this case a cure.[61]

Nicholas Senn reported the treatment of two patients with positive results in 1904.[62] It is noteworthy that all these early applications of the rays to deep-seated and

Fig. 1.13 Treatment of cancer of the breast presented by William Pusey in 1902. The remarkable cosmetic results were followed by a remission in symptoms. (Courtesy of the Center for the American History of Radiology, Reston, Va.)

▼

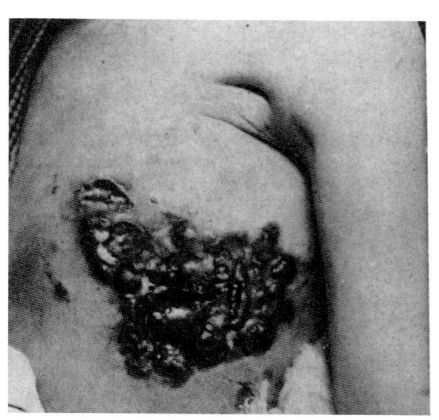
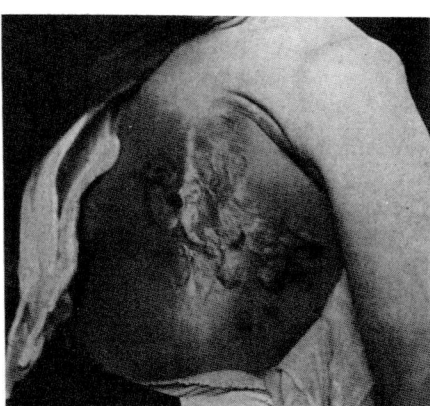
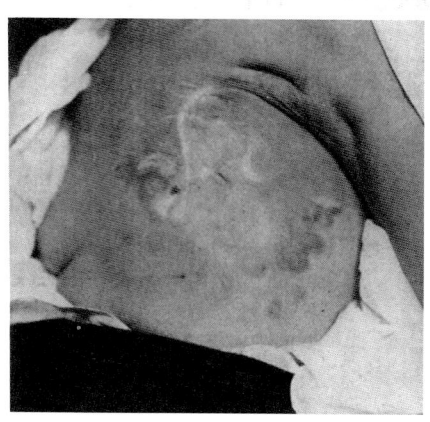

Fig. 1.14 Pusey's first patient treated for Hodgkin's disease, 1902 and 1903. (Courtesy of the Center for the American History of Radiology, Reston, Va.)

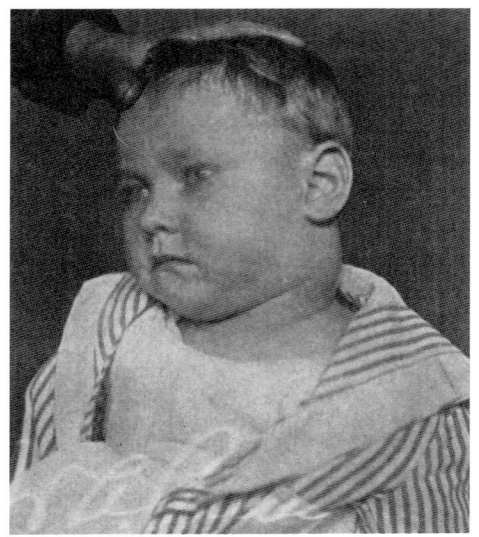
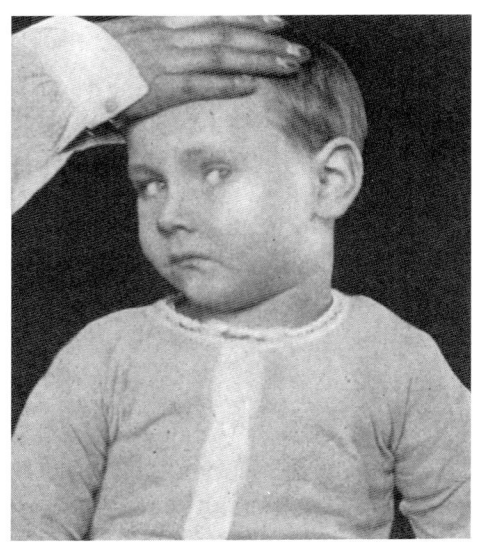

systemic cancers were extended over many days, sometimes into years, with acknowledgement of the possibility of future irradiations on recurrence. The term "therapy" was not lightly applied. This was a commitment for both practitioner and patient, sometimes for as many as one hundred return appointments and gradual improvement.

William Benham Snow described a typical series of treatments for Hodgkin's disease in 1903. Exposures were made with the tube at a distance of 12 to 20 inches, with a medium or high-vacuum tube "for its tonic effects on the structures."[63] The patients were seen daily, with exposures made directly over the involved glands on alternate days from general irradiation of the trunk.

The beneficial results noted do not seem to have given way to immediate widespread use of the technique, perhaps because of the commitment of time and the relative paucity of cases of leukemias compared with cutaneous and site-specific lesions. In 1905 Joseph Clapp and Joseph Smith chaired a session on the treatment of Hodgkin's disease at the meeting of the Chicago Medical Society.[64] Although Senn, Pusey, and others presented promising results, the ensuing discussion centered exclusively on the attending members' own experiences with the rays in dermatological disorders.

Not everyone was ready to see protracted X-ray treatments as beneficial in systemic illness, in part because of a fundamental misunderstanding of the disease process. The work of G. Arneth in Germany, as summarized by Kassabian, illustrated this difficulty. Arneth believed the rays did "not cure the lesion, but they destroyed the parasites which are the cause of the lesions."[65] The action of the X rays was like that of quinine in malaria, "curing the patient by killing off the microorganisms causing the trouble." This explanation and its implications for therapy were not likely to yield promising results, since the assumption was that the "parasites" could be killed off in short order.

The list of cancers treated by irradiation with beneficial results was long in 1910. But there remained little agreement on specific methods, few comparative or long-term follow ups, and no satisfactory explanation for successes in some cases and failures in others.

ANALGESIC AND PALLIATIVE APPLICATIONS

Early in 1897 American medical journals carried tantalizing reports of Russian experiments concluding that X rays had quieting effects on the central nervous system in frogs.[66] Subsequent reports from Germany chronicled surprising success rates in treating "severe neuralgia" with just a few exposures to the rays.[67] In the United States reports of successful treatment of skin lesions in lupus and other diseases were often accompanied with notations on diminution of pain.[68] In one report submitted to the New York Academy of Medicine in 1902 the physician stated that he "had almost invariably noted cessation of pain after the first treatment" of lupus with the X ray.[69] In a survey of potential uses for radiation another physician wrote: "The most astonishing feature is that the X-ray possesses powerful analgesic properties, immediately relieving pain."[70]

The exact mechanism by which the rays appeared to relieve pain in surface lesions was the subject of some dispute. To the electrotherapists, who long had believed in the curative powers of applied current, the results seemed predictable. One electrotherapist even opined that "the relief of pain is due to the action of a high-tension current and connected in no way with the X- or cathode-rays."[71] Others ascribed pain relief to the effects of ozone produced by the generator or to the dessicating properties of the rays.

The positive results could not be denied. In Boston Seabury W. Allen chronicled the relief from pain experienced by patients treated with X rays for maladies ranging from chronic rheumatoid arthritis to suppurating tubercular lesions.[72] Among his notable results was a thirty-six year old woman with arthritis and severely limited flexion of the fingers. After two diagnostic X rays she reported that she had resumed playing the piano, with no pain. Another patient, aged twelve, suffering from what was described as "a painful bilateral tubercular sinus of the neck of one year's duration" felt well enough after two irradiations to go home and throw snowballs. Allen theorized that "the nerve supply of the parts in question" held the answer to these "cures." He noted particularly that the pain-relieving effects often occurred at a distance from the body parts exposed to the light but supplied by nerves which had been exposed. He concluded: "It seems reasonable to suspect that X rays may influence distant parts, either reflexly or through some electric phenomena along the course of the nerves." This was precisely the kind of theory the electrotherapists found most appealing.

The same pain relief noted in benign skin conditions was observed very early in the treatment of cancer with the X ray. Several physicians cited the marked relief of pain as "usually the prompt result of the use of X rays in malignant neoplasms."[73]

After a survey of six case histories in 1902, one doctor noted that the X ray had "a very marked influence upon the pain of nearly all types of malignant tumors, causing entire relief in many cases."[74]

The first application of the rays to malignancies were in surface lesions, where it was found that pain and discharge were substantially reduced.[75] Soon the rays were applied to more deep-seated cancers. First on William Morton's 1902 list of the ten beneficial properties of the X-ray in cancer was "relief from excruciating pain and constant suffering, often immediately."[76] In the same year the editor of the *J.A.M.A.* summarized the reports and views of many physicians:

> In cancers en cuirasse where the chest was imprisoned as in a vise to the constant intense discomfort of the patient, relief was afforded almost immediately after exposure. In the extremely tender cancers of the breast in which often the weight of clothing becomes insupportable, the exaggerated sensitiveness can be made to disappear....It would seem, then, that the X-rays may be resorted to in all inoperable and painful cancers.[77]

The key word in this assessment was "inoperable," for physicians soon came to view the X ray as an analgesic of potentially limitless benefit to terminal cancer patients. One editor wrote: "Even if not curative the X rays have a very decided palliative effect in many cases, reducing discharge, destroying offensive odor, relieving pain, and generally rendering what remains of life more comfortable both to the patients and to those who have the care of them."[78] The X ray, another physician predicted, would enable the afflicted patient to "spend the last few days or months in comfort, compared to the untold suffering heretofore experienced with such growths."[79]

Moreover, the use of the X ray as a palliative in cancer promised to diminish the dual problem of patient addiction to and physician complicity in the overuse of morphine and other opiates. Several authors gratefully acknowledged that their patients could now spend "what remnant of life is left them in comfort and free from the effects of the continuous use of opium."[80] Another physician, in a series of case histories, noted that many patients he treated for cancer of the breast came to him already using large doses of morphine, but "after a few irradiations," the opiates were no longer required.[81]

Perhaps the greatest attraction of the pain-killing effects of the X ray for many practitioners was the seeming congruence with the first tenet of the Hippocratic oath: to do no harm. In detailing his own experience with treating several types of cancer, Pusey concluded that "as X-ray exposures may be given these patients without disturbing them or interfering with their comfort, there seems no reason why they should not have the benefit of the remotest chance of relief."[82] Physicians with X-ray apparatus in their offices found themselves seeing more and more patients who had been rejected as inoperable by surgeons.[83] Prolonged irradiation of the terminally ill was believed to be harmless to all involved—practitioners and patients—and besides relieving pain carried the possibility of improvement in the disease itself. Grubbé noted the beneficial tendency to "prolong life even in hopeless cases."[84] At the turn of the century the X ray was viewed as both a palliative and a last glimmer of hope for sufferers. A series of alarming discoveries and events would soon dim the general optimism and confidence in the "miraculous soothing rays."

"SOME UNTOWARD RESULTS": X-RAY BURNS

Unusual effects on the skin and hair of practitioners and patients were noted almost immediately after X rays came into wide use in the United States. Grubbé later reported that the first of his many overexposures came in the third week of January 1896, with painful lesions on one hand. In April John Daniel, a physicist at Vanderbilt, reported hair loss in a volunteer patient (the dean of the medical school) sitting for a radiograph.[85] In July W. Marcuse in Berlin published an account of a radiographer's public demonstration model, who had experienced "severe skin reactions," including epilation.[86]

By the fall reports of "X-ray burns" or roentgen dermatitis were common in medical journals and newspapers. Described as resembling "a severe sunburn, with the accompanying pain, swelling, blistering, and discoloration," most such burns were discounted as short-lived and the result of something other than the X ray.[87] At this distance the widespread reluctance to suspect the irradiation as the motive factor in these burns seems puzzling. But the notion that any substance, especially an invisible light, might cause this sort of destruction of skin and tissue was entirely foreign to medical thinking in the 1890s.[88]

Many observers were convinced that the burns were caused by ozone from the electrical apparatus or by sparks unnoticed in the excitement of the therapy setting. Others speculated that some persons were more susceptible to the effects of the rays, having "X-ray idiosyncracies" rather like allergies. Still others believed the lesions were caused by unaccustomed handling of photographic chemicals.[89] Others lectured that burns in some patients were the unavoidable results of the sufferers' "existing nervous and mental conditions."[90]

Careful workers with the new rays soon realized that the problem was more complex. Elihu Thomson at the Edison Laboratories reported distressing symptoms among X-ray workers there and even experimented on his own hands.[91] William Rollins, with his brother-in-law and roentgen pioneer Francis Williams, observed a number of deleterious effects. Rollins provoked a torrent of disagreement in his blunt 1901 article "X-Light Kills."[92]

Many practitioners watched as what they had assumed to be temporary burns turn into chronic lesions. In late 1896 G. C. Skinner reported a painful roentgen dermatitis on his hand followed by complete desquamation. Five weeks later the stubborn lesion remained about two inches wide and three inches long (Fig. 1.15).[93] At Johns Hopkins in 1897 T. Gilchrist reported at length on a case of roentgen dermatitis with unfamiliar characteristics and slow healing.[94] Evidence was mounting, but for many practitioners word came too slowly and too late.

The case of Mihran Kassabian (1870–1910) provides an excellent look at the ways in which a full-time radiologist and radiotherapist perceived the dangers of X rays in these earliest years. A medical doctor with an interest in electrotherapy,

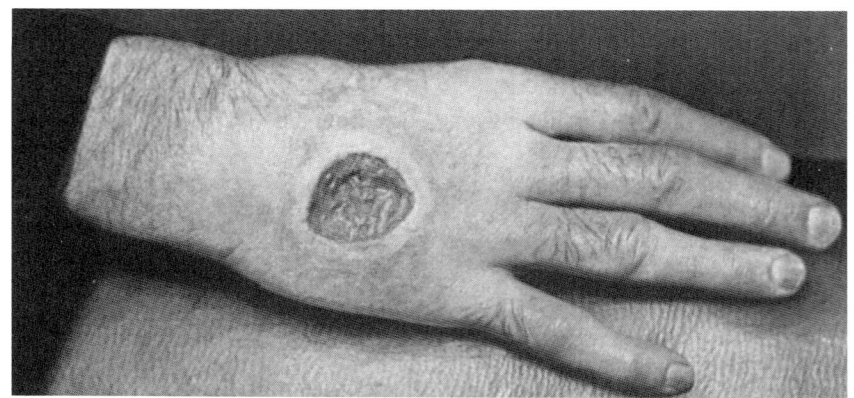

Fig. 1.15 An "X-ray burn" of six months' duration on the hands of a radiological worker in 1898. (Courtesy of the Center for the American History of Radiology, Reston, Va.)

Kassabian began work with the X ray around 1900, and was appointed director of the Roentgen Ray Laboratory of the Philadelphia General Hospital. Early photographs of Kassabian at work in radiography and radiotherapy reveal that, like most of his colleagues, he used little or no shielding.[95] He experienced a series of mild sunburn-like erythemas on his hands and face, with some loss of hair. By 1904 he had noted "chronic dermatitis" with the "nails disfigured and deformed, and they never regained their normal appearance and condition. The skin of the hands remains tough and indurated, with the subsequent occurence of atrophy."[96] Kassabian documented the deteriorating condition of his hands from dermatitis to serial amputations in a number of photographs taken between 1906 and 1909 (Fig. 1.16).

He took a special interest in the explanations for these injuries offered by his colleagues, as well as their suggested treatments. In 1907 he published *Röntgen Rays and Electrotherapeutics*, which was widely read and respected. Here he listed the eight most popular explanations for the cause of X-ray burns, ranging from ultraviolet light to "the flight of minute platinum atoms."[97] In his comprehensive survey of radiotherapists around the world Kassabian asked how his colleagues treated the roentgen dermatitis. Answers ranged from the smug ("Never had any case of dermatitis to treat") to the dismissive ("Treat as any other burn").[98] Kassabian applied salves of zinc and lanolin, took meticulous notes on his deteriorating condition, and died in 1910.

It had been clear for some time to many practitioners that the X-ray dermatitis was not just "any other burn." In 1902 G. S. Johnston followed up an earlier assertion that it was the rays themselves that caused damage with the obervation that the "keratotic peaks" noticed on the hands of radiologists were actually "precancerous conditions prone to epitheliomatous change."[99] Excision and inspection of one of the hard patches of skin revealed "an intense lymphocytosis with a proliferation of fibroblasts." The death of Edison's assistant Clarence Dally in 1904, after a gruesome series of amputations and other surgeries, made the public and the broader medical profession aware of the dangers of X rays. By this time many practitioners like Kassabian

Fig. 1.16 Kassabian documented the deterioration of his hands in diaries and photographs between 1904 and his death in 1910. (Courtesy of the Center for the American History of Radiology, Reston, Va.)

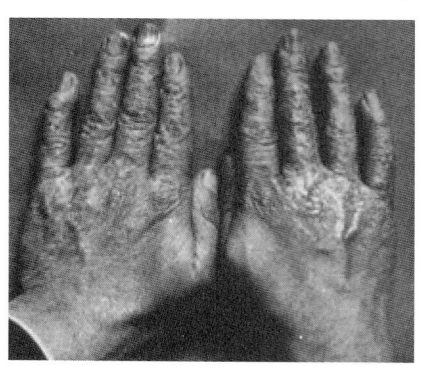
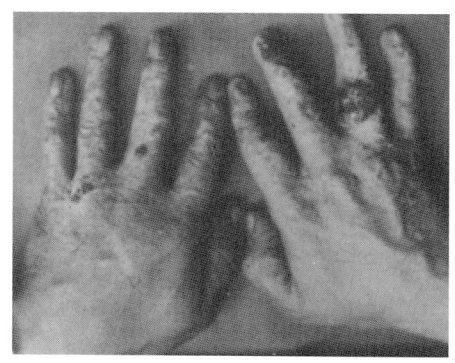
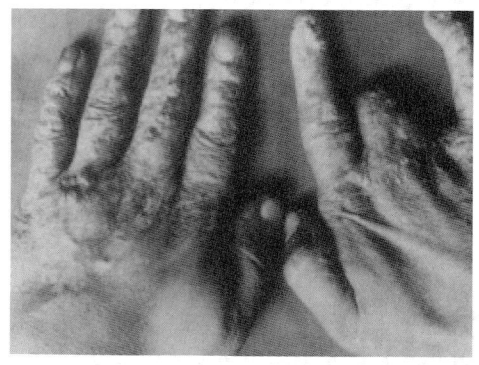

Fig. 1.17 Kassabian designed and recommended a "protective room," where the physician would be removed entirely from the influence of radiation. (Courtesy of the Center for the American History of Radiology, Reston, Va.)

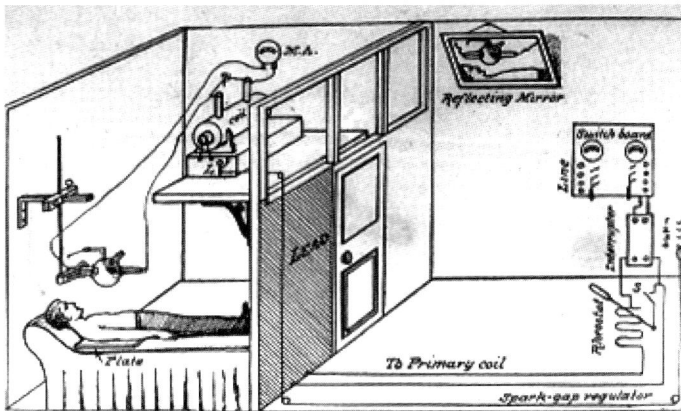

had already noticed alarming symptoms in themselves and their assistants.

Around the world physicians and pathologists looked more carefully at the action of the rays and were forced to conclude that the ray that could cure could also harm. Previous efforts at shielding patients and practitioners from X-ray "sunburn" now took on added significance. Shielded tubes, special filters, and massive screens were added. Determined, if largely ineffective, efforts were made to measure and regulate the amount of radiation given in each treatment. Kassabian recommended that radiation suites be constructed, where the physician worked in a room separated by lead walls from the patient and X-ray tube, monitoring the procedure by a series of wall-mounted mirrors (Fig. 1.17).[100]

By 1910 the enthusiasm of the early years of empirical applications and unlooked-for successes had dimmed. Many of the pioneers were gone, still others ill, from a process they did not entirely understand. Still to come were mysterious and fatal leukemias and anemias associated with X-ray work. Those who took up the field of radiation therapy approached it with great caution, and a growing curiosity about the nature of the action of radiations in the body.

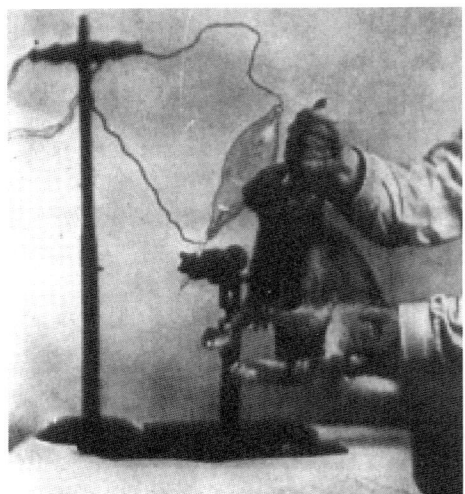

Fig. 1.18 Photographs of early laboratory research are rare. In this 1898 scene the researcher (unwittingly a part of the experiment himself) wished to find out about the effects of X rays on rabbit skin. (Courtesy of the Center for the American History of Radiology, Reston, Va.)

EXPLAINING THE RAYS: EARLY RESEARCH

The rays could heal and they could kill—but how? Many explanations were offered to explain the effects of radiation. Researchers looked for explanations inside and outside the body's cells. They looked in plants and single-celled animals. But as late as 1915 one specialist in the field admitted that "the mechanism by which the disturbances have been brought about remains unexplained."[101]

Early experiments had been simple, training the rays on bacteria to determine "killing" effects and testing the susceptibilities of laboratory animals to varying amounts and exposures to X rays (Fig. 1.18). One of the first of these experimenters was the unfortunate professor who, in February 1896, believed he had raised a drowned lab rat from the dead by the force of X rays—certainly the ultimate therapeutic effect.[102] Most results were not so spectacular, and with wide variations in tubes, currents, and technique, such studies yielded few useful or replicable conclusions.

A number of early investigators in the United States and Europe looked at the effects of radiation on plants and on protozoa. In 1897 Lopriore noted the stimulating effects of short exposures on plant growth, while Schaudinn observed the destructive (coagulating) effects of larger amounts of radiation.[103,104] Schaudinn also reported on a direct correlation between the amount of fluid in certain protozoa and their susceptibility to the rays.[105]

In 1903 Albers-Schönberg in Germany and Bergonié and Tribondeau in France published reports indicating the spermici-

An Organization for Radiation Therapists?

Communication was important in a field where new discoveries seemed to be made every day, and one important venue for such exchanges of information was the medical specialty organization. The American Electrotherapeutic Association (AETA) had been founded in 1891 and included both marginal practitioners and medical luminaries. The AETA prided itself on an ecumenical outlook, including in its ranks a number of different medical specialties, including surgery, dermatology, psychiatry, dentistry, and others. The AETA embraced the therapeutic possibilities of the new ray, featuring lectures on the subject at its October 1896 meeting. Many Fellows of the association became prominent radiologists, while many who wished to work with the new rays sought to join the organization.

Some early specialists in radiation, however, found association with the electrotherapists to be less than desirable and felt that the new field of X rays warranted its own separate organization. The first meeting of the Roentgen Society of America (later the American Roentgen Ray Society [ARRS]) was held in December 1900. Over two thousand invitations to membership were sent out, and full status was accorded to anyone who was already a member of any other medical or scientific society. The official organ, the *American X-Ray Journal,* published numerous articles on X-ray therapy in its short existence, and the ARRS served early radiation therapists with information, collegiality, and a forum for discussion of new methods. In 1905, however, the ARRS dropped seventy-five members whose credentials were considered "dubious." Many of these were members of the AETA without legitimate training, but who had incorporated the use of X rays into their therapeutic practices.

To modern eyes, there seems to have been a substantial portion of journal space (in the *American Journal of Roentgenology*), meeting time, and committee activity devoted to radiation therapy topics within the ARRS after 1905. But the few physicians who practiced radiation therapy exclusively and those in whose work it played a major role sometimes felt as though they had been "tarred by the brush of the ejected members." With the rise in radium therapy and the creation of a domestic American radium product after 1914, a few practitioners and sponsors felt the time was right for an organization devoted to healing with radiation.

On 22 June 1916 during the annual meeting of the American Medical Association, a group of physicians met in Detroit and agreed to found a society in which workers from different disciplines would exchange experiences and contribute to the advancement of the therapeutic uses of radium. The American Radium Society was formally organized in October 1916, dedicated to the promotion of the scientific study of radium, its physical properties, and its therapeutic applications. Members included surgeons, radiologists, gynecologists, physicists, and a few manufacturers of radium. The ARS was a model of interdisciplinary collaboration, and the scientific sessions included groundbreaking clinical reports each year, often including studies in X-ray therapy as well. The ARS would take an active stance on behalf of radiation therapists on legislative and regulatory issues, and its committees would serve as reference points on practice and ethical issues. Radiation therapists would enter the post-World War I medical world with a strong and purposeful organization acting on their behalf, although they were joined in this organization by many other specialists. The small numbers of radiation therapists could not have supported a separate organization in these years and would not for many years to come.

dal properties of the X ray and radium, as well as the ability to induce sterility with the rays.[106,107] This focused additional interest on the action of the rays on dividing cells and on embryos. Perthes's work with *ascaris* eggs and the revelation by Regaud and Blanc that the mitotic phase of the cell was the point of lesser resistance to radiation were widely reviewed in the United States literature.[108,109] Some authors speculated on the meaning this held for the offspring of radiologists, as well as the welfare of their patients.[110] With the publication of their widely accepted "law" in 1906, Tribondeau and Bergonié sought to draw conclusions from the disparate findings of many researchers in the field: "The effects of irradiation on the cells are more intense the greater their [the cells'] reproductive activity, the longer their mitotic phases, and the less their morphology and functions are established."[111] Some observers immediately pointed out inconsistencies in observed research results and the law, but in a field where there were few dependable rules and nothing was ever quite predictable, Tribondeau and Bergonié had provided substantial comfort, however flawed, to many theorists.[112]

In Baltimore Gilman and Baetjer studied the effects of X rays on development in fertilized hens' eggs.[113] Still other researchers looked at the systemic effects of the rays. Heineke observed extensive effects on the lymphoid tissues of rabbits and guinea pigs and was first to note the rapid depletion of irradiated bone marrow.[114]

It was clear by 1910 that radiation had a retarding effect on growth in some organisms and a deadly action in others, that cell division and maturity were crucial factors in determining these effects, and that sensitivity to radiation varied widely. This was of little help in explaining the mysterious healing powers of the rays. In fact, little of the research fit at all into the largely empirical picture of radiation therapy in the period. The competing theories of Schwartz, Hertwig, and Packard, involving foci as diverse as lecithin, chromatin, and enzymes, did little to affect daily work with patients in the years that followed before the first World War.[115] In fact, Hugo de Vries, noted botanist and biologist, in 1915 summarized results thus far and recommended only one possible concrete application: "...we may hope some day to apply the physiological activity of the rays of Roentgen and Curie to experimental morphology."[116] His ideas about "experimental evolution" would have given even the most adventuresome radiation therapist pause.

Conclusion

By 1910 many of the growing number of American roentgenologists included some form of radiation therapy in their practices. All looked with a mixture of fear and awe at the range of results occasioned by their work with the "healing rays." Some limitations had been admitted, while new applications and techniques were explored. With fewer untrained practitioners in the field, public confidence in radiation therapy was on the rise. Better apparatus, greater protection, and the introduction of filtration had improved the clinical picture. Radium would soon be more readily available from domestic sources.

Obstacles to immediate progress remained formidable. The handful of physicians practicing radiotherapy exclusively did not promise soon to constitute a bona fide medical profession. Biological concepts on the action of the rays could provide no consistent explanation of their action which could readily be translated into technique. Dosage remained difficult to control and virtually impossible to compare from clinic to clinic or treatment to treatment. But the hope that radiation therapy held out to sufferers was too great to ignore. Addressing the American Roentgen Ray Society in 1915, President Alfred Gray said, "Today, not withstanding the limitations, the roentgen rays have an established field in the treatment of disease that few, if any, other known agent may enter."[117] The X ray had proven a "light in dark places," offering hope and the promise of cures where there previously had been none.

REFERENCES

1. Several biographies of Röntgen detail his life and the events leading up to the discovery of the X ray. Otto Glasser's many works on Röntgen include: "The Genealogy of the Röntgen Rays," *Am. J. Roent.* 30 (1933):180-200; 349-367; *Wilhelm Conrad Röntgen and the Early History of the Roentgen Rays.* Springfield, Illinois: Charles C. Thomas, 1934; and Dr. W.C. Röntgen. Springfield, Illinois: Charles C. Thomas, 1958. Other worthy summaries include Claxton, K.T., *Wilhelm Röntgen.* Geneva: Heron, 1970; and Nitske, W.R., *The Life of Wilhelm Conrad Röntgen.* Tucson: University of Arizona Press, 1971.

2. Dennis D. Patton's work on the immediate scientific background and events relating to the discovery have shed new light on this remarkable and relatively undocumented event. See "Insights on the Radiological Centennial: A Historical Perspective. Röntgen and the 'New Light'. Part I: Röntgen and Lenard," *Investigative Radiology* 27 (1992):408-414; "Part II: Röntgen's Moment of Discovery," *Investigative Radiology* 28 (1993):954-961; and "Part III. The Genealogy of Röntgen's Barium Platinocyanide Screen," *Investigative Radiology* 29 (1994):836-842; "Röntgen's Inheritance" and "Röntgen and the Discovery" in Gagliardi and McClennan, *A History of the Radiological Sciences,* Vol. I, *Diagnosis* (Reston, Va.: Radiology Centennial, Inc., 1996):1–47.

3. Röntgen, W.C., "Ueber eine neue Art von Strahlen," *Sitzungberichte Physik. Med. Ges. Würzburg,* 1895.

4. "Hidden Solids Revealed! Professor Routgen [sic] Experiments with Crookes Vacuum Tube," *New York Times* (16 January 1896): 9.

5. "Some Startling Results at Yale," *New York Times* (5 February 1896):9.

6. David DiSantis has written extensively on the subject of popular reactions to the rays and radium. See for example, "Radiation and Popular Culture," in Gagliardi and McClennan, *A History of the Radiological Sciences,* Vol. I, *Diagnosis* (Reston, Va.: Radiology Centennial, Inc., 1996):607-617.

7. For a look at the many predicted and attempted uses of the rays see Knight, Nancy, "The New Light: X Rays and Medical Futurism, " in Corn, Joseph T. *Imagining Tomorrow: History, Technology, and the American Future* (Cambridge, Mass: MIT Press, 1986): 1-34.

8. Editorial, *J.A.M.A.* 26 (15 February 1896):337.

9. In New York and Chicago a number of respected physicians, among them William Morton, George Beard, and Margaret Cleaves, had adopted electrotherapy as a standard treatment for any number of illnesses and had static machines installed in hospitals and clinics.

10. The range of illnesses and conditions (presumed and real) at which electrostatic energy was aimed are included in numerous textbooks from the period and in the records of the American Electrotherapeutic Association. One title will suffice to indicate the range of diseases treated. In 1899 William J. Morton published "Electrostatic Currents and the Cure of Locomotor Ataxia, Rheumatoid Arthritis, Neuritis, Migraine, Incontinence of Urine, Sexual Impotence and Uterine Fibroids," *Medical Record* 56 (1899):845-849.

11. Nelson, Paul A., "History of the Once Close Relationship Between Electrotherapeutics and Radiology," *Arch. Phys. Med. Rehab.* 54 supp. (1973):608-640.

12. Among those opening X-ray laboratories were W.C. Fuchs in Chicago, M.E. Parberry in St. Louis, Heber Robarts also in St. Louis, and Samuel Monell in New York. Typically these laboratories were located near (but not in) hospitals, and the patient, whatever his or her condition, was brought out of the hospital for the X-ray examination or treatment.

13. Grubbé would give this explanation as would countless others. The defect in the argument was that they had previously used electricity for the same ailments without similar rationales.

14. White, J. William, "Surgical Application of the Roentgen Rays," *American Journal of the Medical Sciences,* n.s. 115 (1898):14.

15. Letter, G. Pfahler to H. Pancoast, 5 September 1920, Collections of the Center for the American History of Radiology.

16. Grubbé, Émil H., "Priority in the Therapeutic Uses of the X Rays," *Radiology* 21 (1933):158.

17. See, for example, the detailed instructions which appeared throughout January and February 1896 in the *New York Times*. So easy to follow were these directions that, among other groups of hobbyists, a Brooklyn X-Ray Boys' Club was founded for amateur experimentation.

18. Grubbé, E. *X-Ray Treatment. Its Origins, Birth, and Early History* (St. Paul:The Bruce Publishing Company, 1949):45.

19. Ibid., p. 52.

20. Around 1900 Grubbé began to argue at local and national medical meetings that he, and not fellow-Chicagoan Harry Pratt, had pioneered radiation therapy, but it was not until the 1930s that Grubbé began to push his case in the literature.

21. Grubbé, *X-Ray Treatment*, pp. 62-63.

22. Ironically, FBI analysis would lead to the destruction of these notes. The corrosive chemical used for dating darkened and crumpled both slips so that they are today almost illegible. They can be found in the Medical Sciences Collection, National Museum of American History, at the Smithsonian Institution. It should also be noted that proving that the paper and ink were of the right age is not entirely convincing to anyone who has noted that in Grubbé's own collections at the Center for the American History of Radiology he saved every piece of scrap paper, bill, napkin, matchbook, wrapper, and other paper product that came his way during his long life. Had he chosen to postdate the referrals he would have had no problem coming up with paper from 1896.

23. Grubbé, *X-Ray Treatment*, p. 57.

24. In 1986 the authors arranged through the good offices of Mr. Michael Fish to visit the suburban warehouses where older Cook County death certificates are stored and catalogued in large volumes. Every death certificate for 1896 and 1897 was read.
25. Hodges, Paul. *Émil Grubbé* (Chicago: Univ. of Chicago Press, 1966).
26. Taped interview, Paul Hodges to Nancy Knight, through Lloyd Hawes, 15 July 1983.
27. Grubbé, *X-Ray Treatment*, p. 52.
28. See for example Massey, G. Betton, "The Increasing Prevalence of Cancer as Shown in the Mortality Statistics of American Cities," *Am. J. Med. Sciences* n.s. 119 (1900):170-177.
29. "Roentgen or X-Ray Photography," *Scientific American* 74 (1896):103.
30. Burry, James, "A Preliminary Report on the Roentgen or X-Rays," *J.A.M.A.* 26 (1896):402.
31. See for example "Bacilli and X-Rays...," *New York Times* (15 March 1896):9; and Lyon, T. Glover, "Roentgen Rays as a Cure for Disease," letter, *Lancet 1* (1896):328.
32. "The Roentgen Rays in Tuberculosis," 31 *J.A.M.A.* (1898):1437.
33. Brauer, C.H., "Report," in Wagner, R.V., *Some Reports and Talk About X-Ray Therapy* (Chicago: R. V. Wagner & Co., 1900):unpaginated.
34. Jones, Philip Mill, *J.A.M.A.* (1900); and Freund and Schiff, *Archiv fur Derm. und Syph.* 42 (1898).
35. Williams, Francis. *The Roentgen Rays in Medicine and Surgery* (New York: MacMillan, 1901):391-404.
36. Pusey, William, and Caldwell, Eugene. *Practical Applications of the Roentgen Rays in Therapeutics and Disease* (Philadelphia: W.B. Saunders, 1904):Ch. X; 394.
37. This is only a sampling of the types of cases listed by Pusey and Caldwell as "tubercular" and amenable to radiation therapy.
38. Friedlander, A., "Status Lymphaticus and Enlargement of the Thymus, with report of a case successfully treated by the X-ray," *Arch. Pediat.* 24 (1907):490-501. See also Silverman, F., "Thymic Irradiation: A Historical Note," *Am. J. Roent.* 84 (1960):562-564.
39. Freund, L. *Elements of General Radiotherapy for Practitioners.* New Yprk: Rebman Co., 1904
40. Kassabian, Mihran. *Röntgen Rays and Electrotherapeutics* (Philadelphia: J.B. Lippincott Co., 1907): 495-497.
41. "Nikola Tesla Discusses X-Rays," *New York Times* (11 March 1896):16; and "Roentgen Rays and Material Particles," *Electrical World and Engineer* 28 (1896):2.
42. See Berven, Ellis, "Development and Organization of Therapeutic Radiology in Sweden," 79 *Radiology* (1962):829.
43. Johnson, W., and Merrill, W.H., "The X-Rays in the Treatment of Carcinoma," *Phila. Med. J.* 6 (1900):1089-1091.
44. These cases are summarized in the textbooks of the period, many with serial photographs of treatment over time. Surgical intervention for breast cancer was not the norm. In fact, most patients at the turn of the century reported only after the cancers had become inoperable.
45. Sequeria, A., "Report of Cases," *Brit. Med J.* i (1903):1307.
46. Pusey and Caldwell, p. 440.
47. Kassabian, p. 460.
48. Ibid., p. 463.
49. Pusey and Caldwell, p. 499.
50. Ibid., pp. 500-509.
51. Gocht, H., published in *Fortschr. Geb. Röntgenstrahlen* 1 (1897):14.
52. Hopkins, G., "Mammary Cancer Treated with the X Ray," *Phila. Med. J.* 8 (1901):404.
53. Kassabian, p. 465.
54. Pusey and Caldwell, p. 531.
55. Ibid., p. 540.
56. Ibid., p. 552.
57. Skinner, G.C., first published in *Archives of Electrology and Radiology* in October of 1904, then summarized in 45 *J.A.M.A.* (1906): 351.
58. Folllowup case presented in *Transactions of the American Roentgen Ray Society* (1908):101-170.
59. Kassabian, p. 476.
60. Pusey, p. 628.
61. Pusey, pp. 629-631.
62. Senn, Nicholas, report in the *New York Med. J.* (1903).
63. Snow, William Benham. *A Manual of Electrostatic Modes of Application, Therapeutics, Radiography, and Radiotherapy,* 2nd ed. (New York: A.L. Chatterton, 1903):289.
64. "Chicago Medical Society," reported with discussion in *J.A.M.A.* 44 (1905):569-570.
65. Kassabian, p. 486
66. A report of the Russian findings appeared as "Effects of the X-Ray on the Central Nervous System," *J.A.M.A.* 28 (1897):655.
67. A report of the German work appeared as "Pain Soothing Effects of Roentgen Rays," *J.A.M.A.* 35 (1900):328.
68. Report on Cincinnati Academy of Medicine Meeting, "Cutaneous Treatment by Roentgen Ray," *J.A.M.A.* 37 (1901):276.
69. "Report on New York Academy of Medicine Meeting," 38 *J.A.M.A.* (1902):960.
70. Heeve, W.L., "X-Ray as a Therapeutic Agent," *Therap. Gazette* 18 (1902):650-653.
71. Shields, P., "X-Ray Phenomena and Phenomena Not Due to X-Rays," 50 n.s. *Cincinnati Lancet Clinic* (1903):371-377, and continued in a series of articles through p. 388.
72. Allen, S.W., "Notes on the Analgesic Effects of X-Rays," *American Medicine* (1902):461.
73. See for instance Pusey, W.A., "Cases of Sarcoma and of Hodgkins Disease Treated by Exposure to X-Rays," *J.A.M.A.* 38 (1902):166-169.
74. Coley, F., "Sarcoma and the Roentgen Rays," *American Medicine* (1902):380-385.
75. Bondurant, E.O., "Some of the Therapeutic Uses of the X-Ray," *Tr. M. Ass. Alabama* (1902):283-287.

76 Morton, W., "X-Ray Treatment of Malignant Growths," *Medical Record* 62 (1902):533-538.
77 Editorial, "The Roentgen Rays in Therapeutics," *J.A.M.A.* 38 (1902):942-943.
78 Editorial, "The X-Rays and Cancer," *J.A.M.A.* 39 (1902):634-635.
79 Varney, "Results in Radiotherapy," *J.A.M.A.* 40 (1903):1577-1583.
80 Skinner, C.E., "The X-Light in Therapeutics," *Medical Record* 62 (1902):1007-1013.
81 Varney, "Results."
82 Pusey, W.A., "Report of Cases Treated with Roentgen Rays," *J.A.M.A.* 38 (1902):911-919.
83 Pusey, "Cases of Sarcoma."
84 Grubbé, É., "X-Rays in Cancer," *Medical Record* 62 (1902).
85 Daniel, J., "The X-Ray," *Medical Record* 49 (1896):17.
86 Marcuse, W., "Dermatitis und Alopecia nach Durchleuchtungversuchen mit Röntgenstrahlen," *Deutsche Med. Wschr.* 22 (1896):481.
87 "The Effect of the X-Ray on the Skin," *J.A.M.A.* 27 (1896):777.
88 Although there had been several reports linking testicular cancer to the profession of chimney cleaners and thereby to carbon in soot, most nineteenth-century physicians would have balked at the idea that cancer could be caused by contact with substances outside the body.
89 Conard, D., "Dermatitis Roentgenii," *J.A.M.A.* 27 (1896):1014.
90 Skinner, C., "Local Action of the X-Rays," *J.A.M.A.* 27 (1896):1254.
91 Thomson, Elihu, "Roentgen Ray Burns," *Am. X-Ray Journal* 3 (1898):451-453.
92 Rollins, W., "X-Light Kills," *Boston Med. and Surg. J.* 144 (1901):173. A heated and prolonged exchange began in the journal with the reply by E. A. Codman, "No Practical Danger from the X-Ray," *Boston Med. and Surg. J.* 144 (1901):197.
93 Skinner, C., "Dermatitis from X-Rays, " *J.A.M.A.* 27 (1896):1070.
94 Gilchrist, T.C., "A Case of Dermatitis Due to the X-Rays," *Bull. Johns Hopkins Hosp.* 18 (1897):17-23.
95 Photographs of Kassabian at work in his Philadelphia laboratory have been widely published and are on file at the Center for the American History of Radiology in Reston, Virginia.
96 Kassabian, p. 406.
97 Ibid., p. 398.
98 The unnumbered appendices of Kassabian's books provide a wealth of information on the manner, extent, and technique of practice by over thirty of the world's foremost radiotherapists in 1907. The questionnaire was detailed and the responses show a remarkable range of practice and disagreement on the manner in which the X-ray should be applied.
99 Johnson, G., Letter, *Phil. Med. Journal* (1902):220-221.
100 Kassabian, *Röntgen Rays*, p. 408.
101 Richards, A., "The Biological Explanation of X-Radiation Effects," *Am. J. Roent.* 2 (1915):908-911.
102 "Tests with New Plates," *New York Times* (14 February 1896):2.
103 Lopriore, G., "Action of X-Rays on the Living Plant Cell," *Abstract in Botan. Centralbl.* 73 (1898):451.
104 There was a brief period of excitement in Europe when it was believed that the X-ray might cause a form of "perpetual youth" in plants, and by extension, humans. Radium revived this interest; see "Radium and Longevity," *J.A.M.A.* 43 (1904):617.
105 Schaudinn, F., "Uber den Einfluss de Röntgenstrahlen auf Protozoen," *Arch. f. Physiol.* 77 (1899):29.
106 Albers-Schönberg, L., "Uber eine bisher unbekannte Wirkung der Röntgenstrahlen auf den Organismus," *Münch. Med. Wschr.* 50 (1903):1859.
107 Bergonié, J., and Tribondeau, L., "Action des rayons X sur le testicule du rat blanc," *Compt. rend. Soc. de Biol.* 57 (1904):592; and "L'aspermatogenése expérimentale complète obtenue par les rayons X, est-elle definitive?" *Compt. rend. Soc. de Biol.* 58 (1905):678-680.
108 Perthes, G., "Versuche über den Einfluss der Röntgenstrahlen auf die Zellteilung," *Deutsche Med. Wschr.* 30 (1904):632.
109 Regaud, C., and Blanc, J., "Effets généraux produit par les rayons de Röntgen sur les cellules vivantes d'aprés les résultats observés jusqu'à présent dans l'épithélium séminal," *Compt. rend. Soc. de Biol.* 61 (1906):731.
110 Kassabian noted in his textbook that despite growing research that might indicate otherwise, he personally knew six men active in radiology who had fathered healthy children in 1906 and 1907, see Kassabian, p. 414. Such small sample optimism would not last long; later studies would reveal that many early radiologists were receiving sterilizing doses of the rays.
111 Bergonié, J., and Tribondeau, L., "Interpretation de quelques resultats de la radiothérapie et essai de fixation d'une technique rationale," *Compt. Rend. Acad. Sci.* 143 (1906):983-985.
112 del Regato looks briefly at the introduction and subsequent overreliance on the Bergonié and Tribondeau "law" in his biography of Claudius Regaud in *Radiological Oncologists: The Unfolding of a Medical Specialty*. Reston, Va.: Radiology Centennial, Inc., 1993.
113 Gilman, P.K., and Baetjer, F.H., "Effects of Roentgen Rays on Development of Embryos," *Am. J. Physiol.* 10 (1903-4):222.
114 Heineke, H., "Uber die Einwirkung der Röntgenstrahlen auf das Knochenmark nebst einigen Bemerkungden über die Röntgentherapie der Leukemie and Pseudo-leukemia und des Sarcoma," *Deutsch. Ztschr. f. Chirurg.* 78 (1905):196-230; and "Uber die Einwirkung des Röntgenstrahlen aus das Knochenmark," *Verland der Deutsch. Ges. f. Chirurg.* 34 (1905):22-24.
115 These competing theories were summarized in Richards, "Biological Explanations," pp. 908-909.
116 DeVries, H., "Aim of Experimental Evolution," *Carnegie Institute Year Book* 3 (1904):39.
117 Gray, Alfred, "A Brief Review of the Progress of Roentgenology in the Past Decade," *Am. J. Roent.* 2 (1915):869-870.

CHAPTER TWO

CLINICAL PRACTICE

James D. Cox, M.D.

▶ ▶ ▶ ▶ ▶ ▶ ▶ ▶ ▶ ▶ ▶ ▶ ▶

1910–1950

The second decade of the twentieth century was a time of profound and rapid change. The Model T had just been introduced by Ford. The potential of heavier-than-air flight was beginning to be realized. *Psychoanalysis* was published in 1910, and new disciples emerged from each lecture of Freud or Jung. Americans of enormous wealth, whether from railroads, oil, or financial investments, were establishing foundations that would serve the sciences, the humanities, and society at large. In the United States women's quest for the vote was on the verge of success.

Europe hovered on the brink of war, but extraordinary developments in science were occurring. Marie Sklodowska Curie had just been awarded the Nobel Prize for chemistry, the first scientist to receive the award twice and in different fields (physics in 1903 and chemistry in 1910). With the establishment of the Pasteur Pavilion of the Institut du Radium, the medical fruits of Curie's scientific labors had just begun to be realized. Medical applications of ionizing radiations began at the institute under the direction of Claudius Regaud in 1919, on his return from World War I (Fig. 2.1). Regaud characterized the task of the physicians who eventually would be known as radiation oncologists: "Surgeons and radiotherapists who undertake to treat curable cancer assume an exceptionally heavy responsibility because the unique stakes with which they play are the lives of their patients."[1] Throughout the history of the field the men and women who have pursued radiation oncology have been captivated by this challenge to treat diseases which, if left unchecked, would kill their patients. The intriguing intellectual blend of nuclear physics, human tumor biology and pathology, and nascent radiation biology met with progressively effective tools to deliver ionizing radiations to cure cancers.

In the United States therapeutic uses of ionizing radiations were by no means confined to cancer. Cutaneous inflammations of myriad etiologies, arthritis, epilepsy—all were subjected to exposures by poorly penetrating X rays. Only superficial effects were expected, and only they were observed. Roentgen-ray dermatitis and epilation were considered necessary, even desirable.

Fig. 2.1 Claudius Regaud (1870–1940). (Courtesy of the Center for the American History of Radiology, Reston, Va.)

The earliest oncologists, taking on the burden anticipated by Regaud, were physicians from three broad groups: radiological generalists, brachytherapists, and radiation oncologists.

The Generalists

The excitement generated by diagnostic images continued to compel a group of physicians to maintain a practice combining both radiodiagnosis and radiotherapy well into the second half of the twentieth century. During the early years the balance between the two was manageable, and much of the available apparatus could serve either purpose.

Francis Williams (1852–1936), a distinguished Boston physician from a family of Harvard Medical School graduates, followed the reports of Röntgen's discovery with fascination.[2] He quickly became aware not only of the diagnostic power of X rays but also their therapeutic potential. His widely read text, published in 1901 and entitled *The Roentgen Rays in Medicine and Surgery*, was subtitled *As an Aid in Diagnosis and as a Therapeutic Agent*. One of the hallmarks of general radiologists was their preference for fluoroscopy rather than film. Their treatment planning was done in real time, with the tumor first identified fluoroscopically and then treated as so localized. Williams was a proponent of fluoroscopy in his studies of diseases, which he pursued first in a physics laboratory at the Massachusetts Institute of Technology and later in the basement of the Boston City Hospital. He demonstrated cures of basal cell carcinomas of the eyelids and cheek and squamous cell carcinomas of the lower lip. He was among the first radiologists to note the responsiveness of Hodgkin's disease to X rays. Although he retired as head of the department of radiology at Boston City Hospital in 1915, his influence continued well beyond his death two decades later.

Another influential generalist was Mihran Kassabian (1870–1910), an Armenian who emigrated to the United States and eventually became director of the Roentgen Ray Laboratory of Philadelphia Hospital.[3] Although his death occurred at the very beginning of the period under discussion, he provided important documentation of the actual practice of roentgentherapy in the United States in his book *Roentgen Rays and Electro-Therapeutics*, published in 1907.[4] The text itself placed great emphasis on the management of radiation dermatitis. The comprehensive survey of practice throughout the United States documented a host of conditions, mostly benign, for which X rays were being administered. He, too, emphasized fluoroscopy in therapeutic practice but paid dearly for it: he died of widespread metastasis from carcinomas that had developed in the skin of his hands.

George Pfahler of Philadelphia, one of the most highly respected pioneers in the field, reflected on his own early practice after being chosen by the founders of the American College of Radiology (ACR) to be their first president (Fig. 2.2).[5] He had advocated a "saturation" method of radiotherapy in which more brief second and third applications were given at short intervals to maintain the effects of the first prolonged application (of course the roentgen as a unit of exposure was yet to be defined). His basis for doing this has recently been rediscovered: he

Fig. 2.2 George Edward Pfahler (1874–1957) (Courtesy of the Center for the American History of Radiology, Reston, Va.)

assumed tumors had a steady loss of radiosensitivity with time, much like the constant rate of loss seen in biochemical processes, so he compressed the delivery of the irradiation to compensate for it, by "topping-up" with repeated applications after the first one.

Preston M. Hickey (1865–1930) was professor of radiology at the University of Michigan in Ann Arbor.[6] He coined the term "cones" for the pyramidal localizing attachments to the tube heads that dominated roentgentherapy. Light localizers became commercially available in the 1940s.

Although dermatologists were not generalists in the same sense as those who combined roentgentherapy and roentgendiagnosis, they were understandably at the forefront of the practitioners of roentgentherapy in the earliest years of this era. They were the diagnosticians of the vast array of cutaneous conditions for which irradiation was thought a possible remedy, and they assumed expert status in the management of the dermatitis thought to be a desirable result of X-ray treatments. Among the most influential was William A. Pusey of Chicago; with Eugene W. Caldwell, he published the definitive text of this period, *The Roentgen Rays in Therapeutics and Diagnosis*.

The Brachytherapists

Several physicians became intrigued by the potential for placing sources of radiations directly in or immediately adjacent to tumors. The sources contained radium or its gaseous daughter, radon. Francis Williams was as prominent in the use of brachytherapy as he was with roentgentherapy. He purchased his personal supply of radium in Paris and brought it to the United States encapsulated in glass tubes. He and such pioneers as Margaret Abigail Cleaves (1848–1917) of New York placed these tubes in the vagina for cancer of the cervix, applied them to the breast for inoperable mammary cancer, and used them to treat primary carcinomas of the skin such as rodent ulcers.[7]

Many of the early brachytherapists were surgeons. One of them, Robert Abbé (1851–1928) of New York developed a method, now recognized as "afterloading," for cancer of the thyroid.[8] He placed rubber tubes in the neck after thyroidectomy for cancer and then loaded them with glass tubes of radium to enhance the margins of his resection. He also used radium that he had imported from Germany in a specially designed vaginal applicator for the treatment of cancer of the cervix.

Howard A. Kelly (1858–1943), a professor of gynecology and one of the original faculty of the Johns Hopkins Medical School, teamed with James Douglas, an engineer–industrialist, and Charles Parsons, director of the United States Government Bureau of Mines, to buy Colorado mining claims which yielded carnotite.[9] They financed the processing of this uranium-containing mineral to derive radium which could be sold for more than $100,000 per gram. By 1914 they began to donate grams of radium to major cancer centers, such as the Memorial Hospital of New York and Johns Hopkins University Hospital. Cancer of the cervix was the most widely discussed malignant tumor of women at the time, and Kelly, Leda June Stacey, and Margaret Abigail Cleaves were among the first physicians who treated it with intracavitary "curietherapy."

Charles L. Martin (1893–1979) of Dallas was unusual in several ways. He continued to practice general radiology for the early years of his career (he was certified in "radiology" by the American Board of Radiology [ABR] in 1934). However, his therapeutic interests were in brachytherapy. He had the rare if not unique distinction of serving as president of both the American Roentgen Ray Society and the American Radium Society. He did not confine his practice to radiation oncology until 1948.[10]

The Radiation Oncologists

The third group of practitioners included those who quickly confined their practices entirely to the therapeutic uses of radiations for cancer. They were both curietherapists and roentgentherapists. This group was the smallest of the three in the first half of the twentieth century, but was eventually to develop a "critical mass" in the United States. As sophistication in radiodiagnosis developed, imaging procedures were sought for an increasing number of diseases, and diagnostic radiologists, together with pathologists, became consultants to their colleagues. The radiation oncologists, driven by the desire to help their patients, accepted the expertise of their diagnostic colleagues and devoted themselves full time to the search for the physical and biological bases to effect the cure of cancer. After World War II they steadily took over the practice of therapeutic radiology.

One of the original radiation oncologists was Albert Soiland (1873–1946), who was born in Norway and came to the United States at the age of ten. While a medical student at the University of Southern California, he helped build a roentgen-ray generator. In 1904, only four years after his graduation, he was asked to organize the University of California's first department of radiology. Although he was very influential in the rapidly expanding field of radiology (president of the Radiological Society of North America [RSNA] in 1922, founder of the ACR in 1923), Soiland was a pioneer radiation oncologist. He advocated surgical adjuvant radiation therapy for cancer of the breast. He published works on radiation therapy for carcinomas of the oral cavity, cervix, and prostate, as well as leukemia. He even attempted intraoperative radiation therapy (1923).[11]

In his biography of Albert Soiland, Juan del Regato called attention to the birth of supervoltage radiotherapy. In 1930 Soiland asked R. A. Millikan (1868–1954) and C. C. Lauritsen (1892–1968) of the California Institute of Technology to permit a trial of clinical radiotherapy in their laboratory of physical research with their new high-voltage roentgen-ray tube and 750,000 volt generator (Fig. 2.3).[12] Soiland was allowed to bring patients at night for a trial of high voltage (550 kV) radiotherapy of cancer; thus, he was the earliest radiotherapist in the world to enjoy this privilege. The first patient treated, in October 1930, was Dr. C. Edgerton Carter, who suffered from an inoperable carcinoma of the rectum: two years later he was reported "symptomless and working normally having recovered twenty pounds in weight."

By the mid-1920s the Mayo Clinic already was organized into separate sections of radiodiagnosis and radiotherapy. Radiotherapy itself was divided, with Arthur Desjardins as director of Therapeutic Roentgenology and Henry Bowing as chief of radium therapy (Figs. 2.4a and 2.4b).[13] Desjardins, a gentleman in all senses of the word, was a fine representative of early radiation oncologists in an environment where the

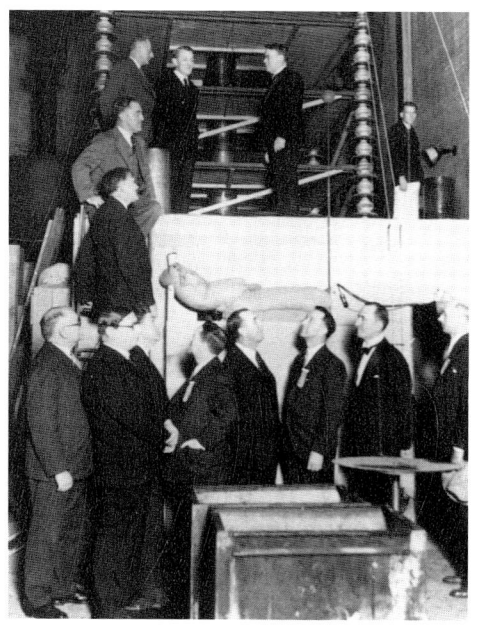

Fig. 2.3 Albert Soiland (1873–1946) standing at top left on platform with Lauritsen and Millikan. The supervoltage generator was at the California Institute of Technology in Pasadena, California. (Courtesy of the Center for the American History of Radiology, Reston, Va.)

Fig. 2.4a Arthur Ulderic Desjardins (1884–1964) (Courtesy of the Center for the American History of Radiology, Reston, Va.)

Fig. 2.4b Harry Herman Bowing (1884–1955) (Courtesy of the Center for the American History of Radiology, Reston, Va.)

skill of his surgical confrères had become world renowned. He contributed numerous reviews of acute and late effects resulting from irradiation of normal tissues. His scholarship and demeanor undoubtedly were important in the Mayo Clinic's recognition of the independent specialty of therapeutic radiology.

In the United States and throughout the world, the major limitation of irradiation with the apparatus available was the magnitude of the effects on the skin and subcutaneous tissues. Except for the handful of experiences with high kilovoltage units such as that available to Soiland, the skin was the dose-limiting normal tissue. When considering irradiation of deep-seated tumors, treatments were delivered to the tolerance of the skin with only a hope that the tumor in depth received sufficient scatter electrons to kill some of its cells. Multiple portals of entry, which allowed cross-firing the site of the tumor, were required; it was common to use at least four portals of entry, but six or eight ports were frequently used.

Radium could be placed in natural body cavities, and eventually it became possible to insert radium needles or radon seeds into tumors. Large lesions of the surface, such as advanced carcinomas of the breast and skin, usually could not be treated with radium because of the large quantities needed. Such quantities were available in a few clinics, but preference was given in these circumstances to the use of X rays.

Henri Coutard, forty-three years old and just home from World War I, was among the small group of physicians who formed the staff of the new Radium Institute of the University of Paris in 1919, under the direction of Regaud, whom he had met during the war (Fig. 2.5).[14] Coutard was intrigued by the work of Regaud on the testes of several animals: a single large application of radium or X rays produced radiodermatitis, the moist desquamation of the skin, which was the dose-limiting phenomenon with roentgentherapy, but spermatogenesis was not greatly affected. When smaller doses were given several days apart, a lesser degree of radiodermatitis resulted, with a dry desquamation of the skin, but spermatogenesis ceased permanently. Coutard applied this approach to patients who were presented to him with inoperable carcinomas of the larynx and pharynx. He prolonged the treatments over several weeks (whereas Regaud considered ten days of treatment to be the maximum). By the time the roentgen unit was defined and internationally accepted in 1928, Coutard already had presented results of a series of patients with carcinomas of the tonsillar fossa who were cured by roentgentherapy. Coutard's protracted-

Table 2.I

Malignant Disease of the Hypopharynx, 1938

		Operable	Prog.* inoperable	Tech. inoperable	Total cured
1929–31	New method	3 of 3	11 of 16 (65%)	16 of 39 (41%)	30 of 59 (50.8%)
1919–29	All cases	0 of 17	0 of 34	0 of 69	0 of 120
1919–29	Surgery	0 of 6	0 of 0	0 of 0	0 of 6
1919–29	Radiotherapy	0 of 11	0 of 34	0 of 69	0 of 114

From Pohle, E.A., ed. *Clinical Roentgen Therapy.* (Philadelphia: Lea and Febiger, 1938):154. Reprinted with permission.

*Prognostically.

fractional method of roentgentherapy was quickly adopted by radiation oncologists around the world.

Maurice Lenz and William Harris, both from New York City, were among many visitors to the Radium Institute of Paris in the 1920s (Figs. 2.6a and 2.6b).[15] Lenz practiced at the Montefiore Hospital and the hospitals affiliated with Columbia University. Harris practiced at Mount Sinai Hospital. In 1931 Coutard made an influential presentation at Mount Sinai Hospital, at the invitation of Harris, on successful roentgentherapy for inoperable cancer of the larynx and oropharynx.

Recognition of the importance of fractionation profoundly affected the practice of radiation oncology, as well as the expectations of both patients and practitioners. Standard textbooks, for example *Clinical Roentgen Therapy*, edited by Ernst A. Pohle, chair of the department of radiology and physical therapy of the University of Wisconsin, reflected this new understanding (Fig. 2.7).[16] Gordon E. Richards of the University of Toronto observed, in the 1938 edition of Pohle's text, "The improvement which has followed upon the adoption of the method introduced by Coutard, of Paris, and known by his name has been little short of spectacular."[17]

Whether fractionated or not, radiation therapy was fully appreciated for its effects on markedly radiosensitive tumors such as malignant tumors of lymphoid structures (lymphosarcomas and Hodgkin's disease), seminoma, myeloma, and Ewing's sarcoma of bone. These malignant diseases were recognized not only as strikingly responsive to irradiation but even potentially curable. In addition, the tumor, which had been recognized

Fig. 2.5 Henri Coutard (1876–1950) (Courtesy of the Center for the American History of Radiology, Reston, Va.)

Fig. 2.6a Maurice Lenz (1890–1974) (Courtesy of the Center for the American History of Radiology, Reston, Va.)

Clinical Practice

Fig. 2.6b William Harris (1894–1963) (Courtesy of the Center for the American History of Radiology, Reston, Va.)

Fig. 2.7 Ernst Albert Pohle (1895–1965) (Courtesy of the Center for the American History of Radiology, Reston, Va.)

concurrently in France by Regaud and in Germany by Schmincke and called lymphoepithelioma, was also recognized as exquisitely sensitive and often curable. The far more common variants of squamous cell carcinoma of the upper respiratory and upper digestive tracts were not consistently curable until the fractionated method of Coutard was adopted (Table 2.I).

Mammary carcinomas were recognized as moderately responsive from the earliest, crude administrations of ionizing radiations. Preference was given to mastectomy for all patients who were operable. However, Lenz reported in 1946 that ten of thirty-one patients deemed inoperable by the criteria of Haagensen and Stout were alive and free of local recurrence or distant metastasis more than five years after roentgen therapy.[18]

Carcinomas of the cervix were the most important conquest of radiation therapy in the era of roentgen therapy, so much so that many radiation oncologists became recognized as leading experts in this disease, contributing to knowledge of clinical evolution, staging, classification, treatment, and prognosis. Successful treatment of cervical cancer, however, was not a consequence of external irradiation, but rather a triumph of brachytherapy. The Stockholm (Radiumhemmet) technique of brief, intensive applications of simultaneous intrauterine and vaginal radium repeated twice; the Paris (Radium Institute) technique of applying a colpostat for four or five days, followed by an intrauterine "tandem" for a similar period; and the Manchester (Holt Radium Institute) variation of the Paris technique all proved strikingly effective for women whose tumors were manifestly inoperable. It was evident that curability by brachytherapy alone was confined to tumors that were not bulky, i.e., which were of lower stage. The success with brachytherapy, however, led to many more innovative and persistent means of external roentgen therapy. Multiple portals of entry were required to achieve maximum doses in depth, since each field was limited by the tolerance of the skin and subcutaneous tissues. There was no clearer example of treating to the tolerance of the skin and hoping the tumor deep within the pelvis received enough total dose.

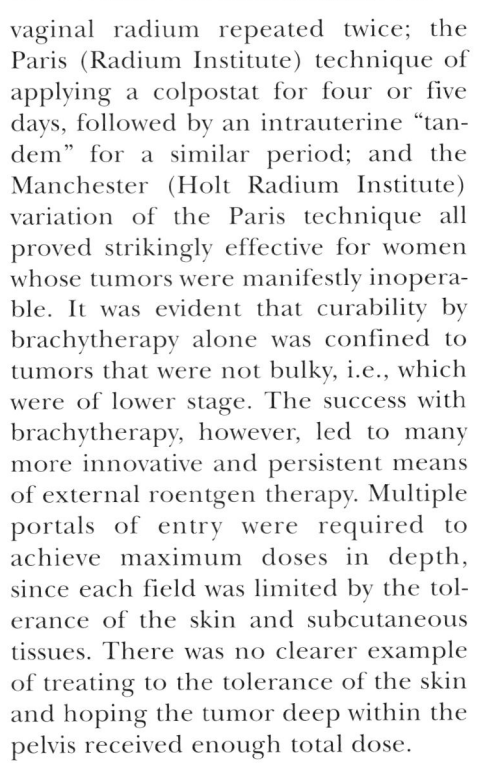

Major forms of cancer, notably carcinomas of the lung, esophagus, prostate, and bladder, were not considered curable with conventional X rays. They required the development of equipment that produced far more penetrating radiations, equipment that would become widely available in the supervoltage era. But the fact that conditions were not considered curable did not prevent physicians from using palliative radiations to treat swelling and to relieve pain.

The X-ray era ended as it began, with optimism and foreboding. The hopes and fears were far more directly related to ionizing radiations than they were at the beginning of the era. The end of World War II was, itself, a cause for vast relief in the United States, but the troubled peace of the Cold War was increasingly evident. The cloud first evident over Hiroshima and Nagasaki, no longer visible but just as threatening, made physicians wonder why a colleague would seek to spend his or her career in such a hostile environment.

In spite of the dramatic therapeutic progress made with X rays and radium, some generalists held that radiation oncology was not a fundamentally different discipline than radiodiagnosis. Throughout the post-World War II era and even into the 1980s, prominent figures in radiology contended that radiation therapy was a relatively trivial enterprise—a sideline to the larger field of radiological diagnosis. They contended that radiation oncology was not a newly emerging specialty and that it should not be recognized with separate administrative or departmental status. They felt it could be practiced quite satisfactorily on a part-time basis by any well-trained radiologist. This viewpoint served as a major impediment to progress in some of the most prestigious institutions in the United States.

There continued to be little interest among medical students in the United States in the fledgling discipline of radiation oncology. The handful of practitioners, no more than a few dozen, most of whom immigrated from Europe or had trained there, were aware that supervoltage radiation therapy was soon to be a powerful instrument in their hands for the care of patients with cancer. As a forerunner of progress to come, Juan A. del Regato, yet another student of Coutard, brought his clinical experience together with the tumor pathology expertise of Lauren Ackerman to publish in 1947, *Cancer: Diagnosis, Treatment, Prognosis*, the first comprehensive textbook in which radiation therapy was presented side by side with surgery (Fig. 2.8).[19] This pre-

Fig. 2.8 Front cover of *Cancer Research*, dedicated to the authors of *Cancer*, showing the first three editions of the English version by Ackerman and del Regato. On the far left is the Polish edition, and on the far right, the Spanish edition by del Regato and Ackerman. (Courtesy of the Center for the American History of Radiology, Reston, Va.)

Fig. 2.9a Franz Julius Buschke (1902–1983) (Courtesy of the Center for the American History of Radiology, Reston, Va.)

Fig. 2.9b Simeon Theodore Cantril (1908–1959) (Courtesy of the Center for the American History of Radiology, Reston, Va.)

saged the enormous contributions radiation oncologists were to make to the care of patients with cancer in the next two decades.

SUPERVOLTAGE ERA: 1950–1970

The supervoltage era actually dawned well before 1950. The clinical exploitation of high-energy X rays by Albert Soiland has already been noted. The first X-ray tube capable of operating at one million volts was developed by another Scandinavian, Charles Lauritsen, a Danish-born engineer. With the assistance of the 1923 Nobel Laureate in physics, Robert A. Millikan, and the General Electric X-ray Corporation, Lauritsen succeeded in establishing a 700,000-volt unit for clinical use at the California Institute of Technology.

One of the earliest systematic explorations of supervoltage roentgentherapy was by Franz Buschke and Simeon T. Cantril at the Tumor Institute of the Swedish Hospital in Seattle (Figs. 2.9a and 2.9b).[20] Buschke was a product of the demanding German educational system. He graduated from the University of Berlin and became one of the rising stars of Hans Schinz at the Röntgeninstitut of the University of Zürich. He was profoundly affected by a brief period as an observer at the Radium Institute in Paris, where he first met Cantril. These two pioneer radiation oncologists thus had come under the influence of Henri Coutard. When they became responsible for the equipment at the Swedish Hospital in 1938, they were among only thirty-nine radiation oncologists in the United States.[21] From the same careful daily observation methods learned from Coutard, they concluded that supervoltage irradiation had definite advantages over orthovoltage roentgentherapy. The record of their observations, presented in a monograph, *Supervoltage Roentgentherapy*, published in 1950, marked a turning point in radiation oncology.[22]

There was considerable disagreement as to the value of supervoltage roentgentherapy. Sherwood Moore, professor of radiology at Washington University in St. Louis, wrote to Juan del Regato, "I have observed these generators in operation at the California Insitute of Technology, Memorial Hospital in New York, Huntington Hospital in Boston and the Massachusetts General Hospital in Boston and elsewhere. My opinion after years of observation is unmistakable against the use of million volt x-radiations in the treatment of disease."[23] However, the weight of opinion among most radiation oncologists by the end of World War II was so great that they sought two-million-volt generators rather than accept lower energies.

Gilbert Fletcher collaborated with Leonard Grimmett, a physicist from England who joined the M. D. Anderson Cancer Center in 1949, in the design of the first cobalt-60 (^{60}Co) teletherapy unit. Grimmet had already designed a teletherapy unit containing 5 grams of radium that became the primary unit for treatment of cancer of the head and neck at the Royal Marsden Hospital in London.[24,25,26] The ^{60}Co containing device he developed with Fletcher was approved by the United States Atomic Energy Commission in 1950 and became a principal treatment unit at M. D. Anderson

Juan del Regato (b. 1909)

Juan Angel del Regato was born in Camaguey, Cuba, the son of Damiana Manzano Nuñez, of Mayan ancestry, and of Juan del Regato Castanedo, of Castillian ancestry. He received his high school education in Santa Clara, Cuba, from 1922 to 1926, then undertook the study of medicine at the University of Havana from 1926 to 1930. When the university was closed due to political unrest in 1930, the Cuban Liga Contra el Cancer sponsored del Regato's continued studies at the University of Paris through 1934. He received his medical degree in 1937 with a medal-winning thesis on successful radiotherapy of inoperable cancers of the maxillary antrum. He also followed a two-year course and received the Diploma of Radiophysiology and Radiotherapy of the University of Paris. He served as an assistant at the Radium Institute of Paris, where he came to know Coutard, Regaud, Marie Curie, Lacassagne, and other founders of the field of radiation therapy.

(Courtesy of the Center for the American History of Radiology, Reston, Va.)

Early in 1937 del Regato came to the United States, bringing with him the Paris tradition of specialization at a time when only a handful of American physicians practiced radiation therapy exclusively. He was a research fellow of the National Cancer Institute (1941–1943); director of radiotherapy at the Ellis Fischel Cancer Center in Columbia, Missouri; and director of the Penrose Cancer Hospital in Colorado Springs (1949–1974). He became professor of radiology at the University of South Florida, Tampa (1974–1981), and emeritus professor of radiology (1981) and distinguished physician of the Veterans Administration. Among numerous awards and honors he has received gold medals from the Radiological Society of North America (1966), the American College of Radiology (1968), and the American Society for Therapeutic Radiology and Oncology (ASTRO) (1977) as well as the American Medical Association Scientific Achievement Award (1993).

Unique in the field, del Regato is noted for his clinical, technical, scholarly, and organizational achievements. In the clinic, his active involvement in ongoing studies and interest in developing new delivery systems led to the development of the del Regato localizer—the first light localizer in the field. His writings include the groundbreaking *Cancer*, with Lauren Ackerman, and a growing body of historical works tracing the personalities and events which have shaped radiation oncology. He has served as an officer in numerous organizations, and it was through his personal efforts that the American Club of Radiotherapists was formed in 1958. Today the organization is ASTRO, the world's largest association devoted to radiation oncology.

del Regato's greatest and most lasting contribution to the field is his ongoing dedication to training young radiation oncologists. The legion of radiation oncologists who trained with del Regato now numbers in the hundreds, and their respect and fondness for the "Chief" is reflected in their support for the activities of the del Regato Foundation, organized to further educational and academic achievement in the field. Dr. del Regato, whose rich and productive life has spanned almost the entire history of the field, remains an active and vital contributor to radiation oncology.

in 1954. A 22-megavolt (MV) Allis Chalmers betatron was installed a year later. With these resources and expanding therapeutic activities, Fletcher was able to divest himself of most of his diagnostic responsibilities.

The clinical observations that were published by Fletcher and his colleagues over the next twenty years changed the practice of radiation oncology not only in the United States but throughout the world. Their results were especially striking for patients with cancer of the upper respiratory and digestive tracts, for whom ^{60}Co teletherapy became a salvation, and for advanced cancer of the cervix, which was shown to be curable in high proportion with skillful use of the betatron.

Although much slower in finding widespread use, the earliest medical linear accelerators were also under development at this time. Henry Kaplan joined the staff of the Stanford University Medical School (then located in San Francisco) in 1948. He was aware that the existing megavoltage accelerators, 1- and 2-mega-electron-volt (MeV) resonance transformers and Van de Graaff electrostatic generators, available to him produced less penetrating photon beams (HVL ≈ 7 mm lead [Pb]) than ^{60}Co gamma rays (11 mm Pb HVL). So, he sought out members of the physics department who had been working for more than a decade on linear accelerators. Edward Ginzton (b. 1915) of W. W. Hansen's group, which had developed a variety of microwave devices between 1938 and 1947, was a major collaborator from the start. Kaplan was the driving force from the medical perspective.[27] Many other scientists, including Russell and Sigurd Varian, C. J. Karzmark, and J. Haimson, contributed to the design, development, and clinical implementation of the first linear accelerator used to treat patients in the United States. It was introduced at Stanford in 1955 and produced 5 MV X rays.

It should be noted that D. W. Fry's group at the Telecommunications Research Establishment (TRE) at Great Malvern, England, was actively investigating the same phenomena and devices at the same time. The first linear accelerator installed in a hospital to treat patients with cancer was introduced in Hammersmith Hospital, London, in 1953. It produced an 8-MV photon beam in a fixed horizontal structure.[28] The first isocentric linac with a 350-degree rotational capability was introduced by Varian in 1962, the first dual photon and electron unit was offered by Applied Radiation in 1965, and the first commercially available treatment unit which produced a photon beam above 15 MV (plus 12 to 32 MeV electrons) was the Sagittaire model of CGR, Paris, 1967.

The most obvious phenomenon in radiation oncology in the 1950s was the rapid acquisition of ^{60}Co teletherapy units or supervoltage X-ray generators. The early proliferation of such units is poorly documented. By the time the first Facilities Master List was generated by the Patterns of Care Studies (PCS) in 1975, there were 970 ^{60}Co teletherapy units and 407 linear accelerators and betatrons.[29]

It is difficult to overestimate the revolution that transpired in cancer care. Whereas the skin had once been the dose-limiting normal tissue and profound effects could be observed quickly, it was now possible to deliver total doses higher than ever before contemplated but with very little effect on the skin. This led to errors in the directions of too much and too little dose. Some of the rapidly proliferating supervoltage units reached well-meaning practitioners of general radiology who delivered too much radiation, as there were none of the usual reactions to follow. The resulting complications led to detailed studies of late radiation effects in deep-seated normal tissues—the heart, lungs, gastrointestinal tract, spinal cord, bone—which previously would have had little relevance. At the other end of the spectrum there were underdosages of tumors resulting from misinterpretations of units of radiation exposure and absorbed dose, understandable in a discipline that was changing from thinking of "roentgens-in-air" to point calcula-

Henry Kaplan (1918–1984)

Henry Seymour Kaplan was born in Chicago in 1918, the son of Sarah Brilliant and Nathan M. Kaplan. He received a B.S. degree from the University of Chicago and his M.D. degree from Rush Medical College in 1940, then served an internship at the Michael Reese Hospital. He went to Minneapolis for training in general radiology under the prestigious radiodiagnostician Leo Rigler at the University of Minnesota. There he received his first exposure to radiation therapy under Karl Stenstrom, Ph.D. (1891–1973). With Rigler, Kaplan made an interesting study of early detection of cancer of the stomach.

He was certified in general radiology by the American Board of Radiology in 1944 and took a position as an instructor and assistant professor of radiology at Yale University Medical School, where he pursued radiodiagnostic interests. A laboratory worker at heart, Kaplan took a fellowship to do experimental research at the National Cancer Institute, where he did original work on leukemia.

(Courtesy of the Center for the American History of Radiology, Reston, Va.)

While in Bethesda he was offered the position of professor of radiology at the Stanford University Medical School. The position called primarily for competence in radiodiagnosis, but Kaplan chose to emphasize the needs of therapeutic radiology with which he was less conversant but in which he achieved undisputed leadership. At that time, most patients with Hodgkin's disease were being inadequately treated by general radiologists under the assumption that small doses of radiation were sufficient to give lasting palliation. Kaplan had learned that it was possible to sterilize the tumor locally with moderately high doses and also that it was advantageous to irradiate the neighboring areas of potential subclinical involvement. He made this his own subject and wrote extensively on it. He recommended vast fields irradiating the cervical, thoracic, and abdominal areas of potential involvement with moderately high doses in relatively short time. This approach was widely adopted, not without untoward effects resulting from the emphasis on total dose and a lesser concern about fractionation. Kaplan went on to recommend staging laparotomies and splenectomies, which led to a vast accumulation of knowledge about the natural history of Hodgkin's disease.

The need for a training program for radiation oncologists was evident, and Kaplan, in his position as a member of the National Advisory Cancer Council, contributed to the extension of federal grants for the establishment of training centers.

When Edward Ginzton undertook to build a linear accelerator at Stanford, Kaplan encouraged him and endeavored to put its originally fixed horizontal beam to the treatment of various forms of cancer with unquestionable success. Kaplan emphasized the role of radiobiological research, putting his hopes successively in electrontherapy, hibaroxyc therapy, radiosensitizers, and other promising approaches.

He was a founder and president of the Radiation Research Society, elected to the National Academy of Sciences, received the Atoms for Peace Award, the Charles L. Kettering Award, and the gold medal of the American Society for Therapeutic and Radiation Oncology—all richly deserved. An ambitious man, he was kind and generous with his associates and residents. As had his father, Kaplan died of cancer of the lung, on 4 February 1984.

—Juan A. del Regato, M.D.

tions of roentgens at depth and more comprehensive dose distributions expressed in rads.

Supervoltage therapy permitted adequate doses to be delivered in depth for diseases thought intractable to radiation therapy: cancer of the lung, esophagus, bladder, and especially carcinoma of the prostate. In a short period of time advanced inoperable adenocarcinomas of the prostate came to be the province of the radiation oncologist. Urologists were astounded to find the palpable evidence of these tumors steadily disappearing, usually over a period of many months. Juan del Regato, Malcolm Bagshaw, and Frederick George each made independent observations of the effectiveness of ^{60}Co gamma rays and 6 MV X rays.

The demonstration that early Hodgkin's disease could be cured with extensive supervoltage irradiation also astounded the medical community. Since the disease afflicted a largely youthful group of patients, the impact of the results was greater than the numbers of patients affected would suggest. The conceptual framework for irradiating lymphatic regions which were not clinically involved had been laid by Rene Gilbert (1892–1962) forty years earlier. Vera Peters (1911–1993) and her colleagues at the Princess Margaret Hospital of Toronto had demonstrated the value of a clinical staging classification and wide-field irradiation; they achieved extraordinary results, especially in patients diagnosed in "early" stages (bipedal lymphography was not available at that time). The results were so extraordinary that they were dismissed by many radiologists and hematologists; patients continued to be treated with brief courses of X rays or individual cytotoxic drugs such as nitrogen mustard. Henry Kaplan and his Stanford University hematologist colleague, Saul Rosenberg (b. 1928), were responsible for the confirmation of the effectiveness of relatively high dose extensive irradiation and especially for the education of the medical community of this efficacy.

Radiation therapy as a postoperative adjuvant to radical mastectomy became widely used in the care of women with carcinoma of the breast. Whereas conventional X rays were suitable for irradiation of the chest wall, the use of ^{60}Co teletherapy permitted more careful modulation of the reactions and, in addition, the peripheral lymphatic regions could be treated with little reaction. The value of postoperative irradiation continued to be the subject of debate between strong proponents and detractors: the former emphasized the reduction in local-regional recurrence, and the latter noted the lack of increase in survival rates.

It quickly became evident that cancer of the lung benefited from megavoltage irradiation far more than it had from conventional X rays. Palliation of most symptoms could be achieved consistently. The questions were whether survival of patients with inoperable tumors was altered and whether it depended on the total dose delivered. The most definitive study during this period was launched in the early 1960s by investigators of the Department of Veterans Affairs. Radiation therapy was compared with supportive care. In spite of the fact that 90 percent of participating institutions did not yet have supervoltage equipment and the total dose prescribed was rather low, and in spite of the fact that small cell carcinomas constituted one-fifth of the patients treated, a small but statistically significant improvement in survival with radiation therapy was reported to the scientific assembly of the RSNA in 1966.

The price that had to be paid for this revolution in cancer treatment was a demand for radiation therapy nationwide far exceeding the available physicians, equipment, and support staff. A compelling solution for this dilemma was to treat patients less often with larger individual doses. Several formulas, based on laboratory data or acute effects on the skin of man and animals, were thought to facilitate fewer treatments with the same results. The most popular of these mathematical expressions in the United States was that developed by

Gilbert Fletcher (1911–1992)

Gilbert Hungerford Fletcher was born in Paris in 1911, the son of Marie Auspel of Auvergne and of Walter Scott Fletcher, an American residing in France who would die when Gilbert was only three years old. In 1929 he graduated from the private Stanislas High School and registered at the Sorbonne to study Latin, Greek, and philosophy. His older brother moved the family business to Belgium, and Gilbert finished a baccalaureate in engineering at Louvain in 1932. He also received a masters degree in mathematics from the University of Brussels in 1935.

His interests turned to medicine, and he received his medical degree in 1941, during the German occupation. As the son of an American citizen, he was entitled to a United States passport, permitting him to leave Belgium. After a residence of only a few months in gynecology at the French Hospital in New York, he entered training in general radiology at the New York Hospital, where he met and married Mary Walker Critz, a resident in pediatrics. He was certified by the American Board of Radiology in 1945.

(Courtesy of the Center for the American History of Radiology, Reston, Va.)

As a captain in the United States Army, Fletcher was assigned as a radiologist to the Veteran's Office in Pittsburgh. He met Randolph Lee Clark, who was recruiting staff for a new cancer institution in Houston. But Fletcher's weakest point in his training and experience had been radiotherapy. He was offered a fellowship and spent several months in Paris, Stockholm, London, and Manchester as an observer. He was particularly impressed by Baclesse and Paterson. On his return in 1948 he was appointed head of the department of radiology at M. D. Anderson Hospital, with responsibilities in radiodiagnosis and radiotherapy.

In the feverish post-war development of radiotherapy he was an unknown, but, blessed with native ingenuity and an abundance of patients, no one rose as rapidly to acquire experience and be recognized as an authority. He developed techniques and pragmatic gadgets that were widely adopted. An indefatigable worker, he made systematic analyses of his own experiences, pointing out causes of failure and complications. His lectures, embellished by his Gallic accent, were remarkably educational presentations of his abundant material. He was also an incisive debater. With his associates he contributed various devices for brachytherapy of cancer of the cervix. He contributed to the development of one of the first cobalt-60 units.

It was not until 1965 that he relinquished his diagnostic obligations to devote himself to the utilization of supervoltage and the investigation of hibaroxyc- and electron-therapies.

Fletcher was a founding member of the International Club of Radiotherapists (1953), president of the American Radium Society (ARS) (1963), and of the American Society of Therapeutic Radiologists (ASTRO) (1967). As chairman of the Committee on Radiological Studies he endeavored to initiate meaningful clinical cooperative studies and trials of combined radiotherapy and chemotherapy. In his brilliant career he earned the Béclère and Janeway medals, as well as the gold medals of the Radiological Society of North America, ASTRO, and the ARS. His associates and former residents founded a Fletcher Society. A man of innate genius and pragmatic resourcefulness, he enjoyed their adulation. On 11 January 1992 Fletcher died of heart failure.

—Juan A. del Regato, M.D.

Fig. 2.10 Victor Marcial (b. 1924) (Courtesy of the Center for the American History of Radiology, Reston, Va.)

Frank Ellis in the 1960s. Ellis's move to this country after his retirement from Oxford University in 1970 gave further impetus to this practice.

Victor Marcial (b. 1924, Fig. 2.10) conducted a survey of fractionation practices in 1965 as an activity of the Committee for Radiation Therapy Studies (CRTS).[30] A questionnaire was sent to the members of the American Society of Therapeutic Radiologists and to selected centers in Canada, the United Kingdom, and France. Of the ninety-five institutions represented by the replies, thirty expressed their units in roentgens (R), twenty-six used rads, and thirty-nine did not specify the units used. He found that most institutions in the United States and Canada treated with single fractions daily, five days per week; the usual fraction size was 200 R or rads, and the total doses ranged from 5,000 to 7,000 R or rads. The British centers usually completed treatment in three or four weeks. There was considerable variability, however. Many institutions treated patients for two weeks and then gave them a rest just at the time mucosal reactions would be expected. This was especially popular when it was unclear whether the patient was being treated with palliative or curative intent; the patient was sent away for two or three weeks, and if he or she returned in improved condition, additional irradiation was given.

In the late 1950s another new trend was seen in the field, less obvious than but of equal importance with supervoltage therapy. This was the expansion of training programs emphasizing radiation oncology. Initially an increase in radiation oncology training was supported by the National Cancer Institute (NCI). The new training programs were designed to emphasize *oncology* at least as much as *radiation* in the emerging specialty. Only a few of the earliest full-time practitioners of radiation oncology had been certified by the ABR in therapeutic radiology, in contrast to the much more common certification in radiology, both diagnostic and therapeutic. This new generation of radiation oncologists, trained in the United States, began to augment the influence of their mentors. The full expression of the training expansion was not seen until the 1970s and 1980s. By the mid-1980s all available residency positions in radiation oncology were filled, mostly with candidates who had graduated with superior records from medical schools throughout the United States.

The third phenomenon, formalized clinical research in oncology, sputtered for more than a decade before gradually establishing a firm foundation.[31] Several prospective comparative clinical trials were begun by the Office of Veterans Affairs, the Eastern Cooperative Group in Solid Tumor Chemotherapy, the Surgery Adjuvant Breast Group, and two groups that were endorsed by the CRTS, one for Hodgkin's disease and another for carcinoma of the prostate. Each suffered. Either the leadership was primarily interested in a modality other than radiation therapy, or there was insufficient infrastructure—data managers, research nurses, biostatisticians, and physicians—to sustain studies which, by their very nature, required many years for completion.

The CRTS, initially chaired by Gilbert Fletcher, had been formed with the encouragement of Kenneth Endicott, director of the NCI in 1963. Endicott wished to have a single set of recommendations from the radiation oncology community rather than the many bits of advice he was receiving. In addition to development of training grants in radiation oncology, CRTS endorsed the concept of a cooperative group devoted to radiation oncology research. The seventeen cooperative groups which had been formed in the mid-1950s had little interest in questions involving radiation therapy. The Hodgkin's and prostate trials that had started in 1967 had required entirely different groups of investigators and statistical centers, and CRTS wanted to develop a cooperative group which would not have to be reinvented with each new study.[32]

Simon Kramer (b. 1918, Fig. 2.11), a member of CRTS, wished to pursue a study of intravenous methotrexate prior to radiation therapy for advanced carcinomas of the upper respiratory and digestive tracts. This was endorsed by CRTS, and he received a grant from the NCI to organize a group, the Radiation Therapy Oncology Group (RTOG). The original award was to the Jefferson Medical College in 1968, but the grantee became the ACR in 1975, the first award that provided direct support to participating institutions.

The period between 1950 and 1970 in the history of the practice of radiation oncology in the United States can best be characterized as having generated much momentum. Supervoltage radiation therapy was immediately superior to conventional X rays with regard to reactions in the skin and subcutaneous tissues. It could be used to relieve symptoms from advanced cancers like skeletal and cerebral metastasis, cancer of the lung, esophagus, pelvic tumors, and it seemed able to eliminate some tumors in the short run: Hodgkin's disease, inoperable carcinomas of the prostate, and advanced cancer of the cervix. The full expression of the supervoltage revolution was seen only after 1970, when improved treatments became available and sufficient numbers of new radiation oncologists were trained to use the equipment. Then it would become evident that fully half of all cancer patients would need radiation therapy and that radiation oncologists would be required increasingly "to assume an exceptionally heavy responsibility" for their lives, as Regaud had challenged decades before.

CLINICAL PRACTICE OF RADIATION ONCOLOGY: 1970–1995

The final quarter of the first hundred years of radiation oncology began with a Cold War between the United States and the Union of Soviet Socialist Republics, an Iron Curtain in Europe, and a gradual awakening of economic superpowers in the Far East. The United States was mired in a civil war in southeast Asia. The major commitment of armed forces to combat in Vietnam influenced the careers of many physicians who were obligated to military service through enlistment or draft.

The care of patients with cancer was dominated by surgeons. Not only was complete resection the most consistent curative approach—albeit frequently with major morbidity—but the wide availability of general surgeons and surgical subspecialists led to their assuming responsibilities for administration of cytotoxic drugs and hormones as well as for the care of many patients who had inoperable or recurrent tumors and needed thoughtful and compassionate care while dying.

Hematologists and those committed to the nascent specialty of medical oncology were becoming aware of striking successes with the use of cytotoxic

Fig. 2.11 Simon Kramer (b.1918) (Courtesy of the Center for the American History of Radiology, Reston, Va.)

drugs for some rare malignant tumors, notably gestational choriocarcinomas and a B-cell lymphoma of African children. Furthermore, there were reports of very preliminary experiences suggesting success with the use of combinations of antineoplastic drugs in patients with less rare malignant disease, such as disseminated testicular carcinomas and Hodgkin's disease.

Radiation oncology was undergoing profound and rapid change. Supervoltage radiation therapy, in the form of ^{60}Co teletherapy units, betatrons, and linear electronic accelerators, was becoming much more widely available. This technology was being used with readily demonstrable effectiveness by general radiologists who elected to confine their activities to the care of patients with cancer and, increasingly, by physicians trained exclusively in radiation oncology. By 1975 training in general radiology, i.e., both diagnostic radiology and therapeutic radiology, was no longer offered. The ABR had ceased certification of physicians in the combined field. By 1980 a majority of universities that offered training in radiology had established separate departments of radiation oncology.

Most groups of physicians who practiced radiology either recruited new associates who were trained exclusively in radiation oncology or agreed that one or more of their colleagues would confine his or her practice to the care of patients with cancer. As a result, by 1983 there were more than 1,600 full-time radiation oncologists, compared with only 111 twenty years earlier.

Schools and Practice Policies

Several individuals and the schools that developed around their approaches to training dominated practice, especially in the 1980s and even today. Each of the schools was characterized by a creative individual with a forceful personality both within the institution and nationally, by the active collaboration of colleagues from other specialties, and by important training programs.

One of the most influential schools developed around Gilbert H. Fletcher and his colleagues at the University of Texas M. D. Anderson Cancer Center. Fletcher's influence on the practice of radiation oncology in the United States was and is to this day a formidable one. It was confined, however, in large part to the treatment of patients with cancer of the upper aerodigestive tract and more broadly the head and neck (excluding brain tumors) and gynecologic cancer—especially cancer of the cervix. The influence of cancer of the head and neck was dependent upon collaborations with surgical colleagues William S. Macomb and Richard H. Jesse. The importance of dental oncology and diagnostic imaging was considerable. Until the mid-1960s there were no full-time residents at M. D. Anderson, but many physicians who had begun training elsewhere took two-year fellowships at the institution. As the residency program expanded in the 1970s, Fletcher's influence in the field became even greater.

Fletcher's influence was extended by his student Rodney Million, who established a department of radiation oncology at the University of Florida in Gainesville in 1964. Million established an effective collaboration with Nicholas Cassisi, a surgeon trained by Joseph Ogura in St. Louis. Million and Cassisi learned from each other, adding to what they had learned from their mentors. In many cases the independent corroboration of the work of Fletcher and his colleagues at M. D. Anderson by the Million–Cassisi practice at the University of Florida emphasized the reproducibility of Fletcher's results and concepts. This was especially true in the management of carcinomas of the glottic and supraglottic larynx and the piriform sinus.

In the field of gynecologic cancer Fletcher is identified especially with the treatment of carcinoma of the cervix. He was an early advocate of high-energy photons using the betatron for whole pelvic irradiation and over time evolved several modifications of the RS (Radium Institute) Manchester approach to intracavitary applications of radium. As the field of after-loading with brachytherapy evolved in the 1960s, the Fletcher–Suit applicator, modified in later years by Luis Delclos, became one of the most popular means of applying brachytherapy with tandem and ovoids.

Two other influences in North America vied with Fletchers's approach to cancer of the cervix. One was promulgated by Victor Marcial at the University of Puerto Rico. He and his colleagues emphasized a slower rate of delivery of external beam radiation therapy, used a single intracavitary application of radium, and attempted to select from the wide range of applicators available that system which most appropriately suited the clinical circumstances of the patients they encountered. The Princess Margaret Hospital in Toronto, promulgating the more rapid fractionation of external radiation common in the United Kingdom, also had a considerable impact on the practice of radiation oncology. All of these influences waned in the 1980s as successful screening with cervical cytology and colposcopy led to consistent treatment of cervical intraepithelial neoplasia and a striking reduction in the incidence of invasive cervical cancer.

Hodgkin's Disease

The treatment of Hodgkin's disease underwent profound changes in the 1970s. As data were analyzed in the 1980s it became clear that treatment changes dating from the early 1970s had resulted in a decrease in the annual mortality of this uncommon malady. Yet standard textbooks that were available a decade before these treatment changes (ca. 1960) depicted Hodgkin's disease as uniformly fatal but acknowledged that some patients could live for prolonged periods of time following surgical or radiotherapeutic intervention.

The fundamental observations that radiation therapy could produce prolonged periods of freedom from all evidence of the disease were made by Rene Gilbert in the 1920s. Very careful observations by this physician from Geneva, Switzerland, revealed that Hodgkin's disease failed in a predictable pattern after local irradiation, namely in immediately contiguous sites. Moreover, he observed that lymphatic regions required treatment beyond the period required for the adenopathy to disappear or it would recur at the same site. This led him to practice the use of more prolonged periods of irradiation, i.e., to higher total doses and to irradiate extensively sites without clinical evidence of involvement in order to prevent contiguous recurrence.

These observations had been corroborated by others, notably Vera Peters of the Princess Margaret Hospital of Toronto, who documented long-term, disease-free survival related to extensiveness of irradiation and clinical stage.

Henry Kaplan, chairman of the department of radiology at Stanford University Medical Center, and Saul Rosenberg, a renowned hematologist, championed the use of extensive prophylactic irradiation of clinically uninvolved areas. They advocated the "mantle" field, which permitted the simultaneous irradiation of all major supradiaphragmatic lymph node-bearing areas so as to avoid multiple adjacent fields. The extent of irradiation was dependent upon the stage. Staging classifications were developed in the mid-1960s and represented an evolution of the classification used by Vera Peters but relied upon bipedal lymphography to investigate infradiaphragmatic nodes. International workshops in Paris; Rye, New York; and eventually Ann Arbor, Michigan, were based on an evolving understanding of sites previously involved at the time of diagnosis and those most likely to have new manifestations of disease following local, regional, or mantle irradiation.

Another important development that led to decreased mortality in Hodgkin's disease was the introduction of combination chemotherapy using MOPP—mechlorethamine (nitrogen mustard), oncovin (Vincristine), procarbazine, and prednisone. MOPP became the most widely recognized effective combination chemotherapy regimen.

Clinical trials were soon underway to investigate comparisons between involved field radiation therapy and more extensive irradiation (mantle or extended field). Kaplan and his colleagues advocated the extensive irradiation but did not actually participate in the clinical trials. Similarly, MOPP, which was developed at the NCI facility in Bethesda, Maryland, was investigated by many cooperative groups nationally and internationally, but the NCI investigators did not lead those trials. Kaplan and Rosenberg, with great credibility among both radiation oncologists and hematologists, convinced a large number of physicians caring for patients with Hodgkin's disease that extended-field radiation therapy was the accepted standard. Carbone, Devita, and their colleagues from the NCI similarly convinced hematologists throughout the nation that MOPP was the standard for patients with advanced disease.

There was considerable potential morbidity with irradiation of very large irregular fields to high total doses. Kaplan recommended a total dose of 44 grays (Gy) in four weeks. Based on a retrospective review of published data, however, questions began to be raised in the mid-1970s as to the morbidity of doses this high. Moreover, an expanding body of data between the mid-1970s and 1992 indicated that the maximal control of Hodgkin's disease was achieved with total dose between 34 and 38 Gy. Concerns about late cardiac effects were increased with the publication of long-term results from Stanford University and other centers. It is now clear that the total dose of radiation administered must be kept to a minimum, 30 Gy for subclinical disease and 38 Gy for tumors greater than 5 centimeters in greatest dimension. Chemotherapy prior to radiation therapy has been used successfully to reduce large masses, especially in the mediastinum, and to permit the protection of more pulmonary tissues. Brief courses of chemotherapy followed by radiation reduced the risk of pulmonary morbidity.

Adenocarcinoma of the Prostate

This most common malignant disease of men is another example of striking benefits from supervoltage radiation therapy. It was not possible to deliver sufficient doses of conventional X rays to the center of the male pelvis to eradicate cancer of the prostate. Unlike cancer of the cervix, which permitted intercavitary brachytherapy as a component of treatment, adenocarcinoma of the prostate required external irradiation alone. The first observation of potential benefit from ^{60}Co teletherapy was published by del Regato in 1962. Malcolm Bagshaw, Fred George, and del Regato each reported small series of patients who remained free of all evidence of cancer of the prostate at least five years after external irradiation in the mid-1960s.

Buoyed by these concurrent observations, urologists and radiation oncologists throughout the nation quickly mounted a prospective comparative trial of radiation therapy compared with radiation and low-dose diethylstilbestrol (DES) treatment. This trial quickly facilitated transfer of concepts and the techniques of the radiation of the prostate to the community at large. Although the clinical trial enrolled nearly four-hundred men, all of whom where followed until death or at least twenty years, the question of adjuvant hormone therapy remained unanswered.

Shifts occurred in the treatment of unresectable carcinoma of the prostate in the early 1970s; the shift was from orchiectomy and/or administration of DES for inoperable patients without distant metastasis to high-dose pelvic dose irradiation. The radiotherapeutic management of cancer of the prostate is very well documented by the PCSs issued initially by Kramer in the early 1970s and continued to the present time. They have shown a high degree of local tumor control for small lesions with total dosage of 65 Gy. They have documented a relatively high control rate for a more advanced (T3/T4) tumors. However, the recent available prostate-specific antigen (PSA) determination has demonstrated that the consistency of control of these more advanced tumors is much less than was thought on clinical grounds alone.

The possible benefit of adjunctive androgen suppression with radiation therapy has resulted in apparent improvement in local control and progression-free survival, although no overall survival benefit has been shown to date.

Nonetheless, a striking improvement in five-year survival rates has been documented by the American Cancer Society. Between the early to mid-1960s and late 1980s, the five-year survival rate increased from 47 percent to 72 percent. Since there is no evidence in the literature to show an influence of hormonal treatment on survival, and since the diagnosis is a result of PSA-based screening which was not a factor until 1989, the only tenable explanation for the striking improvement in survival is the widespread adoption of radiation in the management of men with inoperable but nonmetastatic adenocarcinoma of the prostate.

Cancer of the Breast

More American women are afflicted by cancer of the breast than any other malignant disease. Arguably more intense research efforts and more dollars have been spent on cancer of the breast than any other malignant disease. Unfortunately there is no difference in the annual mortality rate at present than there was fifty years ago. This is despite the increasing availability of well-trained general surgeons, the widespread use of adjuvant radiation therapy since the 1950s, and the even more widespread use of postoperative cytotoxic chemotherapy and hormones since the 1970s.

What is clearly demonstrable is the increasingly widespread adoption of treatments that permit conservation of the breast. First practiced in many institutions in North America, notably the Princess Margaret Hospital in Toronto, Ontario, and the M. D. Anderson Cancer Center in Houston, breast conservation developed the most vocal and effective advocates in France and Italy. Bernard Pierquin, first at the Gustave Roussy Institute in Villejuif and then at the Centre Hôpitalier Henri Mondor in Creteil, France, encouraged excisional biopsy, external irradiation of the entire breast and lymphatics, and then interstitial irradiation with iridium-192. Umberto Veronesi, a highly respected Italian surgeon, launched a prospective trial comparing quadrentectomy plus breast irradiation versus mastectomy at the Italian national cancer institute in Milan. The Milan study was the first comparative trial to show that patients who were afforded breast conservation had the same survival rates as those who had mastectomies.

The most influential group in the United States was that at the Joint Center for Radiation Therapy in Boston. Samuel Hellman enlisted the interest of surgeons at the Harvard teaching hospitals, most prominently William Silen, surgeon-in-chief at the Beth-Israel Hospital, to practice breast conservation. Their collaboration, joined by Jay R. Harris, influenced a large number of surgeons and radiation oncologists around the country through their publications, lectures, and workshops in Boston. Nonetheless, by the early 1990s, only one-quarter of women in the northeastern United States were treated with breast conservation and in some states only one woman in twenty was treated in this manner.

Conclusion

It is not possible to enumerate, let alone elaborate, the contributions of radiation therapy to the care of patients with cancer. As the second century of the therapeutic use of radiation dawns, health care payment reform demands that physicians and patients examine even more seriously the value of treatments offered them. The value of radiation therapy, carefully planned and administered by radiation oncologists and the collaborating healthcare team (physicists, dosimetrists, therapists, nurses), will be recognized as great. Clinical research to increase effectiveness and to expand the indications for radiation therapy will become more formalized and more sophisticated to build upon the enormous scientific and clinical base of the first hundred years.

REFERENCES
1. del Regato, J.A., "Claudius Regaud," *Int. J. Rad. Oncol. Biol. Phys.* 1 (1976):993–1001.
2. del Regato, J.A. *Radiological Oncologists: The Unfolding of a Medical Specialty.* Reston, Va: Radiology Centennial, Inc., 1993.
3. Ibid.
4. Kassabian, M.K. *Electro–Therapeutics and Roentgen Rays.* Philadelphia: J. B. Lippincott Co., 1907.
5. del Regato, *Radiological Oncologists.*
6. Ibid.
7. Ibid.
8. Ibid.
9. Ibid.
10. Ibid.
11. Ibid.
12. Ibid.
13. Ibid.
14. Ibid.
15. del Regato, J.A., "Henri Coutard," *Int. J. Rad. Oncol. Biol. Phys.;* del Regato, *Radiation Oncologists.*
16. Pohle, E.A., ed. *Clinical Roentgen Therapy.* Philadelphia: Lea and Febriger, 1938.
17. Ibid., p. 155.
18. Lenz, "Tumor Dosage and Results in Roentgen Therapy in Cancer of the Breast," *Am. J. Roent.* 56 (1946):67–74.
19. Ackerman, L.V., and del Regato, J.A. *Cancer: Diagnosis, Treatment, and Prognosis.* Springfield: Mosby Company, 1947.
20. del Regato, Radiation Oncologists.
21. del Regato, J.A., "The Unfolding of Therapeutic Radiology," *J.A.M.A.* 262 (1989):1998–2001.
22. Buschke, F.; Cantril, S.T., and Parker, H.M. *Supervoltage Roentgentherapy.* Springfield, Illinois: Charles C. Thomas, 1950.
23. Sherwood Moore to Juan del Regato, personal communication, 1946. Quoted in del Regato, *Radiological Oncologists,* p. 203.
24. For more information about Gilbert Fletcher, refer to the following sources: del Regato, *Radiological Oncologists; The First Twenty Years of the University of Texas–M.D. Anderson Hospital and Tumor Institute.* Houston: University of Texas M.D. Anderson Hospital and Tumor Institute, 1964; Dodd, G.E., "In Memoriam," *Radiology* 185 (1992):287; Peters, L.J., "Memorial Tribute," 1992.
25. Grimmet, L., "A Five–Gramme Radium Unit, with Pneumatic Transference of Radium," *Brit. J. Rad.* 10 (1937):105–117.
26. Lederman, M., "The Early History of Radiotherapy:1895–1939," *Int. J. Rad. Oncol. Biol. Phys.* 7 (1981):639–648.
27. Ginzton, E.L., and Nunan, C.S., "History of Microwave Electron Linear Accelerators for Radiotherapy," *Int. J. Rad. Oncol. Biol. Phys.* 11 (1985):205–216.
28. Karzmark, C.J., and Pering, N.C., "Electron Linear Accelerators for Radium Therapy: History, Principles, and Contemporary Developments," *Phys. Med. Biol.* 18 (1973):321–354.
29. Owen, J.B.; Coia, L.R.; and Hanks, G.E., "Recent Patterns of Growth in Radiation Therapy Facilities in the United States: A Patterns of Care Study Report," *Int. J. Rad. Oncol. Biol. Phys.* 24 (1992):983–986.
30. Marcial, V.A., "The Role of Dose–Fractionation and Time, and Their Relationships in Clinical Radiotherapy," Refresher Course, Radiological Society of North America, 2 December 1966; Marcial, V.A., "Time–Dose–Fractionation Relationships in Radiation Therapy," *National Cancer Institute Monogr.* 24 (1967):187–203.
31. Cox, J.D., "Erskine Lecture."
32. Cox, J.D., "Annual Oration," RSNA.

Early apparatuses for generating X rays were simple, and the components were easily available. In radiation therapy advances in apparatus design would lead to improvements in treatment. (Courtesy of the Center for the American History of Radiology, Reston, Va.)

CHAPTER THREE

Developments in Technology

Martin S. Weinhous, Ph.D., and
Luther W. Brady, M.D.

EARLY DEVELOPMENT: 1910–1950

The years 1910 to 1950 were a time of discovery and rapid progress in radiation oncology. Each year saw investigators outdoing one another, announcing new records for the highest voltage generated or the most intense X-ray beam. Technological advances, at first driven by the desire for better pictures, were soon directed specifically at the need for more penetrating radiation to treat deep-seated tumors. At the same time, new dosimetric methods were introduced to keep pace with the ever-increasing energy of X rays produced by new and improved machines. It was an era of discovery and first-time applications.[1]

In the earliest years there were few or no distinctions between diagnostic and therapeutic X-ray equipment. Eventually, however, the disparate needs of the two applications would result in divergent and highly specialized designs. That is, function dictated form. Therapy tubes, for instance, had to run continuously for many minutes and needed to deliver large (up to 25 centimeters [cm] in diameter) and uniform (flat, symmetric) beams at distances ranging from 15 to 100 cm. In addition, power supplies and control units had to support these operating conditions. The earliest dedicated therapy designs accommodated several different kinds of X-ray therapy, somewhat arbitrarily categorized as Grenz-ray (border-ray), contact, superficial (roentgen), intermediate (medium), deep (orthovoltage), and supervoltage therapies (Table 3.I). The resulting beams had to be collimated and delivered accurately to the intended target. Thus, delivery technology evolved from bare tubes to an interposed lead sheet on the patient, to mountable cones, and, ultimately, to adjustable collimators with light localizers. In addition, beam modifiers such as filters were developed to maximize therapeutic ratios, defined as deep-target-dose to skin-surface-dose.

Meanwhile, dosimetry evolved from describing X rays only by voltage and current to using half-value layers (HVLs) and quantifying the X-ray spectrum. Beam intensity measurement grew from visual assessment of photographic plate darkening to current derived from an ionization chamber. Exposure and dose measurements evolved from "skin erythema" as a

Table 3.I Characteristics of the X-ray Categories

Category	Voltage (kVp)	Filtration	Quality(HVL)
Grenz rays	5-15	0	~1 mm tissue
Contact	30-50	0	~5 mm tissue
Superficial	50-140	0.0-1.0 mm Al	~10 mm tissue
Intermediate (medium)	150-180	0.25 mm Cu	~0.5-2 mm Cu
Deep (orthovoltage)	200-400	0.5 mm Cu	~2-4 mm Cu
Supervoltage	500-3,000	2.0-5.0 mm Cu	~4-13 mm Cu

Table 3.II Significant Events of the 1910–1950 X-ray Era

Year	Event / Technology
1895	Röntgen discovered X rays
1913	X-ray quality first described in terms of half-value layers
1913	Introduction of the hot-cathode roentgen tube
1917	Introduction of self-rectifying X-ray generators
1919	Introduction of oil-immersed shock-proof high-tension generators
1925	Introduction of the air-equivalent wall thimble ionization chamber
1926	Introduction of valve-tube rectified high-voltage transformer generators
1928	Introduction of three-phase generator
1928	Introduction of the condenser dosimeter
1929	Introduction of an X-ray system monitor
1930	Introduction of the single-section supervoltage X-ray tube
1931	Introduction of the multisection cascaded supervoltage X-ray tube
1932	American Board of Radiology (ABR) offered certification in all fields of radiology
1932	Introduction of the radiofrequency supervoltage X-ray generator
1932	Introduction of the "Standard" air chamber for radiation dosimetry
1933	Refinement of electrostatic generators allowing the production of million-volt potential differences
1937	First clinical use of a Van de Graaff accelerator
1937	Acceptance of the "roentgen" by the International Congress of Radiology
1937	Introduction of an oil-immersed rotating tungsten-disk anode tube
1937	Introduction of the resonant-transformer generator
1940	Introduction of the betatron
1942	Introduction of a phototiming circuit
1946	First microwave linear acceleration of electrons
1947	ABR offered certification to radiologic physicists
1948	First clinical use of a betatron

benchmark, to film densitometry, to various kinds of ionization measurements, first in air then in phantoms.

As radiation oncology matured, X-ray tubes and transformer-based power supplies were gradually replaced by other more efficient and effective X-ray production machines. These included electrostatic—also known as Van de Graaff—generators, betatrons, and eventually linear accelerators.

An abbreviated timeline of the era is shown in Table 3.II. While the table simply lists significant events sequentially, we will take a different approach in the text. In discussing progress in technology, we will address separately the two primary topics: X-ray production and delivery, and X-ray dosimetry.

X-ray Production and Delivery

All X-ray producing devices and their power supplies are linked, with changes in one driving advances in others. The common issues were simple. The physics community worked toward producing shorter wavelength X rays to obtain a better tool for investigating the nature of matter. The medical physicists, along with physicians, began to recognize that shorter wavelength X rays were more penetrating and could therefore be used to treat deep-seated targets. In both cases, the production of

Fig. 3.1 Crookes tube. Röntgen worked with a number of shapes and variations on Crookes's widely used experimental physics tube. (Courtesy of the Center for the American History of Radiology, Reston, Va.)

▶

shorter wavelength X rays required that higher energy electron beams strike a target, with the relationship between the resulting X-ray wavelength, λ_{min}, in angstroms (Å) and the applied voltage in kilovolts-peak (kVp), being $\lambda_{min} = 12.4/\text{voltage}$. Thus, considerable effort within both communities was devoted to building tubes that could withstand higher voltages and generators to supply those voltages. In tubes, the limiting factors were size (the minimum required anode and cathode separation increased with increasing voltage) and control of secondary electrons (which could puncture the glass wall of a tube). In generators the limiting factor was often the quality of the available insulation.

Röntgen's discovery of X rays was accomplished with a variant of the cold-cathode Hittorf-Crookes tube as shown in Figure 3.1.[2] This widely produced experimental physics tube usually contained a large aluminum cathode and a small aluminum anode. When high voltage was applied between the anode and cathode, ionization occurred within the residual gas in the tube. The positively-charged oxygen and nitrogen ions striking the cathode liberated many electrons which streamed toward the anode. This electron stream mostly impacted on the glass wall of the tube, where *bremsstrahlung* X rays were produced, albeit with very low efficiency. Röntgen's high-voltage source was a Ruhmkorff coil with a Deprex interrupter. The unit was operated at a voltage that corresponded to a spark gap of about 3 cm.[3]

Many incremental improvements to the cold-cathode tube included concave (focusing) cathodes, platinum foil targets, gas (pressure) regulators, and eventually tungsten-in-massive-copper anodes. These improvements facilitated efficient X-ray production.

The early high-voltage generators— Holtz, Toepler-Holtz, and Wimshurst static machines and Tesla or Ruhmkorff coils—were replaced in 1906 and 1907 respectively by induction coils and the Snook interrupterless (alternating current) transformer.[4] The Snook system included mechanical full-wave rectification and provided near-peak voltage whenever current was flowing in the tube. The first Snook system was installed at the Jefferson Hospital in Philadelphia in 1907, and another was still in use at the Johns Hopkins Hospital in Baltimore as late as 1946. These generators could produce 100 kVp at 100 milliamperes (mA).[5]

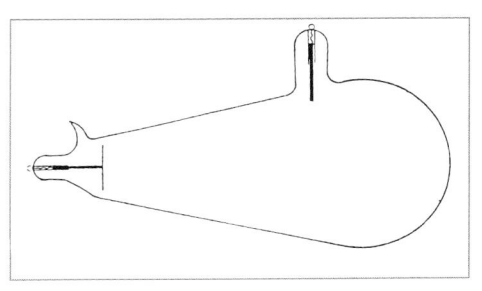

Although relatively inefficient by modern standards, these tubes and generators were enthusiastically applied to a wide range of diseases and conditions. Typically, sheet lead with a suitable aperture was placed on the patient and a bare tube activated a short distance from the intended target.[6] Because of their construction, these tubes were limited to a potential difference of about 100 kV and were only suitable for superficial therapy.

A major breakthrough occurred in 1913 when W. D. Coolidge described the first hot-cathode, high-vacuum roentgen tube.[7] These tubes became commercially available in 1917.[8] Hot-cathode tubes contained an electrically-heated tungsten filament which was used as the source of electrons (thermionic emission) within a cathode assembly (Figs. 3.2a and 3.2b). The rate of electron emission from the cathode depended upon that cathode's temperature rather than upon its bombardment by ions. Because the tube current changed rapidly with increasing filament current, external stabilization devices were required to maintain overall stable operation. Nevertheless, this arrangement provided ready external control over the operation of the tubes.[9] These tubes were the first to provide independent control of beam quality and quantity.

Beam quality was determined by the magnitude and waveform (plot of

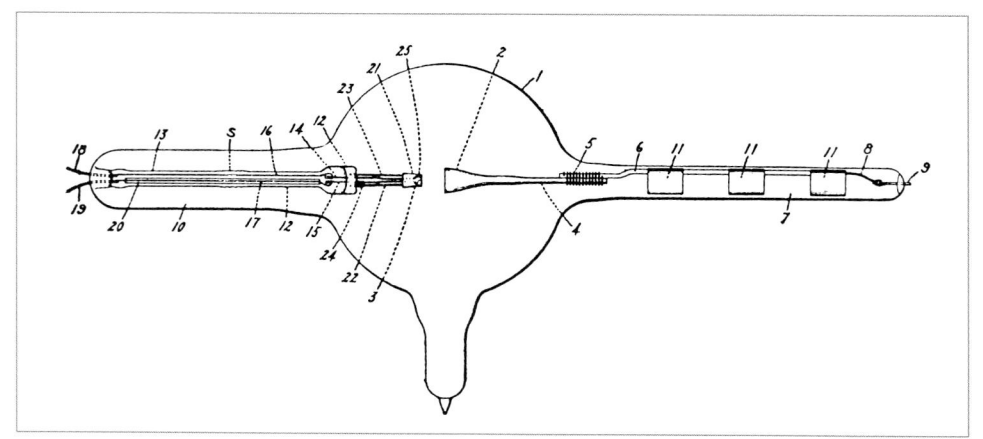

Fig. 3.2a The original design for William Coolidge's improved tube, 1914. (Courtesy of the Center for the American History of Radiology, Reston, Va.)

voltage verses time) of the voltage applied between the cathode and anode, while beam quantity (mA) was determined by the filament current which was, in turn, determined by the temperature of the cathode. The efficiency of hot-cathode tubes, while many times better than that of cold-cathode tubes, was still only on the order of 0.1 percent, with most of the applied energy resulting in heating of the anode.[10] This severe anode heating necessitated both the careful selection of anode material and active cooling of the anode if the tube was to be operated at therapy dose rates. Tungsten, because of its combination of high melting point and relatively high atomic number was (and is) the most often used anode material. Unfortunately, tungsten is only a moderate heat conductor. To conduct away the heat energy deposited in a tungsten anode, a composite anode structure consisting of a tungsten target embedded in a massive copper heat sink (now with external cooling provisions) was usually employed.[11]

In 1915 and 1916 Snook modified his generator to provide an autotransformer with push-button control for easy and accurate selection of kilovolts-peak, a feature that, in function if not in form, is standard on today's generators.[12] Coupled with hot-cathode tubes, beams with reproducible X-ray spectra and intensity became available.

In 1917 the self-rectifying Coolidge radiator tube was introduced.[13] It used half of the high-voltage power supplied by the generator and did so without any need for mechanical rectification. This allowed for considerable reduction in the size and weight of generators (albeit at the cost of reduced beam intensity). The design of roentgen tubes continued to evolve as well. In 1923 Coolidge and Moore described a water-cooled tube designed specifically for therapy applications.[14]

The first X-ray systems were not built with either radiation protection or electrical safety as paramount concerns. Early isolated efforts, such as the internal diaphragm tube designed by Rollins in 1899 and various contraptions for external shielding, were not widely adopted.[15] Eventually, the Philips Metalix tube and others that followed were designed with adequate integral radiation shielding.[16] Electrical safety was also something of an afterthought and was finally well implemented in 1919 with the Waite and Bartlett "shockproof and rayproof" system.[17]

Commercial Grenz-ray tubes were designed specifically for radiation therapy and were put into use in the late 1920s.[18] These were hot-cathode, cooled-anode tubes designed for high-intensity, low-voltage application (5 to 16 kV). In addition, they were designed

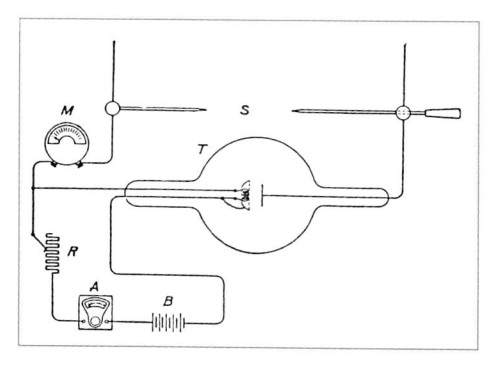

Fig. 3.2b Schematic of tube and apparatus set up. (Courtesy of the Center for the American History of Radiology, Reston, Va.)

Developments in Technology

Fig. 3.3 A deep therapy tube with A, Hooded anode; BW Beryllium window; C cathode; F, filament; TT, tungsten Target; WI cooling water inlet. (Courtesy of the Center for the American History of Radiology, Reston, Va.)

▶

to operate at close range to the skin (~10 cm) for a "reasonable" exposure time. A distinguishing characteristic of these tubes was the very thin window required for the passage of the low-energy (soft) X rays. One successful design was the Linderman window, constructed from the low atomic number (low Z) materials boron, beryllium, and lithium. Unfortunately, this combination of materials was hygroscopic and the windows deteriorated over time (often in spite of a lacquer coating). Otto Glasser built a very thin glass window tube which was stable and allowed the tube's X rays to match those obtained with a Linderman window.[19] Typical HVLs for the X rays from a Grenz-ray tube ranged from 0.007 millimeters (mm) Al to 0.06 mm Al for applied potentials of 4 kV and 12 kV respectively.

In 1921 James T. Case built a generator system designed for therapy that could deliver 8 mA at 200 kVp.[20] Its success led to further production of therapy transformer units. Also in 1921 the sphere gap, a more accurate voltage measuring system, was introduced.[21] Case's generator was coupled with a suitable tube in 1922, and the system operated at 200 kVp.[22] These developments marked the beginnings of the transition from superficial to deep therapy.[23]

True deep/orthovoltage therapy X-ray systems followed soon after and operated at 200 to 400 kV. Tubes for these systems were generally constructed from thick-walled Pyrex, as secondary electrons released when the primary beam impacted the anode could puncture ordinary glass. Most of these tubes also used a hooded-anode design (Fig. 3.3) to minimize the escape of stray electrons. The size of the tube scaled with the maximum allowed voltage—at higher voltages more separation was required to avoid arcing.

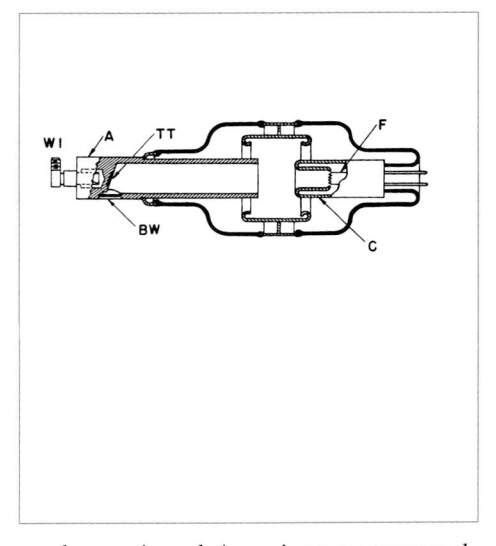

Meanwhile, progress was made in controlling both the quality and quantity of X rays delivered to a patient. As early as 1897 Röntgen described beam hardening by filtration. In 1904 Perthes's depth dose measurements were instrumental in initiating the use of filters in Europe.[24] In 1906 Pfahler's work convinced American roentgenologists that filtration was needed.[25] Soon after, in radium therapy, it was also recognized that there were therapeutic advantages (skin sparing) to "cleaning the radiation by...filtration."[26] However, in the years 1910 to 1920, "Roentgenologists who engaged in therapy were looked upon with suspicion....The sharp reactions produced by soft roentgen rays rendered them hazardous....Not until many months elapsed did the late telangiectases and ulcerations make their appearance...too late to save many patients from these unfortunate sequelae."[27] It was soon recognized that filtration, the removal of soft X rays from a beam, was imperative for any treatments intended for depths greater than a few millimeters. Once X-ray systems (tubes and generators) had been designed for irradiation with filters in place, a new hazard arose. Robert R. Newell in 1938 stated that "There is no blunder in all the practice of medicine quite so black as this: to give 'deep therapy' treatment without the filter."[28] In systems designed for use with filters that removed some 80 percent of the radiation, treating without the necessary filter not only reintroduced soft X rays but also increased the beam intensity some five-fold! Patients so treated would develop deep and often severe ulcerations at depths as great as 2 or 3 cm. The solutions for avoiding these potential catastrophes were first human, then technical, and sometimes both. At some institutions it was required, under pain of dismissal, that two witnesses affirm in writing the presence of the proper filter before the initiation of treatment. Technological solutions ranged from warning devices notifying the user that a filter was missing to interlocks that would prevent the system from producing X rays without the proper filter.[29] Automated systems were employed, where electromechanical devices automatically deployed the filter, unshuttered the treatment portal, and started the timer.[30] Regardless of the means, the recognition of the necessity for proper filtration was an important step in the overall progress of radiation therapy.

In the late 1920s and early 1930s progress was made in filter design by the use of composite materials. Replacing a 2.0 mm copper filter with a so-called Thoraeus filter (0.44 mm tin + 0.25 mm copper + 1.0 mm aluminum) yielded as much hardening and provided 50 percent greater intensity.[31] Along with improvements in filters, improvements in mountable treatment cones were almost continuous during this period, with many different cone designs introduced.[32] Their cross-sections varied from circular to square to rectangular. Their walls were made of materials ranging from leaded glass to steel to lead. Some cones were designed to compress overlying soft tissues so as to bring the X rays more directly to the target. Others even provided simultaneous multi-port compression for pelvic treatments.[33]

The next significant advance in generators, the introduction of vacuum-tube (valve-tube) rectifiers, occurred over many years. The need for more constant high-voltage at increased current was well understood, but valve tube technology was slow to mature. It was first used in 1906, with subsequent notable attempts in 1914 and 1926. By 1930 valve tube systems were well on their way to becoming the standard.[34] A representative generator set up is shown in Figure 3.4. This valve-tube rectification allowed the generators to produce higher average voltages and, thus, higher energy X-ray beams.

In the constant search for greater beam intensity, in 1928 Siemens introduced a three-phase generator capable of providing 2,000 mA at 80 kVp.[35] Three-phase systems have an advantage over single-phase systems in that the output voltage is inherently almost constant with a small superimposed sine-wave "ripple" with a frequency six times the input (line) frequency.[36]

With the introduction of valve tube rectification and three-phase genera-

Fig. 3.4 The Wappler company introduced the first valve-tube rectification into the United States in 1924. This model, the Monex (1927), was used for both diagnosis and therapy, and was especially attractive to the majority of radiologists who performed both functions in their practices. (Courtesy of the Center for the American History of Radiology, Reston, Va.)

Fig. 3.5a Coolidge "cascade" tube, 1928. This two-section tube could generate 600,000 volts and included a sphere gap (above) to measure voltage. (Courtesy of the Center for the American History of Radiology, Reston, Va.)

Fig. 3.5b The cascade principle was soon extended to three sections, which could generate 900,000 volts. (Courtesy of the Center for the American History of Radiology, Reston, Va.)

Fig. 3.6 This 800,000 volt multisection tube was installed at Mercy Hospital in 1933. (Courtesy of the Center for the American History of Radiology, Reston, Va.)

tors, the various therapy units produced widely different high-voltage waveforms. In 1933 Lauriston Taylor and his colleagues at the United States Bureau of Standards reported on a careful and detailed study showing the effects on the quality of X rays resulting from different shapes of high-voltage waveforms applied to tubes.[37]

A therapy tube designed for "contact therapy" made its appearance in the 1930s. Unlike other tubes discussed here, the Chaoul contact therapy tube used a thin, gold-plated nickel anode. Additionally, this transmission target was water cooled. These tubes were typically operated at a potential of 10 to 50 kV. X rays, produced in the gold-plated nickel target, passed through the target (~0.15 mm) and through about 2 mm of water. An external "cone" was used to position the tube reproducibly at a known distance, between 1 and 5 cm, from the skin. An exposure rate of about 100 roentgens (R) per minute at 5 cm was achievable with this device.[38]

Yet another technological advance was required in order to operate a tube beyond 400 to 500 kV (the so-called supervoltage range). New design strategies had to be employed to avoid electrical breakdown either through the residual gas in the tube or through and along the glass walls.[39] One solution to these problems was a cascaded tube as shown in Figures 3.5a and 3.5b. The voltage across each section was 200 to 300 kV; adding additional sections increased the effective first-cathode to last-anode voltage. These cascaded tubes could produce supervoltage X rays but were very large and cumbersome.

An alternative was the multisectioned tube, which used a series of hollow metal electrodes encased within the same glass/metal envelope.[40] Typically, the anode was set at ground potential with each section adding a potential difference of about 80,000 volts. In this fashion, a twelve-section tube could produce one-million-volt X rays (Fig. 3.6).

Multisection tubes were indeed capable of producing high-energy, penetrating X rays but, as was the case for cascaded tubes, were difficult and cumbersome to use. Their high voltage capability, however, served as a harbinger of things to come, particularly linear accelerator waveguides. Despite difficulties with size and operation, a machine of this type, developed in 1928 at the General Electric (GE) Company and powered by a mechanically interrupted induction coil, was installed at Memorial Hospital in 1931 as part of a Works Project Administration project.[41] Its multisection cascade tube delivered

750 kVp X rays and provided clinical service for some two or three years.

Supervoltage therapy developments spanned the world. For example, in about 1932 at the Frauenklinik in Berlin, von Schubert began treating patients with 600 kVp X rays.[42]

The pace of development of devices in the early supervoltage era is well illustrated by activities at the California Institute of Technology during the 1930s.[43] A visit by radiologist Albert Soiland to the physics laboratories of Robert Milliken and C. C. Lauritsen resulted in a collaboration wherein a physics research apparatus would be used for clinical trials. A patient of a Dr. Carter, suffering from inoperable adenocarcinoma of the rectum, was treated with 600 kVp, 4 mA, 6 mm steel-filtered X rays in October 1930 at an SSD of 50 cm.[44] As a result of this and other treatments, W. W. Kellogg donated funds for the construction of the Kellogg Laboratories at the California Institute of Technology.[45] Subsequently, a new high voltage therapy system was built and installed, capable of producing 750 kVp X rays and delivering 20 R per minute at 70 cm SSD. This device included a 30-foot long tube and two 750 kV (r.m.s.) transformers. The tube's water-cooled target was made of gold. The system occupied a multistory structure and could treat four patients at a time with fields ranging up to 15 x 15 cm.[46] From October 1933 treatments were carried out at 900 kV, 3 mA with (6 mm steel + 1 mm lead) filtration, 58 cm SSD, an HVL of 6.5 cm water or 8.1 mm copper at an intensity of 15 R per minute.[47]

The Kelly-Koett Company installed a 900 kV system at Detroit's Harper Hospital in 1931. This durable unit, powered by cascaded Cockroft-Walton generators, stayed in operation until 1956.[48]

A novel high-frequency (radiofrequency at 6 megahertz [MHz]) generator was employed at the San Francisco Hospital in 1935.[49] The tube and transformer were combined into one integral unit, avoiding the then dangerous exposed high-voltage cables. The device was capable of producing 1 megavolt (MV) X rays but could not be used above about 1.2 MV. Additionally, the bulk of the auxiliary equipment (radiofrequency generator) more than made up for the small size of the X-ray generator.

In 1937 a 1 MV X-ray machine, built by Metropolitan Vickers, was first used to treat patients at the St. Bartholomew's Hospital in London. Its X-ray tube was some 10 meters (m) long and was powered by two Cockroft-Walton high-voltage generators.[50] This machine was also noteworthy for two of its numerous technological innovations. It had a parallel-plate transmission monitoring chamber and adjustable collimators. The former assured that the radiation delivered could be monitored directly, and the second provided for easy field size adjustments. It began operations in 1937 at 700 kV, and from 1939 to 1949 was used clinically at 1,000 kV.

The late 1930s saw the development of resonant transformer generators by a GE group headed by E. E. Charlton.[51] These systems consisted of hollow core transformers, typically operated at 180 Hz, with a cascaded Coolidge tube located in the transformer's core. With the entire unit contained within a pressurized tank containing an insulating gas, the system was eventually capable of delivering 1 to 2 MV X rays from a reasonably compact device.[52] A one-million-volt unit was installed at the Memorial Hospital in New York and placed into clinical service in 1938. Further developments at the GE laboratories led to units of more compact size. Systems designed specifically for medical use went into production in 1946. Before they were rendered obsolete by even more compact and higher voltage sources, there were sixteen of these two-million-volt systems installed in the United States.

In 1934, at a private meeting, George Holmes of Massachusetts General Hospital, the physicists Robert Van de Graaff and Edward Lamar, and electrical engineer John Trump of the Massachusetts Institute of Technology (MIT) discussed the feasibility of using a new type of electrostatic generator to power high-voltage X-ray systems. Also

Fig. 3.7 Van de Graaff accelerators. (Courtesy of the Center for the American History of Radiology, Reston, Va.)

in 1934 Dresser of Huntington Memorial Hospital and Trump connected a "standard" 200 kV X-ray tube to a 700 kV Van de Graaff generator with very brief and noisy results.[53] A more rigorous experiment in May of 1935, with suitable dosimetry equipment, easily demonstrated high-intensity, high-voltage X-ray production from the Van de Graaff generator. Subsequently, Dresser obtained a grant from the Hyams Trust for a one-million-volt generator and a building at Huntington Memorial Hospital. The first clinical use of the new system occurred on 1 March 1937. Its specifications included an air-insulated high-voltage terminal 3 m on a side, large enough to walk inside, a water-cooled lead target, and an output of 40 R at 80 cm SSD. The clear advantages of this new device, relative to the common (200 kV and a few 400 kV) tube systems, were greater skin sparing, increased depth doses, and greater intensity.[54] This machine operated until about 1941, when the hospital was closed.

The Van de Graaff generator used a motor-driven belt of nonconducting material to transport negative charge (sprayed onto the belt) to a high voltage electrode (the dome). The entire apparatus (excepting early models such as the one described above) were generally enclosed within a steel tank pressurized with an insulating gas (Fig. 3.7). An accelerating tube with electrodes spaced so as to continuously speed electrons toward the X-ray target was also contained within the tank. Just outside the tank, a water-cooled X-ray target intercepted the high-intensity electron beam, producing copious high-voltage X rays. A generator capable of producing 2 mega-electron-volt (MeV) X rays evolved to be about 1.5 m high and about 60 cm in diameter.

In 1938 the Hyams Trust supported, via a grant to MIT, the development of a 1.25 MV Van de Graaff generator for use at the Massachusetts General Hospital. This machine was used clinically for some sixteen years.[55] With a pressurized vessel, the newer system was much more compact than its predecessor and provided an output of 80 R at 100 cm SSD.[56]

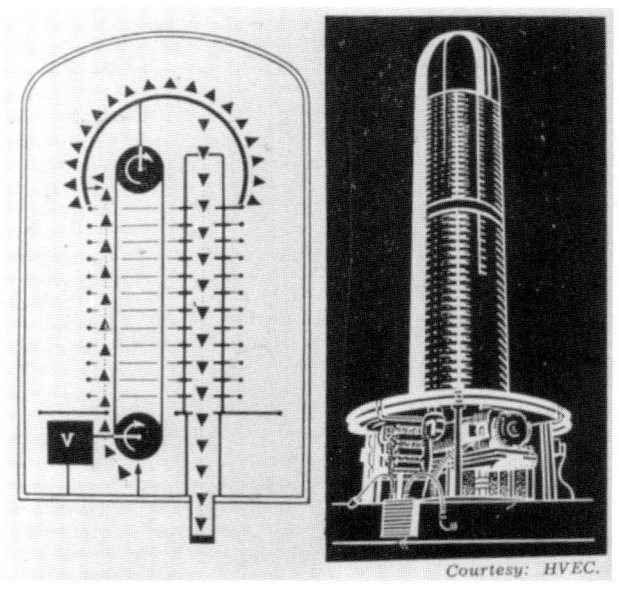

Eventually, an MIT spin-off company, the High Voltage Engineering Corporation, sold forty-three two-million-volt generators for clinical use—thirty-five within the United States. The last was installed in 1969.

Because of insulation and size problems, the practical limit on the X-ray energy for a system that directs acceleration of electrons by an imposed high voltage is about three or four million volts. To achieve higher X-ray energies, the electrons would have to be accelerated stepwise to those higher energies by relatively smaller potential differences. This simple necessity led to the development of both orbital and linear accelerators.

Several investigators independently proposed and attempted to build magnetic induction orbital accelerators, or betatrons.[57] This list includes an electrical engineer, Joseph Slepian in 1922, an engineering student, Rolf Wideröe in 1923, and Steinbeck in 1937.[58] The first working betatron was built at the University of Illinois by Donald Kerst in about 1940 (Fig. 3.8).[59] His first unit

Fig. 3.8 Donald Kerst with two of his prototype betatrons. (Courtesy of the Center for the American History of Radiology, Reston, Va.)

accelerated electrons to 2.3 MeV.

A betatron accelerates electrons in an evacuated "doughnut" inside an AC electromagnet. The changing magnetic flux, in the space occupied by the doughnut, induces an electrical potential difference (voltage) which accelerates any electrons that are injected at the correct time in the alternating magnetic field cycle. The magnetic field not only accelerates the electrons, but also confines them to a stable orbit. Because the electrons have to be injected when the magnetic field is increasing with the correct polarity, only one-fourth of the magnet's AC power cycle is useful for accelerating the electrons. Proper shaping of the magnetic field assures that electrons, even those scattered by gas molecules within the doughnut, will move back toward the equilibrium orbit. When the AC magnet's field reaches its peak, an orbit-expanding magnetic field is added (by another coil), and the electron beam is peeled out of the equilibrium orbit and directed onto a tungsten target. The electrons will have traveled nearly one-hundred miles during the one-quarter-cycle of the AC magnetic field and gained many megavolts of energy.

After his initial success Kerst took a leave of absence from the university to work at GE. While there, he constructed a 20 MeV unit capable of delivering X-ray beams at 20 R per minute at 70 cm SSD. This machine was delivered to the University of Illinois, and basic research and development work continued. In 1948 a twenty-seven-year-old physics graduate student at the university was diagnosed as having a glioblastoma. He was treated by Dr. Quastler using a 22 MeV betatron beginning on 30 April 1948.[60] It should be noted that the fractionated treatment consisted of twenty-five noncoplanar, noncongruent fields, That is, the patient was treated stereotactically.[61] Shortly afterward, in 1949, using a 24 MeV betatron at the Cancer Clinic in Saskatoon, Thomas A. Watson began the first prospective clinical series designed to test the efficacy of multimegavoltage X rays.[62]

Commercial development of the betatron was undertaken first by Allis-Chalmers in North America, and then by Siemens in Germany and Brown-Boveri in Switzerland. By 1975 some fifty units had been installed in North America. The Brown-Boveri unit was capable of reaching 45 MeV and was mounted so that it was more convenient to use. In general, betatrons provided tightly focused monoenergetic electron beams. When used to produce X rays, the focal spot was small enough so that penumbra was essentially negligible. The betatrons were, however, limited to a maximum dose rate of about 100 R per minute at 100 cm SSD and to about 12 x 12 cm fields. Additionally, high-energy betatrons (>25 MeV) were generally too heavy for gantry mounts. These limitations eventually led to the replacement of betatrons by linear accelerators (linacs).

Although linac research and development occurred during the 1910 to 1950 era, clinical implementation did not. The first clinical systems were installed in the early 1950s.[63] Linear accelerator theoretical development began perhaps as early as 1924, when Gustaf Ising suggested acceleration of charged particles by exposure to radiofrequency gaps. Wideröe implemented Issing's suggestion in 1926, accelerating potassium ions through 25 kV gaps to 50 kV. William W. Hansen, working at Stanford in the mid-1930s, developed the rhumbatron *(rhumbra* and *tron* are from Greek, meaning rhythmic and place), a hollow copper cavity in which electrons were accelerated by microwaves. It was only marginally successful, as sources of high-power microwaves were not yet available. In 1937 the Varian brothers, Russell and Sigurd, motivated by the need for high-power microwaves for aircraft detection, became unpaid research associates at Stanford with a $100 budget for equipment and supplies. Working with adaptations of Hansen's rhumbatron, in the late summer of 1937 they produced a two-stage device, the klystron, that could provide high-power microwaves.

Taking a different path, John Randall and Henry Boot of the University of Birmingham in Great Britain developed a 100 kilowatt (kW) magnetron, a circular-beam cavity oscillator, in 1939. By the end of World War II klystron and magnetron development had progressed sufficiently to power linear electron accelerators.

The two most successful of about ten groups operating independently, Hansen's group at Stanford University and Fry's group at the British Telecommunications Research Establishment (later moved to the Atomic Energy Research Establishment [AERE] at Harwell), worked toward the development of a practical, microwave-powered, linear electron accelerator. The AERE group accelerated electrons to 0.5 MeV in November 1946 and the Stanford group to 1.7 MeV in early 1947. These groups continued to achieve higher and higher energies, soon reaching electron energies of 6 MeV.[64] Both groups designed variations of the traveling-wave waveguide with its internal cavities configured to match the phase velocity of the traveling wave to the motion of the electrons.

These systems were still limited by available microwave power. Based on his 1944 work, Edward Ginzton in 1947 proposed the construction of a klystron 1,000 times more powerful than those available at that time. In 1949 Marvin Chodorow, Ginzton, and colleagues demonstrated a new klystron design that, in 1952, reached its goal of producing 30 megawatts (MW).[65]

With the clear success and continuing development of microwave linear accelerators, the British Ministry of Health in 1948 made arrangements for the construction of a stationary-beam, 8 MeV, X-ray linac. That system was placed into clinical use in 1953. In 1949 the ministry coordinated plans for gantry-mounted 4 MeV linacs. These were installed and placed into clinical use in 1953 and 1954 and are discussed below.

X-Ray Dosimetry

X-ray dosimetry includes several topics pertaining to the measurement of the quality and quantity (intensity) of X-ray beams or to the measurement or calculation of exposure or dose. Once again, the years 1910 to 1950 were a time of rapid-paced technological evolution.

X-ray beam quality was, and still is, difficult to quantify. The "proper" method is to completely specify the X-ray spectrum, a full plot of intensity versus energy. This was, and is, difficult and impractical. Initially the quality of a beam was judged according to the fluoroscopic sharpness of the image of the operator's hand. Various devices were quickly invented to replace the operator's hand in testing. These included Kolle's 1896 X-ray meter, a paddle with differing thicknesses of aluminum and a differing number of holes drilled in each thickness.[66] This was followed by Benoist's 1901–1902 penetrometer or radiochromometer, a circular aluminum step wedge, and Beck's 1904 osteoscope, a skeleton arm attached to a board which was inserted into the X-ray beam in place of the operator's hand.[67]

Later, attempts were made to correlate the magnitude of the peak high-voltage with beam quality. This was only marginally successful for several reasons, including: the modification of beam quality by the tube wall and housing, the sometimes large difference between peak voltage and average voltage, and the general uncertainty in measuring voltage (e.g., spark-gap spacing depended upon air quality).

In 1912 Christen described a unified system which attempted to use absorption to quantitate beam quality, in particular by use of the HVL.[68] The HVL is that thickness of a specified material (usually pure aluminum or pure copper) that will reduce the intensity of a transmitted beam to 50 percent of its incident value. However, inserting absorbers into a beam tends to reduce the soft X-ray (low-energy) component of the beam, resulting in an increase of subsequent HVLs relative to the preceding one. A more complete specification of a beam can be accomplished plotting an absorption curve of its intensity versus thickness.[69] If two different systems produce the same absorption curves, then the beam qualities are certainly the same. However, this method did not yield the desired "one simple number" that might be used to compare two different X-ray systems. In the early 1930s Taylor and his colleagues described the constant potential equivalent method for obtaining that one simple number.[70] This enabled anyone to measure the absorption curve for their system and compare it to the standard curves from which the one-number descriptor of the beam could be derived. This greatly facilitated communication of beam quality information among clinicians.

Interestingly, as technology provided higher energies and "purer" beams, the difference between the first and subsequent HVLs became smaller. Thus, the HVL became the desired "one simple number" and eventually replaced a constant potential equivalent for beam characterization. The International Committee on Radiological Units (ICRU) adopted the HVL method for characterizing roentgen-ray beams at their meeting in Paris in 1931.

The determination of X-ray quantity (intensity) began with estimates based upon fluorescence or photographic results. Prior to the introduction of the hot-cathode tube and associated voltage and current measuring apparatus, the intensity of each tube varied widely and results were, at best, uncertain.[71] It was almost impossible to describe a "dose" in meaningful terms so that others might replicate a treatment. Radiochemical measurement methods were attempted, albeit with very limited success, in the earliest days. These methods included observing the discoloration of a fused mixture of HCl and Na_2CO_3 (Holzknecht) and the change in color of a pastille (tablet) of platinobarium cyanide (Sabouraud and Noiré).

This situation began to improve in 1904 when Perthes undertook the first "depth dose," (or more accurately, depth-intensity) measurements and in 1907 when Kassabian published a summary of dosimetric methods.[72] As early as 1905 investigations regarding the use of ionimetric measurements for the determination of intensity, and eventually dose, had been undertaken. Franklin published his results in 1905, and Phillips and Allen both in 1907.[73] In 1908 Villard suggested that radiation intensity be measured by a unit based upon the ionizing effects in a known quantity of air.[74] By 1914 Duane and Szilard had been determining system output with just such a measurement.[75] Early ionization chamber systems suffered from inaccuracies due to perturbations caused by wall materials in the radiation field they were intended to measure. In his 1914 work Duane described a free-air ionization chamber system in which the X-ray beam did not impact upon any solid material (Fig. 3.9). He went on to define intensity, E, in terms of current (charge per unit time) per unit volume of air. It followed that the unit of dose, the ES, was simply intensity multiplied by time.[76] Finally, in 1928 a consensus definition of the unit for measuring radiation (exposure), the roentgen, R, was reported to the second International Congress of Radiology at Stockholm and reaffirmed with minor changes in 1937.[77] This definition continued to depend upon the use of standard free-air chambers for implementation.[78]

More compact, portable, chambers were introduced with enclosed collect-

Fig. 3.9 Duane's original free air chamber was first described in 1914, but did not come into general use. By the 1920s, however, the need for more accurate and standardized measurement of radiation made the profession more receptive to Duane's standard open air ionization chamber, seen here (1928). The principle remained the same in countless chambers for more than fifty years. (Courtesy of the Center for the American History of Radiology, Reston, Va.)

Fig. 3.10 A thimble ionization chamber showing the air cavity from which charge is collected. To collect that charge, a bias voltage is applied between the center electrode and the inner surface of the thimble. (Author's collection)

ing volumes (~1.0 cm3) in the rough shape of a sewing thimble (Fig. 3.10). Since many of the electrons that caused ionization in the air cavity originated in the thimble's wall, there was initially a significant energy dependence, which was a function of the wall's material. In 1924 Fricke and Glasser introduced an "air wall" ionization chamber that overcame the adverse, beam-perturbing effects of the multitude of previously used wall materials.[79] Basically, the chamber walls' composite material was selected to have the same effective atomic number as air. One of the more significant advantages of the air-wall chamber was that it was fairly insensitive to the X-ray beam spectrum. Thus, for example, a chamber could be reliably used to measure HVLs despite the fact that adding absorbers changed the effective beam energy—the chamber (detector) response stayed constant. It should be noted that these were called secondary chambers. They had (and still have) to be calibrated relative to a primary standard free-air chamber. A very commonly used thimble roentgenometer was the condenser electroscope instrument. It combined a condenser thimble chamber with a string electrometer. This device also allowed the chamber to be detached from the reader for exposures and proved quite convenient and accurate.[80] Many other thimble ionization chambers were built and used successfully during this era.[81]

Meanwhile, as the detectors evolved, so did the electronic circuits used to measure ionization current or collected charge. Initially, electroscopes, then string electrometers, served as measuring devices. Semi-electronic devices soon followed with a string or quadrant electrometer serving as a sensitive null device.[82] Vacuum tube amplifiers, used as DC amplifiers, were unsuccessfully used in electrometer circuits. Unfortunately, the static gain of the tubes was not stable. Later, when "choppers" were used to convert the DC signal from the chamber into a pseudo-AC signal, vacuum tube amplifiers could be used in AC mode where stability was far less of an issue.[83]

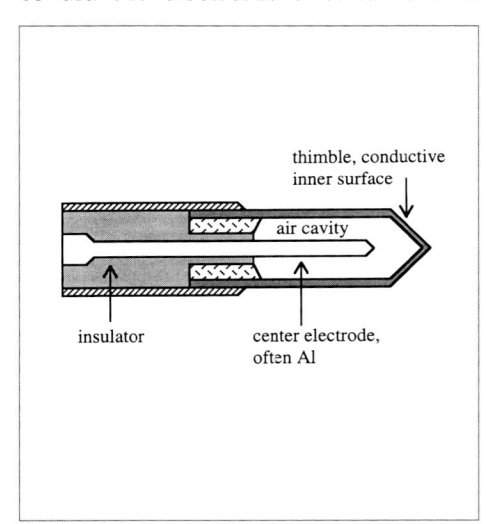

In the earliest days of roentgen therapy, the dose given to the patient was judged by the amount of reddening of the skin, the so-called "erythema" dose. In the 1920s dose was still calculated (rather estimated) by combining quality, intensity, time, and distance factors, an example being:

$$\frac{\text{spark gap(in)} \times \text{current(mA)} \times \text{time(min)}}{\text{distance}^2(\text{in}^2)}$$

A result of about 9/16 implied that the patient would receive about one erythema dose.[84] There were undoubtedly other similar and even more complicated formulae and schemes.

Nevertheless, by 1919 it was clearly recognized that there was a difference between the "dose emitted" and the "dose received."[85] In 1921 Failla spoke of the desirability of expressing dose in terms of energy deposited in tissue.[86]

Table 3.III
Significant Technological Events, 1950–1970

Date	Location, Machine, Manufacturer	New Technology
1950	Cambridge, England	30 MeV Synchrotron[89]
1951	Victoria Hospital London, Ontario Canada	First clinical use of a cobalt-60 teletherapy system[90]
1953		The rad adopted as the unit of absorbed dose by the seventh International Congress of Radiology[91]
1953	Hammersmith Hospital, London, Metropolitan Vickers	First clinical use of a microwave powered linear accelerator for x-ray therapy[92]
1954	St. Bartholomew's Hospital, London, Mullard	Multimodality accelerator, X-rays and electrons[93]
1954	Newcastle, Mullard	First double-gantry orientable unit, pseudoisocentric function, recirculated residual microwave power[94]
1954	Orthotron, Christie Hospital, Manchester, Metropolitan Vickers	First single gantry orientable unit, pseudoisocentric function, no bending magnet[95]
1955	Stanford	Klystron powered, gridded electron gun[96]
1955	U. California, San Francisco	70 MeV Synchrotron[97]
1962	Newcastle Vickers Research	X-band microwave power system, smaller accelerator structures[98]
1962	Clinac 6, Varian	Fully-rotational gantry[99]
1965	Mevatron 8, Applied Radiation	Achromatic beam optics (261° magnet), retractable beamstopper[100]
1965	SL-75, Mullard (now Philips)	Dual dosimetry system, spectrometer-type beam optics (tight energy control), replaceable electron gun filaments[101]
1967	Sagittaire, Paris CSF	Dual traveling-wave structure, novel beam optics, scanned electron beams[102]
1968	Clinac 4, Varian	Standing-wave accelerator, in-line design[103]
1970	Hiroshima, Clinac 35, Varian	Switchable photon energy, compact, high-power beam for research[103]

But the measuring devices described thus far measured, at best, exposure. That is, they did not measure the energy deposited in tissue. Such measurements can, theoretically, be achieved with a calorimeter. Some of the earliest work along these lines was described by Rutherford and Robinson in 1913.[87] It was not to prove practical for X rays until much later.

Meanwhile, it was clear that the roentgen unit for exposure was not usable above about two or three million volts. Very simply, at those high energies, it becomes impossible to build a detector that meets the necessary physical conditions for measurements of exposure. In 1940 Mayneord proposed one possible solution, the "gram-roentgen" as the unit for measurement of integrated dose. In addition, the ICRU commissioned a detailed study which would lead to the definition and initial acceptance of the rad in 1953.

SUPERVOLTAGE, COBALT, AND TECHNOLOGICAL ADVANCEMENT: 1950–1970

The years from 1950 to 1970 saw both improvements in existing technologies and continued development. While progress and discovery continued, they did not occur as rapidly as in the previous era. Technological advances continued to be driven by the search for better dose distributions within the patient. At the same time, dosimetric methods evolved with an improved understanding of the underlying physics and with international standardization of descriptions and

Table 3.IV

Technology Introductions

Item	Early Technology	Follow-on Technology	Benefit
accelerating structure	traveling wave	standing wave	greater efficiency and shorter length
energy gain per m	4 MeV/m	12 to 18 MeV/m	compact linac
bending magnet optics	non-achromatic	doubly achromatic	tighter energy control and greater stability
field size	medium	larger	coverage with fewer fields
dose rate at isocenter	< 200	< 500	shorter treatment times
X-ray energies	< 5 MeV	< 25 MeV	greater skin sparing and greater penetration
dose distributions		better flatness, symmetry and penumbra	higher precision in treatment delivery, normal tissue sparing

Adapted from Ginzton and Nunan[88]

Table 3.V

1968 Distribution of High-Energy Therapy Machines

Type	installed in the US		installed worldwide	
	number	% of total	number	% of total
Cobalt-60	541	89%	1676	88%
Van de Graaff	13	2%	20	1%
Betatron	23	4%	137	7%
Linear Accelerator	33	5%	79	4%

methods. While the evolution of technology continued across the spectrum of therapy related endeavors, perhaps this era is best characterized by the clinical introduction of medical linacs and direct competitors, ^{60}Co teletherapy gamma ray units.[88]

The physics community worked toward producing higher and higher energy accelerators for basic research into the nature of matter. Ancillary technologies developed for these research "atom smashers" enabled the manufacturers of medical accelerators to increase the efficiency and reduce the size of clinical machines.

An abbreviated timeline of the era is shown in Table 3.III. Again, although the table lists events sequentially, we will separately address the two primary topics: X-ray production and delivery, and X-ray dosimetry. The major emphasis for this period will be on X-ray production and delivery, first as technology and then as the clinical implementation of that technology. An alternate view of the advances in technology is shown in Table 3.IV.[104] Here the early and later technologies are compared in terms of their benefit.

In 1968 the International Atomic Energy Agency (IAEA) published their Directory of High-Energy Radiotherapy Centres (Table 3.V). Another use study, conducted for the Bureau of Radiological Health (BRH), Food and Drug Administration (FDA), showed that by 1974 262 medical linacs were operating in the United States and that an additional 61 had been sold but were not yet operating.[105] This and the preceding data are indicators of the worldwide acceptance of high-energy systems during these years.

X-Ray Production and Delivery

As discussed earlier, Van de Graaff systems, with X-ray energies reaching 2 MeV, continued to be manufactured and sold until 1969. High Voltage Engineering Corporation sold forty-three two-million-volt generators for clinical use (thirty-five within the United States). Again, because of insulation and size problems, the practical limit on the X-ray energy for a system that directly accelerates electrons by an imposed high voltage is about three or four million volts. Manufacturers of betatrons included Allis-Chalmers in North America, Siemens in Germany, and Brown–Boveri in Switzerland.[106] By 1975 some fifty units had been installed in North America. The Brown-Boveri unit was capable of reaching 45 MeV. Betatrons were, however, limited to small fields and to a maximum dose rate of about 100 R per minute at 100 cm SSD. Additionally, high-energy betatrons (>25 MeV) were generally too heavy for gantry mounts. These limitations eventually led to the replacement of both Van de Graaff generators and betatrons.

As part of the quest for higher energy and more penetrating radiations, another class of therapy machine—synchotrons—was modeled after those proposed for the high-energy physics research community. Synchrotrons were capable of producing rather high-energy X-ray beams, up to 70 MeV, from a "reasonably" sized device.[107] Three such units were put into clinical trials.[108] Apparently any observed benefits of the very high-energy beams were not sufficient to overcome problems caused by the size, cost, and other practical issues surrounding their use.[109] They were also eventually replaced by smaller moderate-energy machines.

By 1950 most of the elements needed for the manufacture and clinical application of accelerator-based X-ray generators were in place.[110] The significant advantage of these generators over their predecessors was that the electrons (which eventually struck a target to produce X rays) gained energy in devices that, on average, were at near ground potential. Huge air gaps and drastic insulation schemes were simply not necessary for accelerators: there was no constant very high voltage in the system. The medical linac, however, could not become a practical reality until several complex problems were solved. These included: (1) the generation of sufficient radiofrequency (microwave) power; (2) coupling of that power to an accelerating structure capable of efficiently converting some of the microwave energy into increased energy of the electron beam; (3) injecting discrete groups of sufficient numbers of electrons into the accelerating structure at the correct times relative to the phase of the microwave energy, accomplished by the electron gun and the "buncher" at the entrance to the accelerating structure; (4) steering and focusing these "bunches" of accelerating electrons as they traversed the accelerating structure and beyond; (5) bringing the electron bunches to a precise location on a target; and (6) filtering the resultant X rays so as to provide a relatively uniform intensity distribution over the useful treatment field.

The reader is referred to the articles and book by Karzmark and colleagues for detailed descriptions of the physics and engineering principles of accelerator operation.[111] To facilitate further discussions, and to identify the relationship between the components, a greatly simplified block diagram of an electron linear accelerator is shown in Figure 3.11.

The 1930s and 1940s saw the introduction and improvement of more and more powerful microwave devices: the rhumbatron, klystron and magnetron. Ginzton in 1944 and later Chodorow and Ginzton and colleagues from 1949 through 1952 developed a 30 MW klystron.[112] These devices provided the

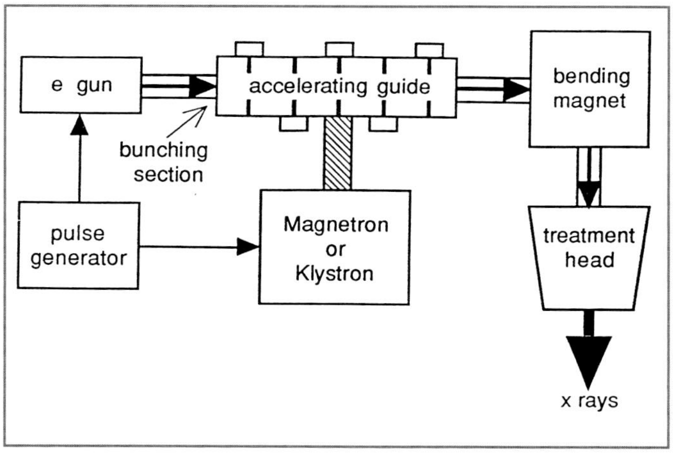

Fig. 3.11 The pulse generator fires (tens to hundreds of times per second) and causes the electron gun and bunching section to sequentially inject many "bunches" of electrons into the evacuated accelerating guide. At the same time the microwave power source (magnetron or klystron) injects radiofrequency energy. The system is designed so that each of the electron bunches arrives in the accelerating section at the correct phase (with respect to the microwaves) to be accelerated. The electron bunches accelerate through the cavities in the guide, are redirected by the bending magnet, and impact upon a target in the treatment head. The X rays produced there are filtered and collimated prior to exiting the head. (Authors' collection)

high-powered microwaves needed for successful electron linacs. In addition, these groups and Fry's group in England accomplished groundbreaking waveguide development work. Much of the design work utilized the readily available wartime microwave radar technology, that is, "S-band"—10 cm wavelength—microwave equipment. By happenstance, microwaves of this wavelength are almost optimally suited for the acceleration of electrons. The resulting accelerator waveguide designs contained "corrugations" that matched the phase velocity of the traveling (or later standing) wave with the speed of the accelerated electrons and thus provided for optimized energy transfer to the electrons.[113]

While the first machines used traveling-wave accelerating structures (physically quite long), newer machines tended to use the shorter standing-wave design.[114] In these standing-wave accelerating structures, every other cavity had zero electric field and did not play a role in transferring energy to the electrons. By moving these zero-filed cavities off-axis, the accelerating structure could be shortened even further, yielding an even higher energy-gain per unit length of structure.[115]

In parallel with accelerating waveguide development, improved electron gun and buncher designs, particularly gridded guns, provided for the necessary high-current and well-timed injection of bunches of electrons into the accelerating structure. Guns were initially diode designs. That is, they consisted of a cathode and anode. Later a control grid was added. This triode design had the advantage of using a relatively low grid voltage (only 2 to 5 percent of the cathode-anode voltage) to gate the gun on and off.[116] This made for sharper switching, better bunching, and a tighter energy range for the accelerated electrons.

In order to produce a clinically usable beam of X rays, the accelerated electron bunches had to be accurately directed onto a specific spot on the target and along a specific path.[117] For straight-beam machines, those with energy below about 8 MeV, the beam transport system was literally straightforward, with the accelerator guide coaxial with the treatment-head (collimator) axis. For higher energy machines, where the accelerating guide was simply too long to use in a straight-through configuration, a beam transport system was used to direct the beam from the roughly horizontal accelerating guide to the axis of the treatment head. Today a beam transport/beam optics system might include several magnets used to steer, bunch, focus, and bend the beam. The first such beam transport systems used 90 degree bending magnets. These nonachromatic systems were able to properly direct only axial, nondivergent, monochromatic beams onto the target. Any electrons entering the magnet off-axis, off-path, or at other than the design energy would strike the target off-center and/or at the wrong angle. The ultimate result would be an unflat and/or asymmetric X-ray field. Many of these difficulties were overcome with the introduction of 270 degree achromatic bending magnets.[118] These magnets can, within reasonable limits such as +10 percent of nominal energy, bring the electrons to the correct point on the target and with the proper trajectory. The 270 degree magnets also have the advantage of being compact relative to other achromatic designs, thus minimizing isocenter and room-ceiling heights. Even with improved beam optics, there were times when the accelerator operating parameters might drift so far from nominal that even a 270 degree bending magnet system could not compensate. To prevent this occurrence active feedback circuits were added to accelerator design. These included, for example, energy-defining "slits" before, in, and/or after a bending magnet.[119] If the beam energy drifted far enough from nominal, so that part of the beam hit the edge of the slit, an electrical signal was obtained from the slit and used to alter the operating parameters and return the energy to nominal. Additional beam monitoring, with active feedback correction, was also implemented through clever use of an accelerator's monitor chambers.[120]

Once the electrons impacted on the target, the resultant X-ray intensity distribution needed to be rendered uniform (i.e., flat and symmetric beam profiles). This was generally accomplished by using a flattening filter. The selection of flattening filter material and shape was heavily dependent on similar selections made for the target in an effort to optimize output, flatness, symmetry, and field-size, while minimizing electron contamination, energy spectrum broadening, etc. A very large body of work has been reported in this area.[121]

Having discussed the enabling technologies, it is worthwhile to look at the landmark clinical systems that applied these innovations to the treatment of cancer. In 1948 the British Ministry of Health gathered together Fry's group at the AERE, Gray's group at the Radiotherapeutic Research Unit (Medical Research Council), and Miller's group at Metropolitan Vickers for the express purpose of building an 8 MeV X-ray linac.[122] The linac was installed at the Hammersmith Hospital in London in 1952. Patient treatments began in September 1953.[123] Incidentally, the Hammersmith department also included another room, a "measurement room," which contained a gantry-mounted diagnostic X-ray system—a simulator!

This first clinical linac did not have a rotating gantry. It did, however, use a 90 degree bending magnet on a swivel so that the head of the machine could rotate 120 degrees, that is, it could deliver X rays within a range of ±60 degrees from the horizontal in a plane perpendicular to the accelerator tube's axis (Fig. 3.12).

The features of this first clinical accelerator included a gold transmission X-ray target, an aluminum flattening filter, a fixed uranium/tungsten-copper primary collimator, an ~8 cm tall adjustable tungsten-copper secondary collimating system, a front and backpointer system, and a light localizer. The machine was capable of providing 4 x 4 cm^2 to 20 x 20 cm^2 fields at 100 cm. Circular fields of various sizes were obtained by adding inserts to the adjustable collimators. 100 rads per minute were available at 2 cm depth in water at 100 cm SSD. The treatment couch was motordriven for rotational treatments, with the couch axis coincident with the beam-down collimator axis. A clever couch positioning mechanism and a treatment room floor with a vertical travel of ~80 cm provided for beam direction adjustments without losing the location of the entry point. That is, adjusting the swivel angle of the collimating system caused the appropriate (pseudo-isocentric) couch and floor motions.[124] Rotational therapy was feasible with this system. A monitor chamber was used to determine the dose delivered and to turn the beam off after a preset value had been reached. A very novel twenty-five chamber dosimetry system for "beam centralizing" was created to assure flat and symmetric beams. This system will be described later. To overcome the limitations of a 90 degree bending magnet and to obtain flat and symmetric fields, this accelerator employed a skewed aluminum flattening filter. The filter's central axis was necessarily displaced from the beam's central axis. The accelerator console included a "dose meter" calibrated in rads to a depth of 2 cm in water on the central axis for an 8 x 8 cm^2 field. From its first use in September of 1953 and for the twelve months of 1964, this linac treated some four hundred patients with only one day of unscheduled down time, an enviable record for any treatment unit.

Anticipating the success of the Hammersmith 8 MeV machine, the British Medical Research Council and the AERE collaborated on the design of 4 MeV gantry-mounted orientable linacs.[125] The resultant designs were made available to British industries. The first clinical use of these new designs occurred in 1954 at Christie Hospital, Manchester, and at Newcastle Upon Tyne. The Christie machine, the "Orthotron," was the first single gantry system. It provided a maximum of 135 degrees of rotation—from 30 degrees

Fig. 3.12 4 MeV linear accelerator installed at Christie Hospital, Manchester, England, in 1954. (Courtesy of the Center for the American History of Radiology, Reston, Va.)

below the horizontal (with a lowering of the floor) to 15 degrees past vertical (Fig. 3.12).

The Newcastle Upon Tyne system used a double gantry capable of directing the beam from over a 210 degree range. The adjustable collimators in this accelerator moved on arcs and kept their inner edges parallel to the edges of the X-ray field. This design minimized the beam penumbra.[126] Additionally, this system used a quadrant dose-rate chamber as part of an automatic centering system. Any deviations of the electron beam caused a difference signal at the chamber which was used to recenter the beam on the target. Otherwise, the features of both of these units were similar to those of the Hammersmith system, that is, ~100 rad/min at the depth of d_{max} at 100 cm SSD with a maximum field size of about 20 x 20 cm².

Meanwhile, in the United States, primarily at Stanford University, work continued on accelerators in general and also on their medical applications.[127] Henry Kaplan, then head of the department of radiology at Stanford, met with Edward Ginzton of the physics department to discuss the possible use of linacs for radiation therapy. A very successful collaboration followed with grants in 1952 from the National Institutes of Health (NIH) and the American Cancer Society (ACS) to build such machines.

Fig. 3.13 The Stanford medical linear accelerator, 1957. Above the patient was the ceiling mounted image amplifier. (Courtesy of the Center for the American History of Radiology, Reston, Va.)

The first Stanford accelerator (Fig. 3.13) was powered by a 1-MW klystron, provided up to 6 MeV X-ray beams, and also had electron beam capability.[128] Installation occurred in 1955, and with the treatment of an infant's retinoblastoma, clinical use commenced in January 1956.[129] As of 1988 that patient still had vision in the treated eye. This accelerator was typically operated at 4 to 5 MeV and provided about 65 rads/min at the depth of d_{max} in water at 100 cm SSD. Its traveling-wave accelerating structure was some 1.6 m long and was contained within a sealed vacuum "bottle." The X rays were produced in a removable gold transmission target (removable so as to allow for electron beam therapy). The X-ray target was followed by a monitor chamber and various filter materials.[130] The head of the machine also contained a light-localizer system and adjustable (lead) collimators. Field sizes were initially limited to 15 x 15 cm² at 1 m. One of the unique features of this machine was the presence of a retractable, 100 kVp, diagnostic, rod-anode, X-ray tube in the head of the machine. It could be quickly inserted so that the diagnostic X-ray target was positioned on the therapy beam axis a scant few centimeters distal to the therapy target. This enabled the production of very high-quality portal images on the accompanying X-ray image intensifier system or on film. An open-field diagnostic X-ray exposure was routinely followed by a second, brief exposure using the collimated therapy beam. This double-exposed film provided an accurate and informative record of the treatment.[131] This machine is now a part of the Medical Sciences Collection of the Smithsonian Institution in Washington, D.C.[132]

In 1958, making use of the British and American experiences, a team began development work on a commercial 6 MeV, fully rotatable, clinical linac. Their intent was the development of a machine and couch that would allow anteroposterior and posterior-anterior irradiations of a patient without having to turn the patient. They were successful, and the first unit was installed in 1962 at Stanford University's new facility in Palo Alto.[133] This machine, a Clinac 6, was fully rotational with an isocentric mount. It was powered by a 2 MW magnetron and used a 1.5 m traveling-wave accelerating guide. A 90 degree magnet directed the electron beam from the horizontal guide to the gold target. In some small part, this fully rotational machine was made possible by the use of a relatively new vacuum pump technology and a sealed-off waveguide.[134] After an initial pump-down of the accelerating guide by conventional pumps, a VacIon pump, with no moving parts and no fluids, was engaged.[135] These physically diminutive sputter-ion pumps were able to maintain high-vacuum, that is, to function effectively in any orientation. They eliminated the need for the relatively dirty oil pumps and for rotating vacuum seals (required for a fixed-position pump and movable gantry). The result was a better vacuum, more efficient transfer of energy from the microwave system to the electron bunches, better focusing of the bunches, a higher dose rate, and a better dose distribution. Other interesting features of this machine included an isocenter height of 46 inches and a modest room ceiling height requirement of only 8 feet. Solid state detectors were located in the distal jaws and provided an indication of beam symmetry. Sealed monitor chambers were combined with an integrator to provide for beam-off at a preset dose with a 1-rad counting accuracy. Some of these Clinac 6 systems were still in use in the United States in the 1980s and even up to 1990.

Another notable design was the Mevatron 8, the first clinical accelerator to use a nominal 270 degree achromatic bending magnet.[136] This accelerator also employed a retractable beam-stop making its use feasible in under-shielded rooms. It was first used clinically in 1965.

The SL-75, introduced in 1965, provided a dual dosimetry system for monitoring the delivered dose.[137] Today, such systems are mandated by several regulatory agencies.

With the invention of the standing wave accelerator guide, commercial development of a small, low-energy, in-line, fullly rotational gantry accelerator was feasible.[138] This design produced the well-known workhorse, the Clinac 4, an 80 cm isocentric machine (Fig. 3.14).

One of the last machines introduced in this era, the Clinac 35 provided two different X-ray energies.[139] This machine, as well as the Sagittaire accelerator, used a two traveling-wave accelerating guide design. Electrons leaving the first accelerating section could or could not be further accelerated depending on the phase relationship of the second section to the first. The significant drawbacks to this design were the overall size (particularly length) of a two traveling-wave accelerating guide construction and its relative inefficiency.

In addition to the new machines introduced during this era, other techniques were modified or introduced as part of linac-based therapy. One of these was moving field, usually rotational, therapy.[140] Then, as now, the intent was to maximize the target dose while sparing normal tissue. "Moving field" was something of a misnomer as, in some cases, the radiation field remained fixed while the patient was rotated around a fixed axis through the target. Some of the early moving therapy units were able to move over any point on the surface of a cylinder (rotate/translate). Others (converging-beam units) could follow a spiral path.[141] As more energetic machines became available, every effort was made to enable their use in a rotational mode.

Fig. 3.14 Varian Clinac 4, which became the workhorse of many radiotherapy departments. (Courtesy of the Center for the American History of Radiology, Reston, Va.)

A technique which developed over all three periods surveyed in this chapter was stereotactic radiosurgery or -therapy. Early uses occurred in 1948 and 1950 and will be discussed later in detail.[142]

X-ray Dosimetry

There were several different aspects to dosimetry work in the period 1950 to 1970. They included defining dose, refining and creating dosimeters and dosimetry systems for treatment machine dose calibrations, refining and creating dosimeters and dosimetry systems for treatment machine beam-parameter measurements, refining and creating real-time dosimetry systems for use as part of negative feedback control systems for treatment machines, and implementation of computerized treatment planning systems.

The beginning of this era was a time for international research and ultimately agreement on the concept of absorbed dose. This was necessitated by observations that dose differences of up to ±20 percent between departments were sometimes found and that the difference between success and failure for some patients might be just a few percent of the delivered dose.[143] It was also clear that the roentgen, the unit for exposure, was not usable for X-ray dosimetry above about two or three million volts. Very simply, at those high energies it became impossible to build a detector to meet the necessary conditions for measurements in R. The ICRU commissioned a detailed study leading to the formal definition and initial acceptance of the rad as the unit of absorbed dose in 1953.[144]

Since absorbed dose is the quotient of absorbed energy and mass, calorimeters and some chemical systems that respond directly to energy deposited per unit mass of material are absolute dosimeters. These observations led to these dosimeters being developed and employed at various national standards laboratories and also led to academic investigations.[145] Building on the absolute dosimeters, better and better relative dosimeters such as ionization-based systems, film systems, and luminescence systems were put into use.[146] Also, several national and international organizations formulated and published detailed protocols for doing dosimetric measurements and calibrations, including the ICRU, the Hospital Physicists Association (HPA), the Nordic Association of Clinical Physicists (NACP), and the American Association of Physicists in Medicine (AAPM).[147] Thus, during these years, more accurate dosimetry was established and better control was achieved over the quantity of radiation delivered by treatment machines.

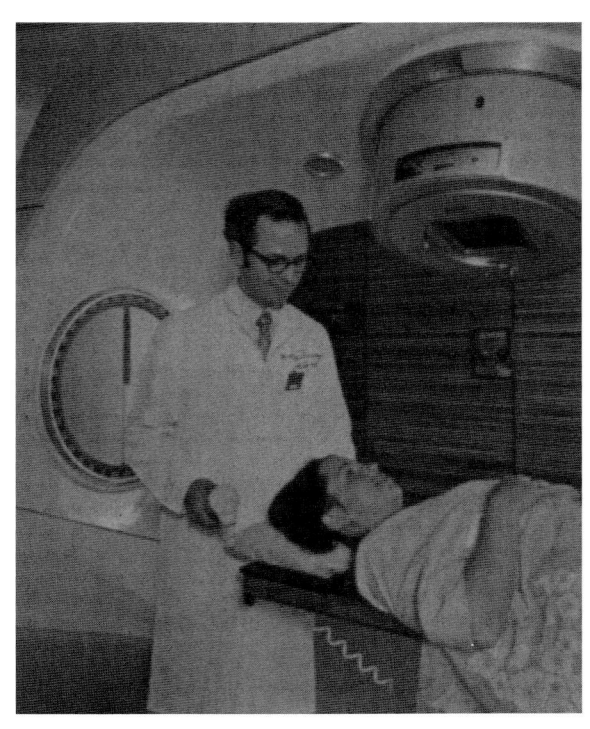

The most significant advances in technology for ionization-based systems were improved chambers and improved electronics. The chamber improvements included recognition of the effect of impurities in the conductive coatings with their subsequent elimination and design standardizations. Improvements in the associated electronics included moving from DC amplifiers to AC amplifiers with a large gain in stability, and moving from vacuum tube systems to solid-state (transistor) systems.

Luminescence systems, particularly thermoluminescence dosimetry (TLD) systems, also profited from a growing understanding of physical fundamentals. Additionally, procedural improvements, especially annealing protocols, improved overall stability. Manufacture of the raw dosimeter materials also improved, as did the capability of the readout electronics with better light sensing tubes and more controlled heating of the TLD materials.

Beam parameter measurements (used for both machine tuning and later for dose distribution calculations) required multiple relative detectors or a single detector that can be easily moved to different parts of a beam (often in a water phantom). The bending magnets, rotating heads or gantry, adjustable collimators, and other features of linacs necessitated many beam parameter measurements. Newbery and colleagues at Hammersmith constructed a twenty-five-way ionization chamber to accomplish their measurements.[148] This 26 cm diameter parallel-plate device was segmented by insulating grooves into a central 2 cm diameter detector and three rings of eight 2 cm diameter detectors each. For any orientation of the collimating head, this 25-way chamber could be rigidly attached at 102 cm and the twenty-five signals read sequentially. This enabled the operator to adjust focusing coils until satisfactory beam flatness and symmetry could be obtained. Other methods for obtaining similar data included film and scanning ion chamber systems.[149]

Real-time accelerator control systems were (and are) necessary to assure that the treatment beam parameters are at any time as they were measured and recorded. Typically, the technological improvements that evolved during this era were the use of "energy slits" to detect changes in electron-bunch energy and to provide a signal used to correct beam steering; multisection monitor chamber(s) to detect flatness or symmetry changes and to provide signals which were again used to correct beam steering; and improved monitor chambers for assuring delivery of the prescribed dose to the patient.[150]

This era just barely saw the genesis of computerized radiation therapy treatment planning, of signal importance in the following era.[151]

1970 TO THE PRESENT

The years 1970 to the present saw a continuing evolution of linacs with most of the technological progress involving beam delivery rather than beam production.[152] During this period, one new type of accelerator, the microtron, made its debut.[153] Another, a lightweight (~280 pound) robot-arm-mounted linac for 6 MV X-ray therapy was also introduced.[154]

Additionally, this era was a time of increasing computerization of all aspects of X-ray therapy.[155] As before, technological advances continued to be driven by the desire for "better" dose distributions within the patient. Also, more and more of the technology research and development work moved to the corporations that supplied treatment machines and treatment planning systems. This is hardly unexpected, as the therapy machines and planning systems have become so complicated that their workings are simply inaccessible to investigators outside the vendor's own research and development groups.

At the same time that treatment machines advanced, national and international efforts in radiation dosimetry improved the radiation dose standards in addition to the standardization of dosimetric methods.[156] This provided for more meaningful results in geographically wide-ranging cooperative studies.

At the beginning of the era, a 1974 study conducted for the BRH showed that 262 medical linacs were in use in the United States and that an additional 61 had been sold but were not yet operating.[157]

In 1976 the IAEA published its Directory of High-Energy Radiotherapy Centres as shown in Table 3.VI. Interestingly, there is an unresolved discrepancy between the accelerator

Table 3.VI
1976 Distribution of High-Energy Therapy Machines

Type	installed in the US number	% of total	installed worldwide number	% of total
Cobalt-60	790	78%	2365	80%
Van de Graaff	18	2%	24	1%
Betatron	44	4%	219	7%
Linear Accelerator	161	16%	336	11%

Table 3.VII
Technology Benefits

Early Technology	Follow-on Technology	Benefit
individual wedges	universal wedge	allows "any" wedge angle
individual wedges	dynamic wedge	allows "any" wedge angle
low melting point alloy blocks	multileaf collimators	eliminates block fabrication and enables dynamic conformal therapy
manual control	computer control	should enable more efficient and more precise therapy
pen and paper	record and verify	should enable more efficient and more precise therapy
localization and verification films	electronic portal imaging devices	catch blocking errors; necessary for verification of dynamic conformal therapy
	linear accelerator stereotactic systems	enables cost-effective and accurate treatment of small lesions in the brain
	microtron	small electron-beam energy spread; accelerated beam can be transported; 1 accelerator can serve more than 1 gantry scanned X rays; enables beam-intensity modulation for missing tissue or dose compensation
	Neurotron 1000	frameless stereotactic robotic therapy system; lack of frame makes treatment easier on the patient; system can compensate for some patient movements during treatment
	tomotherapy	very highly conformal dose distributions, CT-like images from treatment beam

▲ ▲ ▲ ▲ ▲

numbers shown by the two studies. Nevertheless, the trend is clear. Purchasers were moving toward the newer technology, the linacs, in place of the older cobalt units, Van de Graaffs, and betatrons.

Comparing the 1976 and 1968 IAEA studies, one sees a threefold gain for linacs as percent of total for the eight-year span from 1968 to 1976.[158] This and the preceding data are indicators of the increasing worldwide acceptance of linacs as the systems of choice for megavoltage therapy.

A brief overview of the advances in technology is shown in Table 3.VII. Here, the early and later technologies are compared in terms of their benefits.

In discussing the progress in technology, we will again address separately the two primary topics: X-ray production and delivery and X-ray dosimetry. The major emphasis in this chapter will be on X-ray production and delivery. Within that portion of the text, first new subsystems, such as new beam shaping devices, and the new machines, such as microtrons, robotic therapy systems, and tomotherapy systems, will be discussed.

X-ray Production and Delivery

Technological progress in X-ray production and delivery occurred on many fronts during this era. For example, linacs evolved in the sense that there were many refinements in the already invented subsystems such as accelerating waveguides. New treatment machine subsystems, including dynamic wedges and multileaf collimators, were first introduced during this era. In addition to subsystems for existing machines, completely new X-ray therapy machines were introduced. For the sake of brevity, only new subsystems and new machines will be discussed in this section. Those readers seeking further technical descriptions beyond that which can be included here are referred, in particular, to the articles and book by Karzmark and colleagues.[159]

Universal wedges

External wedge filters have long been available for use in compensating for missing tissue and thus for producing better dose distributions. Several technological advances now provide for using internal—to the treatment head—devices to produce these same dose distributions without the necessity for manual insertion and removal of tray-mounted wedges. This technology has the advantage of both saving time and of potentiating more error-free treatment. In 1970 Tatcher suggested a method for combining open and wedged fields to get any desired effective wedge angle.[160] This was experimentally verified by Mansfield and his group in 1974.[161] Later, a "Universal Wedge" was described by Bentel and colleagues in 1982, and detailed methods for obtaining specified angles in dose distributions with their equipment were given by Philips Medical Systems Division.[162] The Philips system automatically inserted an in-the-head 60 degree wedge for the fraction of the treatment time needed to obtain the desired effective wedge. A comprehensive analysis of the methods for obtaining arbitrary wedge angles from the superposition of a 60 degree wedged field and an open field was provided by Petti and Siddon in 1985.[163]

Independent collimators and dynamic wedges

Adjustable collimating jaws on therapy machines have routinely provided for symmetric (about the central axis) rectangular fields. Typically, the opposing jaws were driven in lock-step. By providing for independently acting collimating jaws, it becomes possible, for example, to do away with heavy half-beam blocks and, by moving a jaw during irradiation, to simulate a wedge of almost any effective angle. In the former case, one of a set of opposing jaws is moved to the central axis while the other defines the half-field size. In the second case, known as dynamic wedging, a jaw moves across the field during irradiation such that the integrated dose is larger on one side than on the other side of the field—effectively a wedged field.

Independent collimators potentiate a new and versatile way to design radiation therapy treatment techniques. One compilation of such techniques is given by Chism.[164] He points out the advantages of independent collimation and specifically cites the ability to avoid overlapping beam junctions and the advantages of simplified patient positioning and the potential for compensation.

The simplest type of compensation is wedging a beam. Investigators at the Joint Center made early successful attempts at dynamic wedge operation of an accelerator in the late 1970s.[165] Loshek described the details of a commercial implementation in 1984.[166] A further analysis was given by Leavitt and colleagues in 1990 and by Mohan in 1992.[167] It should also be noted that a computer-controlled collimator jaw can be moved during irradiation to produce a custom dose distribution of any reasonable shape; one is not confined

Fig. 3.15 Multileaf collimator system designed by R. Moeller at Varian Associates. (Courtesy of the Center for the American History of Radiology, Reston, Va.)

to wedge-like sloping distributions.[168]

These independently moving collimator jaws pose significant dosimetry data acquisition difficulties that usually require special apparatus for their solution.[169] These issues will be discussed later in this chapter.

Multileaf collimators

For many years beam shapes have been customized for each treatment portal with low-melting-point-alloy blocks. This is a tedious process for both block construction and block placement. It is further limiting in that a separate block is needed for each field. Multileaf collimators (MLCs) provide for computer-controlled beam shaping from within the head of the treatment machine and should allow for faster, and/or more complex, treatments with less chance of a wrong-block-for-the-field error (Figure 3.15). Additionally, MLCs allow for dynamic conformal therapy where the beam shape and possibly other parameters change with gantry angle. Today, a distinction is made between dynamic and static conformal therapy. Many investigators now believe that multi (6 to 10 or so)-field static conformal therapy is an equal of full dynamic conformal therapy but with fewer complicating issues. This is very much a topic of current study.

In any case, to assure sufficiently accurate delineation of field shapes, the leaves should project to about 1 cm width or less at isocenter. To prevent unintentional irradiation from between the leaves, an interlocking tongue-and-groove design is often used. To assure sufficiently rapid movement of the leaves for dynamic conformal therapy, each leaf may well have its own motor. To minimize penumbra, double-focused leaves may be employed. Some of these features were described as early as 1965 by Takashi.[170] Other early descriptions were given by Mantel and his group in 1977 and by Sofia in 1979.[171] These early experiences were well described in a 1980 conference on conformation and dynamic therapy.[172] More recent work has investigated the dose distributions resulting from MLC (or otherwise conformal) field shaping.[173] MLC systems continue to improve: projected leaf sizes at isocenter get smaller; maximum field sizes increase; cross mid-plane travels increase; and doublefocusing or rounded-end leafs improve penumbra.[174] Last, it must be recognized that full use of a MLC system requires a very sophisticated radiation therapy treatment planning system.

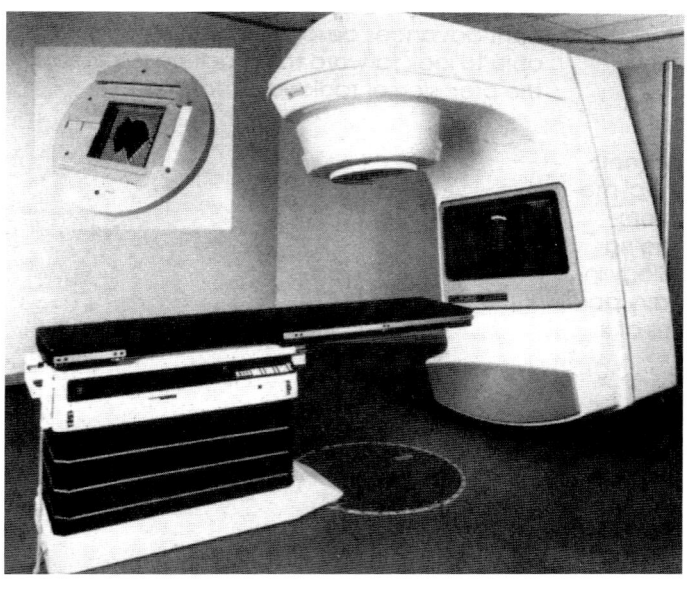

Computer control systems

Computer control of X-ray treatment machines has been a mixed blessing. On the one hand, there was every reasonable expectation that computer controlled machines would be safer, more precise, and more efficient than conventional X-ray treatment machines. Computer control systems are absolutely necessary for the implementation of universal wedges, dynamic wedges, multileaf collimators, and the like. Essentially all of the X-ray treatment machine manufacturers are already using or are developing computer control systems. Many of these are well rec-

Table 3.VIII
Accidents Timeline

Date	Event
3 June 1985	overdose in 10 MeV e⁻ treatment at Marietta, Georgia, facility; estimated delivered-dose of 150 to 200 Gy
26 July 1985	overdose at Hamilton, Ontario, Canada facility; estimated delivered-dose of 130 to 170 Gy
September 1985	AECL modifies hardware (turntable position sensor)
8 November 1985	Canadian Radiation Protection Bureau (CRPB) requests hardware additions and changes
December 1985	overdose at Yakima, Washington, facility
21 March 1986	overdose in 22 MeV e⁻ treatment at Tyler, Texas, facility; "Malfunction 54"; estimated delivered-dose of 165 to 250 Gy to 1 cm2 area in less than 1 second, machine taken out of service. The patient died from the overdose about five months after the accident.
7 April 1986	Tyler machine returned to service after no problems were found.
11 April 1986	overdose 10 MeV e⁻ treatment at Tyler, Texas, facility; "Malfunction 54"; estimated delivered-dose of about 250 Gy, machine taken out of service. The patient died from the overdose about three weeks after the accident.
17 January 1987	overdose at Yakima, Washington, facility; estimated delivered-dose of about 80 to 100 Gy. The patient died from the overdose about three months after the accident.

ognized as safe, precise, and efficient.

Unfortunately, one early implementation of a computer controlled treatment machine was involved in a series of accidental overdoses and deaths between June of 1985 and January of 1987. A detailed discussion of these Therac-25 accidents was authored by Leveson and Turner in 1993.[175] The following short discussion of those accidents will illustrate the care that must be taken when designing today's newest, more sophisticated, and more complicated X-ray therapy machines.

The treatment machines involved in the accidents were produced by Atomic Energy of Canada Ltd. (AECL), a Crown corporation. The Therac-25 used a magnetron-powered double-pass accelerator and could produce 25 MV photon beams or any of several electron beams. The Therac-25 was designed as a computer controlled machine (using a DEC PDP-11 minicomputer) with many of the traditional safety interlocks subsumed into their software equivalents. Apparently, the Therac-25 software was developed by one person and was based on the earlier (1972) Therac-6 software. Therefore, it was written in PDP-11 assembly language—a so-called low-level computer language which is very difficult to understand and maintain. The first commercial version of the machine became available in 1982. Eventually five machines were installed in the United States and six in Canada. Six accidents were recorded between 1985 and 1987 with the FDA, the Canadian Radiation Protection Bureau (CRPB), and the AECL taking actions resulting in a series of safety modifications to the accelerator. Many of the problems were rooted in the software that controlled the accelerator and were very difficult to find—hence the delay in overcoming the problems. An abbreviated accidents timeline is shown in Table 3.VIII, adapted from the Leveson and Turner article.[176]

It was subsequently determined, primarily by the Tyler physicist and machine operator, that the speed with which one typed data into the controlling computer and the speed with which one edited that data could affect the state of the machine. In the worst case, high-current electron beams were injected into the accelerating structure (as though for photon treatment) with no X-ray target in place and without the electron-beam scanning system operating. This resulted in an extremely high dose rate and narrow beam incident on the patient. Doses of about 25,000 rads could have been delivered to small areas within a second or so. The only error message displayed at the console was the cryptic "Malfunction 54." Different software problems were involved in some of the other accidents. The reader is referred to additional information about these accidents and about safety critical systems in general available from the various other sources.[177,178]

In some part because of these accidents, computer-based medical systems are much safer today.[179] Moreover, since the accidents medical physicists have endeavored to become more familiar with computer control systems and have taken part in some rigorous testing of those systems.[180] Done properly, computer control systems are beneficial if not essential to the implementation of universal wedges, dynamic wedges, multileaf collimators, and various other treatment tools. The present generation of computer-controlled X-ray therapy machines seems to be living up to the expectations of the radiation therapy community.

Linac-based stereotactic radiosurgery/radiotherapy

Pre-linac-based stereotactic radiosurgery and radiotherapy (single high-dose treatment and fractionated treatment respectively) began as early as 1948 using betatron-generated X rays.[181] Other early uses occurred in about 1951 with 200–300 kVp X rays.[182] Systems using ^{60}Co trays and heavy charged particle beams were also evolving at this time.[183]

The early, high-energy X-ray techniques, developed at several centers, have many features in common. Typically, a frame, e.g., the commonly used Brown-Roberts-Wells (BRW) head ring, is rigidly attached to the patients skull early in the day.[184] A device often called a computed tomography (CT)-localizer is attached to the head ring for the purpose of displaying fiducial markers on a subsequent CT scan. The localizer is removed, and the patient waits while treatment planning is accomplished. A true three-dimensional planning system uses the CT data to produce a plan of treatment relative to the coordinate system of the ring. Meanwhile, the accelerator is fitted with an external collimating system that provides small beams of circular cross section, typically 1.0 to 6.0 cm diameter. With all planning done, and with the machine set up according to the plan, the patient's head ring is bolted to a support which is adjusted to place the target at the machine's isocenter. Then for all of the prescribed couch and gantry motions, for all of the non-coplanar beams, each beam is always directed at the target.[185]

Further advances in the technology of the frames and localizers have led to the use of data from planar angiography, digital subtraction angiography, and magntice resonance (MR) imaging for stereotactic cases.[186] A caveat, however, is that it is difficult to use MR-compatible frames and localizers. The frames and localizers must fit into the scanner's head coil, and they must not distort the magnetic fields. Some very recent work by Kooy and colleagues has removed the need for use of a head ring and localizer. Rather, an image processing technique is used to register a freely taken MR data set with a stereotactically constrained CT data set.[187]

The traditional circular collimators used in stereotactic radiosurgery/radiotherapy pose another problem. They make it difficult to treat other than roughly spherical targets. Such treatments have been accomplished by setting multiple isocenters, that is, overlapping spherical dose distributions within the patient. This limitation may be overcome by the use of adjustable collimators or even miniature multileaf collimators. McGinley has recently described an adjustable collimator system that significantly improves dose distributions for some nonspherical targets.[188] A very detailed modeling study of the effects/advantages of dynamic field shaping was carried out by Nedzi and colleagues. They concluded that there was a real gain from simple shaping devices and a considerable potential gain from multileaf dynamic collimation.[189]

One of the remaining difficulties of stereotactic radiosurgery treatments is the head ring. Pinning a ring to a patient's skull induces much discomfort, and, additionally, the ring gets in the way of imaging and of treatment. There are now a few techniques available for minimizing these difficulties. One example of a relocatable frame was designed by Gill, Thomas, and Cosman (the GTC frame). It uses dental and cranial molds to establish a coordinate system fixed to the patient's skull. There are no pins, the system is non-invasive and can be reattached to the patient in minutes. A clever quality assurance device, the depth helmet, can be quickly and easily used to verify proper positioning of the frame.[190] Because of the ease with which this device can be used, fractionated stereotactic radiotherapy is completely feasible. Beyond the easy-to-apply GTC frame are some frameless techniques for stereotactic treatments. One of these techniques uses gold-filled titanium cortex screws to attach radiographic markers rigidly to the patients skull. A treatment plan is done in a coordinate system derived from the marker locations. At the time of treatment, new images are obtained from which the target location is determined. That information is used to move the target to the treatment machine isocenter.[191] Finally, and to be described in a bit more detail later, a new commercial robotic therapy system uses a comparison of calculated radiographic images with pairs of images taken during treatment to determine the patient's head position and to correct the system's aim.[192]

Treatment planning for linac-based, X-ray, stereotactic radiosurgery/radiotherapy has also progressed during this era. Systems now offer full three-dimensional graphics and beam's eye view planning to spare normal tissues.[193] Optimization routines are also finding their way into stereotactic planning.[194]

Real-time imaging

Several new technologies for obtaining live portal images were developed during this period.[195] The early systems were those of Baily and colleagues in the late 1970s and Meertens's group in the early 1980s.[196] The devices went by a variety of names, among them on-line imager, real-time portal imager, and electronic portal imaging device (EPID). Names aside, the devices have the same basic features. They obtain and display a portal image during treatment. The various techniques include an image receptor on the far side of the patient and a means for displaying that image on a monitor (typically located at the control console). The image receptors range from fluorescent screens viewed by a video camera (using a mirror or a light-pipe bundle) to a matrix of liquid-filled ionization chambers (Fig. 3.16).

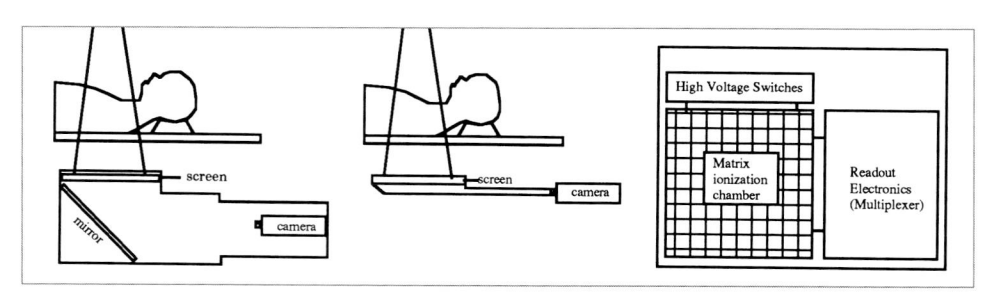

Since these systems are capable of producing a new image every few seconds, the number of images per treatment can be overwhelming. There are considerable efforts underway to provide tools for dealing with the data. Meanwhile, the real-time imagers tend to be used to check for proper blocking at the beginning of irradiation for each field, and the real-time images are also used, as traditional TV monitors are used, to watch for gross patient move-

Fig. 3.16 Left, a mirror-box system, the image on the bottom of the fluorescent screen is reflected by the mirror to the camera lens; Center, a (now no longer available) fiber reducer system, the image on the bottom of the fluorescent screen enters a bundle of light pipes, each pipe shrinks in size on its way to the camera. Right, a matrix-ionization-chamber cassette, the signal from each tiny ionization chamber, picture element (pixel) is sequentially read by an attached computer. (Authors' collection)

Fig. 3.17 Schematic illustration of a circular microtron. The extraction tube can be positioned to intercept different orbits resulting in the extraction of beams of different energies. The beam transport system can switch a beam between different treatment rooms (gantries). (Authors' collection) ▶

ment. More sophisticated uses of the systems range from providing after-treatment movie loops to computer-assisted matching of treatment to prescription images, attempts at automatic image comparison, and using the multitude of images for dose estimations.[197, 198, 199]

Electronic portal imaging is perceived as sufficiently important to radiation therapy to have been the topic of several workshops, namely those at Chapel Hill in 1987, Las Vegas in 1989, and Newport Beach in 1992. The third International Workshop on Electronic Portal Imaging occurred in San Francisco in October 1994. When mature, this portal imaging technology is likely to significantly improve the accuracy and precision of radiation therapy.[200]

Microtrons

One new and significant accelerator, the microtron, evolved into a practical clinical device during this era. Microtrons differ from linacs in that microtrons are cyclic accelerators in which each electron bunch passes through the same accelerating structure many times before being extracted, transported to, and striking an X-ray target (Fig. 3.17). The electrons follow a well-defined path of expanding discrete circular orbits, because their energy increases by the same increment on each pass through the accelerating structure. Thus energy selection is accomplished by physically moving the extraction tube to intersect different orbits. This method was first suggested by Veksler in 1944.[201] In addition, the energy spread is necessarily much smaller for microtrons than for linacs (in order for the circulating electrons to remain in phase with the microwaves in the accelerating structure). Thanks to this narrow energy spread, it is feasible to transport the beam to semi-remote locations, that is, one microtron can serve multiple treatment rooms (i.e., gantries). However, microtron develop-

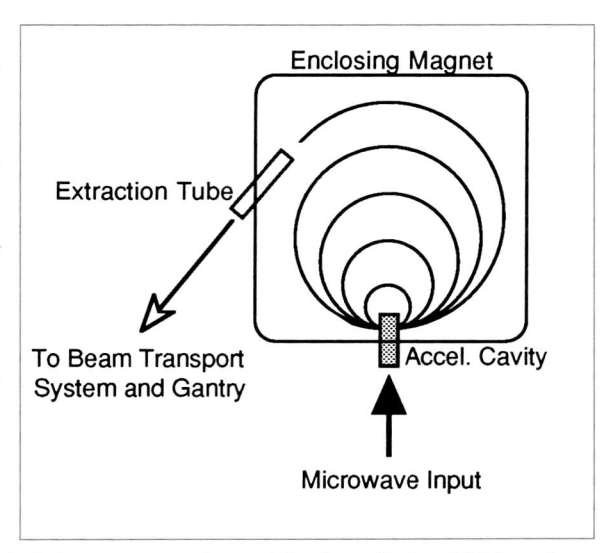

ment languished until the 1960s, when better electron guns and injection systems overcame the earlier low-beam-current limitations.[202] The first clinical microtron, a 10 MeV machine, was described in 1972 by Reistad and Brahme and put into routine clinical service in 1976.[203] A commercial system, the MM 22, was then developed by AB Scanditronix for the University of Umeå and was delivered in 1977. This circular microtron could deliver X-ray beams of either 10.2 or 20.9 MV and nine different energy electron beams.[204]

The original microtron design, using one accelerating cavity and roughly circular orbits, was superseded by the "racetrack" microtron, which used a multicavity accelerating guide and racetrack-shaped orbits (Fig. 3.18). Note that the circular microtron was limited to one cavity because the orbit was curving through the cavity. Because the electron bunches in the racetrack microtron have a straight path between the semicircular orbit ends, multiple accelerating cavities can be used. This system provides much greater acceleration (energy gain) per orbit and results in a more compact machine. The racetrack design was proposed in 1946 by Schwinger.[205] Early implementations included an accelerator at the University of Western Ontario and another at the Royal Institute of Technology, Stockholm.[206] Present models can provide X-ray energies of up to 50 MeV in 5 MeV steps.

Fig. 3.18 Schematic illustration of a racetrack microtron. Once again, extraction-tube positioning is used to control beam energy selection. The entire two-magnet structure is contained within an evacuated vessel. (Authors' collection)

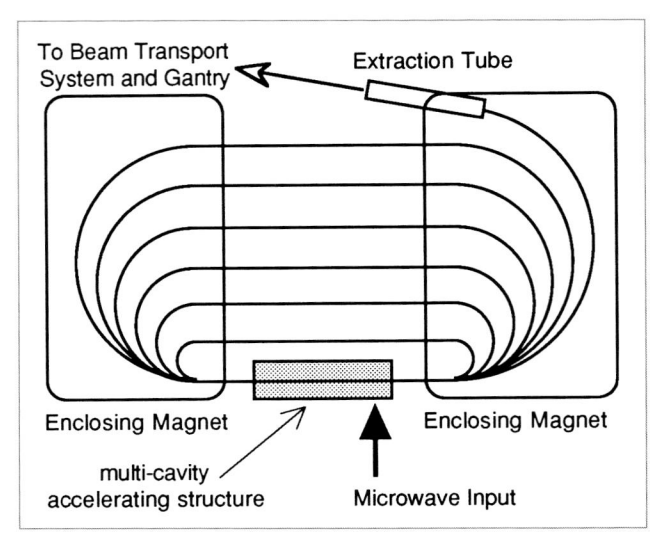

The present generation of commercial clinical microtrons have some unique features.[207] They provide a scanned X-ray beam, an X-ray beam purging magnet, helium in place of air in the treatment head, multileaf collimators as standard equipment, and a beam's eye view video system. The first three of these features are intended to work together to provide the highest quality photon beams. That is, they reduce beam contamination by electrons and low energy photons (maintaining the greatest possible depth of d_{max} and provide consistent beam energy across a treatment field). The scanning system also does the beam flattening out to a 50 x 50 cm² field. The scanning feature potentiates dose modulation within a treatment field, i.e., dose compensation.[208] The multileaf collimators work as described earlier, and the beam's eye view video system serves as a quality assurance device for multileaf operation.

The present commercial system is already noteworthy for its innovative technology in both accelerator design and in beam delivery design (the treatment head devices). Microtron technology is still evolving, and we can anticipate interesting future developments.

Robotic therapy

Within a few weeks of this writing, in June of 1994, a frameless stereotactic robotic therapy system (an investigational device), should treat its first patient at the Stanford University Medical Center. There are several technological innovations in this system.[209]

The accelerator is quite small and compact. It uses X-band rather that S-band microwaves to power its accelerating waveguide and to provide a 6 MV X-ray beam. Since the wavelength of X-band microwaves is 3 cm rather than the S-band's 10 cm, all of the microwave components are about one-third "normal" size. Also, this straight-beam accelerator is designed to use simple external circular collimators (0.5 to 6.0 cm diameter); thus there is no treatment head. Overall, these design choices result in a roughly 125 kg (280 pound) system. The system specifications are presently 6 MV with, at 80 cm SSD and for a 4 cm diameter collimator, dmax=1.5±0.2 cm and 55 percent at d=10 cm, a spot size of less than 0.2 cm, a maximum dose rate of greater than 3 Gy per minute at 80 cm at the depth of dmax, flatness across the central 80 percent of better than 4 percent at 7 cm depth, and symmetry across the central 80 percent of better than 2 percent at 7 cm depth.

The accelerator is mounted on a robotic manipulator similar to the type used, among other applications, to weld cars together. This industrial robot provides automatic positioning and aiming of the accelerator relative to the patient. There is no "isocenter" in this system. The manipulator is capable of positioning the accelerator to an absolute accuracy of ±0.05 cm. The corresponding pointing accuracy for the accelerator's X-ray beam is ±0.0001 radians (0.006 degrees). Although the manipulator is capable of moving the accelerator over the full surface of a sphere (pointing inward), practical considerations do limit somewhat the available pointing directions. To avoid contact with the patient's body and with some of the auxiliary equipment, the system is limited to about 3π of the 4π steradians of a full sphere.

In practice the system is a point-and-shoot device, meaning the X-ray beam is engaged only when the system is stationary. It is anticipated that most treatments will be accomplished using pseudo-arcs whose totality will be represented by up to 200 point-and-shoot nodes. Taken together, the accelerator, manipulator, and their control systems provide for from 0.1 to 50.0 MU per node in increments of 0.1 MU. In addition, since there is no isocenter and the robot can vary the SSD from node to node, field sizes can also vary from node to node without changing the physical collimator. This potentiates conformal therapy as the field size can be adjusted to the target size node by node.

This system's third technological innovation, which in particular allows for frameless intracranial stereotactic treatments, is the use of real-time images for patient motion compensation. This is not portal imaging. Rather, two diagnostic X-ray sources, with their axes widely separated but intersecting at the treatment target, provide a pair of images as the accelerator is moved to each node. The image receptors are similar to portal imaging mirror-box systems in that a camera looks at a fluorescent screen. The digitized images are immediately compared to a set of images previously calculated from a high-resolution CT scan of the patient. These precalculated images, some four hundred or so, represent the imaging system's views of a patient at or near the nominal treatment position. Thus, the correlation of the real-time images with some images from the precalculated set determines the patient's present head position and that position's deviation from nominal. An immediate correction for the accelerator's position is calculated and the accelerator moved to the proper place, relative to the patient's head, prior to irradiation from that node. The system's patient motion compensation provides an overall spatial accuracy of ±0.1 cm.

The advantages of this treatment system over conventional systems is that little patient restraint is required, there is a constant cycle of check-and-correct for any patient motion, and even with circular collimators conformal dose distributions are possible and practical. Once this investigational device succeeds for intracranial targets, there is every intent to extend its reach to other anatomical sites.

Tomotherapy

At this time, tomotherapy exists only as a very interesting proposal put forward by Mackie and colleagues.[210] Tomotherapy, or "slice therapy," would be accomplished with a small, straight-line, 6 to 10 MV linac mounted in a CT-like ring gantry. A spatially and temporally modulated fan beam from the rotating gantry would irradiate the patient while the patient moves through the ring. This is achieved with a multileaf collimator whose leaves individually open and close during irradiation. Simulations of this type of therapy show very highly conformal dose distributions. It might seem that the treatment design problem would be overwhelming for this type of treatment, but, it is essentially a solved problem. The same algorithms used in emission CT image reconstruction can be used to provide dose optimization for tomotherapy.[211]

The proposed system has other attractive features. By mounting a standard CT X-ray source and detector system some 90 degrees around the ring from the therapy X-ray source, CT pretreatment positioning images can easily be obtained. Also, CT verification images can be obtained during treatment. By mounting megavoltage image detectors opposite the therapy X-ray source, it also becomes possible to determine the dose distribution within the patient for each treatment. The CT image and dose distributions thus obtained are easily correlated because they were taken with the same hardware at the same time. The verification images for each treatment are a series of CT images with delivered-dose information superimposed on the anatomy.

Mackie and colleagues have also shown that tomotherapy dose distributions are almost independent of beam energy, with very high energy beams providing just somewhat better skin sparing. In part, this independence of energy comes about because the system would not need a flattening filter. The basic slit-beam can be modulated as needed by the MLC system. The omission of the filter provides for lower beam contamination by electrons and low energy photons, hence better skin sparing than conventional machines of the same energy.

Collision avoidance, a problem for some other systems attempting three-dimensional conformal therapy, is simply not an issue for this system. If the patient can fit through the ring gantry, then he or she can be treated without any fear of collision. Also, the ring-gantry and slit-beam make the installation of a beam-stopper very feasible. With a beam stopper, an approximately 10 MV tomotherapy system could likely use a room that had been previously shielded for a ^{60}Co system

All in all, the physics and engineering studies of Mackie's group strongly suggest that this technology could provide stereotactic-like precision throughout the body and at a cost comparable with that of conventional therapy systems.

X-ray Dosimetry

There are several different aspects to the dosimetry work of this era. They include refining the definition of absorbed dose, the dosimetry protocols for treatment machine dose calibrations, the dosimeters and dosimetry systems for treatment machine beam-parameter measurements, and the computerized treatment planning systems.

Absolute dosimetry

While absorbed dose had been previously defined, the various national laboratories relied on ^{60}Co exposure standards for dose calibrations.[212] This changed in the mid-1970s when Domen and Lamperti described their graphite absorbed dose calorimeter.[213] This instrument provided a direct measure of absorbed dose to graphite, measured by observing the temperature rise in an irradiated disk of graphite. While this device provided an improvement relative to exposure methods, it still required a conversion from dose-to-graphite to dose-to water. This transfer is considered to be accurate to about 1 percent.[214] Beginning in 1980 Domen showed that the dose transfer problem could be bypassed by using a water calorimeter to directly measure absorbed dose to water.[215] The first Domen calorimeters were large bulky devices. Other investigators redesigned the system to provide for a portable device.[216] Unfortunately, water exhibits a "heat defect" whereby the temperature rise is not strictly proportional to dose. This resulted in a 3.5 percent discrepancy between water and graphite calorimeter measurements. The problem was thoroughly investigated and the cause and solutions found.[217]

With continuing improvements in the absolute calorimeter and ferrous sulfate dosimetry standards, more serious efforts are underway to move from ^{60}Co exposure calibrations of the equipment used to calibrate X-ray therapy machines. We can expect to see absorbed dose calibrations of those systems as soon as the national standards laboratories (and the ICRU) finalize their devices and reach a consensus on some details.[218] The expected impact on clinical dosimetry is a ±1 percent accuracy rather than the current ±3 to 4 percent and simpler, less error-prone dosimetry protocols.

Calibration protocols

During this era several groups published new or replacement protocols for determining the dose to water from high-energy X-ray beams.[219] These are the protocols used in the field by physicists to calibrate treatment machines. With work beginning in the 1970s, first the NACP and then the AAPM (TG-21) published their protocols in 1980 and

1983 respectively.[220] While we will not go into details here, it is worth noting that these new protocols did introduce a new ion chamber dependent parameter, the cavity-gas calibration factor N_{gas}, which represents the dose to the gas in the chamber per charge collected (e.g., in Gy/C). Unlike previous protocols, these explicitly took chamber design into account through N_{gas}.

More recently the IAEA published their protocol.[221] It is very similar to the AAPM TG-21 protocol and seems to agree to within 1 percent for the vast majority of comparisons.[222] All of these protocols are currently based on exposure calibrations of the users' ionization chambers. Again, we are nearing a time when exposure calibrations will be replaced with dose calibrations. The protocols will then become both simpler and more accurate.

Beam parameter measurements

The advent of new accelerator subsystems and new techniques has necessitated the development of new dosimetry approaches for measuring beam parameters. For example, traditional beam scanning methods are inappropriate for obtaining profiles for universal or dynamic wedges. The small-diameter beams used for stereotactic radiosurgery also require new methods for beam scanning.

In the case of universal or dynamic wedges, where the intensity at a point along a scan path changes with time, the traditional method of moving a chamber along a scan path will not work. Film certainly will work but is considered less accurate than a scanning ion chamber or scanning diode measurement. The solution is found in multidetector arrays, where the many detectors stay in place during the operations in the treatment head (e.g. mixing an open field with a 60 degree wedge field or moving a jaw). One example of a multidetector array was recently discussed by Leavitt.[223] He described dynamic wedge beam profile measurements with an eleven-diode linear array. Many of the commercial manufacturers of beam scanning systems are now offering similar array detectors.

Small diameter beams pose a severe challenge for conventional, Farmer-chamber dosimetry systems, in that the field is sometimes smaller than the chamber and in that lateral equilibrium is not achieved. Descriptions of appropriate dosimetry techniques are referenced in the forthcoming AAPM task group report on stereotactic radiosurgery.[224] A particularly interesting description of the problems, along with a method for correcting one's data, is given by Rice and colleagues.[225] The almost universal recommendation for obtaining beam data for small fields is to use more than one system (e.g. film, thermoluminescent, and ionometric), compare the results, and if they differ, find out why. Note that even film dosimetry techniques must be modified for small field work. Most available film densitometers have too large an aperture and will report too large a beam penumbra. Some physicists have resorted to visits to astronomy departments to use microdensitometers in order to overcome this difficulty.

Another new method for the dosimetry of stereotactic radiosurgery systems involves measuring the end result, the dose distribution. Of course this does an end-to-end check on the system from CT system geometric fidelity, to accelerator mechanical systems, to beam data, and to the treatment planning system throughout treatment delivery without identifying any specific problem area. Nevertheless, it is a very useful check. This technique uses MR to image dose in a ferrous sulfate gel-filled anthropomorphic head phantom. The techniques was first described by Gore, Kang, and Schulz in 1984.[226] Subsequently, several investigators have contributed to the technique.[227] There are several remaining problems. MR scanners are subject to their own geometric distortions, and ferrous sulfate gels are prone to both postirradiation diffusion of ferric ions (blurring the image) and autooxidation of ferrous ions (decreasing the signal to noise ratio).

Treatment planning systems

None of the several techniques described in this chapter could have had practical implementation without computerized radiation therapy treatment planning (RTP). This era saw tremendous evolution of such systems, from the programmed console to today's three-dimensional RTP systems.[228] It is unfortunately well beyond the scope of this chapter to describe these systems in any detail. The reader is referred to the 1991 special issue of the *International Journal of Radiation Oncology, Biology, Physics* for a snapshot of the state of the art in 1989 and 1990.[229]

Much, if not most, of today's treatment planning system development has shifted to commercial vendors. In large part this is due to the software engineering requirements for a system that will meet today's expectations for image display and graphics display. Various investigators now tend to break new ground in specialty areas of treatment planning. The successful ideas are eventually engineered into the commercial systems.

For example, RTP systems are even now evolving beyond simple geometric and physical determinants. Some investigators are attempting to use biological endpoints in the process of optimizing a plan of treatment.[230]

In the pursuit of increased computing power to more quickly solve radiotherapy treatment planning and dose display problems, Rosenman and colleagues ported a planning program to a remote supercomputer which they accessed over a gigabit network.[231]

There are other examples too numerous to mention. It is safe to say that today's and tomorrow's RTP systems will take advantage of ever increasing computer power. We should expect to be provided with more detailed patient information and to receive optimization aid from the system. This might take the form of automatic beam size, shape, and orientation advisories or "just" the real-time display of dose-volume histograms. Suffice it to say that three-dimensional planning systems are very much on the beginning of the steep upward reach of their development curve.

Acknowledgments:
Thanks go to Bill Minowitz and Richard Morse, Ph.D., for providing historical information from their archives, and to Leonard Stanton, M.S., and Pamela G. Duncan for their critical reading of the manuscript.

▷ ▷ ▷ ▷ ▷

REFERENCES

1   Ewing, J., "Early Experiences in Radiation Therapy," *Am. J. Roent.* 31 (1934):153-163; Pohle, E.A., and Chamberlain, W.E., eds. *Theoretical Principles of Roentgen Therapy* (Philadelphia: Lea and Febiger, 1938):271; Glasser, O., ed. *Medical Physics*. vol 1. (Chicago: The Year Book Publishers, 1944):1744; Glasser, O., ed. *Medical Physics*. vol. 2. (Chicago: The Year Book Publishers, 1950):1227; Krabbenhoft, K.L., "A History of Roentgen Therapy," *Am. J. Roent.* 76 (1956):859-865; Glasser, O., ed. *Medical Physics*. vol. 3. (Chicago: The Year Book Publishers, 1960):754; Buschke, F., "Radiation Therapy: The Past, the Present, the Future," *Am. J. Roent.* 108 (1970):235-246; Lederman, M., "The Early History of Radiotherapy: 1895-1939," *Int. J. Radiat. Oncol. Biol. Phys.* 7 (1981):639-648; Eisenberg, R.L. *Radiology: An Illustrated History*. 1st ed. (St. Louis: Mosby Year Book, 1991):606.

2   Glasser, O., ed. *The Science of Radiology* (Baltimore: Charles C. Thomas, 1933):450; and Coolidge, W.D., and Atlee, Z.J., "Roentgen Rays: Tubes," in Glasser, *Medical Physics*, vol. 1, pp. 1395-1401.

3 Feldman, A., "A Sketch of the Technical History of Radiology from 1896 to 1920," *RadioGraphics* 9 (1989):1113-1128.
4 Jerman, E.C., "Roentgen-Ray Apparatus," in Glasser, *Science of Radiology*, p. 450; Krabbenhoft, "A History"; and Feldman, "A Sketch."
5 Grigg, E.R.N. *The Trail of the Invisible Light*. Springfield, Ill.: Charles C. Thomas, 1965.
6 Eisenberg, *Radiology*.
7 Coolidge, W.D., "Powerful Roentgen-Ray Tube with a Pure Electron Discharge," *Phys. Rev.* 2 (1913):409.
8 Krohmer, J.S., "Radiography and Fluoroscopy: 1920 to the Present," *RadioGraphics* 9 (1989):1129-1153.
9 Langmuir, I., "Effect of Space Discharge and Residual Gases on Thermionic Currents in High Vacuum," *Phys. Rev.* 2 (1913):450; and Krohmer, "Radiography."
10 Rutherford, E., and Barnes, J., "Efficiency of Production of X rays from a Coolidge Tube," *Phil. Mag.* 30 (1915):361.
11 Gross, M.J., and Atlee, Z.J., "Progress in the Design and Manufacture of X-ray Tubes," *Radiology* 21 (1933):365; and Coolidge and Atlee, "Roentgen Rays: Tubes."
12 Glasser, *The Science of Radiology*.
13 Coolidge, W.D., "The Radiator Type of Tube," *Am. J. Roent.* 6 (1919):175.
14 Coolidge, W.D., and Moore, C.N., "A Water Cooled High Voltage X-ray Tube," *Am. J. Roent.* 10 (1923):884.
15 Feldman, "A Sketch."
16 Newell, R.R., "Roentgen Therapy Apparatus," in Pohle and Chamberlain, *Theoretical Principles of Roentgen Therapy*, p. 61-120.
17 Jerman, "Roentgen-ray Apparatus," and Krohmer, "Radiography."
18 Glasser, O., and Fortmann, W., "The Physical and Clinical Foundations of Oversoft Roentgen-ray (Grenz-ray) Therapy," *Am. J. Roent.* 19 (1928):442-452; Bucky, G. *Grenz-ray Therapy*. New York: Macmillan, 1929; Glasser, O., "The Physical Foundation of Grenz-ray Therapy," *Radiology* 18 (1932):713-726.
19 Glasser, O., and Beasley, I.E., "An Improvement of Grenz-ray Tubes," *Strahlentherapie* XL (1930):389; and Glasser, "The Physical Foundation."
20 Glasser, *The Science of Radiology*.
21 Darnell, C., "A Sphere Gap for Measuring X-ray Voltage: Service Suggestions," *Victor* 21 (1921):1.
22 Case, J.T., "History of Radiation Therapy," in Buschke, F., ed. *Progress in Radiation Therapy*. vol. 1. (New York: Grune and Stratton, 1958):1-18; and Trout, E.D., "History of Radiation Sources for Cancer Therapy," in Buschke, *Progress in Radiation Therapy*.
23 Buschke, "Radiation Therapy," and Lederman, "The Early History."
24 Perthes, G., "Experiments to Determine the Transparency of Human Tissues to Roentgen Rays and the Significance of this Transparency of Tissues in Radiation Therapy," *Fortschr. Geb. d. Roent.* 8 (1904):1; Krabbenhoft, "A History"; and Glasser, *The Science of Radiology*.
25 Pfahler, G.E., "A Roentgen Filter and a Universal Diaphragm and Protecting Screen," *Trans. Amer. Roent. Ray. Soc.* (1906):217.
26 Lacassagne, A., "Idées directrices et principes de techniques actuels de la curiethérapie des cancers à l'Institut du Radium de Paris," *Radiophysiolique et Radiotherapie* I (1927):162-206; and Buschke, "Radiation Therapy."
27 Ewing, "Early Experiences."
28 Newell, "Roentgen Therapy Apparatus."
29 Pfahler, G.E., "A Device to Prevent the Omission of a Filter in Deep Roentgen Therapy," *Am. J. Roent.* 10 (1923):562-563; Fried, C., "Ein neues bestrahlungsgerät in verbindung mit der Metalix-Therapieröhre," *Strahlentherapie* 33 (1929):160-168.
30 Janker, R., "Eine automatische vorrictung zur filterbetätigung und einhaltung der bestrhalungszelt," *Strahlentherapie* 52 (1935):349-352.
31 Thoraeus, R., "New Filter for Roentgen Deep Therapy," *Acta Radiol.* Supplement 3, Part 2 (1929):207-208; and Pohle and Chamberlain, *Theoretical Principles*.
32 Newell, "Roentgen Therapy Apparatus."
33 van Roojen, J., "Practical Applications of Geometric Principles in Cross-Beam Radiation," *Brit. J. Rad.* 1 (1928):454-456.
34 Moran, E.F., "Roentgen Rays: Generators," in Glasser, *Medical Physics*, vol. 2, pp. 879-884; Atlee, Z.J., "Roentgen Rays: Generators; Rectifier Tubes," in Glasser, *Medical Physics*, vol. 2, pp. 884-887; and Jerman, "Roentgen-ray Apparatus."
35 Krohmer, "Radiography."
36 Glasser, O.; Quimby, E.H.; Taylor, L.S.; and Weatherwax, J.L. *Physical Foundations of Radiology* (New York: Paul B. Hoeber, 1944):426; Meredith, W.J., and Massey, J.B. *Fundamental Physics of Radiology*. 3rd ed. (Chicago: Year Book Medical Publishers, 1977):710.
37 Taylor, L.S.; Singer, G.; and Stoneburner, C.F., "A Basis for the Comparison of Roentgen Rays Generated by Voltages of Different Wave Form," *Am. J. Roent.* 30 (1933):368-379.
38 Mayneord, W.V., "Measurements of Low Voltage X rays (Chaoul Technique)," *Brit. J. Rad.* 2 (1936):215; Robertson, J.K. *Radiology Physics*. 2nd ed. (New York: D. Van Nostrand Company, 1948):323.

39 Coolidge, W.D., "Cathode-ray and Roentgen-ray Work in Progress," *Am. J. Roent.* 19 (1928):313-321.
40 Charlton, E.E.; Westendorp, W.F.; Dempster, L.E.; and Hotaling, G., "A New Million Volt X-ray Outfit," *J. Applied Physics* 10 (1939):374-385.
41 Coolidge, W.D.; Dempster, L.E.; and Tanis, H.E., "High Voltage Cathode Ray and X ray and their Operation," *Physics* 1 (1931):230; Coolidge, W.D.; Dempster, L.E.; and Tanis, H.E., "High Voltage Induction Coil and Cascade Tube," *Am. J. Roent.* 27 (1932):405; Coolidge, W.D., "Production of X rays of Very Short Wavelengths," *Radiology* 30 (1938):537; Failla, G., "Discussion of Coolidge Paper," *Radiology* 30 (1938):537; Schulz, M.D., "The Supervoltage Story," *Am. J. Roent.* 124 (1975):541-559; and Trout, "History of Radiation Sources."
42 von Schubert, E., "Vorläufage erfahrungen mit der karzinomtherapie mit extrem harten rontgenstrahlen," *Strahlentherapie* 44 (1932):293-310.
43 Soiland, A., "Experimental Clinical Research Work with X-ray Voltages above 500,000," *Radiology* 20 (1933):99; Mudd, S.; Emery, C.K.; Meland, O.M.; and Costolow, W.E., "Data Concerning 3 Years Experience with 600 kVp Roentgen Therapy," *Am. J. Roent.* 31 (1934):520; Mudd, S.G.; Emery, C.K.; and Levi, L.M., "Clinical Observations in the Treatment of Cancer by Supervoltage X rays," *Radiology* 30 (1938):489-492; Soiland, A., "Clinical Experience in Use of Supervoltage Roentgen Rays," *Am. J. Roent.* 40 (1938):52; Costolow, W.E., "Clinical Aspects of Supervoltage Radiation Therapy in Cancer," *Radiology* 34 (1940):28; and Lauritsen, C.C., "The Development of High Voltage X-ray Tubes at the California Institute of Technology," *Radiology* 31 (1938):354-361.
44 Soiland, "Experimental Clinical Research."
45 Holbrow, C.H., "The Giant Cancer Tube and the Kellogg Radiation Laboratory," *Physics Today* (July 1981):42-49.
46 Mudd, Emery, Meland, and Costolow, "Data Concerning."
47 Mudd, Emery, and Levi, "Clinical Observations," and Lauritsen, "The Development of High Voltage."
48 Schulz, "The Supervoltage Story."
49 Stone, R.S.; Livingstone, M.S.; Sloan, D.H.; and Chaffee, M.A., "A Radio Frequency High Voltage Apparatus for X-ray Therapy (Parts I and II)," *Radiology* 24 (1935):153-159 and 298-302.
50 Jones, A., "The Development of Megavoltage X-ray Therapy at St. Bartholomew's Hospital," *Brit. J. Rad.* suppl. 22(1988):3-10; Innes, G.S., "The One Million Volt X-ray Therapy Equipment at St. Bartholomew's Hospital," *Brit. J. Rad.* suppl. 22 (1988):11-16; and Laughlin, J.S., "Development of the Technology of Radiation Therapy," *RadioGraphics* 9 (1989): 1245-1266.
51 Trout, "History of Radiation"; Schulz, "The Supervoltage Story"; and Laughlin, "Development of the Technology."
52 Charlton, E.E.; Hotaling, W.F.; Westendorp, W.F.; and Demster, L.E., "Oil Immersed X-ray Outfit for 500,000 Volts and Oil Immersed Multisection X-ray Tube," *Radiology* 29 (1937):327; Charlton, E.E.; Westendorp, W.F.; Demster, L.E.; and Hotaling, G., "Million Volt X-ray Unit," *Radiology* 35 (1940):585; and Charlton, Westendorp, Dempster, and Hotaling, "A New Million Volt X-Ray Outfit."
53 Trump, J.G., "Radiation for Therapy: In Retrospect and Prospect," *Am. J. Roent.* 91 (1964):22-30.
54 Trump, J.G.; Moster, C.R.; and Cloud, R.W., "Efficient Deep Tumor Irradiation with Roentgen Rays of Several Million Volts," *Am. J. Roent.* 57 (1947):703.
55 Trump, "Radiation for Therapy."
56 Schulz, "The Supervoltage Story."
57 Kerst, D.W., "Betatron," in Glasser, *Medical Physics*, vol. 1, pp. 27-32; and Schulz, "The Supervoltage Story."
58 Kerst, D.W., "Development of Betatron," *American Scientist* 36 (1947):57; and Wideröe, R., "Der Strahlentransformator," *Arch. f. Electrotechnik* 21 (1928):400.
59 Kerst, D.W., "The Betatron," *Radiology* 40 (1943):115; Kerst, D.W., and Skaggs, L.S., "Accelerators: High-Energy; Betatron," in Glasser, *Medical Physics.*, vol. 2, p. 1120; Laughlin, J.S., "Accelerators: High-Energy; Betatron," in Glasser, *Medical Physics.*, vol. 3, p. 10-14; and Kerst, "Development of Betatron."
60 Schulz, "The Supervoltage Story," and Laughlin, "Accelerators: High-Energy."
61 Quastler, H.; Adams, G.D.; Almy, G.M.; et al., "Techniques for Application of the Betatron to Medical Therapy," *Am. J. Roent.* 61 (1949):591-625; and Laughlin, "Development of the Technology."
62 Watson, T.A., and Brukell, C.E., "Betatron in Cancer Therapy," *J. Canad. A. Radiologists* 2 (1951):60-64.
63 Karzmark, C.J.; Nunan, C.S.; and Tanabe, E. *Medical Electron Accelerators*. 1st ed. (New York: McGraw-Hill, 1993):316.
64 Ginzton, E.L., and Nunan, C.S., "History of Microwave Electron Linear Accelerators for Radiotherapy," *Intl. J. Radiat. Oncol. Biol. Phys.* 11 (1985):205-216; Schulz, "The Supervoltage Story"; and Karzmark and Nunan, "History of Microwave Electron."
65 Karzmark and Nunan, "History of Microwave Electron."
66 Kolle, F.S., "A New X-ray Meter," *Elect. Engineer* 22 (1896):602; and Eisenberg, *Radiology.*
67 Benoist, L., "The Radiochromater," *Comptes rendus Acad. Paris* 13 (1902):225; Beck, C. *Roentgen-Ray Diagnosis and Therapy*. New York: Appleton, 1904; Eisenberg, *Radiology*;

Glasser, *The Science of Radiology*; and Glasser, Quimby, Taylor, and Weatherwax, *Physical Foundations*.

68. Christen, T., "An Absolute Measure for Quality of Roentgen Rays and Its Use in Roentgen Therapy," *Verhandl. Deutsch. Röntgen Gesellsch.* 13 (1912):119-122.
69. Victoreen, J.A., "Roentgen Rays: Measurement of Quality," in Glasser, *Medical Physics*, vol. 2, pp. 887-894.
70. Taylor, L.S.; Singer, G.; and Stoneburner, C.F., "Effective Applied Voltage as an Indicator of the Radiation Emitted by an X-ray Tube," *Am. J. Roent.* 30 (1933):221; and Taylor, L.S., and Singer, G., "Standard Absorption Curves for Specifying the Quality of X-radiation," *Radiology* 22 (1934):445.
71. Portman, U.V., "Roentgen Therapy," in Glasser, *Science of Radiology*, pp. 210-241.
72. Kassabian, M.K., "A Résumé of the Radiometric Dosage of Roentgen Rays," *Amer. Quart. Roent.* 1 (1907):14; Glasser, *The Science of Radiology*; and Perthes, "Experiments to Determine."
73. Franklin, M., "Measurement of the X ray," *Phila. Med. J.* 4 (1905):22; Phillips, C.E.S., "The Standardization of Radiations," *Amer. Quart. Roent.* 1(3) (1907):1; and Allen, S.J., "A Null Instrument for Measuring Ionization," *Trans. Amer. Roent. Ray Soc.* (1907):237.
74. Villard, P., "The Radiosclerometer," *Arch. d. elec. Med.* 14 (1908):692.
75. Duane, W., "X rays and Gamma Rays," address, American Roentgen Ray Society, Cleveland, 1914; and Szilard, B., "The Absolute Measurement of Roentgen and Gamma Rays in Biology," *Strahlentherapie* 5 (1914):742.
76. Duane, W., "Roentgen Rays of Short Wavelengths and their Measurement," *Am. J. Roent.* 9 (1922):167; Duane, W., "The Scientific Basis of Short Wavelength Therapy," *Am. J. Roent.* 9 (1923):781; Duane, W., "Measurement of Doses by Means of Ionization Chambers," *Am. J. Roent.* 10 (1923):399; and Duane, W., and Lorenz, E., "The Standard Ionization Chamber for Roentgen Ray Dosage Measurements," *Am. J. Roent.* 19 (1928):461.
77. "Recommendations of International Committee for Radiological Units," *Radiology* 29 (1937):634.
78. Taylor, L.S.," Measurement of X-Ray Quantity (Dose)," in Glasser, Quimby, Taylor, and Weatherwax, *Physical Foundations*, pp. 146-181.
79. Fricke, H., and Glasser, O., "Standardization of the Roentgen-ray Dose by Means of the Small Ionization Chamber," *Am. J. Roent.* 13 (1925):462.
80. Glasser, O.; Portmann, W.; and Seitz, V.B., "The Condenser Dosimeter and its Use in Measuring Radiation over a Wide Range of Wave Lengths," *Am. J. Roent.* 20 (1928):505.
81. Victoreen, J.A., "Roentgen Rays: Measurement of Quantity," in Glasser, *Medical Physics*, vol. II, pp. 894-901.
82. Failla, G., "A New Instrument for Measuring X-radiation," *Radiology* 15 (1930):437; and Taylor, L.S., "Accurate Measurement of Small Electric Charges by a Null Method," *Radiology* 17 (1931):294.
83. Taylor, "Measurement of X-ray Quantity."
84. Grigg, E.R.N.; Bagshaw, M.A.; Chamberlain, W.E.; et al., "The RSNA Historic Symposium on American Radiology: Then and Now," *Radiology* 100(1971):126.
85. Quimby, E.H., "The Specification of Dosage in Radium Therapy," *Am. J. Roent.* 45(1941):1-16.
86. Failla, G., "Dosage in Radium Therapy," *Am. J. Roent.* 8 (1921):674-685.
87. Rutherford, E., and Robinson, H., "Heating Effect of Radium and its Emanation," *Phil. Mag.* 25 (1913):312.
88. Newbery, G.R., "Microwave Linear Accelerator," *Brit. J. Rad.* 22 (1949):473; Glasser, *Medical Physics*, vol. 2., p. 1227; Glasser, *Medical Physics*, vol. 3, p. 754; Johns, H.E.; Bates, L.M.; and Watson, T.A., "1,000 Curie Cobalt Units for Radiation Therapy: 1. The Saskatchewan Cobalt-60 Unit," *Brit. J. Rad.* 25 (1952):296; Krabbenhoft, "A History," pp. 859-865; Miller, C.W., "The Continuing Evolution of Linear Accelerators," *Brit. J. Rad.* 35 (1962):182; McGinty, G.K., "Modern Trends in Accelerator Design for Therapy," *Brit. J. Rad.* 35 (1962):196; Morrison, R., "History of Linear Accelerator Development," in *Modern Trends in Radiotherapy*. London: Butterworth, 1967; Buschke, "Radiation Therapy," pp. 235-246; Karzmark, C.J., and Pering, N.C., "Electron Linear Accelerators for Radiation Therapy: History, Principles and Contemporary Developments," *Phys. Med. Biol.* 18 (1973):321-354; Schulz, "Supervoltage Story,"pp. 541-559; Lederman, "Early History," pp. 639-648; Karzmark, C.J., "Advances in Linear Accelerator Design for Radiotherapy," *Med. Phys.* 11 (1984):105-128; Ginzton, E.L. *An Informal History of the Microwave Electron Accelerator for Radiotherapy*. Proceedings of Tenth Varian Users Meeting, Palm Springs, Ca., 1984; Ginzton and Nunan, "History of Microwave Electron Linear Accelerators," pp. 205-216; Eisenberg, *Radiology: An Illustrated History*, p. 606; and Karzmark, Nunan, and Tanabe, *Medical Electron Accelerators,* p.316.
89. Mitchell, J.S.; Smith, C.L.; Allen-Williams, D.J.; and Broom, R., "Experiences with 30 MeV Synchrotron as a Radiotherapeutic Instrument," *Acta Radiol. Ther. Phys. Biol.* 39 (1953):419.
90. Green, D.T., and Errington, R.F., "Design of Cobalt-60 Beam Therapy Unit," *Brit. J. Rad.* 25 (1952):309; Johns, H.E., "The Physicist in Cancer Treatment and Detection," *Intl. J. Radiat. Oncol. Biol. Phys.* 7 (1981):801-808; Johns, Bates, and Watson, "1000 Curie Cobalt Units"; and Schulz, "Supervoltage Story."
91. National Bureau of Standards. *Handbook 62—Report of the International Commission on Radiological Units and Measurements (ICRU)*. Washington, D.C.: U.S. Government Printing Office, 1956; National Bureau of Standards. *Handbook 78—Report of the International Cominission on*

Radiological Units and Measurements (TCRU). Washington, D.C.: U.S. Government Printing Office, 1959; and National Bureau of Standards. *Handbook 84—Report 10a of the International Commission on Radiological Units and Measurements (ICRU)*.Washington, D.C.: U.S. Government Printing Office, 1962.

92 Howard-Flanders, P., "The Development of the Linear Accelerator as a Clinical Instrument," *Acta. Radiol. Suppl*. 116 (1954):649-655; Miller, C.W., "Traveling Wave Linear Accelerator for X-ray Therapy," *Nature* 171 (1953):297; Morrison, R.; Newbery, G.R.; and Deeley, T.J., "Preliminary Report on the Clinical Use of the Medical Research Council 8 MeV Linear Accelerator," *Brit. J. Rad*. 29 (1956):177; Newbery, G.R., and Bewley, D.K., "The Performance of the Medical Research Council 8 MeV Linear Accelerator," *Brit. J. Rad*. 28 (1955):241-251; and Wood, C.A.P., and Newbery, G.R., "Medical Research Council Linear Accelerator and Cyclotron," *Nature* 173 (1954):233.

93 Atherton, L., "Design Evolution of an Advanced Linear Accelerator for Supervoltage Therapy," *Medicamundi* 10 (1965):77; and Rotblat, J., "The 15 MeV Linear Accelerator at St. Bartholomew's Hospital," *Nature* 175 (1955):745.

94 Day, M.J., and Farmer, E.T., "4 MeV Linear Accelerator at Newcastle on Tyne," *Brit. J. Rad*. 31 (1958):669; and McGinty, "Modern Trends."

95 Fry, D.W.; Shersbie-Harvie, R.B.R.; Mullet, L.B.; and Walkinshaw, W., "A Traveling Wave Linear Accelerator for 4-MeV Electrons," *Nature* 162 (1948):859; Greene, D., and Nelson, K.A., "Performance of Linear Accelerators in Clinical Service," *Brit. J. Rad*. 33 (1960):336; Greene, D., and Stephenson, S.K., "The Design of a Treatment Room to House a Gantry-Mounted 4 MeV Linear Accelerator," *Brit. J. Rad*. 34 (1961):640; Murison, C.A., and Hughes, H.A., "Physical Measurements on a 4-MeV Linear Accelerator," *Radiology* 68 (1957):367; Sutherland, W.H., "Dose Monitoring Methods in Medical Linear Accelerators," *Brit. J. Rad*. 42 (1969):864; Miller, C.W., "Linear Accelerators for X-ray Therapy," in *Proceedings of Eighth International Congress of Radiology*. Mexico City: Metropolitan Vickers Co. Ltd., Manchester, England, 1956; Miller, C.W., "Continuing Evolution"; and Ginzton and Nunan, "History of Microwave Linear Accelerators."

96 Ginzton, E.L; Mallory, K.B.; and Kaplan, H.S., "The Stanford Medical Linear Accelerator I. Design and Development," *Stanford Med. Bull*.16 (1957):123-140; and Weissbluth, M.; Karzmark, C.J.; Steele, R.E.; and Selby, A.H., "The Stanford Medical Linear Accelerator II. Installation and Physical Measurements," *Radiology* 72 (1959):242-253.

97 Adams, G.D., "Use of 70 MeV Synchrotron for Cancer Therapy. I. Physical Aspects," *Radiology* 83 (1964):785; and Stone, R.S., and Lowie, R.V., "Use of 70 MeV Synchrotron for Cancer Therapy. Part II. Clinical Aspects," *Radiology* 83 (1964):797.

98 Beadle, R., and Kelliher, M.G., "Recent Developments in Linear Accelerators at X-band for Radiotherapy," *Brit. J. Rad*. 35 (1962):188; and Bentley, R.E.; Jones, J.C.; and Lillicrap, S.C., "X-ray Spectra from Accelerators in the Range 2-6 MeV," *Phys. Med. Biol*. 12 (1967):301.

99 Gardner, A.; Bagshaw, M.A.; Page, V.; and Karzmark, C.J., "Tumor Localization, Dosimetry, Simulation and Treatment Procedures in Radiotherapy: The Isocentric Technique," *Am. J. Roent*. 114 (1972):163; Haimson, J., and Karzmark, C.J., "A New Design 6 MeV Linear Accelerator System for Supervoltage Therapy," *Brit. J. Rad*. 36 (1963):650; Horsley, R.J.; Price, R.H.; Saunders, J.E.; and Dingwall, P.W., "Performance of a 6 MeV Varian Linear Accelerator," *Brit. J. Rad*. 41 (1968):312; Karzmark, C.J., and Rust, D.C., "A Time and Motion Study of the Delivery of Radiation Treatment," *Brit. J. Rad*. 45 (1972):276.

100 Karzmark and Pering, "Electron Linear Accelerators."

101 Naylor, G.P., and Chiveralls, K., "The Stability of the X-ray Beam from an 8 MV Linear Accelerator Designed for Radiotherapy," *Brit. J. Rad*. 43 (1970):414; Karzmark, C.J., and Morton, R.J. *A Primer on Theory and Operation of Linear Accelerators in Radiation Therapy*. Vol FDA 828181. (Rockville, Md.: U.S. Department of Health and Human Services, 1981):48; Atherton, "Design Evolution"; and Bentley, Jones, and Lillicrap, "X-ray Spectra."

102 Aucounturier, J.; Huber, H.; and Jaouen, J., "Système de transport du faisceau d'électrons dans le Sagittaire," *Rev. Tech. Thomson* CSS 2 (1970):655.

103 Hansen, H.H.; Connor, W.G.; Doppke, K.; and Boone, M.M.L., "A New Flattening Filter for the Clinac-4," *Radiology* 103 (1972):443; Sabel, M.; Gunn, W.G.; Penning, D.; and Gardner, A., "Performance of a New 4 MeV Standing-Wave Linear Accelerator," *Radiology* 97 (1970):169; Karzmark, "Advances in Linear Accelerator Design"; and Karzmark, Nunan, and Tanabe, *Medical Electron Accelerators*.

104 Ginzton and Nunan, "History of Microwave Accelerators."

105 Safety Sciences Inc. *The Use of Electron Linear Accelerators in Medical Radiation Therapy; Overview Report No. 2; Market Use and Characteristics: Current Status and Future Trends*. National Technical Information Service, 1975.

106 Laughlin, J.S., "Accelerators: High-Energy; Betatron," in Glasser, *Medical Physics*, vol. 3, pp. 10-14.

107 Fry, D.W., "Synchrotron Accelerator: Its Possibilities as a Generator of X Rays in 10 to 50 MeV Energies for Medical Use," *Brit. J. Rad*. 22 (1949):642; and Schulz, "Supervoltage Story."

108 Shulz, "Supervoltage Story"; Mitchell, Smith, Allen-Williams, and Broom, "Experiences with 30 MeV Synchrotron"; Adams, "Use of 70 MeV Synchrotron:" and Stone and Lowie, "Use of 70 MeV Synchrotron."

109 Laughlin, J.S.; Mohan, R.; and Kutcher, G.J., "Choice of Optimum Megavoltage for Accelerators for Photon Beam Treatment," *Intl. J. Radiat. Oncol. Biol. Phys*. 12 (1986): 1551-1557.

110 Karzmark, Nunan, and Tanabe, *Medical Electron Accelerators*.
111 Karzmark and Fering, "Electron Linear Accelerator Design"; Karzmark, "Advances in Linear Accelerator Design"; Karzmark, Nunan, and Tanabe, *Medical Electron Accelerators*; and Karzmark and Morton, *A Primer*.
112 Ginzton, E.L.; Hansen, W.W.; and Kennedy, W.R., "A Linear Electron Accelerator," *Rev. Sci. Instrum*. 19 (1948):89; and Chodorow, M.; Ginzton, E.L.; Neilsen, I.R.; and Sonkin, S., "Design and Performance of a High-Power Pulsed Klystron," *Proc. Inst. Radio Engrs*. 41 (1953):1584.
113 Chodorow, M.; Ginzton, E.L.; Hansen, W.W.; et al., "Stanford High-Energy Linear Electron Accelerator (Mark III)," *Rev. Sci. Instrum*. 26 (1955):134-204; Fry, D.W.; Shersbie-Harvie, R.B.R.; Mullet, L.B.; and Walkinshaw, W., "Traveling Wave Linear Accelerator for Electrons," *Nature* 160 (1947):315; Fry, D.W., "The Linear Electron Accelerator," *Philips Tech. Rev.*14 (1952):1; Morrison, "History of Linear Accelerator"; Karzmark, "Advances in Linear Accelerator Design"; Karzmark, Nunan, and Tanabe, *Medical Electron Accelerators*; Fry, Sherabie-Harvey, Mullettm, and Walkinshaw, A Traveling Wave Linear Accelerator"; and Ginzton, Hansen, and Kennedy, "A Linear Electron Accelerator."
114 Karzmark and Pering, "Electron Linear Accelerators."
115 Karzmark and Pering, "Electron Linear Accelerators"; Karzmark, "Advances in Linear Accelerator Design"; Karzmark, Nunan, and Tanabe, *Medical Electron Accelerators*; and Karzmark and Morton, "A Primer."
116 Karzmark, Nunan, and Tanabe, *Medical Electron Accelerators*.
117 Karzmark, "Advances in Linear Accelerator Design"; Ginzton, "An Informal History"; Ginzton and Nunan, "History of Microwave Electron Linear Accelerators"; and Karzmark, Nunan, and Tanabe, *Medical Electron Accelerators*.
118 Petersilka, E., and Schiegl, W.E., *Electromedica* 2-3 (1975):99; Enge, H.A., "Particle Accelerator Provided with An Adjustable 270° NonDispersive Magnetic Charged-Particle Beam Bender," U.S. Patent No. 3,379,911, 1968; LeBoutet, H., "Magnetic Deflecting and Focusing Device for a Charged Particle Beam." U.S. Patent No. 4,056,728, 1977; Enge, H.A., "Achromatic Magnet Mirror for Ion Beams," *Rev. Sci. Instrum*. 34 (1963):385-389; Karzmark, Nunan, and Tanabe, Medical Electron Accelerators; and Aucounturier, Hubert, and Jaouen, "Système de transport."
119 Brown, E.L., and Turnbull, W.G., "Achromatic Magnetic Beam Deflection System." U.S. Patent No. 3,867,635, 1975; Sutherland, W.H., *Brit. J. Rad*. 49 (1975):262; and Tronc, D., "Device for the Achromatic Magnetic Deflection of a Beam of Charged Particles and an Irradiation Apparatus Using such a Device." U.S. Patent No. 4,322,622, 1982.
120 Boux, R., "System for Monitoring the Position, Intensity, Uniformity and Directivity of a Beam of Ionizing Radiation." U.S. Patent No. 3,942,102, 1974.
121 Flock, S.T., and Shragge, P.C., "A Semianalytical Method for the Design of a Linac X-ray Beam Flattening Filter," *Med. Phys*. 14 (1987):202-209; Lane, R.G., and Paliwal, B.R., "Extended Field Treatment Flatness Filter for 4 MV Linear Accelerators," *Radiology* 115 (1975):478-479; Larson, R.D.; Brown, L.H., and Bjarngard, B., "Calculations of Beam Flattening Filters for High-Energy X-ray Machines," *Med. Phys*. 5 (1978):215-220; McCall, R.C.; McIntyre, R.D.; and Turnbull, W.G., "Improvements in Linear Accelerator Depth-Dose Curves," *Med. Phys*. 5 (1978):518-524; Nair, R.; Menon, N.S.K.; Bauer, C.; and Batley, F., "Dosimetric Aspects of 4 MeV X rays from a Linear Accelerator Equipped with a Uranium Flattening Filter," *Appl. Radiol*. (May-June 1978):81-83; Nair, R., and Wrede, D.E., "Depth Dose Data for 4 MeV Linear Accelerators with Lead or Uranium Field Flatteners," *Acta Radiol. Ther. Phys. Biol*. 19 (1980):107-109; Podgorsak, E.B.; Rawlinson, J.A.; Glavinovic, M.I.; and Johns, H.E., "Design of X-ray Targets for High Energy Accelerators in Radiotherapy," *Am. J. Roent*. 121 (1974):873-882; and Karzmark, Nunan, and Tanabe, *Medical Electron Accelerators*.
122 Newbery, "Microwave Linear Accelerator"; Morrison, "History of Linear Accelerator Development"; Karzmark, Nunan, and Tanabe, *Medical Electron Accelerators*; Morrison, Newbery, and Deeley, "Preliminary Report"; and Newbery amd Bewley, "The Performance of the Medical Research Council 8 MeV."
123 Miller, "Traveling Wave Linear Accelerator"; Newbery and Bewley, "The Performance of the Medical Research Council 8 MeV"; and Wood and Newbery, "Medial Research Council Linear Accelerator."
124 Howard-Flanders, P., and Newbery, G.R., "The Gantry Type of Mounting for High Voltage X-ray Therapy Equipment," *Brit. J. Rad*. 23 (1950):355-357.
125 Ibid., and Howard-Flanders, "The Development of the Linear Accelerator."
126 Day and Farmer, "4MeV Linear Accelerator."
127 Ginzton and Nunan, "History of Microwave Linear Accelerators"; Ginzton, Mallory, and Kaplan, "The Stanford Medical Linear Accelerator"; and Weissbluth, Karzmark, Steele, and Selby, "The Stanford Medical Linear Accelerator II."
128 Chodorow, Ginzton, Hansen, et al., "Stanford High-Energy Linear Electron Accelerator."
129 Kaplan, H.S., and Bagshaw, M.A., "The Stanford Medical Linear Accelerator III. Application to Clinical Problems of Radiation Therapy," *Stanford Med. Bull*. 15 (1957):141-151.
130 Weissbluth, Karzmark, Steele, and Selby, "The Stanford Medical Linear Accelerator II."
131 Kaplan and Bagshaw, The Stanford Medical Linear Accelerator III."
132 Ginzton, "An Informal History."
133 Ibid.

134 Haimson and Karzmark, "A New Design 6 MeV."
135 Hall, L.D.; Helmer, J.C.; and Jepsen, R.L., "Electrical Vacuum Pump Apparatus And Method." U.S. Patent No. 2,993,638, 1957/1961.
136 Karzmark and Pering, "Electron Linear Accelerators."
137 Naylor, G.P., and Williams, P.C., "An Instrument for Measuring Small Differences in Dose Rate at Points in a Radiation Beam," *Phys. Med. Biol*. 16 (1971):525; Naylor, G.P., and Williams, P.C., "Dose Distribution and Stability of Radiotherapy Electron Beams from a Linear Accelerator," *Brit. J. Rad*. 46(1972):603; Atherton, "Design Evolution"; and Naylor and Chiveralls, "The Stability of the X-Ray Beam."
138 Knapp, E.A.; Knapp, B.C.; and Potter, J.M., "Standing Wave High Energy Linear Accelerator Structures," *Rev. Sci. Instrum*. 39 (1968):979-991; Karzmark and Pering, "Electron Linear Accelerators"; and Sabel, Gunn, Penning, and Gardner, "Performance of a New 4 MeV."
139 Karzmark and Pering, "Electron Linear Accelerators"; Karzmark, "Advances in Linear Accelerator Design"; and Karzmark, Nunan, and Tanabe, *Medical Electron Accelerators*.
140 Johns, H.E., "Radiation Therapy: Rotation," in Glasser, *Medical Physics,* vol. 3, pp. 515-523.
141 Wachsmann, F., "Various Forms of Moving Field Therapy and Its Possibilities," *Acta Radiol. Ther. Phys. Biol*. 116 (1954):524.
142 Kerst, D.W., "Betatron-Quastler Era at the University of Illinois," *Med. Phys*. 2 (1975):297-300; Leksell, L.T., "The Stereotactic Method and Radiosurgery of the Brain," *Acta. Chir. Scand*. 102 (1951):316-319; and Leksell, L., "Stereotactic Radiosurgery," *J. Neurol. Neurosurg. Psychiatry* 46 (1983):797-803.
143 Worsnop, B.R., "Dose Delivered to the Mediastinum: An Intercomparison Between Centers," *Radiology* 82 (1964):1062; Worsnop, B.R., "Phantom Thermoluminescent Dosimeter Comparison for a Cooperative Radiotherapy Trial," *Radiology* 91 (1968):545; and Golden, R.; Cundiff, J.H.; Grant, W.H, III; and Shalek, R.J., "A Review of the Activities of the AAPM Radiological Physics Center in Interinstitutional Clinical Trials Involving Radiation Therapy," *Cancer* 29 (1972):1468.
144 National Bureau of Standards. *Handbook 62—Report of the International Commission on Radiological Units and Measurements (ICRU)*.Washington, D.C.: U.S. Government Printing Office, 1956; National Bureau of Standards. *Handbook 78—Report of the International Cominission on Radiological Units and Measurements (TCRU)*.Washington, D.C.: U.S. Government Printing Office, 1959; and National Bureau of Standards. *Handbook 84—Report 10a of the International Commission on Radiological Units and Measurements (ICRU)*.Washington, D.C.: U.S. Government Printing Office, 1962.
145 Tubiana, M., and Dutreix, J., "High Energy Radiation Dosimetry," *Proceedings of Symposium on Quantities, Units, and Measuring Methods of Ionizing Radiation* (Rome: Ulrico Hoelphied, 1958):149-181; Genna, S., and Laughlin, J.S., "Absolute Calibration of a Cobalt-60 Gamma Ray Beam," *Radiology* 65 (1955):394; Laughlin, J.S., and Genna, S., "Calorimetry," in Attix and Roesch, *Radiation Dosimetry,* 2nd ed. (New York: Academic Press, 1966) vol. 2.; Fricke, H., and Morse, S., "The Actions of the Rays on Ferrous Sulfate Soluitions, " *Phil. Mag*. 7 (1929):129; and Fricke, H., and Hart, E.J., "Chemical Dosimetry," in Attix and Roesch, *Radiation Dosimetry*, vol. 2, 167-239.
146 Reese, H., Jr., "Design of a Vibrating Capacitor Electrometer," *Nucleonics* 6 (3) (1950):40; Farmer, F.T., "A Sub-Standard X-ray Dose-Meter," *Brit. J. Rad*. 28 (1955):304-330; Boag, J.W., "Ionization Chambers," in Hine, G.J., and Brownell, G.L., eds. *Radiation Dosimetry*. New York: Academic Press, 1956; Boag, J.W., "Ionization Chambers," in Attix and Roesch, *Radiation Dosimetry*, vol. 2., pp. 1-72; Dutreix, J., and Dutreix, A., "Etude comparée d'une série de chambres d'ionisation dans des faisceaux d'électrons de 20 et 10 MeV," *Biophysik* 3 (1966):249-258; Fowler, J., and Attix, F., "Solid State Integrating Dosimeters," Attix and Roesch, *Radiation Dosimetry*, vol. 2; Lovenger, R., "Precision Measurement with the Total-Feedback Electrometer," *Phys. Med. Biol*. 11 (1966):267-279; Mauderli, W., and Bruno, F.P., "Solid State Electrometer Amplifier," *Phys. Med. Biol*. 11 (1966):543; Aird, E.G., and Farmer, F.T., "The Design of a Thimble Chamber for the Farmer Dosimeter," *Phys. Med. Biol*. 17 (1972):169-174; Ehrlich, M. *National Bureau of Standards: Handbook 57 Photographic Dosimetry of X and Gamma Rays*. Washington, D.C.: U.S. Government Printing Office, 1954; Karzmark, C.; White, J.; and Fowler, J., "Lithium Fluoride Thermoluminescence Dosimetry," *Phys. Med. Biol*. 9 (1964):273; and Cameron, J.; Suntharalingam, N.; and Kenney, G. *Thermoluminescent Dosimetry*. Madison, WI: University of Wisconsin Press, 1968.
147 ICRU. *Report No. 10d: Clinical Dosimetry*. Washington, D.C.:1963; ICRU. *Report No. 14, Radiation Dosimetry: X-Ray and GammaRays with Maximum Photon-Energies between 0.6 and 50 MeV*. Washington, D.C.:1969; "HPA. Hospital Physicists Association," *Phys. Med. Biol*. 14 (1969):1; "NACP. Nordic Association of Clinical Physicists," *Acta Radiol. Ther. Phys. Biol*. 1971; AAPM. Subcommittee on Radiation Dosimetry (SCRAD): Protocol for the dosimetry of high energy electrons. *Phys. Med. Biol*. 1966; 11:505-520; and "AAPM. Subcommittee on Radiation Dosimetry (SCRAD): Protocol for the dosimetry of X and gamma-ray beams with maximum energies between 0.6 and 50 MeV," *Phys. Med. Biol*. 16 (1971):379.
148 Newbery and Bewley, "The Performance of the Medical Research Council 8 MeV."
149 "AAPM. Code of practice for X-ray therapy linear accelerators," *Med. Phys*. 2 (1975):110-121; and Weissbluth, Karzmark, Steele, and Selby, "The Stanford Medical Linear Accelerator II."
150 Karzmark and Pering, "Electron Linear Accelerators"; Karzmark, "Advances in Linear Accelerator

Design"; Karzmark, Nunan, and Tanabe, *Medical Electron Accelerators*; Sutherland, "Dose Monitoring Methods"; Karzmark and Morton, "A Primer"; and Boux, "System for Monitoring."

151 Cox, J.R.; Gallagher, T.L.; Holmes, W.F.; and Powers, W.E., "Programmed Console: An Aid to Radiation Treatment Planning," in *Proceedings of the Eighth IBM Medical Symposium*. BME 15 (1968):128-129; Cox, J.R, " Recollections on Biomedical Computing. in Computing at Washington University," January/February 1988: 20-31; Onai, Y.; Irifune, T.; and Tomaru, T., "Calculation of Dose Distributions in Radiation Therapy by a Digital Computer. I. The Computation of Dose Distributions in a Homogeneous Body for Cobalt-60 Gamma-rays and 4.3 MeV X-rays," *Nippon Acta Radiologica* 27 (1967):653.

152 Karzmark and Pering, "Electron Linear Accelerators"; Schulz, "Supervoltage Story"; Lederman, "Early History"; Karzmark, "Advances in Linear Accelerator Design"; Ginzton, "An Informal History"; Ginzton and Nunan, "History of Microwave Electron Linear Accelerators"; Eisenberg, *Radiology,* p. 606; Mohan, R., "Secondary Field Shaping, Asymmetric Collimators and Multileaf Collimators," in Purdy, J.A., ed. *Advances in Radiation Oncology Physics*. Woodbury, NY: American Institute of Physics, 1992; Bova, F.J., "Stereotactic Radiosurgery," in Purdy, *Advances in Radiation Oncology Physics;* Karzmark, Nunan, and Tanabe, *Medical Electron Accelerators*; and Schell, M.C.; Bova, F.J.; Larson, D.A.; et al., "Stereotactic Radiosurgery: Report of Task Group 42 of the Radiation Therapy Committee Report," 1994-in press.

153 Brahme, A., "Investigations on the Application of a Microtron Accelerator for Radiation Therapy," Ph.D. thesis, University of Stockholm, 1975; Svensson, H.; Jonsson, L.; Larson, L-G.; et al., "A 22 MeV Microtron for Radiation Therapy," *Acta Radiol. Ther. Phys. Biol*. 16 (1977):145-156; Brahme, A., "Design Principles and Clinical Possibilities with a New Generation of Radiation Therapy Equipment," *Acta Oncol*. 26 (1987):403-412; Masterson, M.E.; Mageras, G.S.; LoSasso, T.; et al., "Preclinical Evaluation of the Reliability of a 50 MeV Racetrack Microtron," *Intl. J. Radiat. Oncol. Biol. Phys*. 28 (1994):1219-1227.

154 *The Neurotron 1000,* Accuray, Inc., Santa Clara, California, 1994.

155 "Computers in Radiation Therapy," in Glicksman, A.; Cederlund, J.; and Cohen, M., ed. *Proceedings of the Fourth International Conference on the Use of Computers in Radiation Therapy*. Uppsala, August 7-11, 1972 (Uppsala: Tofters tryckeri ab, 1973); Cox, "Recollections"; Zink, S., ed. *Future Directions of Computer-Aided Radiotherapy: Joint U.S.-Scandinavian Symposium on Future Directions of Computer-Aided Radiotherapy.* San Antonio, August 13, 1988. Bethesda: National Cancer Institute, 1988; and Rosen, I., and Purdy, J.A., "Computer Controlled Medical Accelerators," in Purdy, *Advances in Radiation Oncology Physics*.

156 Rogers, D.W.O., "New Dosimetry Standards," in Purdy, *Advances in Radiation Oncology Physics*, and Rogers, D.W.O., "Fundamentals of High Energy X-ray and Electron Dosimetry Protocols," in Purdy, *Advances in Radiation Oncology Physics*.

157 Safety Sciences Inc. *The Use of Electron Linear Accelerators in Medical Radiation Therapy; Overview Report No. 2; Market Use and Characteristics: Current Status and Future Trends*. National Technical Information Service, 1975.

158 *Directory of High-Energy Radiotherapy Centres*. Vienna: International Atomic Energy Agency, 1976; and *Directory of High-Energy Radiotherapy Centres*. Vienna: International Atomic Energy Agency, 1968.

159 Karzmark and Morton, "A Primer," p. 48; Karzmark and Pering, "Electron Linear Accelerators"; Karzmark, "Advances in Linear Accelerator Design"; and Karzmark, Nunan, and Tanabe, *Medical Electron Accelerators.*

160 Tatcher, M., "A Method for Varying the Effective Angle of Wedge Filters," *Radiology* 97 (1970):132.

161 Mansfield, C.M.; Suntharalingam, N.; and Chow, N., "Experimental Verification of a Method for Varying the Effective Angle of Wedge Filters," *Am. J. Roent*. 120 (1974):699-702.

162 Bentel, G.C.; Nelson, C.E.; and Noell, K.T. *Treatment Planning and Dose Calculation in Radiation Oncology*. New York: Pergamon, 1982; and *Philips Medical Systems Division: Product Data*. Eindoven, The Netherlands: 1983.

163 Petti, P.L., and Siddon, R.L., "Effective Wedge Angles with a Universal Wedge," *Phys. Med. Biol*. 30 (1985):985-991.

164 Chism, S.E.; Chism, D.B.; Kalsched, M.; et al. *Treatment Techniques Using Independent Collimators*. Shelton, Conn.: Philips Medical Systems North America, 1993.

165 Kijewski, P.K.; Chin, L.M.; and Bjarngard, B.E., "Wedge-Shaped Dose Distributions by Computer-Controlled Collimator Motion," *Med. Phys*. 5 (1978):426-429; and Levene, M.B.; Kijewski, P.K.; Chin, L.M.; et al., "Computer-Controlled Radiation Therapy," *Radiology* 129 (1978):769-775.

166 Loshek, D.D., "Applications and Physics of the Independent Collimator Feature of the Varian 2500," *Proceedings of Tenth Varian Users Meeting*. California, 1984.

167 Leavitt, D.D.; Martin, M.; Moeller, J.H.; and Lee, W.L., "Dynamic Wedge Field Techniques through Computer Controlled Collimator Motion and Dose Delivery," *Med. Phys*. 17 (1990):87-91; and Mohan, R., "Secondary Field Shaping."

168 Mohan, R., "Secondary Field Shaping."

169 Leavitt, D.D., "Evaluation of a Diode Detector Array for Measurement of Dynamic Wedge Dose Distributions," *Med. Phys*. 20 (1993):381-382; Dunscombe, P., and Ross, C., "Jaws 2: A Description of In-Air Beam Profiles for Half Blocked and Dynamically Wedged Radiation Fields," *Med. Phys*. 20 (1993): 1705-1707.

170 Takahashi, S., "Conformation Radiotherapy, Rotation Techniques as Applied to Radiography and Radiotherapy," *Acta Radiol*. Suppl. (1965):242.

171 Mantel, J.; Perry, H.; and Weinkam, J.J., "Automatic Variation of Field Size and Dose Rate in Rotation Therapy," *Intl. J. Radiat. Oncol. Biol. Phys*. 2 (1977):697-704; and Sofia, J.W., "Computer Controlled Multileaf Collimator for Rotational Radiation Therapy," *Am. J. Roent*. 133 (1979):956-957.

172 "Conformation and Dynamic Therapy.," in Umegaki, Y., ed. *Proceedings of the Seventh International Conference on the Use of Computers in Radiation Therapy*. Tokyo, 1980.

173 Maleki, N., and Kijewski, P., "Analysis of the Field Defining Properties of a Multi-Leaf Collimator," *Med. Phys*. 11 (1984):390; Brahme, A., "Optimal Setting of Multileaf Collimators in Stationary Beam Radiation Therapy," *Strahlenther. Onkol*. 164 (1988):343-350; Cheng, C-W.; Kijewski, P.W.; and Langer, M., "Field Shaping by Multileaf Collimators (MLC): A Comparison with Convential Blocks," *Med. Phys*. 16 (1989):671; Biggs, P.; Capalucci, J.; and Russell, M., "Comparison of the Penumbra between Focused and Nondivergent Blocks—Implications for Multileaf Collimators," *Med. Phys*. 18 (1991):753-758; Masterson, M.E.; LoSasso, T.; Larson, A.; et al., "Design and Performance of the Multileaf Collimator on the Scanditronix 50 MeV Racetrack Microtron," *Med. Phys*. 19 (1992):816; and Zhu, Y.; Boyer, A.L.; and Desorbry, G.E., "Dose Distributions of X-ray Fields as Shaped with Multileaf Collimators," *Phys. Med. Biol*. 37 (1992):163-173.

174 Galvin, J.M.; Smith, A.R.; Moeller, R.D.; et al., "Evaluation of a Multileaf Collimator Design for a Photon Beam," *Intl. J. Radiat. Oncol. Biol. Phys*. 23 (1992):789-801; Mohan, "Secondary Field Shaping"; and Brahme, A., "Optimization of Radiation Therapy and the Development of Multileaf Collimation," *Intl. J. Radiat. Oncol. Biol. Phys*. 25 (1993):373-375.

175 Leveson, N.G., and Turner, C.S., "An Investigation of the Therac-25 Accidents," *Computer* 26 (1993):18-41; and "Therac-25 Revisited," Correspondence, *Computer* 26 (10) (1993):45.

176 Leveson and Turner, "An Investigation."

177 Kivel, M., ed. *Radiological Health Bulletin*.Vol XX. U.S. Federal Food and Drug Administration, 1986; Miller, E., "The Therac-25 Experience," *Proceedings of Conference of State Radiation Control Program Directors*, 1987; "Medical Device Recalls, Examination of Selected Cases." U.S. Government Accounting Office Report GAO/PEMD-90-6, 1989; and Bowsher, C.A., "Medical Devices: The Public Health at Risk," U.S. Government Accounting Office Report GAO/T-PEMD-90-2, 046987/139922, 1990.

178 Houston, M.F., "What Do the Simple Folk Do?: Software Safety in the Cottage Industry," *Proceedings of IEEE Computers in Medicine Conference*, 1985; Houston, M.F., "What Do the Simple Folk Do?: Software Safety in the Cottage Industry," *Proceedings of COMPAS '87 Computer Assurance Conference,* 1987: S20-S24; Levenson, N., "Software Safety: Why, What and How," *Computing Surveys* 18 (2) (1986):125-163; and Leveson, N.G., "Software Safety in Embedded Computer Systems," *Comm. ACM* (February 1991):34-46.

179 Peterson, M., "Advances Cited in Computer-Based Medical Systems since Therac-25 Accidents," *Computer* 26 (9) (1993):109; and in Kriewall, T.J., ed. *Proceedings of IEEE Symposium on Computer-Based Medical Systems*. Ann Arbor, Michigan: Computer Society Press, 1993.

180 Weinhous, M.S.; Purdy, J.A.; and Granda, C.O., "Acceptance Testing of a Computer-Controlled Medical Linear Accelerator," in Povilat, C., ed. *1987-1988 Radiation Oncology Center Scientific Report*. (St. Louis: Mallinckrodt Institute of Radiology, Washington University, 1988):195-200; Weinhous, M.S.; Purdy, J.A.; and Granda, C.O., "Testing of a Medical Linear Accelerator's Computer-Control System," *Med. Phys*. 17 (1990):95-102; Weinhous, M.S.; Purdy, J.A.; and Granda, C.O., "Testing of a Medical Linear Accelerator's Computer-Control System," *Varian's Centerline* 13 (1990):1-2; Weinhous, M.S., "Quality Assurance of Radiotherapy-Accelerator Computer-Control Systems," in Starkschall, G., and Horton, J., ed. *Proceedings of Quality Assurance in Radiotherapy Physics*. (Galveston, Texas: Medical Physics Publishing Corporation, 1991):45-60; Masterson, Mageras, LoSasso, et al., "Preclinical Evaluation"; Rosen and Purdy, "Computer Controlled Medical Accelerators"; and Masterson, LoSasso, Larson, et al., "Design and Performance."

181 Kerst, "Betatron-Quastler Era."

182 Leksell, L.T., "The Stereotactic Method and Radiosurgery of the Brain," *Acta. Chir. Scand*. 102 (1951):316-319; and Leksell, L., "Stereotactic Radiosurgery," *J. Neurol. Neurosurg. Psychiatry* 46 (1983):797-803.

183 Pike, B.; Podgorsak, E.B.; Peters, T.M.; and Pla, C., "Dose Distributions in Dynamic Stereotactic Radiosurgery," *Med. Phys*. 14 (1987):780-789; and Schell, Bova, Larson, et al., *Stereotactic Radiosurgery*.

184 Heilbrun, M.P.; Roberts, T.S.; Wells, T.H.; et al. *Instruction Manual for the BRW Brown-Roberts-Wells CT Stereotaxic System*. Burlington, Mass.: Radionics, Inc., 1983.

185 Columbo, F.; Benedetti, A.; Pozza, F., et al., "Stereotactic Radiosurgery Utilizing a Linear Accelerator," *Appl. Neurophysiol*. 48 (1985):133-145; Pike, Podgorsak, Peters, and Pla, "Dose Distribution"; Houdek, P.V.; Fayos, J.V.; Van Buren, J.M.; and Ginsberg, M.S., "Stereotaxic Radiotherapy Technique for Small Intracranial Lesions," *Med. Phys*. 12 (1985):469-472; Hartmann, G.H.; Schlegel, W.; Sturm, V.; et al., "Cerebral Radiation Surgery Using Moving Field Irradiation at a Linear Accelerator Facility," *Intl. J. Radiat. Oncol. Biol. Phys*. 11 (1985):1185-1192; Siddon, R.L., and Barth, N.H., "Stereotaxic Localization of Intracranial Targets," *Intl. J. Radiat. Oncol. Biol. Phys*. 13 (1987):1241-1246; Lutz, W.; Winston, K.R.; and Maleki, N., "A System for

Stereotactic Radiosurgery with a Linear Accelerator," *Intl. J. Radiat. Oncol. Biol. Phys.* 14 (1988):373-381; Winston, K.R., and Lutz, W., "Linear Accelerators as a Neurosurgical Tool for Stereotactic Radiosurgery," *Neurosurgery* 22 (1988):454-464; Podgorsak, E.B., "Stereotactic Radiosurgery," *On Target* 2 (2) (1988):1-4; Podgorsak, E.B.; Olivier, A.; Pla, M.; et al., "Dynamic Stereotactic Radiosurgery," *Intl. J. Radiat. Oncol. Biol. Phys.* 14 (1988):115-125; and Podgorsak, E.B.; Pike, G.B.; Olivier, A.; et al., "Radiosurgery with High Energy Photon Beams: A Comparison Among Techniques," *Intl. J. Radiat. Oncol. Biol. Phys.* 16 (1989):857-865.

186 Olivier, A.; de Lotbiniere, A.; Peters, T.; et al., "Combined Use of Digital Subtraction Angiography and MRI for Radiosurgery and Stereoencephalography," *Appl. Neurophys.* 50 (1-6) (1987):92-99; Peters, T.M., "Principles of Stereotactic Imaging," *On Target* 2 (2) (1988):6-8; Lutz, Winston, and Maleki, "A System."

187 Kooy, H.M.; van Herk, M.; Barnes, P.D.; et al., "Image Fusion for Stereotactic Radiotherapy and Radiosurgery Treatment Planning," *Intl. J. Radiat. Oncol. Biol. Phys.* 28 (1994):1229-1234.

188 McGinley, P.H.; Butker, E.K.; Crocker, I.R.; and Aiken, R., "An Adjustable Collimator for Stereotactic Radiosurgery," *Phys. Med. Biol.* 37 (1992):413-419.

189 Nedzi, L.A.; Kooy, H.M.; Alexander, E., III; et al., "Dynamic Field Shaping for Stereotactic Radiosurgery: A Modeling Study," *Intl. J. Radiat. Oncol. Biol. Phys.* 25 (1993):859-869.

190 Gill, S.S.; Thomas, G.T.; Warrington, A.P.; and Brada, M., "Relocatable Frame for Stereotactic External Beam Radiotherapy," *Intl. J. Radiat. Oncol. Biol. Phys.* 20 (1991):599-603; and Kooy, H.M.; Dunbar, S.F.; Tarbell, N.J.; et al., "Adaptation and Verification of the Relocatable Gill-Thomas-Cosman Frame in Stereotactic Radiotherapy," 1994.

191 Jones, D.; Reike, J.; and Hafermann, M., "Frameless Stereotactic Multiple Arc Radiotherapy," *Administrative Radiology* (May 1993):64-68.

192 Hanemann, W.P.; Cox, R.S.; and Brain, S.W., "Treatment Planning for a Multi-Degree-of-Freedom Robotic Linear Accelerator," *Med. Phys.* 20 (1993):889; Cox, R.S.; Hanemann, W.P.; and Brain, S.W., "Dose Distributions Produced by a Robot-Mounted Linac," *Med. Phys.* 20 (1993):889; also see *The Neurotron 1000*.

193 Kooy, H.M.; Nedzi, L.A.; Loeffler, J.S.; et al., "Treatment Planning for Stereotactic Radiosurgery of Intra-Cranial Lesions," *Intl. J. Radiat. Oncol. Biol. Phys.* 21 (1991):683-693; Goetsch, S.J.; Holly, F.E.; Solberg, T.D.; et al., "Treatment Planning for Stereotactic Radiosurgery with Acceleratorbased Systems," *Med. Phys.* 19 (1992):788; and Bova, "Stereotactic Radiosurgery."

194 Weeks, K.J.; Marks, L.; Ray, S.K.; et al., "3-Dimensional Optimization of Multiple Arcs for Stereotactic Radiosurgery," *Intl. J. Radiat. Oncol. Biol. Phys.* 26 (1993):147-154.

195 Boyer, A.L.; Antonuk, L.; Fenster, A.; et al., "A Review of Electronic Portal Imaging Devices (EPIDs)," *Med. Phys.* 19 (1992):1-16; Wong, J.; Munro, P., and Fenster, A., "On-Line Radiotherapy Treatment Verification Systems," in Purdy, *Advances in Radiation Oncology Physics*; and Weinhous, M.S., "Portal Imaging," in Weller, M., ed. *Proceedings of Third Annual Dosimetry/Physics Course*. Philadelphia: Hahnemann University, 1994.

196 Baily, N.A.; Horn, R.A.; and Kampp, T.D., "Fluoroscopic Visualization of Megavoltage Therapeutic X-ray Beams," *Intl. J. Radiat. Oncol. Biol. Phys.* 6 (1980):935-939; Meertens, H.; van Herk, M.; and Weeda, J., "A Liquid Ionization Chamber Detector for Digital Radiography of Therapeutic Megavoltage Photon Beams," *Phys. Med. Biol.* 30 (1985):313-321; and van Herk, M., and Meertens, H., "A Matrix Ionization Chamber Imaging Device for Online Patient Setup Verification during Radiotherapy," *Radiotherapy and Oncology* 11 (1988):369-378.

197 Weinhous, M.S., "Treatment Verification Using a Computer Workstation," *Intl. J. Radiat. Oncol. Biol. Phys.* 19 (1990):1549-1554; Halverson, K.J.; Leung, T.C.; Pellet, J.B.; et al., "Study of Treatment Variation in Radiotherapy of Head and Neck Tumors Using a Fiber-Optic On-Line Radiotherapy Imaging System," *Intl. J. Radiat. Oncol. Biol. Phys.* 21 (1991):1327-1336; Ezza, A.; Munro, P.; Porter, A.T.; et al., "Daily Monitoring and Correction of Radiation Field Placement Using a Video-Based Portal Imaging System: A Pilot Study," *Intl. J. Radiat. Oncol. Biol. Phys.* 22 (1992):159-165; Leszczynski, K.W.; Shalev, S.; and Gluhchev, G., "Verification of Radiotherapy Treatments: Computerized Analysis of the Size and Shape of Radiation Fields," *Med. Phys.* 20 (1993):687-694; Michalski, J.M.; Wong, J.W.; Bosch, W.R.; et al., "An Evaluation of Two Methods of Anatomical Alignment of Radiotherapy Portal Images," *Intl. J. Radiat. Oncol. Biol. Phys.* 27 (1993):1199-1206; and Balter, J.M.; Chen, G.T.; Pelizzari, C.A.; et al., "Online Repositioning during Treatment of the Prostate: A Study of Potential Limits and Gains," *Intl. J. Radiat. Oncol. Biol. Phys.* 27 (1993):137-143.

198 Balter, J.M.; Pelizzari, C.A.; and Chen, G.T., "Correlation of Projection Radiographs in Radiation Therapy Using Open Curve Segments and Points," *Med. Phys.* 19 (1992):329-334.

199 Michalski, J.M.; Wong, J.W.; Gerber, R.L.; et al., "The Use of On-Line Image Verification to Estimate the Variation in Radiotherapy Dose Delivery," *Intl. J. Radiat. Oncol. Biol. Phys.* 27 (1993):707-716.

200 Boyer, Antonuk, Fenster, et al., "A Review."

201 Veksler, V.J., "A New Method for Acceleration of Relativistic Particles," *Dokl. Akad. Nauk. SSSR* 43 (1944):329.

202 Brahme, "Investigations"; and Svensson, Jonsson, Larson, et al., "A 22 MeV Microtron."

203 Reistad, D., and Brahme A., "The Microtron: A New Accelerator for Radiation Therapy," *Proceedings of Third International Conference on Medical Physics*. Gothenburg, 1972.

204 Svensson, Jonsson, Larson, et al., "A 22 MeV Microtron."

205 Schwinger, J., *Phys. Rev.* 75 (1949):1912.
206 Brannen, E., and Froelich, H., *J. Appl. Phys.* 32 (1961):1179-1180; and Rosander, S.; Sedlacek, M.; and Wernholm, D., "The 50 MeV Racetrack Microtron at the Royal Institute of Technology," *Nucl. Instr. Meth. Stockholm* 204 (1982):120.
207 Brahme, "Design Principles."
208 Brahme, "Optimization."
209 Promotional materials provided by Accuray Inc., Santa Clara, California, 1994; Brain, S.; Haneman, W., and Cox, R., "The UNIX Workstation as a Treatment Planning Platform," *Med. Phys.* 20 (1993):863; Hahneman, Cox, and Brain, "Treatment Planning"; and Cox, Hahneman, and Brain, "Dose Distribution."
210 Mackie, T.R.; Holmes, T.; Swerdloff, S.; et al., "Tomotherapy: A New Concept for the Delivery of Dynamic Conformal Radiotherapy," *Med. Phys.* 20 (1993):1709-1719.
211 Holmes, T.W., "A Model for the Physical Optimization of External Beam Radiotherapy," Ph.D. thesis, University of Wisconsin, 1993.
212 National Bureau of Standards. *Handbook 62—Report of the International Commission on Radiological Units and Measurements (ICRU)*. Washington, D.C.: U.S. Government Printing Office, 1956; National Bureau of Standards. *Handbook 78—Report of the International Commission on Radiological Units and Measurements (ICRU)*. Washington, D.C.: U.S. Government Printing Office, 1959; National Bureau of Standards. *Handbook 84—Report 10a of the International Commission on Radiological Units and Measurements (ICRU)*. Washington, D.C.: U.S. Government Printing Office, 1962; and AAPM, "Subcommittee on Radiation Dosimetry (SCRAD): Protocol for the Dosimetry of X- and Gamma-ray Beams with Maximum Energies between 0.6 and 50 MeV. *Phys. Med. Biol.* 16 (1971):379.
213 Domen, S.R., and Lamperti, P.J., "A Heat-Loss-Compensated Calorimeter: Theory, Design and Performance," *J. Res. NBS* 75A (1974):595610.
214 Rogers, "New Dosimetry Standards."
215 Domen, S.R., "Absorbed Dose Water Calorimeter," *Med. Phys.* 7 (1980):157-159; and Domen, S.R., "An Absorbed Dose Water Calorimeter: Theory, Design and Performance," *J. Res. of NBS* 87 (1982):211-235.
216 Schulz, R.J., and Weinhous, M.S., "Calorimetric Determination of the Cavity Gas Calibration Factor, N_{gas}," *Med. Phys.* 12 (1985):166-168; and Schulz, R.J., and Weinhous, M.S., "Convection Currents in a Water Calorimeter," *Phys. Med. Biol.* 30 (1986):1093-1099.
217 Ross, C.K.; Klassen, N.V.; and Smith, G.D., "The Effect of Various Disolved Gases on the Heat Defect of Water," *Med. Phys.* 11 (1984):653-658; Schulz, R.J.; Wuu, C.S.; and Weinhous, M.S., "The Direct Determination of Dose-To-Water Using a Water Calorimeter," *Med. Phys.* 14 (1987):790-796; Domen, S.R., "The Role of Water Purity, Convection and Heat Conduction in a New Water Calorimeter Design," in Ross, C.K., and Klassen, N.V., ed. *Proceedings of NRC Workshop on Water Calorimetry, NRC Report—29637*. Ottawa: NRC, 1988; and "NRC Report—29637," in Ross and Klassen, *Proceedings of NRC Workshop*.
218 Rogers, "New Dosimetry."
219 Rogers, "Fundamentals."
220 NACP, "Procedures in External Beam Radiotherapy Dosimetry with Photons and Electron Beams with Maximum Energies between 1 and 50 MeV," *Acta Radiol. Oncol.* 19 (1980):55; Schulz, R.J.; Almond, P.R.; Cunningham, J.R.; et al., "A Protocol for the Determination of Absorbed Dose from High-Energy Photon and Electron Beams," *Med. Phys.* 10 (1983):741-771; Schulz, R.J.; Almond, P.R.; Cunningham, J.R.; et al., "Erratum: A Protocol for the Determination of Absorbed Dose from High-Energy Photon and Electron Beams, "*Med. Phys.* 10, 741 (1983). *Med. Phys.* 1984;11(2):213; and Schulz, R.J.; Almond, P.R.; Kutcher, G.; et al., "Clarification of the AAPM Task Group 21 Protocol," *Med. Phys.* 13 (1986):755-759.
221 IAEA. *Absorbed dose determination in photon and electron beams: An international code of practice*. IAEA Report 277, 1987.
222 Rogers, "Fundamentals."
223 Leavitt, "Evaluation."
224 Schell, Bova, Larson, et al, *Stereotactic Radiosurgery*.
225 Rice, R.K.; Hansen, J.L.; Svensson, G.K.; and Siddon, R.L., "Measurements of Dose Distributions in Small Beams of 6 MV X-rays," *Phys. Med. Biol.* 32 (1987):1087-1099.
226 Gore, J.C.; Kang, Y.S.; and Schulz, R.J., "Measurement of Radiation Dose Distributions by Nuclear Magnetic Resonance (NMR) Imaging," *Phys. Med. Biol.* 29 (1984):1189-1197.
227 Olsson, L.E.; Peterson, S.; Ahlgren, L.; and Mattsson, S., "Ferrous Sulphate Gels for Determination of Absorbed Dose Distributions Using MRI Techniques: Basic Studies," *Phys. Med. Biol.* 34 (1989):43-52; Schulz, R.J.; deGuzman, A.F.; Nguyen, D.B.; and Gore, J.C., "Dose-Response Curves for Fricke-Infused Agarose Gels as Obtained by Nuclear Magnetic Resonance," *Phys. Med. Biol.* 35 (1990):1611-1622; Olsson, L.E.; Fransson, A.; Ericsson, A.; and Mattson, S., "MR Imaging of Absorbed Dose Distributions for Radiotherapy Using Ferrous Sulphate Gels," *Phys. Med. Biol.* 35 (1990):1623-1631; Prasad, P.V.; Nalcioglu, O.; and Rabbani, B., "Measurement of Three-Dimensional Radiation Distributions Using MRI," *Rad. Res.* 128 (1991):113; and Schulz, R.J.; Maryanski, M.J.; Ibbott, G.S.; and Bond, J.E., "Assessment of the Accuracy of Stereotactic Radiosurgery Using Fricke-Infused Gels and MRI," *Med. Phys.* 20 (1993):1731-1734.

228 Cox, Gallagher, Holmes, and Powers, "Programmed Console"; Onai, Y.; Irifune, T.; and Tomaru, T., "Calculation of Dose Distributions in Radiation Therapy by a Digital Computer. I. The Computation of Dose Distributions in a Homogeneous Body for Cobalt-60 Gamma-rays and 4.3 MeV X-rays," *Nippon Acta Radiologica* 27 (1967):653; "Three-Dmensional Photon Planning: Report of the Collaborative Working Group on the Evaluation of Treatment Planning for External Photon Beam Radiotherapy," *Intl. J. Radiat. Oncol. Biol. Phys*. 21 (1991):1-266; and Cox, "Recollections."
229 "Three-Dimensional Photon Planning."
230 Niemierko, A., "Random Search Algorithm (RONSC) for Optimization of Radiation Therapy with Both Physical and Biological End Points and Constraints," *Intl. J. Radiat. Oncol. Biol. Phys*. 23 (1992):89-98; Niemierko, A.; Urie, M.; and Goitein, M., "Optimization of 3-D Radiation Therapy with Both Physical and Biological End Points and Constraints," *Intl. J. Radiat. Oncol. Biol. Phys*. 23 (1992):99-108; and Sodertrom, S., and Brahme, A., "Optimization of the Dose Delivery in a Few Field Techniques Using Radiobiological Objective Functions," *Med. Phys.* 20 (1993): 1201-1210.
231 Rosenman, J.G.; Chaney, E.L.; Cullip, T.J.; et al., "VISTAnet: Interactive Real-Time Calculation and Display of 3-Dimensional Radiation Dose: An Application of Gigabit Networking," *Intl. J. Radiat. Oncol. Biol. Phys*. 25 (1993):123-129.

William Coolidge's contribution of an improved tube for treatment revolutionized radiation therapy and emphasized the importance of physics research in medical applications of X rays. (Courtesy of the Center for the American History of Radiology, Reston, Va.)

CHAPTER FOUR

MEDICAL PHYSICS AND RADIATION ONCOLOGY

R.J. Shalek, Ph.D., J.D.

X-RAY ERA: 1910–1950

In retrospect, much of the history of medical physics related to radiation therapy has been directed toward one goal: to provide the basis for safely increasing the radiation dose to a target volume while reducing the radiation dose elsewhere. This simply stated idea has required extensive and diverse scientific and engineering efforts. Medical physicists borrow and create the appropriate science and engineering ideas or stimulate others, particularly in industry, to do so.

But this knowledge has flowed in both directions; findings by medical physicists have increased general scientific understanding. Radiation oncologists have contributed greatly in articulating clinical problems requiring physical solutions and suggesting methodology and testing innovations. In addition to efforts to improve radiation treatment quantitatively with X rays, gamma rays, and electrons, there have been and are repeated experimental efforts to change radiation interactions qualitatively with heavy particles or by altering the chemical environment.

The scientific and technical developments beginning in the first decades of this century led to a great expansion in the effectiveness and use of radiation for therapy. The availability of large amounts of radium (and therefore, radon) from ores mined in Colorado allowed the advantages inherent in brachytherapy (namely, high radiation doses near the sources and low doses elsewhere) to be exploited in North America. External beam treatment progressed more slowly, as developments in X-ray tubes, potential generators, other accelerators, and measurement standards were translated into commercially available apparatus permitting radiation treatment of deeper and deeper target volumes within patients.

The period between 1910 and 1950 is sometimes described as the golden age of radiology, including advances in both therapeutics and diagnosis. However, these years were also a golden age for atomic and nuclear physics, a period during which the boundaries of classical physics were exceeded in order to understand radioactivity, atomic structure, nuclear constituents, and nuclear transmutations, including fission. Medical physicists were contributors to this explosion of physical understanding, and the field of radiology was a beneficiary.

Several people have written comprehensive histories of radiation therapy and radiological physics starting with the discovery of X rays and radioactivity and continuing up to the date of their respective publications. Brecher and Brecher wrote a well-organized history of diagnostic and therapeutic radiology in the United States and Canada which gave considerable attention to physics.[1] Grigg wrote a less organized account of world-wide events which remains valuable for its remarkable detail, photographs, and memorabilia relating to physicians, physicists, engineers, and commercial companies.[2] Schulz has produced a careful and highly readable account of supervoltage history, and Laughlin wrote clearly on the development of technology of radiation therapy and its related physics.[3,4] Finally, del Regato wrote about medical and other physicists in radiation therapy as though he were an intimate but objective friend of each.[5]

Status in 1910

The general understanding of the nature of X rays and radioactive sources with their emitted radiations progressed rapidly between 1895 and 1910. In 1910 it was known that alpha particles were helium ions, that beta particles were identical to cathode rays and to electrons liberated by ionization, and that gamma rays and X rays were both high-energy electromagnetic radiation. The uranium radioactive series of which radium is a member was understood, as were the ideas of exponential decay, half life, average life, and disintegration constant. Relative measurements of X-ray intensity were made by observing the fluorescence of barium platinocyanide screens and by gauging various chemical reactions, some observable by color changes. These methods, however, were quite dependent on the individual performing the measurement. The strength of radium sources, even when specified by the weight of the radioactive source material, was not fully reliable as a measure of activity because of impurities in the sources and variable amounts of hydration. Thus in 1910 a great need existed for X-ray and radium standards so that results of experiments and particularly of clinical efforts could be communicated, compared, and replicated reliably.

By 1910 the state of practice in radiation therapy had advanced from the first makeshift experimental efforts following Röntgen's discovery. Gas-filled X-ray tubes were still somewhat unpredictable in performance. Transformers to 100 kilovolts (kV) were available with mechanical rectification only in institutions served by reliable alternating current; for institutions with direct current power, the interrupterless transformer, consisting of a motor-AC generator, transformer, and mechanical rectifier, provided reliable potentials to 120 kV. The biological effects of X rays were correctly attributed to the passage of the rays through tissue, and various competing theories of biological action of X rays by ozone formed in the air or by static charge on the skin had been discredited. It was understood that aluminum or leather filters for X rays removed the less penetrating rays in an X-ray beam, that increasing the electrical potential across an X-ray tube increased the dose at a depth relative to the dose at the surface, and that for radium sources the use of material around the source reduced the dose to tissue immediately near the source by absorbing beta particles and low-energy gamma rays. For both X rays and radium it was known that increasing the treatment distance increased the relative dose at depth. In order to protect patients, radiotherapists used diaphragms to limit X-ray beams to the region under treatment. It was known that X rays could induce skin cancer. An early attempt to measure radiation received by medical workers involved the darkening of film. Table 4.I lists the major advances in physics related to radiation therapy that would occur between 1910 and 1950.

Radiation Interactions

From the time of the discovery of X rays, radioactivity, and radium, there

Table 4.1
Major Advances from 1910 to 1950[6]

Advancement	Date	By whom	Where
International radium standard	1911	M. Curie	France[7]
Atomic nucleus discovered from α-ray scattering	1911	Rutherford	UK[8]
Nuclear atom model	1913	Bohr	Denmark[9,10,11]
Heated cathode X-ray tube	1913	Coolidge	USA[12]
Standard free-air chamber	1914	Duane	USA[13,14]
X-ray cutoff energy; effective energy of X ray	1915	Duane and Hunt	USA[15]
Glass radon seeds	1915–1917	Duane	USA[16]
Radium irradiator	1922	Burnam and Ward	USA[17]
		Lysholm	Sweden[18]
Compton scattering theory and confirmation	1923	Compton	USA[19]
Gold radon seeds	1926	Failla	USA[20]
Definition of exposure (roentgen unit)	1928, 1937	Committee	Intl.[21,22]
Victoreen R meter	1928	Glasser and Seitz	USA[23]
Van de Graaff accelerator	1929	Van de Graaff	USA[24]
Discovery of neutron	1932	Chadwick	UK[25]
Quimby brachytherapy system	1932 et seq	Quimby	USA[26]
Manchester brachytherapy system	1934 et seq	Paterson and Parker	UK[27]
Transformation of nuclei by neutron capture	1934	Fermi	Italy[28]
Bragg-Gray formula	1936	Gray	UK[29]
1 MeV resonant transformer X ray	1937	Charlton et al.	USA[30]
Neutron therapy	1938	Stone	USA[31]
Nuclear fission	1939	Hahn and Strassman	Germany[32]
Betatron	1940	Kerst	USA[33]
Nuclear reactor	1942	Fermi	USA[34]
Electronic computer	1942	Atanasoff	USA[35]
Linear accelerator	1946	Fry et al.	UK[36]
Linear accelerator	1948	Ginzton, Hansen,	

were individuals who were both academic physicists, interested in the nature of the radiation and in the structure of matter as revealed by the interaction of radiation with matter, and medical physicists, interested in the application of extant knowledge to medical treatment. An outstanding example of such a person is William Duane, an American who trained over an extended period of time with the best-known physicists in Germany, France, and England and worked at Harvard and the Huntington Memorial Hospital in Boston. We will see his involvement in the understanding of X-ray spectra, defining the X-ray energy limit, the quality of X-ray beams, and a radiation standard; inventing the free-air chamber to measure the radiation standard; improving radon generating plants; introducing interstitial radon sources; and participating in a debate on the scattering of X rays and gamma rays. Duane published widely in both physics and medical journals.

Photoelectric effect and X-ray cutoff

As early as 1909 it was known that some photoelectrons ejected by X rays in matter had speeds up to those of the electrons accelerated in the X-ray tube. If the X rays were material particles, it would be a reasonable explanation that an electron striking a target gave all its energy to an X-ray particle, which in turn gave all its energy to a photoelectron. The difficulty, however, was that it was also recognized, principally from Barkla's 1905 work on polarization of X rays, that X rays were

WILLIAM DUANE (1872-1935)

Duane was born in 1872 into a noted Philadelphia family, and attended the University of Pennsylvania (B.A. 1892) and Harvard (a second B.A. in 1893). He stayed on at Harvard as an assistant in the physics department, earning his M.A. In 1895, the year of Röntgen's discovery, Duane left for Europe as a Tyndall Fellow. He received his doctorate from the University of Berlin in 1897, showing special interests in magnetic fields and thermal chains.

On returning to the United States, Duane accepted a position as the first professor of physics at the University of Colorado. On a year's sabbatical from his teaching and research, Duane studied with the Curies in Paris in 1905. He worked there on his apparatus for radium emanation extraction and spent an additional three months studying with J.J. Thomson in England. Shortly after his return to Colorado, Duane was contacted by Marie Curie with a request and funding to return to the University of Paris to resume his studies of radioactivity. He would remain there for six years, contributing research which would eventually combine with that of others to culminate in the discovery of artificial radioactivity and nuclear fission.

Duane returned to head a new radium institute at Harvard in 1913. There his continued work with radium emanations, spectrometers, ionization chambers, radiations in crystals, and his work with Franklin Hunt on wavelength contributed to his fame as an eminent scientist. He was actively interested in the application of his researches and innovations to clinical practice in radiation therapy, participating with enthusiasm as both a presenter and commentator at meetings of the American Radium Society and the American Roentgen Ray Society. Duane died in 1935, after a long and fruitful life in academic, research, and organizational physics, working at the pinnacle of the profession. It was not until some years after his death, however, that the extent of his contributions in what would be known as nuclear physics was fully realized.

—Juan A. del Regato, M.D.

(Courtesy of the Center for the American History of Radiology, Reston, Va.)

---

electromagnetic radiation with wavelength, like visible light. In 1915 Duane and Hunt demonstrated experimentally and confirmed a calculated high-energy cutoff for tungsten targets, suggested a definition for the effective wavelength of X rays, and calculated values of Planck's constant, h, from the X-ray cutoff observed for various applied potentials.[38] Other researchers demonstrated that, as predicted, this short-wave limit depended on the potential impressed across the X-ray tube and not on the material of the X-ray target. This line of inquiry resulted in two important findings for radiation therapy: determining how the high-energy limit of X-ray spectra changed with potential across the X-ray tube and understanding the photoelectric effect for X rays, one of the major ways X rays and gamma rays interact and deposit energy in matter.

Compton scattering

Compton scattering is a second major way in which X rays and gamma rays interact with matter and is of great importance in radiation therapy because it accounts for the backscatter in X-ray beams and the similar absorption on a mass basis for materials of different atomic numbers for X rays and gamma rays greater than 140 kV. Again, medical physicists in north America played significant roles in the scientific arguments and competition that led to the acceptance and demonstration of this phenomenon. The works of British researchers Eve in 1904 and Florance in 1910 in England and the work of Canadian researcher J. A. Gray in 1913 foreshadowed later theories and demonstrations.[39] Gray noted that "when gamma rays are scattered, there is a change in quality, the scattered rays being less penetrating the greater the angle of scattering....the quality and quantity of the scattered radiation is approximately independent of the nature of the radiator.... a similar explanation must be given of the scattering of X and gamma rays."[40]

These early observations were based on the measurement of absorption of primary and scattered gamma rays. The Bragg X-ray spectrometer allowed more precise measurements. The story of Arthur Holly Compton's (Fig. 4.1) experience at Washington University in St. Louis and the ensuing controversy in 1922 to 1924 is an example of the competition and criticism sometimes seen in science. From absorption measurements Compton estimated an approximately constant increase in wavelength at a given scattering angle for various primary wavelengths. He verified this observation with an X-ray spectrometer. Compton explained, "this led me to examine what would happen if each quantum of X-ray energy were concentrated in a single particle and would act as a unit on a single electron."[41] Compton went on to write, "This idea, of an X-ray quantum losing energy by collision with an electron must have been already in the mind of Peter Debye, then working at Zürich, for immediately upon the appearance of my report in the bulletin of the National Research Council, he published a paper in the *Physikalische Zeitschrift* in which he presented an explanation of the change in wavelength of the scattered rays identical in principal with my own hypothesis...." Later in the same paper Compton said, "The results, confirming accurately the theoretical predictions, immediately became a subject of the most lively scientific controversy that I have ever known." The American Physical Society arranged a debate between Compton and Duane during its December 1923 meeting. Duane had observed a different scattered spectrum that he attributed to tertiary X rays excited by photoelectrons in the scattering material. Compton concluded that, "It

▶

Fig. 4.1 Arthur Holly Compton (1892–1962), whose work in both theoretical and applied physics led to innovations in radiation treatment. (Courtesy of the Center for the American History of Radiology, Reston, Va.)

Gioacchino Failla (1891–1961)

Failla was born on 19 July 1891 in Sicily and in 1906 moved with his widowed mother to New York. After graduating with honors from the prestigious Science High School in 1911, he entered Columbia University to study engineering, receiving his E.E. in June 1915. His interest in physics grew in graduate school, and in 1915 he was recommended for a part-time position at the Memorial Hospital, assisting Henry Harrington Janeway with the operation of Duane's newly-installed radon plant.

Failla applied himself to the task with enthusiasm and innovation. Only two years later the young physicist's name appeared as coauthor with Janeway and Benjamin Barringer on *Radium Therapy in Cancer*. During the first World War Failla served as a science attaché at the United States embassy in Rome, then returned via Paris and the Radium Institute. Marie Curie and others encouraged him, and he returned to obtain a doctorate from the University of Paris.

With Edith Quimby, Failla would spend forty years providing innovations and insights in the physics of radiation therapy. Their investigations into filters, sizes of fields, dose, phantoms, and ionization chambers changed practice in the field. Their work on delivery systems, biophysics, and radiation protection are of lasting importance.

Failla retired from Memorial in 1960, with numerous testimonials and awards. He accepted a position as senior physicist emeritus at the Argonne National Laboratories, where his former student, John E. Rose, was head of the radiological physics division. Before new researches at Argonne could begin, Failla was killed in a car accident on 15 December 1961.

—Juan A. del Regato, M.D.

(Courtesy of the Center for the American History of Radiology, Reston, Va.)

---

became evident that though X rays moved and did things as particles, they nevertheless have also the characteristic optical properties that identify them as waves....It may be fair to say that these experiments were first to give, at least to physicists in the United States, a conviction of the fundamental validity of the quantum theory."[42]

One of the early experimental agreements with Compton's theory was by C. T. R. Wilson with cloud chamber photographs of recoil electrons from Compton scattering. Highly convincing evidence for Compton's theory was soon provided by Bothe and Geiger, who observed coincidences between scattered photons and recoil electrons in counters. Hugo Fricke and Otto Glasser, then medical physicists at the Cleveland Clinic Foundation, corroborated the amount of energy given to recoil electrons by measurements with a series of ionization chambers constructed of different materials.[43]

The theory and demonstration of particle-like properties for X rays and gamma rays was followed in 1924 by de Broglie's prediction of wave-like properties for particles. The duality of particle and wave properties, important for electrons and photons, and later shown for nuclear constituents, laid the foundation for quantum mechanics, a theory critical to modern physics.

Pair production

Pair production, a third major way in which photon radiation interacts with matter (in addition to the photoelectric effect and Compton scattering), was first observed in 1930. Explanations of the phenomena followed in the 1930s and early 1940s.[44] Medical physicists during this period were not very concerned with the interactions that occurred with photons above 1.02 megaelectron volts (MeV) and thus did not contribute substantially to the understanding of pair production.

Relative measurement

In 1910 radiation could be measured by ionization, fluorescence, or chemical changes. The Curies had used electroscopes to separate more radioactive from less radioactive preparations during the purification of radium salts. These instruments measured ionization in the air within the instrument by noting the decrease in potential indicated by the position change of a gold leaf. Radioactive sources could be placed within or near the electroscope. Until recently the electroscope in various forms continued to be an important tool for comparing gamma ray intensity from radioactive sources. References from scientists in the period indicate that ionization methods were used to measure the relative intensities of X rays well before 1908, but information about the specific methodology was not found.[45]

Fluorescent and chemical methods for measuring X rays were reported most often. The measurement of free iodine in a dilute iodoform-chloroform solution and the precipitation of calomel from a solution of ammonium oxalate and bichloride of mercury required expertise in chemistry and was not very practical. More successful efforts used the fluorescence of barium platinocyanide and zinc sulfide. Visually, the fluorescence was compared to the luminosity of a standard lamp or that of a radioactive preparation. As late as 1926 the fluorescent method was still being refined. Perhaps the most used method of X-ray measurement in those early days was that of a color change caused by X rays in a pastille (i.e., a small mass or little loaf) of platinobarium cyanide. (Fig. 4.2) These methods are discussed by Hudson.[46]

Relative measurements for X rays and radium sources in experienced hands were probably quite reproducible; however, it was difficult to compare sources and measurements between institutions. Even an amount of radium that could be stated by its weight was suspect because of variabilities in the purity of the radium salt and the degree of hydration, and even whether the specified weight referred to the radium compound or the radium element. A story told by Curtis Burnam, an associate of Howard Kelly in the gynecology department at Johns Hopkins Medical School in Baltimore, illustrates how tenuous source calibrations were in early 1911.[47] A shipment of radium ordered from Paris was compared by electroscope ionization measurements to a radium source at Yale University, which had been compared to a standard source in Vienna. Burnam and his coworkers concluded that the Paris shipment contained only 60 milligrams (mg) of radium instead of the 100 mg as ordered—an error worth thousands of dollars at the time. Upon notification, Paris sent an additional 40 mg of radium. The discrepancy was

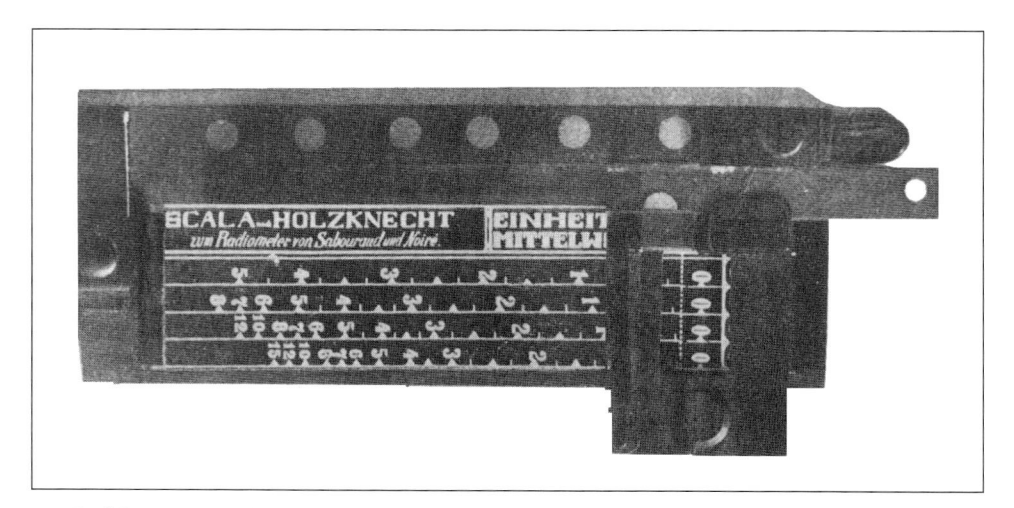

Fig. 4.2 Holzknecht's chromoradiometer, first presented in 1902 in an early effort at radiation dosimetry. This later model made use of the color changes radiation induced in the barium platinocyanide pastilles of Sabouraud and Noiré. (Courtesy of the Center for the American History of Radiology, Reston, Va.)

probably a mistake rather than an experimental error, but the story illustrates that the knowledge of radium source strength was insecure in its dependence on informal channels.

Brachytherapy

Standardization of radium sources

Before 1910 the need for agreed standards for measuring X rays and stating of source strength for radium was apparent. The method of stating the weight of radium in a closed container was in common use, but with flaws like those discussed above. In 1910 the International Congress of Radiology (ICR) and Electricity asked Marie Curie to prepare an international radium standard. The same congress defined the curie as the activity of the amount of radon in radioactive equilibrium with one gram of radium. In 1911 Curie completed the preparation of a standard of 21.99 mg of pure radium chloride. An ionization comparison with sources prepared independently by Honigschmid in Vienna showed all the sources to be in acceptable agreement. Subsequently, substandards were prepared for other countries by ionization comparison with Curie's standard. This system of readily available standards has provided a stable base for radium and radon treatment and for brachytherapy even to this day.

Relation of exposure rate to radium source strength

The knowledge of the strength of a radium source in units of milligrams of radium contained in that source did not relate directly to energy absorbed in tissue. It was necessary to determine the exposure rate at a specified distance from a standardized radium source. Because of the range of electrons generated from the most energetic gamma rays from radium (2.45 MeV), a free-air chamber at atmospheric pressure would have to be quite large to accommodate the range of recoil electrons generated. Friedrich, at the University of Berlin, actually tried such a measurement in an armory, 100 x 50 x 22 meters[3] in size. Other measurements were made with cavity ionization chambers and free-air chambers under pressures of up to 50 atmospheres. Mayneord and Roberts in 1934 used a cavity chamber to determine that the exposure rate at 1 centimeter (cm)

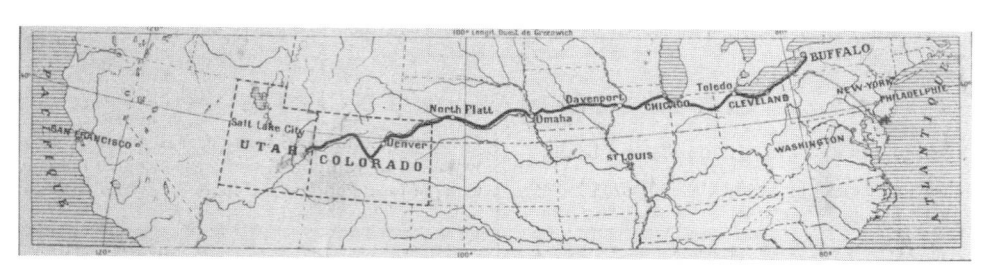

Fig. 4.3 This 1910 map shows the route of radium ore from Utah and Colorado to Buffalo, New York. The ore began the trip in oxcarts, was transferred to rail cars, partially refined in Buffalo, and then shipped to France for final reduction to radium. The Standard Chemical Company shortened this laborious process by refining the ore in Pittsburgh and producing a reliable supply of radium for the United States. (Courtesy of the Center for the American History of Radiology, Reston, Va., from the E.D. Trout Collection)

Medical Physics and Radiation Oncology

from a point source of radium filtered by 0.5 millimeter (mm) of platinum was 8.3 r cm^{2}hr^{-1}mg^{-1}.[48] An authoritative value of 8.25 r cm^{2}hr^{-1}mg^{-1} was based on measurements of 8.26 ± 0.05 by Attix and Ritz in 1957 at the United States National Bureau of Standards (NBS).[49] Cavity ionization chambers, gamma standardization, and measurement of high-energy X-ray and electron beams are discussed in a later section.

Availability of radium and radon

As mentioned previously, Howard Kelly in Baltimore in early 1911 was able to purchase 100 mg of radium from France. Soon after that purchase, however, the success of brachytherapy, particularly in treating carcinoma of the uterine cervix, increased the demand for radium so sharply that the mines in Austria (Bohemia) could not fill all the orders. The small supply of radium available before 1913 would have precluded progress in radium therapy in north America had not the Standard Chemical Company of Pittsburgh commenced mining carnotite ore in Colorado and Utah. This ore was shipped to Pittsburgh for refining (Fig. 4.3). Refining one gram of radium required 500 to 600 tons of ore, 10,000 tons of distilled water, 1,000 tons of coal, and 500 tons of chemicals. By 1922 the world supply of radium was 175 grams, of which the United States had produced 120.[50] In 1923 mines in the Belgian Congo began producing very high-grade uranium ore, causing a drop in price from $120,000/gram of radium metal to $70,000/gram.[51] As a result of this overwhelming competition, the United States mines closed in 1923.

Radon implants

Interestingly, Alexander Graham Bell in a 1903 letter suggested the use of interstitial radium sealed in glass tubes.[52] By 1914 most of the radium in the United States was used for the production of radon.[53] Adapting the features of radon plants in France and England, Duane introduced the idea of collecting radon gas into glass capillaries which could be sealed into small sections for interstitial use.[54] The glass, however, was not thick enough to stop the beta particles. Penetration through the walls by the beta particles was anticipated and thought to be favorable, but the high radiation dose near the glass radon sources produced tissue necrosis. Barium paste introduced around the sources to absorb beta particles penetrating the glass was too difficult to control. Gioacchino Failla, working at the Memorial Hospital in New York, addressed the beta particle problem. His early approach was to enclose the glass tube in an outer platinum tube. However, this method, which necessitated sources of greater diameter, proved to be too traumatic to the tissue under treatment. Failla then considered introducing the radon gas directly into gold or platinum tubing. He found that pinched and cut gold tubing produced a permanent seal for the radon gas within the tube.[55] The choice of wall thickness for the gold radon seeds required (a) understanding that biological effects depended not on the radiation passing through tissue but on the radiation energy absorbed by the tissue; (b) understanding that because sources were spaced at 1 cm or more from one another, the treatment depended on the gamma rays and not electrons; and (c) measuring radiation at and near the surface of the source. The radiation near the source consisted of about 49 gamma rays with an average energy of 0.83 MeV from the radium daughter products and also the associated beta particles. In addition, photoelectric and Compton recoil electrons were ejected from the gold wall of the seed by gamma ray interactions. Failla measured the mix of these radiations by using the bleaching effect of radiation on ordinary table butter. "When a tube containing a sufficient amount of radon is placed on the smooth surface of a block of butter (which is kept on ice), one can observe an area of discoloration which increases in size for a number of days. The outline of this region is quite sharp and can be measured with fair precision."[56] From the butter experiments

William David Coolidge (1873-1975)

Coolidge was born in Hudson, Massachusetts, on 3 October 1873, the only child of Martha Alice Shattuck and Albert Edward Coolidge, a shoemaker. After grade and high schools in Hudson he received a B.S. degree from the Massachusetts Institute of Technology (MIT) in 1896. After a study fellowship in Leipzig, Germany, he received his Ph.D. in 1899 and was appointed to the staff of MIT.

In 1905 he took a position in the General Electric Research Laboratory at Schenectady, New York, where he remained for the rest of his long life. He worked at first on ductile tungsten as a replacement for carbon filament in electric lamps. In 1913, using the new tungsten filament, he developed his famous hot cathode tube. The Coolidge tube was perhaps the single most important innovation for radiology in the first half of this century. For radiation therapists the hot cathode tube provided reliable and predictable X rays for treatment of a variety of ailments. More important, it promised for the first time to provide results which could be easily reproduced and compared among practitioners.

Widely honored for his many inventions and innovations, Coolidge obtained eighty-three patents during his working life, which continued until his death at 102.

—Juan A. del Regato, M.D.

(Courtesy of the Center for the American History of Radiology, Reston, Va.)

---

and experiments in animals, Failla concluded that a wall thickness of 0.3 mm of gold, yielding an outside diameter of 0.75 mm, was preferable. A gold seed of those dimensions removed 99.6 percent of the beta radiation and 18 percent of the gamma radiation (mostly the low-energy gamma radiation). The seemingly small amount of beta radiation remaining was nonetheless significant and affected the choice of 0.3 mm wall thickness instead of 0.2 mm, which removed 99.2 percent of the beta radiation, or 0.5 mm, which removed all of the beta radiation but resulted in a seed judged to have an unacceptably large outside diameter. Typically, the gold or glass radon seeds were a few millimeters in length (Fig. 4.4).

Brachytherapy systems

Radium and radon have been used clinically in four major ways: as, external applicators, intracavitary sources, interstitial sources, and sources in external beam treatment devices. The first three uses were organized into evolving systems beginning about 1932. The Manchester System, built on the French and British practice of using sources of low linear activity (e.g., 0.33 and 0.66 mg/cm), culminated in a collection of journal articles published in book form as the Manchester System.[57] Edith Quimby, a colleague of Failla at Memorial Hospital in New York, published regularly on the characteristics of radium and radon sources from 1922 onward. In 1932 she published the first papers on the so-called Quimby System, which codified the usual American practice of using equal strength sources of high linear intensity (e.g., 1 mg/cm or more).[58] With these sources "treatments had to be quite short, reactions were often severe and the techniques did not become popular....In America Dr. Charles Martin was the leader in the employment of low intensity needles, and achieved such good results that others followed his lead."[59]

Radium systems specified the strength and geometry of sources for various treatment arrangements. Then, based on the resultant geometry and total strength of the sources, a single exposure rate was calculated from tables. That single exposure rate and the cumulative total exposure characterized an implant and was the basis for determining the removal time of a temporary implant. Clinicians related the treatment result to the calculated total exposure and adjusted future prescriptions accordingly. Thus, the radium systems were of enormous importance in regularizing treatment on the basis of radiation absorbed and in permitting the dissemination of clinical experience. The older methods of characterizing treatment in terms of milligram-hours, that is, total source strength multiplied by the total treatment time, disregarded source and target volume geometry and did not address dose effects on tissue. However, milligram-hour calculations retained some usefulness in setting safe upper treatment limits.

External Beams

Radium irradiators

The first radium irradiator was built in 1912 and was used briefly in Germany to treat pelvic cancer.[60] This irradiator is notable both because it was first and because it was called a "radium cannon." Radium packs were

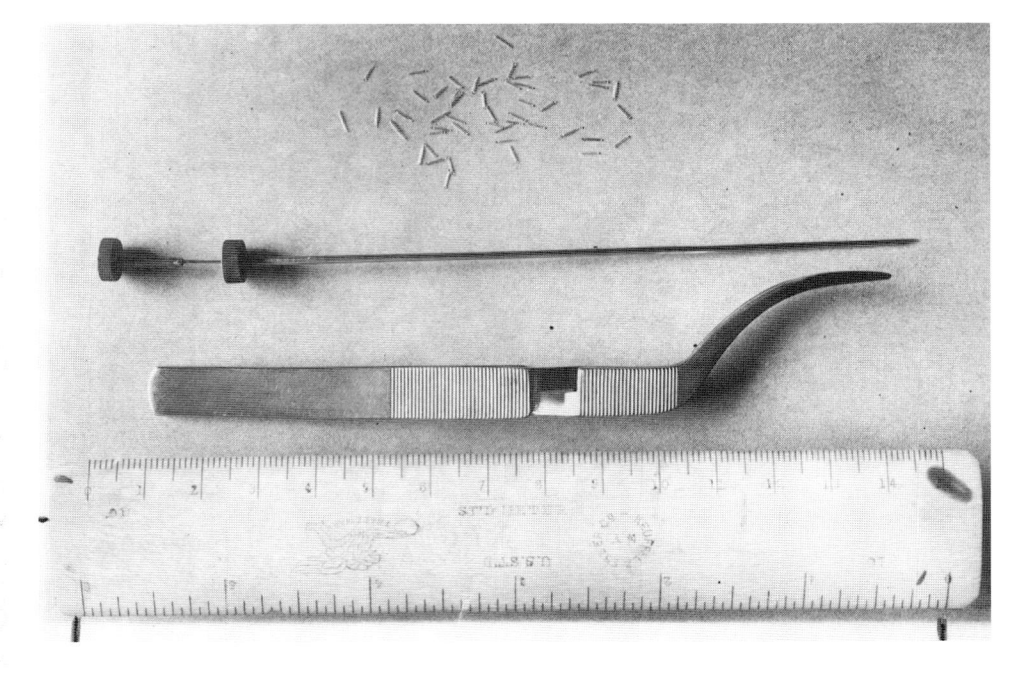

Fig. 4.4 Henry Janeway's glass radon seeds for "buried emanation." The tiny glass seeds were inserted into the hollow steel needle with small forceps and then injected into cancerous growths. The tubes were not removed, as they were said by Janeway and colleagues to become "extinct" after two or three weeks. (Courtesy of the Center for the American History of Radiology, Reston, Va.)

Fig. 4.5 left Gioacchino Failla's 4-gram radium packs devised for the Memorial Hospital (1928). (Courtesy of the Center for the American History of Radiology, Reston, Va.)

Fig. 4.5 right The packs were raised and lowered for patient contact using a heavy metal container maneuvered by controls in an adjacent room. (Courtesy of the Center for the American History of Radiology, Reston, Va.)

portable protected boxes which contained available radium or radon for treatment at distances set by fillers of cork or balsa wood. As the supply of radium became more abundant, efforts to build radium irradiators in Baltimore by Howard Kelly's group and in Stockholm by Eric Lysholm, who called his apparatus a "radium howitzer," came to fruition about the same time in 1922 (Fig. 4.5).[61,62] The Baltimore unit had a collimator that could be placed at a variable distance from the source, and the Stockholm unit had a variable source-treatment distance. Neither group mentioned source strength, suggesting that it may have been whatever was available.

By 1929 there was a radium unit in Baltimore, one in New York, and at least four in Europe.[63] Other radium units came into being after that; some found niches, particularly for head and neck treatment, which allowed their continued use into the 1960s.[64] The era for the radium irradiator, however, ended with the building of a 50 gram multisource radium irradiator by Failla for the Roosevelt Hospital in New York in the early 1950s.[65] The unit was still operating in 1956. The advent of cobalt-60 irradiators in 1951 ended plans for developing new radium irradiators.

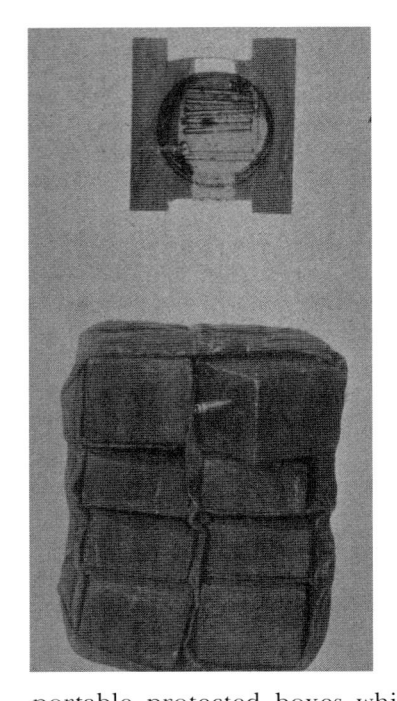

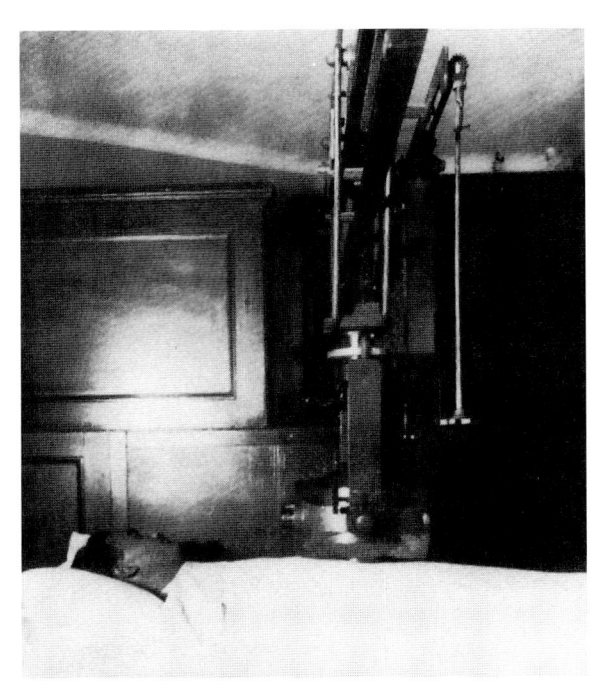

X-ray tubes

Early gas X-ray tubes were difficult to control. Electrons for acceleration were produced when positive ions from the gas in the tube struck the negative cathode. Successful operation employed cathodes with the capability of adsorbing gas (e.g., hydrogen gas in aluminum cathodes). These tubes required enough gas in the tube to form positive ions and electrons but not so much gas as to preclude acceleration of the electrons to the anode. The state of the vacuum tended to vary during operation, causing instability. There was considerable variation from tube to tube. Many tube designs were tried and used, including tubes with means of replenishing the gas in the tube.

A series of scientific observations led to the invention of the hot cathode high-vacuum X-ray tube in 1913 by William D. Coolidge at the General Electric (GE) Company in Schenectady, New York. In the 1880s Thomas Edison had shown that in an evacuated incandescent bulb current would flow from the hot filament to another electrode.[66] This was called the "Edison effect." The relationship between the emission of electrons and temperature was described by Richardson in 1902.[67] In Germany Wehnelt and Trenkle tried a

Fig. 4.6 Irving Langmuir (left) and William Coolidge (right) show visitor J.J. Thomson one of the improved X-ray tubes during a visit to the General Electric Laboratories in Schenectady. (Courtesy of the Center for the American History of Radiology, Reston, Va.)

▶

lime-coated hot cathode in a tube with a rather low vacuum, resulting in an X-ray tube limited to a potential of 1 kV because of positive ion bombardment and destruction of the cathode. In 1912 Lilienfeld and Rosenthal introduced a hot cathode into a gas tube with reduced pressure to aid in controlling the conductivity of the tube. No doubt some of the electrons striking the anode originated in the hot cathode, but the successful operation of the tube required some gas in the tube to produce electrons. In 1903 Irving Langmuir, also at GE, produced stable and reproducible emission of electrons in high vacuum.[67] In 1913 Coolidge successfully tried a similar method in an X-ray tube in a high vacuum with much higher voltages.[68]

One of Coolidge's major contributions was the production of ductile tungsten. In 1905 work began at the GE laboratories investigating and altering the mechanical properties of metallic tungsten. This normally brittle and unworkable metal was rendered ductile by a carefully controlled process in which purified tungsten powder was formed into rods under heavy pressure in a hydrogen atmosphere just below the tungsten melting point; the rods so formed were worked mechanically with heat, thus elongating the tungsten crystals and producing the desired ductile tungsten.

Ductile tungsten was first used in incandescent bulbs and later for anodes in gas-filled X-ray tubes, replacing platinum. Tungsten embedded in copper (a complex process) produced anodes with a higher melting point, a lower vapor pressure, a greater heat conductivity and only a slightly lower atomic number than platinum, permitting X-ray tubes to operate at a higher potential and tube current. Tungsten targets soon replaced platinum targets in gas tubes for applications that required high potential and high tube current. However, if these tubes were overloaded electrically, the vaporized tungsten united with nitrogen in the gas tubes, depositing as a solid nitride on the tube walls and thus reducing the gas pressure in the tube below operating levels. Other gases such as hydrogen, argon, and helium worked somewhat better but were slowly lost to the gas in the tube. An additional difficulty in the gas tubes operating at high potentials and tube currents was the destruction of the aluminum cathodes by positive ion bombardment. The gas in the aluminum cathodes was important in starting the current flow in gas tubes. Substitution of tungsten as a cold cathode in gas tubes produced instability in the tube operation because of the lack of occluded gas in the tungsten metal. Coolidge's solution was to produce a cathode of coiled tungsten wire heated by a separate filament-heating circuit in which all the electrons constituting the current through the tube were emitted from or boiled off the filament (Fig. 4.6). The production of the electrons for acceleration in the tube in high vacuum without the need of gas solved the complicated problems of gas pressure control and cathode preservation. The new tubes were long-lasting and functional in a stable, reproducible way. As stated by Grigg, "Coolidge's high vacuum, hot cathode X-ray tube, more than any other single development ushered in the Golden Age of Radiology."[69]

The Coolidge tube was characterized by stability, high output, and accuracy of adjustment. When potentials greater than 200 to 300 kV were applied, current sometimes flowed through the tube even with an unheated filament, a phenomenon called the cold cathode effect. If the potential was raised higher, there might be arcs in the

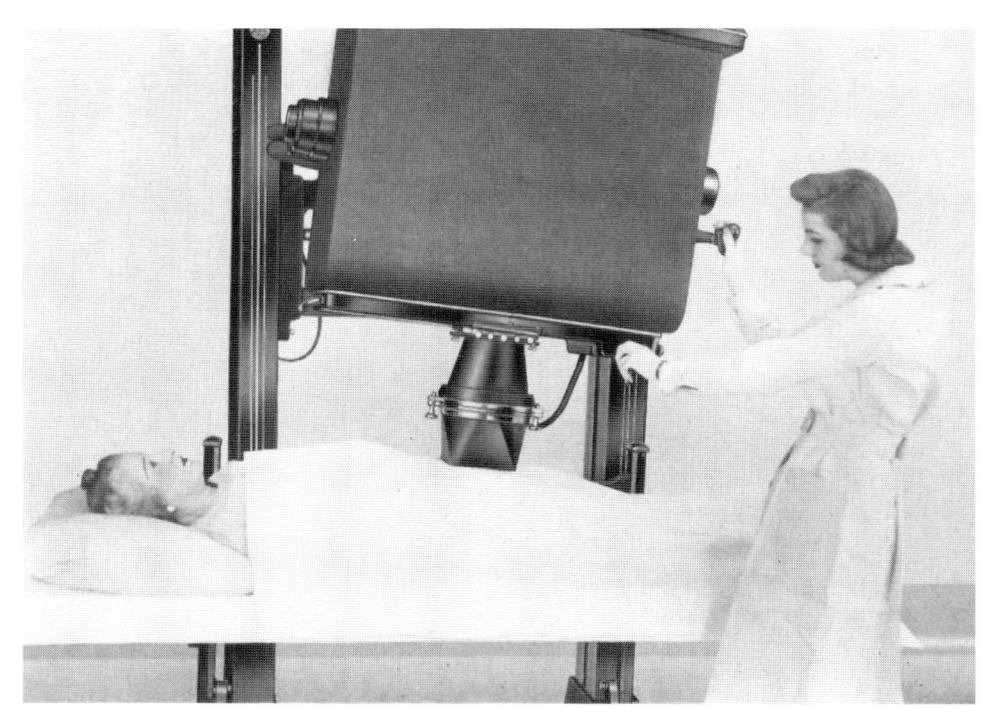

Fig. 4.7 General Electric Maximar 250, originally introduced as a 200 kVp unit in 1936, became the workhorse of radiation therapy. (Courtesy of the Center for the American History of Radiology, Reston, Va.)

tubes or even punctures in the tube wall. X-ray tubes in which potentials in increments of 200 to 300 kV were applied serially overcame this limitation. The size required by the multistage tubes provided a practical limit to the potential possible in X-ray equipment where a fraction of the full accelerating potential appeared across sections of the tube. Self-rectification by the X-ray tube and vacuum tube rectifiers became available in the early 1920s. By 1921 transformers capable of about 200 kV and 8 milliamperes (mA) had become available (Fig. 4.8).

High-energy X-ray machines

A brief summary of salient developments in high-energy X-ray machines suggests the range of apparatus and clinical issues with which medical physicists were faced as the century progressed.

The late 1920s and 1930s brought a variety of ingenious ways to achieve higher accelerating potentials. Lauritsen provided a way to generate 750 kV X rays at the California Institute of Technology in 1928 (Fig. 4.9).[70] The potential appeared between a long inner electrode functioning as a cathode and reaching close to a grounded anode; the glass wall of the tube was protected by a series of four corona shields held at intermediate potentials, which prevented the

Fig. 4.8 Many of these machines, with their impressive control panels, remained in routine use well into the 1970s. (Courtesy of the Center for the American History of Radiology, Reston, Va.)

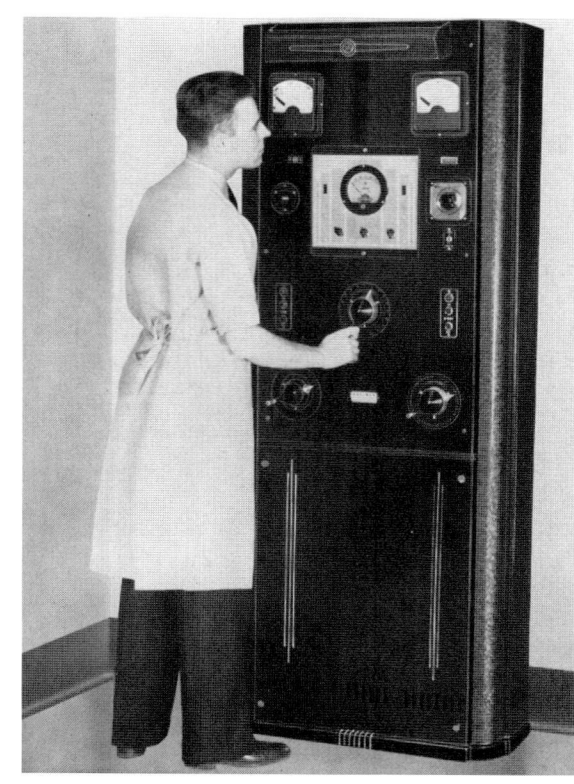

Medical Physics and Radiation Oncology

Fig. 4.9 Radiologist Albert Soiland and C.C. Lauritsen in the early 1930s, with the 600 kV unit Soiland acquired for his own institution after conducting limited clinical trials with Lauritsen's 750 kV machine. (Courtesy of the Center for the American History of Radiology, Reston, Va.)

▶

Fig. 4.10 800 kV therapy unit installed in Mercy Hospital in 1933. The control panel, located in an adjacent room, occupied an entire wall. (Courtesy of the Center for the American History of Radiology, Reston, Va.)

tube from being exposed to the full potential difference. This device required a 50-foot ceiling for transformers and a derrick-like structure 14 feet high to support the tube. A descendent of Lauritsen's tube was installed in Harper Hospital, Detroit, in 1932. It was 13 feet in length. Six transformers in cascade provided up to 900 kV (Fig. 4.10).[71]

In 1931 a two-section X-ray tube powered by an induction coil built at GE under the direction of Coolidge was installed at Memorial Hospital, New York.[72] This unit was usually operated at 700 kV but could be operated up to 840 kV. The separation between the cathode and the target was about 10 feet, and, because of the separation, both electrostatic and electromagnetic focusing of the electrons in the tube were employed to avoid loss of tube current. Collaborators Failla and Quimby did many classical physical and biological tests on this machine.

Particularly interesting was a 1 MV device invented by David Sloan of the University of California Radiation Laboratory in 1934. This device was 40 inches high and 42 inches in diameter.[73] A 6 megahertz oscillator stepped 15,000 V to a range of 700 to 1,000 kV in an enclosed transformer. It was a short-wave radio oscillator operating at high power and sending its power into a high voltage resonant circuit consisting of about fifteen turns of copper tubing. This transfer of power occurred within the vacuum. The vacuum container was of thick steel. Clinician Robert Stone in San Francisco used this machine for many years (Fig. 4.11).

At GE a group working with E. E. Charlton developed a 1 MV resonant transformer X-ray set about 1937.[74] In a resonant transformer, at the resonant frequency (180 cycles per second in this machine) determined by the number of turns in the transformer secondary and the geometry around the transformer, no core for the transformer is required. Instead, a multisection X-ray tube is placed within the coils with terminal and intermediate connections (and thus intermediate potentials) between the tube and transformer secondary. An insulating gas fills the enclosing tank. A 2 MV unit followed. Both the 1 MV and the 2 MV units were used industrially during World War II. Production of medical machines commenced in 1946.[75]

Van de Graaff accelerators, invented by Robert Van de Graaff in 1929, developed high potentials by carrying charge on motor-driven insulating belts to and from a central electrode.[76] These machines were first developed to accelerate positive ions for experiments in nuclear physics and were later adapted for medical purposes by accelerating electrons. Efforts of John Trump and Robert Van de Graaff of the Massachusetts Institute of Technology resulted in a 1 MV medical Van de Graaff accelerator in 1937. By 1940 a 1.25 MV medical unit had been built.[77] This machine evolved into the 2 MV medical Van de Graaff units popular after World War II. The date of the first such machine is not clear from the literature. As produced in the 1950s and 1960s, the medical units were 5.5 feet long and 3 feet in diameter (Fig. 4.12).

It is fair to say that the 1920s brought 200 kV, the 1930s 1 MV, and the 1940s 2MV to medical practice. By the end of the 1940s the prospect for higher energies was well on its way and treatments had been done with 22 to 24 MeV X rays and with high-energy electrons. A characteristic of the betatron and linear accelerator machines developed in the 1940s was that the electrons were not subject to the full electrical potential commensurate with their final energy.

In a three-year period Donald Kerst, working at the University of Illinois at Urbana, improved the theory for accelerating electrons by magnetic induction and produced the first working betatron of 2.3 MeV in 1940. Within a year, a 20 MeV betatron had been built; it was followed by a 300 MeV unit (Fig. 4.13).[78] In reviewing the development of the betatron, Kerst carefully and generously credited a dozen or so physicists and engineers in the United States and Europe who, starting in 1922, worked on the theory of electron acceleration by magnetic induction and the daunting technical problems of electron injection, capture, and extraction from a stable orbit.[79,80] In these machines electrons are accelerated and trapped in a stable orbit by an increasing magnetic field during a quarter cycle of the applied alternating potential. Kerst had produced a leap to multi-MeV for X rays and electrons in therapy. The first medical betatrons manufactured by Allis-Chalmers Manufacturing Company were installed at the Saskatoon Cancer Clinic in 1948, where Harold Johns was the principal physicist, and a second unit was installed in 1949 at the University of Illinois College of medicine in Chicago. These betatrons were 22 to 24 MeV machines. Electrons extracted through a window in the

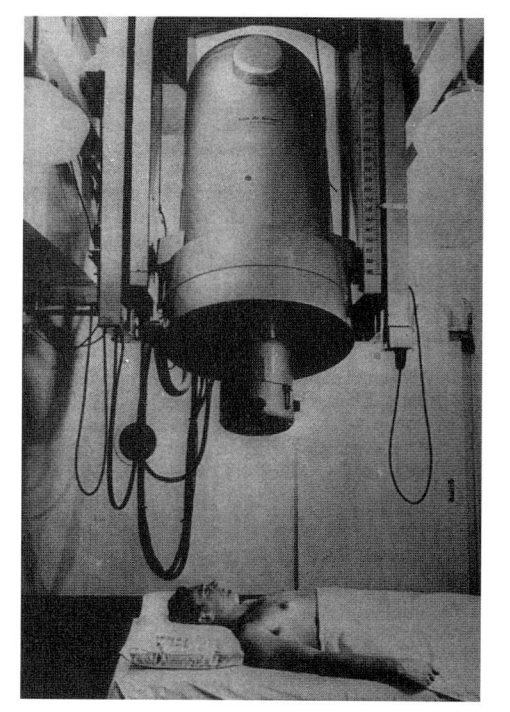

Fig. 4.11 Robert S. Stone (1895–1966) (Courtesy of the Center for the American History of Radiology, Reston, Va.)

Fig. 4.12 2-million volt Van de Graaff generator X-ray unit positioned for single portal treatment. (Courtesy of the Center for the American History of Radiology, Reston, Va.)

Fig. 4.13 Professor D. W. Kerst at the University of Illinois, placing doughnut-shaped vacuum tube in position between pole faces of magnet of 20-million volt betatron. (Courtesy of the Center for the American History of Radiology, Reston, Va.)

round, doughnut shaped accelerating tube provided early physical and biological experience with electron therapy, which John Laughlin began on the Chicago unit in 1950.[81] However, betatrons are limited in the number of electrons which can be captured in a stable orbit and thus the output for X-ray production is limited. Limitation in output means restricted size for flattened X-ray fields, because the amount of beam that can be discarded in field flattening is determined by the acceptability of a low dose rate. The largest practical field for X rays from a 22 MeV Allis-Chalmers betatron was about 18 x 18 cm^2 and thus not sufficient for some large fields used for chest or pelvic treatment.

A group of physics graduate students working on the betatron problems in the early 1940s with Kerst and Henry Quastler, a radiologist from Urbana, became medical physicists who contributed enormously in all areas of the field over their professional lifetimes. These include Gail Adams, Lawrence Lanzl, John Laughlin, Jacques Ovadia, Lester Skaggs, and Rosalyn Yalow. Adams worked at the University of California and University of Oklahoma, Lanzl at the University of Chicago and Presbyterian-St. Luke's in Chicago, Laughlin at the University of Illinois and Memorial Sloan Kettering in New York, Ovadia at Michael Reese Hospital in Chicago and the University of Chicago, Skaggs at the University of Chicago, and Yalow at the Veterans Administration in New York. Yalow shared a Nobel Prize with Solomon Berson for their work on radioimmunoassay. It is interesting that the physics department of the University of Illinois produced few or no medical physicists before or after this distinguished group.

Microwave technology, developed during World War II for radar, had a direct carryover to the development of linear accelerators soon after the war. Two groups working quite independently developed the principles. The frequencies and power possible from radar technology permitted the acceleration of electrons in very closely machined

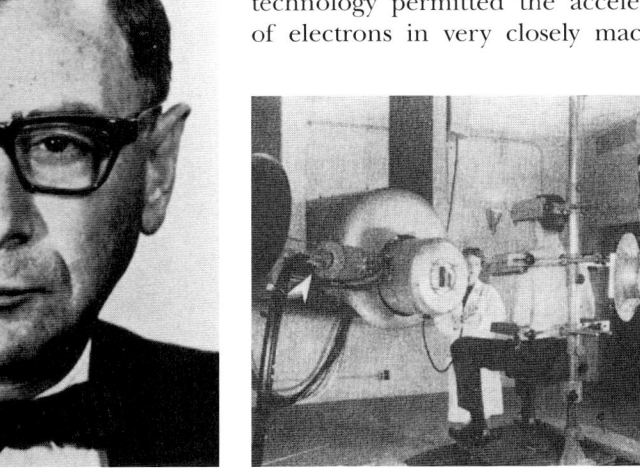

Fig. 4.14 left Henry Kaplan (1918–1984) (Courtesy of the Center for the American History of Radiology, Reston, Va.)

Fig. 4.14 right Stanford medical linear accelerator, 1957. The arrow designates the side-mounted diagnostic tube used for alignment and localization. (Courtesy of the Center for the American History of Radiology, Reston, Va.)

Shalek

OTTO JULIUS ALEXANDER GLASSER (1895–1964)

Glasser, born in Germany, performed seminal work on dosimetry in association with W. Friedrich of Freiburg in Breisgau. In 1929 he devised a condenser dosimeter, later marketed as the r-meter by John Victoreen. In 1931, in Berlin, Glasser published his classic and still definitive biography of Röntgen with a bibliographical list of over one thousand publications on X rays from 1896.

In 1932 he was engaged by Howard Kelly of Johns Hopkins to come to Baltimore. After a short term of duty there, he became the physicist of the Cleveland Clinic, a position he kept for his lifetime. In 1933 he was editor of a book of contributed chapters, *The Science of Radiology*, summarizing in detail the development of all aspects—clinical and technological—of the specialty. Glasser contributed to the production of charts of intensity distribution of doses in depth, which he named "isodose curves." He was one of the earliest trustees of the American Board of Radiology and, with Edith Quimby, established the requirement of physics competence of candidates for certification long before their clinical competence could be adequately tested.

—Juan A. del Regato, M.D.

(Courtesy of the Center for the American History of Radiology, Reston, Va.)

---

wave guides. One group at the Telecommunications Research Establishment at Great Malvern, England, had the first working device in late 1946.[82] The other group, led by W. W. Hansen and E. L. Ginzton at Stanford University in California, used a similar but different technology to develop an operating electron accelerator, which they reported on in 1948.[83] Both groups then produced medical electron accelerators. In 1953 the British group designed medical accelerators which produced X rays of 4 and 8 MeV.[84] In 1957 the Stanford group produced a medical machine to operate at less than 6 MeV.[85] This machine was installed at the Stanford Medical School under the direction of Henry Kaplan (Fig. 4.14).

With the stimulation of World War II, technological advances in the decade of the 1940s led to the develop-

Fig. 4.15 Duane's standard open-air ionization chamber (1928) consisted of two parallel plates, one of which, A, could be moved to any desired distance from the other. An insulated portion of the opposite plate B was connected to a galvanometer C to measure ionization. (Courtesy of the Center for the American History of Radiology, Reston, Va.)

ment of radiation sources that have since dominated radiation therapy. The betatron played a war-time role in the industrial radiography of heavy metal pieces; the generation of microwaves or radar was a step toward electron linear accelerators; and sustained nuclear fission in an atomic pile in 1942 demonstrated the feasibility of a nuclear bomb but also permitted the production of many radioactive isotopes, including cobalt-60, cesium-137, iridium-193, and iodine-125.

Standardization of X-ray measurement

In France in 1908 Villard proposed a unit of quantity for X rays as "that which liberates by ionization one electrostatic unit of electricity per cubic centimeter of air under normal conditions of temperature and pressure."[86] Air and tissue were understood to have similar atomic numbers and would absorb energy in approximately proportionate amounts as X-ray quality was varied. However, there was no experimental way to make such an ionization measurement at the time of Villard's suggestion. William Duane used the same unit, noting that the electrical charge should be collected under saturation conditions (i.e., all ions collected). The standard free-air chamber proposed and built by Duane in about 1914 consisted of a large parallel-plate air chamber in which the collimated X-ray beam struck nothing but air (Fig. 4.15).[87,88] In principle, the spacing of the movable electrodes should be great enough to allow the energy of secondary electrons to be absorbed fully in air before they reach the plates; in practice, the plates could be closer without loss of measured ionization but adequate plate spacing was an important parameter. In the years that followed, other minor changes were made to Duane's free-air chamber, but none were fundamental. Chambers of larger size accommodated higher-energy X-ray beams. By the mid-1920s units based on the Villard-Duane formulation and the Duane method of measurement were in use in France and Germany. An international unit of exposure was defined at the second ICR held in Stockholm in 1928 as "the quantity of X radiation which, when the secondary electrons are fully utilized and the wall effect of the chamber is avoided, produces in 1 cubic centimeter of atmospheric air at 0° C and 760 mm mercury pressure such a degree of conductivity that one electrostatic unit of charge is measured at saturation current. That the international unit of X radiation shall be called the 'Roentgen' and designated by 'r'" (Fig. 4.16).

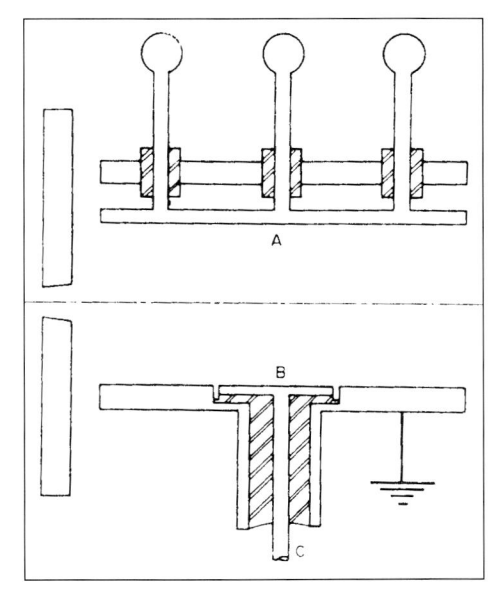

In 1928 this definition of exposure with the unit of roentgen relied on the existence of a free-air chamber which could satisfy the experimental conditions in the 1928 definition. The definition was revised in 1937 at the fifth ICR in Chicago to read: "The roentgen shall be the quantity of X or gamma radiation such that the associated corpuscular emission per 0.001293 gram of air produces, in air, ions carrying 1 e.s.u. of quantity of electricity of either sign."[89] Thus, the roentgen was redefined in a more general way, separating the definition from the method of measurement. The new definition included gamma rays as well as X rays and opened the possibility of a standard measurement with cavity ionization chambers. Free-air chambers allowed for absolute X-ray measurement depending on measurement of charge,

Fig. 4.16 Executive Committee of the International Congress of Radiology, 1928. This group was responsible for seriously pursuing the appointment and follow-through of a committee to adopt and implement radiation standards. Members included, from left to right: Martin Haudek (1880–1931), Austria; Walter Friedrich (1883–1968), Germany; Pasquale Tandoja (1870–1934), Italy; Congress President Gösta Forssell (1876–1950), Sweden; Charles Thurston Holland (1863–1941), England; George E. Pfahler (1874–1957), United States; and Antoine Béclère, France. (Courtesy of the Center for the American History of Radiology, Reston, Va.)

length, and air pressure and not other radiation measurements. The new exposure definition also allowed for the absolute measurement of high-energy gamma rays with cavity ionization chambers, depending on physical parameters and not other radiation measurements.

In 1937 the dose to tissue was to be stated in roentgens, but at that time it was recognized that this "dose" was not to be confused with the energy actually absorbed by tissue. At the sixth ICR in London in 1950, dose was defined as "the energy imparted per unit mass of material at the place of interest." At the seventh ICR in Copenhagen in 1953, the unit for absorbed dose, the rad, was defined as 100 ergs per gram of any irradiated material. The ambiguous use of "dose" to mean charge in air or energy absorbed in any material ended in 1962 by defining exposure as ionization in air.[90]

Cavity ionization chambers for X-ray and gamma ray measurement

Measuring the quantity of radiation by ionization in gas is the primary method for therapeutic radiology and radiation protection. The ionization chambers and the electrical systems used to measure charge or current can be precise, stable over long periods of time, and robust for portability. Ionization chambers can be fashioned into many configurations to measure exposure as a free-air chamber does or to measure absorbed dose as a cavity chamber does. Clearly Duane holds the strongest claim for inventing the free-air chamber, but the origins of cavity ionization chambers are a little murky. Hudson mentioned that Villard utilized the ionizing effect for measurements in 1908, and Szillard constructed a 1 cm^3 chamber with three lead walls and one aluminum wall in 1914.[91] Glasser wrote that in 1914 Duane made small ionization chambers using carbon walls, but the reference is based on a verbal presentation by Duane to the American Roentgen Ray Society (ARRS) in that year.[92] In 1922 Duane reported on cavity chambers made from aluminum sheets and confirmed his reports on other chambers in 1914.[93] By 1926 small secondary chambers were available both in Europe and the United States.[94]

Glasser and Fricke at the Cleveland Clinic and Failla at Memorial Hospital in New York were early leaders in developing cavity ionization chambers. In 1925 Fricke and Glasser, utilizing chambers that were forerunners of those associated with the Victoreen r-meter, appreciated that electrons crossing the chamber arose almost completely in the chamber wall.[95] Accordingly, their recommendation was to make a cavity chamber with an air-like wall with an effective atomic number of 7.69 (the same as air). Such a chamber "will always give an ionization current which is proportional to its volume, and equal to the current produced in the same volume of an infinitely extended air space." Later that year Glasser, Portmann, and Seitz described a condenser dosimeter which functioned like the well-known Victoreen r-meter with a built-in electrometer. They compared the performance of the condenser chamber for X-ray measurement, concluding that correction factors were required at some wavelengths for the condenser chamber.[96] The Victoreen r-meter, which was widely used for over fifty years, was constructed by Glasser and Seitz in 1928 (Fig. 4.17). They were awarded a patent in 1932.[97] Shortly thereafter in England, Gray related the ionization measured in the gas of a cavity to the energy absorbed in the wall.[98,99] With the resulting Bragg-Gray theory, the foundation was laid for utilizing cavity ionization chambers for the absolute measurement of high-energy gamma rays.[100,101] Bragg had considered electrons crossing a cavity in a qualitative way in 1912.[102]

Radiation Protection

Radiation oncologists, diagnostic radiologists, dentists, medical physicists, engineers, and their associates were the principal actors in using ionizing radiation for a very long time, at least until the early 1940s. Their observations of patients and radiation injuries that they themselves incurred were a major source of data for recommendations on radiation protection. These early workers also brought an understanding of benefit versus risk (that is, potential benefit to the patients weighed against possible risks to both patients and practitioners) to the formulation of protection recommendations.

Soon after the discovery and use of X rays in medicine, deterministic effects of X rays, such as skin burns and eye irritations were observed. Sinclair defined deterministic effects as those direct radiation effects in organs or tissues that impair the function of that organ or tissue mainly because of the loss of individual cells by killing.[103] These effects have also been called somatic effects. They are characterized by a threshold dose below which the effect is not observed and above which the severity of the effect increases with radiation dose. A definition consistent with these characteristics was contained in a 1993 report by the National Council on Radiation Protection and Measurement (NCRP).[104]

As previously mentioned, it was known early on that radiation induced skin cancer and by 1911 it was reported that radiation also induced leukemia. In 1927 Muller demonstrated that genetic mutations in fruit flies were caused by X rays. Cancer and the hereditary defects in offspring are examples of stochastic effects caused by ionizing radiation. Sinclair defined stochastic effects as those that occur with a frequency (or probability) dependent on the dose (and usually assumed to be proportional to dose in the low-dose region).[105] The severity of the effect, however, is the same at all doses, and it is assumed that there is no dose threshold; i.e., it is an all-or-none effect. A consistent definition was given in

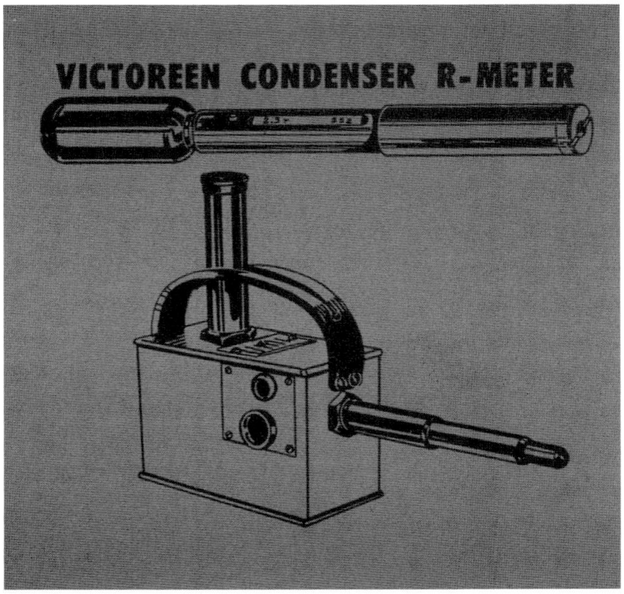

▶ Fig. 4.17 Victoreen r-meter, based on the condenser dosimeter devised by Glasser and Fricke in 1929 and first marketed by John Victoreen in the 1930s, was for over fifty years a mainstay of medical physics. (Courtesy of the Center for the American History of Radiology, Reston, Va.)

the 1993 NCRP report, which also stated the objectives of radiation protection:

> The specific objectives of radiation protection are:
> (1) to prevent the occurrence of clinically significant radiation-induced deterministic effects by adhering to dose limits that are below the apparent threshold levels and
> (2) to limit the risk of stochastic effects, cancer and genetic effects, to a reasonable level in relation to societal needs, values, benefits gained, and economic factors.[106]

From 1910 to 1950 only deterministic radiation effects which would affect the radiation worker in the near term were considered in the development of radiation protection recommendations. Spiers pointed out that in 1921 the first-ever recommendations came from the British X-ray and Radium Protection Committee.[107] They stated in part, "that nurses and attendants should not remain in close proximity to patients undergoing radium treatment with quantities of radium exceeding 0.5 gram." The basis for quantitative recommendations was the suggestion that 1/1000 of the skin erythema dose might be tolerated safely if spread over five days. In 1925 a committee appointed by the ICR adopted a recommendation of 0.2 R per day or 1 R per week based on the erythema dose. The International Commission on Radiological Protection (ICRP) was formed in 1928, and in 1934 reaffirmed the recommendation of 0.2 R per day.[108] The Advisory Committee on X-ray and Radium Protection was established in 1929 becoming the National Committee on Radiation Protection and Measurements in 1946 and the National Council on Radiation Protection and Measurements with a charter from the United States Congress in 1964. This organization in the period 1931 to 1934 recommended an exposure limit of 0.1 R/day. The international recommendation of 0.2 R/day was to be measured at the skin surface (therefore with backscatter) while the American recommendation of 0.1 R/day was to be measured in air (therefore without backscatter).[109] Acknowledging this measurement difference, the two recommendations are closer than a bare statement indicates, because the backscatter factor could exceed 50 percent at some X-ray energies. The enunciation of radiation protection recommendations in the late 1920s occurred at about the same time as the health histories of radiologists became similar to those of other physicians.[110]

With the advent of nuclear reactors and the development of the atomic bomb in the early 1940s there was a great increase in the number of professionals devoted to the problems posed in the radiation protection of workers. A new profession of health physics studied the underlying scientific and technical questions and created training programs for radiation inspectors. Many heath physicists are employed in federal and state agencies that regulate radiation. Brodsky and Kathren have discussed these developments more completely.[111]

SUPER VOLTAGE ERA: 1950–1970

The basis for modern radiation therapy was provided by scientific and technical advances made between 1940 and 1950, many of which resulted from war research and production. The betatron, the electron linear accelerator, and electronic computers were all invented during this period, and large quantities of manmade radioactive isotopes became available from nuclear reactors, the first of which went critical in 1942. In the period 1950 to 1970 medical physicists, radiation oncologists, and many commercial researchers participated in developing innovations from the 1940s. In their clinical responsibilities, physicists and oncologists learned how to calculate dose distributions more completely and more rapidly with electronic computers and how to control, measure, and use new radiation sources for treatment. During this period heavy particles reentered the armamentarium at a few academic centers.

Status in 1950

Except at a dozen or so centers in

Table 14.II

Major Advances from 1950 to 1970

Advancement	Date	By whom	Where
Electron beam above 6 MeV	1951	Laughlin and Haas	USA[112]
Cobalt irradiator	1951	Johns	Canada[113]
	1951	Errington and Green	Canada[114]
	1954	Grimmett	USA[115]
Calorimetry for high-energy X-ray	1951	Laughlin and Beattie	USA[116]
Heavy particle therapy capability	1952	Tobias, Anger, Lawrence	USA[117]
X-ray calculations with computer	1955	Tsien	USA[118]
Brachytherapy calculations with computer	1958	Nelson and Meurk	USA[119]
Iridium-192 brachytherapy	1958	Henschke	USA[125]
Thermoluminescent dosimeters, development for therapy	1961	Cameron and Daniels et al.	USA[120]
Cobalt calibration extended to high-energy X rays	1964 and 1969	Hospital Physicists Association	UK[121] UK[122]
Remote afterloading-cervix treatment	1965	O'Connell et al.	UK[127]
	1965	Henschke, Hilaris, Mahan	USA[128]
Iodine-125 brachytherapy	1966	Lawrence et al.	USA[126]
Cobalt calibration extended to high-energy electrons	1967	Almond	USA[123]
		Svensson and Petterson	Sweden[124]

North America in 1950, radiation therapy was a part-time occupation for general radiologists, surgeons, and gynecologists. There were probably fewer than two dozen full-time medical physicists in addition to part-time medical physicists holding other academic positions. The professions of radiotherapy and related medical physics were poised for expansion, stemming in part from new technology, and in part from hard-won success and recognition of radiation therapy as a treatment for cancer.

In 1950 radium needles were readily available and radon seeds could be purchased for use with patients. Radioactive sources made from radionuclides produced in nuclear reactors existed, but were not commercially available for radiation therapy. X-ray machines operating at 250 or 400 kV were available from several manufacturers, and X-ray machines operating at 1 and 2 MV, utilizing resonant transformers, had become commercially available for medical purposes. Medical betatrons operating at 20 to 24 MeV were in use, and medical Van de Graaff accelerators were operating and soon would be available commercially. Electron linear accelerators were operating, and medical prototypes would be tested soon. Work was proceeding on planning and building the initial cobalt irradiators, and sources were being activated. The advantages of greater depth dose, maximum dose beneath the skin, and similar energy absorption for bone and muscle were known for high-energy X-ray and gamma-ray therapy. Table 4.II lists the major advances occurring between 1950 and 1970.

Brachytherapy

Computer calculations and brachytherapy systems

In 1958 Nelson and Meurk of Memorial Hospital in New York were the first to demonstrate the usefulness of electronic computers for calculating isodose patterns around brachytherapy implants.[129] They summed doses to

Edith Hinkley Quimby
(1891–1982)

Quimby was born in Rockford, Illinois, on 10 July 1891. With a bachelors degree from Whitman College in Walla Walla, Washington, she taught high school science at Nyassa, Oregon in 1912, then obtained an M.Sc. degree from the University of California at Berkeley. There she met and married Shirley Leon Quimby. They moved to New York, where he was to do graduate work as an instructor in physics at Columbia University. She took a position as an assistant physicist at the Memorial Hospital in 1919. As an assistant, she developed activities of her own that brought her personal recognition and a wide following but performed the bulk of her work with Gioacchino Failla. In addition to their studies of filtration and depth doses, she published pioneer work in the study of time-dose relationships and pioneered in the trials of radioactive sodium and iodine.

An educator by choice and a didactician by nature, she began teaching physics to residents in radiology before there were any requirements that they have such training. At Cornell, and later at Columbia, she taught a course of weekly lectures on radiation physics which, over time, was attended by over one thousand physicians, industrialists, physicists, and residents-in-training. With Glasser, Taylor, and Weatherwax, she wrote a compendium on the fundamentals of radiation physics that was widely used by students and practitioners.

(Courtesy of the Center for the American History of Radiology, Reston, Va.)

She became a member of the American Radium Society (ARS) and was its president in 1954, writing a valuable history of the organization in 1966. As a trustee of the American Board of Radiology, she established physics as part of the examination of residents in training. She received numerous awards and honors, including the Janeway Medal and the gold medal of the Radiological Society of North America. Quimby and her husband lived in Greenwich Village and were avid theatre goers. She died in 1982, leaving behind hundreds of well-trained hospital physicists and a legacy of research and education.

—Juan A. del Regato, M.D.

points using a computer that did double duty as an accounting machine. Others soon followed in the computer field. For multisource implants, the manual calculation of isodose curves in three dimensions required too much time and effort to be practical for individual implants. In a sense, the possibility of constructing specific isodose patterns for individual implants constituted an entirely new system. However, at M. D. Anderson Hospital in Houston, clinicians determined early that the discipline imposed by the Manchester System rules for source strength and placement with the calculation of a single dose rate was important in maintaining a tie to previous clinical experience. Relying too heavily on the isodose patterns led to overprescription in a few instances. The merit of the more complete knowledge of the isodose distributions was demonstrated by Gilbert Fletcher and Marilyn Stovall.[130] They showed that of fifty patients who were treated with interstitial implants and had experienced complications (recurrence or necroses), approximately 80 percent of the complications could be explained by local regions of low or high dose. Some institutions continue to use both computer isodose and single dose rate calculations with an existing system to maintain a consistency with earlier clinical experiences and to confirm the general correctness of an individual computer calculation.

The availability of isodose distributions around implants showed that the Manchester and the Quimby systems for interstitial and externally applied sources were not compatible and should not be used interchangeably.[131] Single doses derived from each system differed widely for some implants that satisfied the rules of both systems. The study of isodose patterns for typical implants supported the internal consistency of the Manchester system.

In the 1960s Pierquin and Dutreix described a new brachytherapy system, called the Paris system, which used wires of iridium-192 (^{192}Ir).[132] The single dose rate characterizing the implant could be calculated manually by superimposing isodose curves or could be calculated by computer. Gillin and colleagues, using extensive studies, contrasted the Paris and Manchester systems.[133]

The use of iodine-125 (^{125}I) seeds, which today has achieved wide use, was pioneered at Memorial Hospital in New York beginning in the early 1960s. ^{125}I emits 5 gamma and X rays with an average energy of 28.3 keV and has a half life of 59.56 days. ^{125}I seeds, when used as a replacement for radon-222 (0.83 MeV average energy and 3.824 days half life) or gold-198 (0.412 MeV and 2.696 days half life) seeds in permanent implants, are a large departure from past clinical experience. Thus, the Memorial system was an innovation in major physical and biological respects.[134]

External Beams

Cobalt irradiators

Development of the cobalt-60 (^{60}Co) irradiator was a North American affair, with two of the three first cobalt irradiators from Canada and the remaining one from the United States. The idea for using ^{60}Co in teleirradiators may well have occurred independently to a number of people. Various undocumented claims have been made, but the first published suggestion for using ^{60}Co as a substitute for radium was in 1946 by J. S. Mitchell of Cambridge, in the United Kingdom.[135] Mitchell, however, did not specify whether the substitution of ^{60}Co for radium was for teletherapy or brachytherapy. Because his ^{60}Co suggestion was followed by other possible substitutions of nucleotides with short half lives, one can speculate that he had brachytherapy in mind.

To implement the possibility of a cobalt irradiator, Gilbert Fletcher was instrumental in bringing physicist Leonard Grimmett to M. D. Anderson in 1949. Grimmett had previously designed teleradium units in Britain, and of twelve plans submitted by universities for a ^{60}Co irradiator, the Grimmett design was selected by the United States

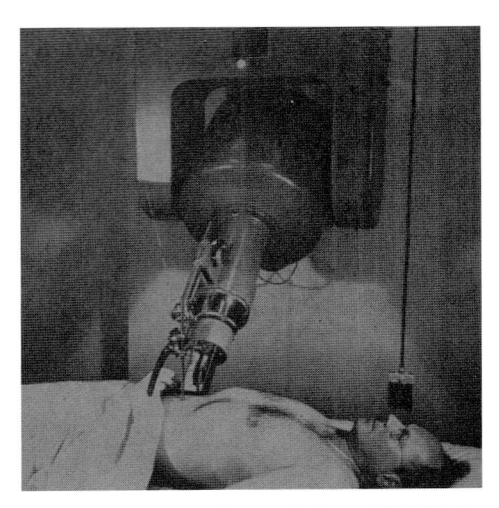

Atomic Energy Commission (AEC) in February 1950.[136,137]

While the early activity at M. D. Anderson may have influenced Canadian counterparts, the Canadian cobalt units were built and placed in clinical service long before the Texas unit.

The first kilocurie cobalt unit used in clinical treatment was designed by R. F. Errington and D. T. Green of the Eldorado Mining and Refining Company in Ottawa, Canada, and was built by the same company.[138] Radiation oncologist Ivan Smith treated the first patient on 27 October 1951, at Victoria Hospital, London, Ontario. The design utilized lead shielding and pumped liquid mercury to form a shutter or open a portal. Eldorado Mining and Refining then manufactured the unit for commercial use.

Shortly afterward, on 8 November 1951, the first patient was treated by Thomas Watson on the Harold Johns unit, which was built by John MacKay, owner of the Acme Machine Shop and Electric Company in Saskatoon, Saskatchewan, Canada.[139] This unit had lead shielding and a rotating source wheel. A commercial model by the Picker X-Ray Corporation followed Johns's design (Fig. 4.18).

The Texas unit treated its first patient on 22 February 1954, with Gilbert Fletcher as radiation oncologist (Fig. 4.19). This unit was designed by Leonard Grimmett and engineer E. Bailey Moore.[140] It was built by GE with tungsten alloy for shielding and utilizing a rotating source wheel. This unit did not serve as a prototype for further commercial manufacture. By 1960 there were nineteen companies worldwide manufacturing cobalt units.

Electron beam

As early as 1934 Braasch and Lange in Germany wrote about the outlook and feasibility of therapy with 2 to 2.5 MeV electrons.[141] In 1940 Trump, Van de Graaff, and Cloud at MIT reported on physical studies with 1.5 MeV electrons issuing from a Van de Graaff accelerator.[142] In 1953 this group, along with Hugh Hare and other physicians from the Lahey Clinic, reported on the use of 2.5 MeV electrons to treat a patient with mycosis fungoides.[143] Meanwhile in Germany, treatment with electrons of up to 6 MeV was reported in 1950.[144]

Fig. 4.18 Harold E. Johns's cobalt unit, installed in the Saskatchewan Cancer Clinic in August, 1951. (Courtesy of the Center for the American History of Radiology, Reston, Va.)

Fig. 4.19 Mock-up of M. D. Anderson cobalt unit displayed at Oak Ridge in May 1951. From left to right, Marshall Brucer, Gilbert Fletcher, R. Lee Clark, and Leonard Grimmett. This photograph was shortly before Grimmett's sudden death. (Courtesy of the Center for the American History of Radiology, Reston, Va.)

Skaggs and co-workers at Urbana devised a magnetic shunt to withdraw an electron beam from the betatron and reported this advance in 1948.[145] In 1951 a 7 to 22 MeV electron beam from a betatron was used for treatment at the University of Illinois College of Medicine in Chicago.[146]

Extension of calibrations to high energy X rays and electrons

It is not feasible for national or regional agencies to calibrate field cavity ionization instruments at all the X-ray and electron energies that users may require. In their codes of practice, the British Hospital Physicists Association in 1964 and 1969 recommended factors with which to make transitions from 2 MV X-ray (or cobalt) calibration, available from the national calibration agencies, to higher X-ray energies.[147,148] North American codes of practice or protocols followed.

Electron beams posed a similar problem. Independently, in 1967 Peter Almond at M. D. Anderson Hospital and H. Svensson and C. Petterson in Sweden recommended factors with which to make a transition from the chamber calibration energy (^{60}Co) to various electron energies.[149,150] Before these recommendations (which were later codified into protocols and codes of practice), the output of electron beams was measured using cavity ionization chambers with ^{60}Co calibration factors. It was understood that output was not the same as absorbed dose.

The theoretical basis for the high-energy X-ray and electron factors depends on the use of the Bragg-Gray theory for cavity ionization chambers at the calibration energy and again at the high-energy X-ray or electron energy. At the calibration energy the mass of air in the chamber is derived implicitly as part of the chamber calibration factor; then, using that fact, the absorbed dose at higher X-ray or electron energy is measured and calculated. A great merit of this system is the tie of high-energy electron and X-ray measurements to the national ^{60}Co standards of Canada, Mexico, and the United States. These national ^{60}Co standards are compared to the standards of other countries ensuring worldwide consistency.

Calorimetry

Measurement of radiation energy absorbed by temperature increase in an absorber was an important method during 1950 and 1970, particularly for verifying measurement of high energy X-ray and electron beams by ion chambers. The energy deposited in an absorber is measured in a direct way with little interpretation of the absorption processes required.[151] Calorimetry has a long history. Perhaps the first effort at calorimetry was in 1900 by Rutherford.[152] In 1903 Pierre Curie and A. Laborde measured the temperature rise caused by a source of radium in a calorimeter.[153] This heating of an absorber around a radioactive source can be observed by touch in high-activity cobalt irradiators. Other successful calorimetric measurements were made in the 1920s.

Energy in a beam (total absorption calorimetry) or energy absorbed from a transmitted beam (local absorption calorimetry) may be measured.[154] The latter is of great interest since the figures may be compared directly with ion chamber or chemical measurements.[155] Johns and others at Saskatoon reported on an early local dose calorimeter that used water as the absorber.[156] This possibility is particularly interesting because water is usually the substance in which dose is stated; despite difficulties such as radiochemical reactions with water as an absorber, the development has continued.[157]

Thermoluminescent dosimeters

In a 1953 article in *Science* Farrington Daniels, a physical chemist at the University of Wisconsin, and his co-workers pointed out that many natural rocks, including limestones, granites, and other minerals are thermoluminescent, emitting an easily seen bright white or orange light if they are heated strongly.[158] However, upon reheating, these rocks and minerals do not again emit light. This phenomena was known and reported by geologists as early as 1913.[159] By the early 1930s the association of radioactivity with thermoluminescent minerals was noted. Once heated, thermoluminescent samples will regain thermoluminescence if they are irradiated by X rays or gamma rays.

John Cameron, also at the University of Wisconsin, recognized the potential usefulness of thermoluminescent dosimeters (TLD) in radiation therapy. Small amounts (around 25 mg) of powder or small millimeter crystals could be subjected to radiation and analyzed later with controlled heating and observation with a phototube. Cameron investigated the response characteristics of TLD to radiation and developed the methodology to a practical state.[160] TLD has proved valuable for monitoring radiation measurements and for making measurements where the small size or flat shape of the dosimeter is important.

Computer calculations for X-ray beams

In 1955 K. C. Tsien at Memorial Hospital, New York, was the first to use computers to calculate treatment plans for external beam treatment.[161] In the years that followed until 1970, research workers in the United Kingdom, Germany, France, Holland, Canada, the United States, and elsewhere diligently tried new computer methods. Communication between groups was vital. Researchers, such as John Cunningham of the Ontario Cancer Institute, organized international meetings to report on progress and exchange ideas about computer use in radiotherapy. In the joint chairman's address at the third International Conference on Computers in Radiotherapy, held in Glasgow in September 1970, Cunningham stated, "although this is the 'third' international conference on the subject, it is at least the fifth meeting of this general type. The first was a panel meeting held at the I.A.E.A. in Vienna in October, 1965."[162]

Initially, computer use to calculate two-dimensional compounding of iso-dose curves for multiport treatment was of moderate interest because manual methods produced a similar result in a reasonable time for a single treatment plan. Two-dimensional calculations on dedicated computers were first performed around 1970. The ease of calculation on readily available computers resulted in a willingness by radiation oncologists to order several or many treatment plans for the same case, thus allowing greater optimization than was customary with manual calculations. By 1970, however, it was not clear whether computer calculation had yet had much impact on the practice of external-beam radiotherapy.

Radiation Protection

In 1950 the ICRU substituted maximum permissible dose (MPD) for tolerance dose in the statement of radiation protection recommendations. This small change in words acknowledged risk even within recommended boundaries. The NBS *Handbook 59*, published in 1954 for the NCRP, stated that a "permissible dose may then be defined as the dose of ionizing radiation that, in the light of present knowledge, is not expected to cause appreciable bodily injury to a person at any time during his lifetime."[163] The drafting committee recognized that ionizing radiation, in amounts less than the threshold dose, received by gonads of parents could cause genetic mutation. However, they did not believe that there was sufficient knowledge at that time to address the genetic consequences for individuals; instead, they stated, "under present conditions and for some time to come, genetic damage to the population as a whole in future generations is not a limiting factor in setting up a permissible level for occupational exposure to ionizing radiation." An MPD of 300 mR/week or 15 R/year of X rays was set for whole body or gonad exposure for occupational workers. Rather soon, in an addendum to NBS handbook 59 dated 15 April 1958, this MPD was lowered for workers consistently receiving high occupational radiation doses. Comments in that document included:

> The risk to the individual is not precisely determinable but, however small, it is believed not to be zero. Even if the injury should prove to be proportional to the amount of radiation the individual receives, to the best of our present knowledge, the new permissible levels are thought not to constitute an unacceptable risk. Since the new rules are designed to limit the potential hazards to the individual and to the reproductive cells, it is therefore necessary to control the radiation dose to the population as a whole, as well as to the individual. For this reason, maximum permissible doses are set for the small percentage of the whole population who may be occupationally exposed, in order that they not be involved in risks greater than are normally accepted in industry.

In the addendum, the recommended limits for external exposure to whole body, head and trunk, active blood-forming organs, or gonads were limited to 5 (N-18) rem/year, which is 5 rem/year for a worker consistently receiving the maximum recommended. N is the worker's age. For the first time, a recommended MPD for radiation exposure to the public was made in this addendum.

The United States government invested substantially in radiation protection and related activities between 1950 and 1970. The AEC operated sizable national laboratories with missions relating to the uses of nuclear energy for war and peace. Radiation biology was high on the agenda for both the work at the laboratories and work contracted out to other research groups. The United States Public Health Service funded basic and clinical radiation investigations and operated the Bureau of Radiological Health, which was highly active in collating survey materials and presenting radiation data in a usable form.

MEGAVOLTAGE ERA: 1970 TO PRESENT

Is it better to be correct or to be consistent? Correctness builds a scientific base; consistency provides a base for clinical evaluation of treatment results. In response to this question the medical physics community in 1970 seems to have answered that it is best to be both correct and consistent (increasingly between institutions) when dealing with radiation matters. Tireless committees have generated many protocols and codes of practice ensuring the quality of procedures in or related to therapeutic radiology. A notable example is the work of Task Group 21 of the American Association of Physicists in Medicine (AAPM), whose members, under the chairmanship of Robert Schulz of Yale University, reviewed in detail the process of utilizing cavity ionization chambers calibrated at ^{60}Co energy for the measurement of high-energy X-ray and electron beams.[164] The protocol accommodated a number of small corrections to the straight-forward application of the Bragg-Gray theory, as well as differences expected from a variety of commercially available chambers. AAPM reports of relevance to radiation therapy are listed in the reference to this chapter.[165] Through the stimulus of interinstitutional clinical trials, the AAPM provides direction and oversight to the Radiological Physics Center at the M.D. Anderson Hospital, which does detailed on-site reviews of physics practices at institutions participating in the trials.[166]

The major research thrust during this period is related. It has been directed at understanding radiation therpy dose distributions more completely, more precisely, and with greater accuracy. Important in this effort is the diagnostic precision of localizing treatment volumes afforded by simulators, computed tomography (CT), and magnetic resonance (MR) imaging. The intellectual and technical challenges are substantial.

Status in 1970

High-energy X-ray and electron beams up to energies greater than required, as well as ^{60}Co units, were commercially available by 1970. Cesium-137, ^{192}Ir, gold-198, and ^{125}I sources for brachytherapy were also available. Three-dimensional calculations were

Table 4.III Major Advances from 1970 to 1993

Advance	Date	By whom	Where
Computed tomography (CT)	1972	Hounsfield	UK[167]
		Cormack	USA[168]
Magnetic resonance imaging (MRI)	1973	Lauterbur	USA[169]
Use of CT in treatment planning	1977	Goitein, Munzenrider	USA[170,171]

available for brachytherapy sources but not for external beam treatments. Dedicated and interactive computers were generally not available, nor were CT and MR imaging. High dose-rate brachytherapy, though demonstrated, had limited clinical use. Table 4.III lists major advances occurring between 1970 and 1993.

Brachytherapy

High dose-rate brachytherapy

Manual afterloading in brachytherapy has had a long history starting in 1910 in the hands of Robert Abbé and continuing in the 1950s and 1960s.[172] These later efforts reduced to some extent the irradiation of health care workers and thus reduced a disadvantage of brachytherapy. The advent of machine-controlled source loading opened the possibility of using sources of higher activity with reduced treatment times.[173,174] Treatment could then be on an outpatient basis, with treatment times approaching those for external beam therapy. However, the change in radiation dose rate from conventional brachytherapy to high dose-rate changed the biological response to the radiation. In a comprehensive discussion, Hall and Brenner considered the effect of the higher dose-rate.[175] They recommended that the number of fractions of high dose-rate not be reduced to less than five.

There are at least eight manufacturers of brachytherapy remote afterloading devices in North America and Europe.[176] Some of the manufacturers, particularly those in Europe, have models that deliver low, medium, and high dose rates. Low dose-rate units deliver radiation dose rates comparable to those of conventional interstitial or intracavitary implants, whereas the high dose-rate units deliver dose rates which may be thirty times higher than conventional implants.

In addition to convenience for the patient and the physician in most treatments, the absence of radiation while locating the source guides and the relatively short treatment times allow clinicians to apply brachytherapy in new sites (or old sites again) such as the uterus, vagina, rectum, esophagus, bronchus, trachea, breast, chest, and head and neck.[177]

The measurement, safety, and quality assurance tests for physicists are more demanding with the remote afterloading apparatus than for conventional brachytherapy. Special rooms are required to accommodate 10 to 12 curies of ^{192}Ir; tests for the positioning accuracy of sources and the measurement of the source strength require ingenuity.[178] The chance for loss or mix up of sources is reduced with remote afterloading devices, but while the possibility of mislocated sources or sources coming loose is low, either is a serious event. Orton warns that with the new devices "time" must be explicitly considered and translated into fractionation schedules and that source-to-healthy-tissue distance needs to be respected as before.[179]

External Beams

Optimum energy for X-ray beams

The drive for higher energy X-ray beams has given rise to many discussions of the optimum energy of these beams for radiation therapy. An excellent analysis was given by Laughlin, Mohan, and Kutcher with editorial comments by Suit.[180,181] Laughlin and colleagues considered the depth dose, electron build-up, and related effects at interfaces with the patient, beam penumbra, bone dose, and the production of photo-neutrons for beams of various energies, as well as the advantages conferred by different beams at different treatment sites. They concluded that for facilities with a single unit, a 6 MeV X-ray apparatus is close to the optimal machine; for two-unit facilities they recommended one X-ray unit at 4 to 6 MeV and the other at 10 to 18 MeV. Suit commented that in his view ^{60}Co beams still have considerable utility, particularly if sources of small diameter and secondary collimation are used.[182]

Computer calculations for external-beam treatments

The availability of dedicated computers during the 1970s for two-dimensional calculations permitted a major advance in the possibility of calculating multiple plans for the same patient. John Cunningham at the Princess Margaret Hospital in Toronto contributed considerably to computer applications since their inception and has stated, "Another step, that is part of treatment planning, includes choosing the best distribution. This part of the process is frequently referred to as 'optimization.' It is usually done visually by trial and selection. There has been only very limited success in applying computers to these procedures."[183]

Even with manual calculation of dose distributions, there is the possibility of displaying some three-dimensional information, particularly on a line at the intersection of central rays perpendicular to the plane defined by the central rays. However, this calculation was seldom done because the field sizes used were usually larger than the probable tumor boundary. Around 1970 it was sometimes said that knowing the radiation dose pattern within a centimeter or so was adequate, because the uncertainty in the tumor position was that large or greater. Extension of tumors not visualized radiographically was and is another important concern. The availability of commercial CT in 1972 altered aspects of tumor visualization.[184] Multisection CT could place the visualized tumor within the body contour with more certainty than could perpendicular radiographs.[185]

During the 1980s much effort was invested in three-dimensional planning for external-beam treatments. Incorporation of CT diagnostic data plus the possibility of conformal treatment with or without multileaf collimators allowed the target volume to be established more precisely and enabled the X-ray treatment field to be tailored to the target volume. In the last several years this enormously complex process appears to be yielding to systematic use in the clinic.[186] At Memorial Hospital in New York, prostate treatment volumes are closely defined by these techniques, and the total dose is increased in steps in a planned program. The diagnostic process is repeated during treatment to allow for anatomic change of the treated volume during treatment. This program is meant to test whether the treatment volume can be reduced safely, and if so, whether the tumor dose can be increased safely. Herman Suit and Michael Goitein, with the group at the Massachusetts General Hospital, are pursuing various components of the problem of precise dose delivery, including diagnostics and calculation, patient monitoring during treatment, proton treatment where appropriate, evaluation of uncertainties, and clinical logic.[187] A number of centers have worked on three-dimensional dose computation.[188,189,190] A volume of the *International Journal of Radiation Oncology, Biology, and Physics* in 1991 was dedicated to reports of the Collaborative Working Group on the Evaluation of Treatment Planning for External Photon Beam Radiotherapy funded by the National Cancer Institute.[191] This volume contained reports on methods, heterogeneities, tolerance of normal tissue, and considerations for various sites.

An international symposium, entitled "3-D Radiation Treatment Planning and Conformal Therapy," was held at Washington University School of Medicine in April 1993. This symposium revealed substantial improvements in tumor visualization using CT and simulators with automatic input to three-dimensional treatment planning computers. The magnitude and technical difficulty of this effort to limit radiation to the intended treatment volume is considerable. Clinical studies are underway to evaluate conformal radiation therapy to the prostate, lung, liver, and head and neck. Most of the studies include increasing the total dose stepwise as experience accumulates. Clinical experience with conformal planning should allow conclusions about required treatment margins beyond indicated tumor boundaries. Predictably, surgeons will continue to remove accessible tumors with well-defined boundaries and to believe that radiation therapy should take care of unrevealed or regional spread of disease. A long period of evaluation between surgery and radiation therapy by conformal radiation treatment is probable.

For perspective on this issue, portions of two editorials are quoted. Lichter has stated that:

> ...additional developments in machine technology, localization techniques, treatment planning, and dose calculation will likely produce new developments that will improve upon today's technology in the same way that linear accelerators improved upon ^{60}Co technology. Such has been the history of our field and other fields of technical endeavor over the past fifty years, and this history will likely continue.[192]

Words of caution, however, have come from Smith, who has indicated that gains in patient radiation treatment from imaging and calculation technology are likely to be obtained in increments that require heavily time-consuming and expensive efforts. Speaking for the journal editors he states,

> As a general guideline, we would like studies to include: (a) a clear statement of the clinical problem and aim of the study; (b) a study design; (c) if new technology is being evaluated and compared to existing technology, a complete description of each and how they were applied; (d) data tables or graphs showing the types of differences found, their magnitude, and frequency of occurrence; (e) conclusions based on data, on both the benefits (if any) and disadvantages (if any) of the new technology....It is generally recognized that most new software and hardware technology may be expensive to implement in terms of capital investment and manpower. In times of increasing pressure to limit or even decrease the costs of health care, there should be a demonstrable gain in quality of patient care, efficiency, or safety to justify the expense.[193]

Stereotactic radiation treatment

Stereotactic methods for radiosurgery of the brain to treat a variety of conditions, including arteriovenous malformations and benign and malignant tumors, have been pursued diligently by Leksell in Sweden since the late 1940s, culminating in the commercial Leksell gamma unit or "gamma knife" which uses 201 ^{60}Co sources in a helmet-like apparatus.[194] A unit was installed in the United States in 1987 after overcoming trying regulatory and import problems.[195] In North America interest in these and related techniques has increased sharply during the 1980s (Fig. 4.20).

Stereotactic methods have required careful physics and engineering. Houdek et al. stated the problem: "In particular, it was recognized that while brain lesions as small as 2-3 mm could be localized in nearly every medical imaging department, they could not be optimally treated with correspondingly small radiation fields because, in the majority of radiotherapy departments, the errors associated with data transfer and patient setups were commonly greater than the lesion itself."[196] To localize radiation placement to within 1 mm requires techniques and apparatus not used for conventional radiation therapy.[197] Means of applying radiation, in addition to the Leksell gamma knife, have included linear accelerator photon

Fig. 4.20 Leksell gamma knife, a 201-source cobalt unit for radiosurgery. (Courtesy of Elekta Radiosurgery, Inc., of Atlanta, Georgia)

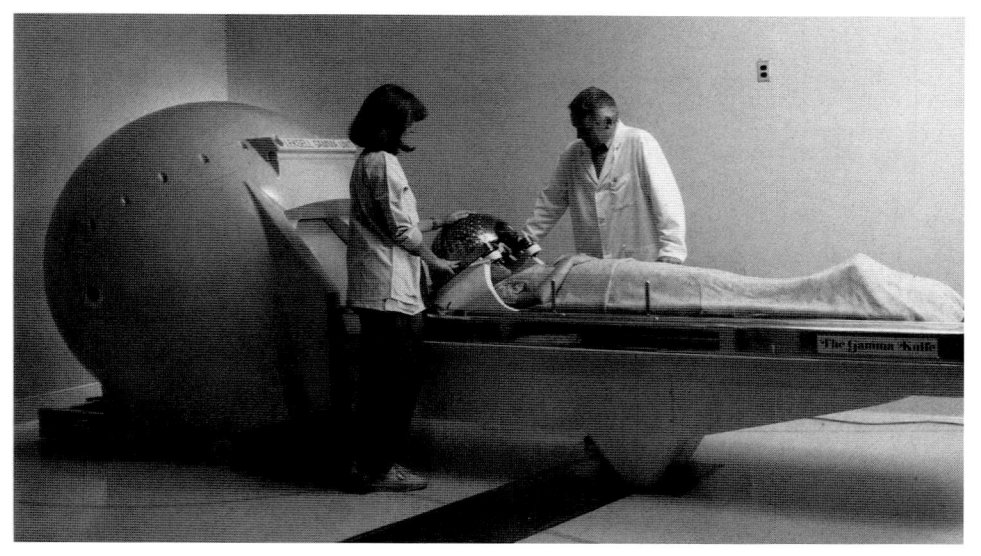

beams (such beams modified by additional collimation to elliptical shape or by motor driven circular and rectangular collimators to shape the instantaneous field to the target shape continuously for each change in arc angle) and positive ion beams with sharp side and end delineation.[198]

Radiation Protection

The increase in the body of knowledge related to radiation biology was spurred on by the inventions of the nuclear reactor and the atomic bomb. Information from exposed human populations has been focused on radiation protection recommendations. Groups such as the National Academy of Sciences have published comprehensive reports of a review nature since 1956. Beginning in 1972 a series of very influential reports by the Advisory Committee on the Biological Effects of Ionizing Radiations were published by the National Academy of Sciences and the National Research Council. Also in that year, a series of reports on similar subjects was prepared by the United Nations Scientific Committee on the Effects of Atomic Radiation. Reviewing this information is beyond the scope of this chapter. Fortunately, there are highly readable reviews with extensive references by W. K. Sinclair that relate to radiation protection recommendations.[199,200]

The most recent recommendations from the NCRP are based on the risk of stochastic effects (i.e., genetic changes and cancer induction). An interesting few sentences from that publication follow:

> By the early 1950s, the emphasis had shifted to chronic or late effects. The maximum permissible dose then employed was designed to ensure that the probability of the occurrence of injuries was so low that the risk would be readily acceptable to the average individual. In that decade, based on the results of genetic studies in drosophila and mice, the occupational limit was substantially reduced and a public limit introduced. Subsequently, the genetic risks were found to be smaller and cancer risks larger than were thought of at the time.[201]

In the 1993 recommendations, in accord with the understanding that cancer induction and genetic mutation are all-or-none phenomena, a linear extrapolation to low radiation doses was taken; a linear extrapolation results in lower recommended limits to radiation exposure than other extrapolations. For radiation workers, the current (since 1987) dose limit in one year is 50 mSv (5 rem) compared to 15 rem in 1958; the cumulative effective dose limit is 10 mSv/yr (1 rem) x age (years) compared with 5 (age - 18) rem in 1958. Thus, present limits are one-third to about one-half what they were in 1958. The recommendations for occupational radiation exposure are translated to expected worker fatalities and compared to the fatal accident rate in other industries. Assuming that an average radiation worker receives one-fourth to one-sixth of the maximum permissible dose over a lifetime (about 0.15 Sv or 15 cGy of X-ray, gamma-ray, and electron radiation), the risk to that worker is about the same as to workers in safe industries (trade, manufacture, service, government). If a radiation worker received the maximum permissible dose over a lifetime (0.7 Sv or 70 cGy of X-ray, gamma-ray, and electron radiation), that worker's risk of death is about the same as that for workers in the more hazardous jobs within safe industries (construction, mines, agriculture).[202]

PHYSICISTS IN RADIOTHERAPY : FUTURE ACTIVITIES

Physicists continue to define and improve the basis for safely increasing the radiation dose to a target volume while reducing the radiation dose elsewhere. Four major areas in that effort are: (a) three-dimensional external-beam dose calculation combined with the intent to cause the radiation pattern to conform to the tumor volume, (b) stereotactic localization and treatment of small volumes in the brain with small multiple fields of X rays or positive ions, (c) remote afterloading high-dose rate brachytherapy, and (d) positive ion beams. The application of increasingly sophisticated diagnostic methods, computer dose calculation algorithms and hardware, physical knowledge of radiation interactions, and engineering precision cause these programs to be heavily dependent upon the physicist and his mathematical and engineering colleagues. Suite and Verhey believe that precision radiation therapy requires the treatment technique that achieves the closest feasible approximation of the treatment volume to the target volume.[203] Acknowledging that imaging methods such as CT and MR imaging do not necessarily display the spread of subclinical disease from solid tumor, they propose beginning with a treatment volume large enough to enclose the spread of subclinical disease. During treatment, with additional diagnostic data, the treatment volume is decreased—probably more than once. This plan is an elaboration of widely practiced boost techniques, in which treatment is concluded with a smaller treatment field. The Suit and Verhey logic reduces the possibility of a geographic miss in treatment by using explicit margin allowances, permits an increase in tumor dose because the volume is limited, and spares normal nearby structures with shrinking fields. To implement precision therapy, there must be an effective technique and discipline in defining targets and in setting and maintaining patient treatment position. Repeated diagnostics during treatment and on-line monitoring of treatment portals are indicated. Suit and Verhey state that "present experience shows that the target can be defined and enclosed by a treatment volume within a few millimeters. For some sites, such as the head, this can be 1-2 mm with procedures which are well tolerated by the patient for highly fractionated treatment."[204]

The drive toward precision radiotherapy (or, as stated at the beginning, to provide the basis for safely increasing the radiation dose to a target volume while reducing the dose elsewhere) continues to be the major task of radiation therapy physicists. In 1996 the process seems complicated and expensive, requiring rather large amounts of professional time per case. However, with expanded, automated linking of steps as is done between diagnostic and computational procedures, the time required and the difficulty of each case should decrease. The example of the ease of operating very complex military technology to diagnose and respond to a situation within a short time should give encouragement to the practicality of the precision radiation therapy enterprise.

There must be some level of precision beyond which no gain can be expected. This boundary will probably be established empirically over an extended period of time. Studies within an institution in which step-wise escalated dose prescriptions are used may be the most effective method for testing precision radiation therapy at various sites.

The complexity of dose calculation, the reliance upon diagnostic and calculation methods by computers using proprietary programs, and the difficulty of verifying the correctness of dose delivered in a simple way in treatments in which beam movement and collimator movement are computer controlled are features and challenges of contemporary medical physics practice. The technical complication of these procedures can only increase. In addition, the regulatory and litigious atmosphere within which radiotherapy operates is likely to continue despite political discussions of the problem.

As the push towards increased precision in radiotherapy continues, there will be a need for continuing refinements in methodology. Innovative breakthroughs usually are not foreseen; with that caveat, this writer sees present types of radiation sources being adapted to new treatment sites and new fractionation schedules which utilize existing physics and an increasingly complex technology for control of radiation beams. If this is correct, it suggests that medical physicists in radiation therapy will need to exercise a growing range of management skills to protect the integrity of complex systems in which machines and humans interact, and will continue to negotiate the pitfalls of regulatory and legal constraints.

▶ ▶ ▶ ▶ ▶ ▶

REFERENCES

Complete refrence citations are given once; additional references to the same citation are abbreviated.

1. Brecher, R., and Brecher, E. *The Rays*. Baltimore: Williams and Wilkins, 1969.
2. Grigg, E.R.N. *The Trail of the Invisible Light*. Springfield, Ill.: Charles C. Thomas, 1965.
3. Schulz, M.D., "The Supervoltage Story," *Am. J. Roent.* 124 (1975):541-559.
4. Laughlin, J.S., "Development of the Technology of Radiation Therapy," *Radiographics* 9 (1989):1245-1256.
5. del Regato, J.A. *Radiological Physicists*. New York: American Institute of Physics, 1985.
6. The form of this table and selected references are taken from Evans, R.D. *The Atomic Nucleus* (New York: McGraw-Hill, 1955):4-5.
7. Failla, G., and Quimby, E.H., "Radium Physics," in Glasser, O., ed. *Science of Radiology* (Springfield, Ill.: Charles C. Thomas, 1933):242-256.
8. Rutherford, E., "The Scattering of Alpha and Beta Particles by Matter and the Structure of the Atom," *Phil. Mag.* 21 (1911):669-688.
9. Bohr, N., "Part : On the Constitution of Atoms and Molecules," *Phil. Mag.* 26 (1913):1-25.
10. Bohr, N., "On the Constitution of Atoms and Molecules. Part II: Systems Containing Only a Single Nucleus," *Phil. Mag.* 26 (1913):476-502.
11. Bohr, N., "On the Constitution of Atoms and Molecules. Part III: Systems Containing Several Nuclei," *Phil. Mag.* 26 (1913):857-875.
12. Coolidge, W.D., "A Powerful Roentgen Ray Tube with a Pure Electron Discharge," *Phys. Rev.* 2 (1913):409-430.
13. Duane, W., "X Rays and Gamma Rays," Address before the Cleveland Meeting of the American Roentgen Ray Society, 1914. Discussed in Hudson, J.R., "Roentgen-Ray Dosimetry," in Glasser, *Science of Radiology*, pp. 120-138.
14. Duane, W., and Lorenz, E., "The Standard Ionization Chamber for Roentgen-ray Dosage Measurements," *Am. J. Roent.* 19(1928):461-469.
15. Duane, W., and Hunt, F.L., "On X-ray Wave Lengths," *Phys. Rev.* 6 (1915):166-171.
16. Duane, W., "Methods of Preparing and Using Radioactive Substances in the Treatment of Malignant Disease and of Estimating Suitable Dosages," *Boston Med. Surg. J.* 177 (1917):787-799.
17. Burnam, C.F., and Ward, G.E., "Recent Developments in Protective Methods and Appliances as Used in Radium Therapy," *Am. J. Roent.* 10 (1923): 625-632.
18. Lysholm, E., "Apparatus for the Production of a Narrow Beam of Rays in Treatment by Radium at a Distance," *Acta Radiol.* 2 (1923):516-519.
19. Compton, A.H., "A Quantum Theory of the Scattering of X Rays by Light Elements," *Phys. Rev.* 21 (1923):483-502.
20. Failla, G., "The Development of Filtered Radon Implants," *Am. J. Roent.* 16 (1926):507-525.
21. International X-Ray Unit Committee, "Second International Congress of Radiology at Stockholm. International X-ray Unit of Intensity," *Brit. J. Rad.* 1 (1928):363-364.
22. ICRU, "Recommendations of the International Committee for Radiological Units (Chicago, 1937)," *Am. J. Roent.* 39 (1938):295-298.
23. Glasser, O., and Seitz, V.B., "Method and Apparatus for the Measurement of Radiation Intensity," U.S. Patent 1,855,669, 26 April 1932.
24. Brecher and Brecher, *The Rays*.
25. Chadwick, J., "The Existence of a Neutron," *Proc. Roy. Soc.* A136 (1932):692-708.

26 Quimby, E.H., "Dosage Calculations with Radioactive Materials," in Glasser, O.; Quimby, E.H.; Taylor, L.S.; et al., eds. *Physical Foundations of Radiology*. 3rd ed. (New York: Harper and Row, 1961):336-381.
27 Meredith, W.J., ed. *Radium Dosage: The Manchester System*. Edinburgh and London: Livingstone, 1947.
28 Fermi, E., "Radioactivity Induced by Neutron Bombardment (Letter)," *Nature* 133 (1934):757.
29 Gray, L.H., "An Ionization Method for the Absolute Measurement of γ Ray Energy," *Proc. Roy. Soc.* A156 (1936):578-596.
30 Charlton, E.E.; Westendorp, W.F.; Dempster, L.E.; and Hotaling, G., "A Million-Volt X-ray Unit," *Radiology* 35 (1940):585-597.
31 Stone, R.S., "Neutron Therapy and Specific Ionization. Janeway Memorial Lecture," *Am. J. Roent.* 59 (1948):771-785.
32 Hahn, V.O., and Strassmann, F., "Über den nachweis und das verhalten bei der Bestrahlung des urans mittels neutronen entstehenden erdalkalimetalle," *Naturwiss.* 27 (1939):11-15.
33 Kerst, D.W., "The Acceleration of Electrons by Magnetic Induction," *Phys. Rev.* 60 (1941):47-53.
34 del Regato, *Radiological Physicists*.
35 Burks, A.R, and Burks, A.W. *The First Electronic Computer: The Atanasoff Story* (Ann Arbor: The University of Michigan Press, 1988):1.
36 Fry, D.W.; R-S-Harvie, R.B.; Mullett, L.B.; and Walkinshaw, W., "Traveling-Wave Linear Accelerator for Electrons," *Nature* 160 (1947):351-353.
37 Ginzton, E.L.; Hansen, W.W.; and Kennedy, W.R., "A Linear Electron Accelerator," *Rev. Sci. Instrum.* 19 (1948):89-108.
38 Duane and Hunt, "On X-ray Wave Lengths."
39 Florance, D.C.H., "Primary and Secondary γ Rays," *Phil. Mag.* 20 (1910):921-938.
40 Gray, J.A., "The Scattering and Absorption of the γ Rays of Radium," *Phil. Mag.* 26 (1913):611-623.
41 Compton, A.H., "The Scattering of X Rays as Particles," *Am. J. Phys.* 29 (1961):817-820.
42 Ibid.
43 Fricke, H., and Glasser, O., "The Secondary Electrons Produced by Hard X Rays in Light Elements," *Proc. Natl. Acad. Sci.* (USA) 10 (1924):441-447.
44 Evans, *The Atomic Nucleus*.
45 Hudson, J.R., "Roentgen-ray dosimetry," in Glasser, *Science of Radiology*, pp. 120-138.
46 Ibid.
47 Burnam, C.F., "Early Experiences with Radium," *Am. J. Roent.* 36 (1936):437-452.
48 Mayneord, W.V., and Roberts, J.E., "The Ionization Produced in Air By X Rays and Gamma Rays," *Brit. J. Rad.* 7 (1934):158-175.
49 Attix, F.H., and Ritz, V.H., "A Determination of the Gamma-Ray Emission of Radium," *J. Res. Natl. Bur. Std.* 59 (1957):293-305.
50 Davis, K.S., "The History of Radium," *Radiology* 2 (1924):287-342.
51 Press notice from U. S. Geological Survey, 17 Nov. 1922, "Large Deposits of Radium Ore Discovered in Africa," *Radium* 1 (1923):308-309.
52 Davis, "The History of Radium."
53 Burnham, "Early Experiences with Radium."
54 Duane, "Methods of Preparing."
55 Failla, "The Development of Filtered Radon Plants."
56 Ibid.
57 Meredith, *Radium Dosage*.
58 Quimby, "Dosage Calculations."
59 Quimby, E.H., "The Background of Radium Therapy in the United States, 1906-1956," *Am. J. Roent.* 75 (1956):443-450.
60 Schulz, "The Supervoltage Story."
61 Burnham and Ward, "Recent Developments."
62 Lysholm, "Apparatus for the Production."
63 Wilson, C.W., "Thirty Years of Telecurietherapy," *Brit. J. Rad.* 33 (1960):69-81.
64 Grimmett, L.G., "A Five-Gramme Radium Unit, with Pneumatic Transference of Radium," *Brit. J. Rad.* 10 (1937):105-122.
65 Quick, D., and Richmond, J.D., "50 Gram Convergent Beam Radium Unit," *Am. J. Roent.* 74 (1955):635-650.
66 Coolidge, W.D., "The Development of Modern Roentgen-ray Generating Apparatus," *Am. J. Roent.* 24 (1930):605-620.
67 Ibid.
68 Coolidge, "A Powerful Roentgen-ray Tube."
69 Grigg, *The Trail of the Invisible Light*.
70 Lauritsen, C.C., "The Development of High Voltage X-ray Tubes at the California Institute of Technology," *Radiology* 31 (1938):354-361.
71 Leucutia, T., and Corrigan, K.E., "The Present Status of Roentgen Therapy with Voltages Above 200 kV," *Am. J. Roent.* 31 (1934):628-662.
72 Coolidge, W.D.; Dempster, L.E.; and Tanis, H.E., Jr., "A High Voltage Induction Coil and Cascade Tube Roentgen-ray Outfit," *Am. J. Roent.* 27 (1932):405-414.

73. Stone, R.S.; Livingston, M.S.; Sloan, D.H.; and Chaffee, M.A., "A Radio Frequency High Voltage Apparatus for X-ray Therapy," *Radiology* 24 (1935):153-159.
74. Charlton, Westendorp, Dempster, and Hotaling, "A Million Volt X-ray Unit."
75. Schulz, "The Supervoltage Story."
76. Brecher and Brecher, *The Rays*.
77. Trump, J.D.; Van de Graaff, R.J.; and Cloud, R.W., "Compact, Supervoltage, Roentgen-ray Generator Using a Pressure-Insulated Electrostatic High Voltage Source," *Am. J. Roent.* 44 (1940):610-614.
78. Laughlin, "Development of the Technology."
79. Kerst, D.W., "Historical Development of the Betatron," *Nature* 157 (1946):90-95.
80. Kerst, D.W., "Development of the Betatron and Applications of High Energy Betatron Radiations," *Am. Sci.* 35 (1947):57-84.
81. Haas, L.L.; Laughlin, J.S.; and Harvey, R.A., "Biological Effectiveness of High Speed Electron Beam in Man," *Radiology* 62 (1954):845-851.
82. Fry, Harvey, Mullet, and Walkinshaw, "Traveling-Wave Linear Accelerator."
83. Ginzton, Hansen, and Kennedy, "A Linear Electron Accelerator."
84. Miller, C.W., "Traveling-Wave Linear Accelerator for X-ray Therapy," *Nature* 171 (1953):297-298.
85. Gintzon, E.L.; Mallory, K.B.; and Kaplan, H., "The Stanford Medical Linear Accelerator. Part I: Design and Development," *Stanford Med. Bull.* 15 (1957):123-140.
86. Hudson, *Roentgen Ray Dosimetry*.
87. Duane, "X Rays and Gamma Rays."
88. Duane and Lorenz, "The Standard Ionization Chamber."
89. ICR, 1928 *Recommendations*; ICRU, 1937 *Recommendations*.
90. Wyckoff, H.O., "From 'Quantity of Radiation' and 'Dose' to 'Exposure' and 'Absorbed Dose' — A Historical Review," Lauriston S. Taylor Lectures in Radiation Protection and Measurement, Lecture No. 4. Washington: National Council on Radiation Protection and Measurements, 1980.
91. Hudson, *Roentgen-Ray Dosimetry*.
92. Glasser, O., "Radiation-Measuring Instruments," in Glasser, Quimby, Taylor, et al., *Physical Foundations of Radiology* 3rd ed., pp. 187-224.
93. Duane, W., "The Scientific Basis of Short Wave-Length Therapy," *Am. J. Roent.* 9 (1922):781-791.
94. Hudson, *Roentgen-Ray Dosimetry*.
95. Fricke, H., and Glasser, O., "A Theoretical and Experimental Study of the Small Ionization Chamber," *Am. J. Roent.* 13 (1925):453-461.
96. Glasser, O.; Portmann, U.V.; and Seitz, V.B., "The Condenser Dosimeter and Its Use in Measuring Radiation Over a Wide Range of Wave Lengths," *Am. J. Roent.* 20 (1928):505-513.
97. Glasser, "Radiation Measurement Instruments."
98. Gray, "An Ionization Method."
99. Gray, L.H., "The Absorption of Penetrating Radiation," *Proc. Roy. Soc.* A122 (1929):647-668.
100. Mayneord and Roberts, "The Ionization Produced in Air."
101. Attix and Ritz, "A Determination of the Gamma Ray Emission."
102. Bragg, W.H. *Studies in Radioactivity* (London: MacMillan, 1912):95.
103. Sinclair, W.K., "Radiation Protection: Recent Recommendations of the ICRP and the NCRP and Their Biological Bases," *Adv. Rad. Biol.* 16 (1992):303-324.
104. NCRP. *Report 116: Limitation of Exposure to Ionizing Radiation*. Bethesda: National Council on Radiation Protection and Measurements, 1993.
105. Sinclair, "Radiation Protection."
106. NCRP, *Report 116*.
107. Spiers, F.W., "Forty Years of Development in Radiation Protection," *Phys. Med. Biol.* 29 (1984):145-151.
108. Sinclair, "Radiation Protection."
109. NBS. *Handbook 59. Permissible Dose from External Sources of Ionizing Radiation: Recommendations of the National Committee on Radiation Protection*. Washington: National Bureau of Standards, 1954.
110. Sinclair, W.K., "Risk, Research, and Radiation Protection," *Radiat. Res.* 112 (1987):191-216.
111. Brodsky, A., and Kathren, R.L., "Historical Development of Radiation Safety Practices in Radiology," *RadioGraphics* 9 (1989):1267-1275.
112. Laughlin, J.S.; Ovadia, J.; Beattie, J.W.; et al., "Some Physical Aspects of Electron Beam Therapy," *Radiology* 60 (1953):165-185.
113. Johns, H.E.; Bates, L.M.; and Watson, T.A., "1000 Curie Cobalt Units for Radiation Therapy I. The Sakatchewan Cobalt-60 Unit," *Brit. J. Rad.* 25 (1952):296-302.
114. Green, D.T., and Errington, R.F., "1000 Curie Cobalt Units for Radiation Therapy III. Design of a Cobalt Beam Therapy Unit," *Brit. J. Rad.* 25 (1952):309-313.
115. Grimmett, L.G., "A 1000-Curie Cobalt-60 Irradiator," *Texas Reports on Biology and Medicine* 8 (1950):480-490.
116. Laughlin, J.S., and Beattie, J.W., "Calorimetric Determination of the Energy Flux of 22.5 MeV X Rays," *Rev. Sci. Instrum.* 22 (1951):572-574.
117. Tobias, C.A.; Anger, H.O.; and Lawrence, J.H., "Radiological Use of High Energy Deuterons and Alpha Particles," *Am. J. Roent.* 67 (1952):1-27.
118. Tsien, K.C., "The Application of Automatic Computing Machines to Radiation Treatment

Planning," *Brit. J. Rad.* 28 (1955): 432-439.
119 Nelson, R.F., and Meurk, M.L., "The Use of Automatic Computing Machines for Implant Dosimetry (Abstract)," *Radiology* 70 (1958):90.
120 Cameron, J.R.; Daniels, F.; Johnson, N.; and Kenney, G., "Radiation Dosimetry Utilizing the Thermoluminescense of Lithium Flouride," *Science* 134 (1961):333-334.
121 Hospital Physicists Association, "A Code of Practice for the Dosimetry of 2 and 8 MV X-ray and Caesium-137 and Cobalt-60 Gamma-ray Beams," *Phys. Med. Biol.* 9 (1964):457-463.
122 Hospital Physicists Association, "A Code of Practice for the Dosimetry of 2 to 35 MV X-ray and Caesium-137 and Cobalt-60 Gamma-ray Beams," *Phys. Med. Biol.* 14 (1969):1-8.
123 Almond, P.R., "The Physical Measurements of Electron Beams from 6 to 18 MeV, Absorbed Dose and Energy Calibration," *Phys. Med. Biol.* 12 (1967):13-24.
124 Svensson, H., and Pettersson, C., "Absorbed Dose Calibration of Thimble Chambers with High Energy Electrons at Different Phantom Depths," *Arkiv för Fysik* 34 (1967):377-384.
125 Henschke, U.K., "Interstitial Implantation in the Treatment of Primary Bronchogenic Carcinoma," *Am. J. Roent.* 79 (1958):981-987.
126 Lawrence, D.C.; Sondhaus, C.A.; Feder, B.; and Scallon, J., "Soft X-ray 'Seeds' for Cancer Therapy (Abstract)," *Radiology* 86 (1966):143.
127 O'Connell, D.; Howard, N.; Joslin, C.A.F.; et al., "A New Remotely Controlled Unit for the Treatment of Uterine Carcinoma," *Lancet* ii (1965):570-571.
128 Henschke, U.K.; Hilaris, B.S.; and Mahan, G.D., "Intracavitary Radiation Therapy of Cancer of the Uterine Cervix by Remote Afterloading with Cycling Sources (Presented April, 1965)," *Am. J. Roent. Rad.* 96 (1966):45-51.
129 Nelson and Meurck, "The Use of Automatic Computing Machines."
130 Fletcher, G.H., and Stovall, M.A., "A Study of the Explicit Distribution of Radiation in Interstitial Implantations. II: Correlation with Clinical Results in Squamous-Cell Carcinomas of the Anterior Two-Thirds of Tongue and Floor of Mouth," *Radiology* 78 (1962):766-782.
131 Shalek, R.J., and Stovall, M., "Dosimetry in Implant Therapy," in Attix, F.H., and Touchlin, E., eds. *Radiation Dosimetry* (New York: Academic Press, 1969):743-807.
132 Pierquin, B., and Dutreix, A., "Towards a New System in Curietherapy (Endocurietherapy and Pleseocurietherapy with Non-Radioactive Preparation)," *Brit. J. Rad.* 40 (1967):184-186.
133 Gillin, M.T.; Kline, R.W.; Wilson, J.F.; and Cox, J.D., "Single and Double Plane Implants: A Comparison of the Manchester System with the Paris System," *Int. J. Radiat. Oncol. Biol. Phys.* 10 (1984):921-932.
134 Anderson, L.L., "Spacing Nomograph for Interstitial Implants of Iodine-125 Seeds," *Med. Phys.* 3 (1976):48-51.
135 Mitchell, J.S., "Application of Recent Advances in Nuclear Physics to Medicine," *Brit. J. Rad.* 19 (1946):481-487.
136 Grimmett, "A 1000-Curie Cobalt-60 Irradiator."
137 Grimmett, L.G., "A Five-Gramme Radium Unit, with Pneumatic Transference of Radium," *Brit. J. Rad.* 10 (1937):105-122.
138 Green and Errington, "1000 Curie Cobalt Units."
139 Johns, Bates, and Watson, "1000-Curie Cobalt Units."
140 Grimmett, "A 1000-Curie Cobalt-60 Irradiator."
141 Brasch, A., and Lange, F., "Aussichten und möglichkeiten einer therapie mit schnellen kathoden-strahlen," *Strahlentherapie* 51 (1934):119-128.
142 Trump, J.G.; Van de Graff, R.J.; and Cloud, S.M., "Cathode Rays for Radiation Therapy," *Am. J. Roent.* 43 (1940):728-734.
143 Trump, J.G.; Wright, K.A.; Evans, W.W.; et al., "High Energy Electrons for the Treatment of Extensive Superficial Malignant Lesions," *Am. J. Roent.* 69 (1953):623-629.
144 Bode, H.G.; Paul, W.; and Schubert, G., "Elektronentherapie menschlicher hautkarzinome mit einem betatron von 6 millionen elektronen-volt," *Strahlentherapie* 81 (1950):251-266.
145 Skaggs, L.S.; Almy, G.M.; Kerst, D.W.; and Lanzl, L.H., "Development of the Betatron for Electron Therapy," *Radiology* 50 (1948):167-173.
146 Haas, L.L.; Harvey, R.A.; Laughlin, J.S.; et al., "Medical Aspects of High Energy Electron Beams," *Am. J. Roent.* 72 (1955):250-259.
147 HPA, "A Code of Practice," 1964.
148 HPA, "A Code of Practice," 1969.
149 Almond, "The Physical Measurements."
150 Svensson and Pettersson, "Absorbed Dose Calibration."
151 Laughlin, J.S., and Genna, S., "Calorimetry," in Attix and Roesch, *Radiation Dosimetry*, pp. 389-441.
152 Webster, D.L., "Roentgen-ray Physics," in Glasser, *Science of Radiology*, pp. 39-63.
153 Curie, P., and Laborde, A., "Sur la chaleur dégagée spontonément por les sels de radium," *Comptes rendus* 136 (1903):673-675.
154 Laughlin and Beattie, "Calorimetric Determination."
155 Bewley, D.K., "The Measurement of Locally Absorbed Dose of Megavoltage X Rays by Means of a Carbon Calorimeter," *Brit. J. Rad.* 36 (1963):865-878.
156 Bernier, J.P.; Skarsgard, L.D.; Cormack, D.V.; and Johns, H.E., "A Calorimetric Determination of the Energy Required to Produce an Ion Pair in Air for Cobalt-60 Gamma Rays," *Radiat. Res.* 5 (1956):613-633.

157 Domen, S.R., "Advances in Calorimetry for Radiation Dosimetry," in Kase, K.R.; Bjarngard, B.E.; and Attix, F.H., eds. *The Dosimetry of Ionizing Radiation*. vol. II (Orlando: Academic Press, 1987):245-320.
158 Daniels, F.; Boyd, C.A.; and Saunders, D.F., "Thermoluminescence as a Research Tool," *Science* 117 (1953):343-349.
159 Ibid.
160 Cameron, J.R.; Suntharaligam, N.; and Kenney, G.N. *Thermoluminescent Dosimetry*. Madison: The University of Wisconsin Press, 1968.
161 Tsien, "The Application of Automatic Computing Machines."
162 Cunningham, J.R., "Computers in Radiation Therapy," in Glicksman, A.S.; Cohen, M.; and Cunningham, J.R., eds. *Computers in Radiotherapy*. London: *Brit. J. Rad.* Special Report Series No. 5 (1971):7.
163 NBS. *Handbook 59. Permissible Dose from External Sources of Ionizing Radiation: Recommendations of the National Committee on Radiation Protection*. Washington: National Bureau of Standards, 1954.
164 AAPM, "A Protocol for the Determination of Absorbed Dose from High-Energy Photon and Electron Beams. Task Group 21. American Association of Physicists in Medicine," *Med. Phys.* 10 (1983):741-771.
165 AAPM. *Report No. 41. Remote Afterloading Technology*. Published for American Association of Physicists in Medicine. New York: American Institute of Physics, 1993; AAPM. *Report No. 7. Protocol for Neutron Beam Dosimetry*. Published for American Association of Physicists in Medicine. New York: American Institute of Physics, 1980; AAPM. *Report No. 11. A Guide to the Teaching of Clinical Radiological Physics to Residents in Radiology-1982*. Published for the American Association of Physicists in Medicine. New York: American Institute of Physics, 1982; AAPM. *Report No. 13. Physical Aspects of Quality Assurance in Radiation Therapy*. Published for the American Association of Physicists in Medicine. New York: American Institute of Physics, 1984; AAPM. *Report No. 16. Protocol for Heavy Charged-Particle Therapy Beam Dosimetry*. Published for the American Association of Physicists in Medicine. New York: American Institute of Physics, 1986; AAPM. *Report No. 17. The Physical Aspects of Total and Half Body Photon Irradiation*. Published for the American Association of Physicists in Medicine. New York: American Institute of Physics, 1986; AAPM. *Report No. 18. A Primer on Low-Level Ionizing Radiation and Its Biological Effects*. Published for the American Association of Physicists in Medicine. New York: American Institute of Physics, 1986; AAPM. *Report No. 19. Neutron Measurements Around High Energy X-ray Radiotherapy Machines*. Published for the American Association of Physicists in Medicine. New York: American Institute of Physics, 1987; AAPM. *Report No. 21. Specification of Brachytherapy Source Strength*. Published for the American Association of Physicists in Medicine. New York: American Institute of Physics, 1987; AAPM. *Report No. 23. Total Skin Electron Therapy: Technique and Dosimetry*. Published for the American Association of Physicists in Medicine. New York: American Institute of Physics, 1988; AAPM. *Report No. 24. Radiotherapy Portal Imaging Quality*. Published for the American Association of Physicists in Medicine. New York: American Institute of Physics, 1988; AAPM. *Report No. 26. Performance Evaluation of Hyperthermia Equipment*. Published for the American Association of Physicists in Medicine. New York: American Institute of Physics, 1989; AAPM. *Report No. 27. Hyperthermia Treatment Planning*. Published for the American Association of Physicists in Medicine. New York: American Institute of Physics, 1989; AAPM. *Report No. 28. Quality Assurance Methods and Phantoms for Magnetic Resonance*. Published for the American Association of Physicists in Medicine. New York: American Institute of Physics, 1990; AAPM. *Report No. 32. Clinical Electron-Beam Dosimetry*. Published for the American Association of Physicists in Medicine. New York: American Institute of Physics, 1991; AAPM. *Report No. 34. Acceptance Testing of Magnetic Resonance Imaging Systems*. Published for the American Association of Physicists in Medicine. New York: American Institute of Physics, 1992; AAPM. *Report No. 36. Essentials and Guidelines for Hospital Based Medical Physics Residency Programs*. Published for the American Association of Physicists in Medicine. New York: American Institute of Physics, 1992; AAPM. *Report No. 38. The Role of the Physicist in Radiation Oncology*. Published for the American Association of Physicists in Medicine. New York: American Institute of Physics, 1993; AAPM. *Report No. 39. Specification and Acceptance Testing of Computed Tomography Scanners*. Published for the American Association of Physicists in Medicine. New York: American Institute of Physics, 1993.
166 Hanson, W.F.; Shalek, R.J.; and Kennedy, P., "Dosimetry Quality Assurance in the U.S. from the Experience of the Radiological Physics Center," Proceedings of an American College of Medical Physics Symposium, May 1991. Starkschall, G., and Horton, J.L., eds. *Quality Assurance in Radiotherapy Physics* (Madison: Medical Physics Publishing, 1991):255-269.
167 Hounsfield, G.N., "Computerized Transverse Axial Scanning (Tomography). Part I: Description of the System," *Brit. J. Rad.* 46 (1973):1016-1022.
168 Cormack, A.M., and Doyle, B.J., "Algorithms for Two-Dimensional Reconstruction," *Phys. Med. Biol.* 22 (1977):994-997.
169 Lauterbur, P.C., "Image Formation by Induced Local Interactions: Examples Employing Nuclear Magnetic Resonance," *Nature* 242 (1973):190-191.
170 Goitein, M., "Computed Tomography in Planning Radiation Therapy," *Int. J. Radiat. Oncol. Biol. Phys.* 5 (1979):445-447.
171 Munzenrider, J.E.; Pilepich, M.; Rene-Ferrero, J.B.; et al., "Use of Body Scanner in Radiotherapy Treatment Planning," *Cancer* 40 (1977):170-179.
172 Henschke, U.K.; Hilaris, B.S.; and Mahan, G.D., "Afterloading in Interstitial and Intracavitary Radiation Therapy," *Am. J. Roent.* 90 (1963):386-395.

173 O'Connell, Howard, Joslin, et al., "A New Remotely Controlled Unit."
174 Henschke, Hilaris, and Mahan, "Intracavitary Radiation Therapy of Cancer."
175 Hall, E.J., and Brenner, D.J., "The Dose-Rate Effect Revisited: Radiobiological Considerations of Importance in Radiotherapy," *Int. J. Radiat. Oncol. Biol. Phys.* 21 (1991):1403-1414.
176 AAPM, *Report No. 41*.
177 Ibid.
178 Ezzel, G.A., "Evaluation of Calibration Techniques for a High Dose Rate Remote Afterloading Iridium-192 Source," *Endocurietherapy/Hyperthermia Oncology* 6 (1990):101-106.
179 Orton, C.G., "HDR: Forget Not 'Time' and 'Distance,.'" *Int. J. Radiat. Oncol. Biol. Phys.* 20 (1992):1131-1132.
180 Laughlin, J.S.; Mohan, R.; and Kutcher, G.J., "Choice of Optimum Megavoltage for Accelerators for Photon Beam Treatment," *Int. J. Radiat. Oncol. Biol. Phys.* 12 (1986):1551-1557.
181 Suit, H.D., "What's the Optimum Choice?" *Int. J. Radiat. Oncol. Biol. Phys.* 12 (1986):1711-1712.
182 Ibid.
183 Cunningham, J.R., "Development of Computer Algorithms for Radiation Treatment Planning," *Int. J. Radiat. Oncol. Biol. Phys.* 16 (1989):1367-1376.
184 Hounsfield, "Computerized Transverse Axial Scanning."
185 Steward, J.R.; Hicks, J.A.; Boone, M.L.M.; and Simpson, L.D., "Computed Tomography in Radiation Therapy," *Int. J. Radiat. Oncol. Biol. Phys.* 4 (1978):313-324.
186 Personal communication, C.J. Kutcher, 1993.
187 For several reports see Goitein, M.; Abrams, M.; Rowell, D.; et al., "Multidimensional Treatment Planning. II: Beams Eye View Back Projection, and Projection through CT Sections," *Int. J. Radiat. Oncol. Biol. Phys.* 9 (1983):789-797; Leong, J.C., and Stracher, M.A., "Visualization of Internal Motion within a Treatment Protocol during a Radiation Therapy Treatment," *Radiother. Oncol.* 9 (1987):153-156; Suit, H.D.; Goitein, M.; Munzenrider, J.E.; et al., "Clinical Experience with Proton Beam Radiation Therapy," *J. Can. Assoc. Radiologists* 31 (1980):35-39; Goitein, M., "Calculation of the Uncertainty in the Dose Delivered during Radiation Therapy," *Med. Phys.* 12 (1985):608-612; and Suit, H., and Du Bois, W., "The Importance of Optimal Treatment Planning in Radiation Therapy," *Int. J. Radiat. Oncol. Biol. Phys.* 21 (1991):1471-1478.
188 Fraass, B.A.; McShan, D.L.; Diaz, R.F.; et al., "Integration of Magnetic Resonance Imaging into Radiation Therapy Treatment Planning: I. Technical Considerations," *Int. J. Radiat. Oncol. Biol. Phys.* 13 (1987):1897-1908.
189 Roseman, J.; Sherouse, G.W.; Fuchs, H.; et al., "Three-Dimensional Display Techniques in Radiation Therapy Treatment Planning," *Int. J. Radiat. Oncol. Biol. Phys.* 16 (1989):263-269.
190 Vijayakumar, S.; Low, N.; Chen, G.T.Y.; et al., "Beams Eye View-Based Photon Radiotherapy I," *Int. J. Radiat. Oncol. Biol. Phys.* 21 (1991):1575-1586.
191 Twenty-one papers on three-dimensional photon treatment planning can be found in *Int. J. Radiat. Oncol. Biol. Phys.* 14 (1988):373-381;
192 Lichter, A.S., "Three-dimensional conformal radiation therapy: a testable hypothesis," *Int. J. Radiat. Oncol. Biol. Phys.* 21(1991):1-265.
193 Smith, A.R., "Evaluation of New Radiation Oncology Technology," *Int. J. Radiat. Oncol. Biol. Phys.* 18 (1990):701-703.
194 Leksell, L., "A Stereotaxic Apparatus for Intercerebral Surgery," *Acta Chir. Scandinav.* 99 (1949):229-233.
195 Lunsford, L.D.; Maitz, A.; and Lindner, G., "First United States 201 Source Cobalt-60 Gamma Unit for Radiosurgery," *Appl. Neurophysical* 50 (1987):253-256.
196 Houdek, P.V.; Schwade, J.G.; Serago, C.F.; et al., "Controlled Stereotaxic Radiotherapy System," *Int. J. Radiat. Oncol. Biol. Phys.* 22 (1992):175-180.
197 Goitein, M.; Abrams, M.; Gentry, R.; et al., "Planning Treatment with Heavy Charged Particles," *Int. J. Radiat. Oncol. Biol. Phys.* 8 (1982):2065-2070; Lutz, W.; Winston, K.R.; and Maleki, N., "A System for Stereotactic Radiosurgery with a Linear Accelerator," *Int. J. Radiat. Oncol. Biol. Phys.* 14 (1988):373-381; Siddon, R.L., and Barth, N.H., "Stereotaxic Localization of Intracranial Targets," *Int. J. Radiat. Oncol. Biol. Phys.* 13 (1987):1241-1246.
198 Serago, C.F.; Lewin, A.A.; Houdek, P.V.; et al., "Improved Linac Dose Distributions for Radiosurgery with Elliptically Shaped Fields," *Int. J. Radiat. Oncol. Biol. Phys.* 21 (1991):1321-3125; Leavitt, D.D.; Gibbs, F.A.; Heilbrun, M.P.; et al., "Dynamic Field Shaping to Optimize Stereotactic Radiosurgery," *Int. J. Radiat. Oncol. Biol. Phys.* 21 (1991):1247-1255; and Philips, M.H.; Frankel, K.A.; Lyman, J.T.; et al., "Comparison of Different Radiation Types and Irradiation Geometries in Stereotactic Radiosurgery," *Int. J. Radiat. Oncol. Biol. Phys.* 18 (1990):211-220.
199 Sinclair, W.K., "Risk, Research, and Radiation Protection," *Radiat. Res.* 112 (1987):191-216.
200 Sinclair, W.K., "Radiation Protection: Recent Recommendations of the ICRP and the NCRP and Their Biological Bases," *Advances in Radiation Biology* 16 (1992):303-324.
201 NCRP. *Report 116. Limitation of Exposure to Ionizing Radiation*. Bethesda: National Council on Radiation Protection and Measurements, 1993.
202 Ibid.
203 Suit, H.D., and Verhey, L.J., "Precision in Megavoltage Radiotherapy," *Brit. J. Rad.* 22 (suppl.) (1988):17-24.
204 Ibid.

CHAPTER FIVE

Clinical and Radiobiological Research

Eric J. Hall, D.Sc.

▸ ▸ ▸ ▸ ▸ ▸ ▸ ▸ ▸ ▸ ▸ ▸ ▸ ▸

Although this volume focuses mainly on developments on this continent, radiobiological research became a "global market" at a very early date; ideas, and indeed scientists, crossed the Atlantic at great speed and with great frequency. As a consequence, no attempt will be made here to impose geographical boundaries on the evolution of new ideas.

There are four principal themes in the evolution of radiobiological research related to radiation therapy:

(1) The oxygen effect. This was first discovered in the laboratory and eventually influenced much of clinical research in the form of high pressure oxygen tanks, neutrons, radiosensitizers, and bio-reductive drugs.

(2) Fractionation. Laboratory experiments over the years shaped the protocols used in clinical practice.

(3) Radiation quality. Radiobiological experiments with neutrons and alpha particles led ultimately to the introduction of exotic particles in the treatment of cancer patients.

(4) The development of experimental biological endpoints. In the early years, plant systems were widely used, later replaced by mammalian cells cultured *in vitro*, clever assays for radiation damage in normal tissues of experimental animals, and a variety of transplantable tumors. Recombinant techniques in molecular biology have gradually been introduced into radiation research beginning in the 1980s. The development of these research themes will be emphasized at the end of the chapter.

The X-Ray Era: 1910–1950

Leading the Way

One of the first documented uses of X-ray therapy was the treatment of a hairy nevus by the Viennese dermatologist Leopold Freund in 1896. Priority for the first X-ray therapy has been claimed by others, including Grubbé, Despeignes, Williams, and Voigt, though these were largely one-of-a-kind efforts rather than systematic attempts to apply a new modality.[1,2]

The first planned experiment in radiobiology was performed by Pierre Curie, who applied a tube of radium to his own forearm and described in detail the various phases of the resulting moist radio-epidermitis and his recovery from it (Fig. 5.1).

Radiobiological research in the early years of the twentieth century emphasized the importance of the proliferative potential in determining the radiosensitivity of cells. Kienbock in 1901 commented that organs in which active cell proliferation occurred were especially sensitive to radiation.[3] In 1903 Albers-Schönberg showed that radiation could induce azoospermia after irradiation of the testes, while three years later, Bergonié and Tribondeau used the same biological system to develop their "law," which proposed a proportionality between radiosensitivity and reproductive activity of cells.[4,5] In these earliest years the German school dominated radiation therapy and favored the view that single dose "caustic" treatments were superior, while fractionated treatment was judged to be "weak irradiation." As Wintz, one of the proponents of caustic irradiation, argued:

> The cells of the human body are endowed with variable radiosensitivity and capacity for recovery from radiation damage. It is also reasonable to assume that recovery from radiation injury depends on cellular metabolism, and further that a rapidly growing tumor cell is better able to effect recovery from injury than a connective-tissue cell with its comparatively slow metabolism. Therefore, the difference in response will favor the tumor if the cancericidal dose is not applied in the first treatment.[6]

This is the first but not the last example to be found in our history where radiobiological ideas were used retrospectively as a rationale for clinical protocols decided arbitrarily.

The controversy between "concentrated" and fractionated treatments continued into the 1920s, each based on proposed radiobiological principles. The supporters of concentrated treatments argued that recovery from injury was more efficient in rapidly growing tumor cells so that one big dose was best, while advocates for fractionation evoked the chance of irradiating cells in a more radiosensitive phase if the dose was divided. The controversy was unresolved when Coutard (Fig. 5.2) began treating head and neck tumors with fractionated low dose-rate (LDR) beam therapy in 1919, attempting to mimic the radium technique of Regaud.

Biological support for the superiority of fractionated treatment came from the experiments of Regaud and Ferraux, published in 1927 (Fig. 5.3).[7] It was found that rabbits could not be sterilized by exposing their testes to a single dose of radiation without extensive skin damage to the scrotum, whereas if the radiation was spread out in a series of daily fractions, sterilization was possible without producing unacceptable skin damage. It was postulated that the testes were a model of a growing tumor, whereas the skin of the scrotum represented a dose-limiting normal tissue. The reasoning may have been flawed, but the conclusion proved valid: fractionation of the radiation dose produces, in most cases, better

Fig. 5.1 Pierre Curie performed the first radiobiology experiment when he used a radium tube to produce a reaction on his own arm. This illustration depicts his experiment. (From *Radiobiology for the Radiologist*. Courtesy of E. J. Hall and Lippincott)

Fig. 5.2 Henri Coutard treated head and neck tumors with fractionated low dose-rate beam therapy. (Courtesy of Dr. Juan del Regato)

Fig. 5.3 Claudius Regaud pioneered protracted radiotherapy and showed the biological basis for it in animal experiments. (Courtesy of Dr. Juan del Regato)

tumor control for a given level of normal tissue toxicity than a single large dose.

In the years that followed, the conversion to fractionated treatments became essentially complete. The success of Coutard's technique was recognized when he presented it at the second International Congress of Radiology in Stockholm in 1928, reporting better tumor control with less normal tissue damage than for single dose treatments for head and neck cancer.[8]

Meanwhile, treating cancer of the uterine cervix by means of radium tubes inserted into the uterus yielded success. In other sites radium needles placed interstitially were successful for some highly-localized cancers. These methods of treatment, called brachytherapy, made use of the good radiobiological properties of continuous LDR irradiation which have only been explained more recently.

Isoeffect Curves and Overall Treatment Time

During the 1920s and 1930s a number of researchers tabulated what were termed "recovery factors" for fractionated protocols derived from daily treatments. Recovery factors were defined to be the ratio of biologically effective doses in a fractionated protocol (over a specific number of days) to the biologically effective dose given in a single sitting. Most notable were the efforts of Reisner; Duffy, Arnesen, and Edward; and MacComb and Quimby.[9,10,11] In retrospect, the most influential work was that of Strandqvist in 1944, who based the analysis of his data on a biological principle that we now know to be incorrect, namely "...radiation effect depends only on the total dose and overall time...it matters very little if the fractional doses are of variable size...."[12] He found that the total dose required to produce a given level of skin damage (either erythema or necrosis) was related to the overall treatment time by a simple power law:

$$\text{Total Dose} = \text{Constant } T^{0.33}$$

As the X-ray era drew to a close, a most significant contribution was the publication of the nominal standard dose (NSD) concept of Ellis and his colleagues at Oxford in the late 1960s.[13,14]

This came about from a chance circumstance. Frank Ellis was appointed as external examiner for the doctoral thesis of Lionel Cohen of the University of the Witwaterstrand in South Africa. This thesis contained extensive clinical data on acute skin reactions and the cure of squamous cell carcinoma of the skin (Fig. 5.4). The burden of the thesis was that the isoeffect curve for squamous cell carcinoma was different than for normal skin. The isoeffect curve followed a power law for both normal skin reactions and tumor cure, but the slope was less (0.22) for tumor than for skin (0.33).

Ellis reasoned that normal skin was under homeostatic control, whereas the tumor was not, and concluded that the difference in slope (0.11) could be taken as the time factor for normal tissue recovery. The formula therefore became:

$$\text{Total Dose} = \text{NSD } N^{0.24} T^{0.11}$$

where N = number of fractions, and T = overall time in days. For all its problems, the NSD concept had a significant impact on clinical practice. In retrospect, the biggest contribution it made was to separate the effects of fraction size and overall time and to emphasize

Fig. 5.4 Frank Ellis developed wedge filters and tissue compensators and also was responsible for the nominal standard dose (NSD) system. Shown here (center) on his eightieth birthday with students who had trained with him at Oxford, including Herman Suit, Eric Hall, and Martin Brown. (Author's collection)

that fraction size or number was by far the most important variable. The NSD concept received considerable support from the animal experiments of Fowler and colleagues, who showed that for skin reactions in pigs, the number of fractions was of greater importance in determining isoeffect doses than time, at least up to twenty-eight days.[15] In retrospect, we now know that this overall time was too short to show much repopulation in either pig or human skin.[16]

The Oxygen Effect in Radiotherapy

The first radiation experiments in which oxygen was mentioned as a factor were performed by Wald Schwarz in 1910.[17] He applied a radium applicator to the skin of his own forearm and observed that the subsequent reaction was reduced if the applicator was pressed hard against skin during the exposure. He concluded that metabolism was involved in the radioresistance of the skin cells but excluded oxygen as a factor because he found that if the blood flow to the arm was temporarily interrupted by the application of a tourniquet this had no effect on the radiation response. The oxygen effect could have been discovered in 1910, but it wasn't! However, this work was the stimulus for subsequent experiments by Holthusen and Petry.

Holthusen, in 1921, studied the effect of radiation on *ascaris* eggs and synchronized the division of cells by depriving them of oxygen (Fig. 5.5).[18] He observed that cellular radiosensitivity was diminished under hypoxia, but attributed this resistance to the absence of cell division, not to the absence of oxygen. He did the right experiment but drew the wrong conclusion. Soon after, Eugene Petry compared the radiosensitivity of various grains by measuring the length of irradiated roots.[19] Hypoxia resulted in radioresistance, from which Petry concluded that oxygen was directly or indirectly related to the damage produced by X rays. He was, therefore, the discoverer of the oxygen effect, but since both he and Holthusen published in German, their work appears to have gone unnoticed in the English-speaking world at the time.

Crabtree and Cramer in 1933 irradiated thin slices of animal tumors and

Fig. 5.5 Holthusen performed some of the early experiments involving the oxygen effect. He subsequently became well known as a clinical radiation oncologist. (Courtesy of Dr. Juan del Regato)

Fig. 5.6 Gioacchino Failla and Hal Gray. Both had been trained in physics, but led groups on opposite sides of the Atlantic involved in radiation biology; Gray in London, Failla in New York. (Courtesy, archives of the Center for Radiological Research, Columbia University, New York)

scored the percent of positive transplants and subsequent tumor growth rate.[20] Their primary concern was to show that respiration played a key role in the radiation response, their working hypothesis being that radiation damage was mediated through the respiratory mechanism. They clearly showed that radioresistance could be conferred by hypoxia, and they were aware of the radioresistance of tumors in badly vascularized tissues and the implications of this in radiotherapy. Thus, the oxygen effect was rediscovered in Britain ten years after its discovery in Germany.

Meanwhile, in 1935 Mottram, using seedlings of *vicia faba*, demonstrated the modifying influence of oxygen and discussed its importance in radiotherapy.[21] Eleven years earlier he had shown that the skin reaction produced by X rays on a rat's tail was reduced if blood circulation was stopped by a ligature applied at the base of the tail, but attributed the resistance to reduced blood flow, not to reduced oxygen.[22] Again, the right experiment but wrong conclusion. Years later, in light of the report of Crabtree and Cramer, he reinterpreted his data in terms of oxygen. It was Mottram who introduced Gray and Read to the use of the bean root as an experimental system (Fig. 5.6). There followed the classic series of papers published in the *British Journal of Radiology* between 1942 and 1952 in which Read, Gray, and colleagues investigated X rays, neutrons, alpha particles, mixtures of high- and low-linear energy transfer (LET) radiations, and the effect of oxygen.[23] They published the first numerical estimate of the oxygen enhancement ratio. In this way the stage was set for the subsequent era, in which the presumed limitation of radiotherapy was the radioresistance of hypoxic cells.

One lesson we can learn from the early experiments with the oxygen effect is the danger of preconceived ideas. The dogma of the day, epitomized in Bergonié and Tribondeau's so-called law, was that radiosensitivity depended on cell division. The discovery of the oxygen effect was delayed time and time again because investigators interpreted their data to support the fashionable ideas of the day and thereby missed a new discovery. How often do we do the same thing today?

Neutrons and Heavy Ions

The medical application of heavy particles was to blossom and reach its peak in the 1970s and 1980s at a time when physics research *per se* had lost some of its glamour; research directed toward more humanitarian goals became more fashionable. Nevertheless, it must be remembered that particles for radiation therapy are a spin-off from the incredible achievements of the halcyon era of nuclear physics. By 1919 Rutherford had demonstrated artificial transmutation of atomic nuclei. Alpha particles were used as projectiles to bombard nitrogen nuclei; the absorption of an alpha particle and ejection of a proton caused a nitrogen nucleus to become an oxygen nucleus. The alpha particles came from the only source then available, which were naturally occurring radioactive elements. As president of the Royal Society in 1927, Rutherford (Fig. 5.7) expressed a wish for a supply of "atoms and electrons which have an energy far transcending that of the alpha and beta particles from radioactive bodies."[24]

The race was on. From the outset, two of the big rivals were the Cavendish Laboratory in Cambridge and the Lawrence Berkeley Laboratory in California. At Cambridge, Cockroft and

Walton used a voltage multiplier and won the first race when they demonstrated transmutation using an artificially accelerated particle. Their generator and most similar devices of the day relied on high potentials, which are difficult to contain. The cyclotron, invented by Ernest Lawrence in 1931, was based on the principle of using the same electrical potential over and over again—an idea in fact demonstrated by Wideröe in 1928. Lawrence's contribution was the circular format (Fig. 5.8).

The first successful cyclotron, built by Lawrence and his graduate student, could fit in the palm of a hand and accelerated a few hydrogen ions to 80,000 electron volts (Figs. 5.9 and 5.10). In the following year, 1932, James Chadwick demonstrated the existence of the neutron at the Cavendish Laboratory (Fig. 5.11).[25]

In spite of the difficulties of the depression years, Lawrence succeeded in financing the design and construction of a series of cyclotrons of increasing size and energy. The 37-inch cyclotron was used as a source of neutrons to treat cancer patients from September 1938 to June 1939; figure 5.12 shows the first patient treated by John Lawrence and Robert Stone. The 60-inch cyclotron became available in December 1939, accelerating 16 megaelectron-volt (MeV) deuterons that were used to generate neutrons. Several hundred cancer patients were treated before the effort was terminated by the entry of the United States into World War II. The trial, not based on any particular radiobiological principle, was conducted at a time when little was known about the biological properties of neutrons and led to the conclusion that neutrons were not suitable for radiotherapy because of the disproportionately serious late effects produced in normal tissues.[26]

Meanwhile, on the other side of the Atlantic, radiobiological experiments by Gray and his colleagues were setting the scene for a post-World War II resurgence of interest in the neutron as a particle for radiotherapy, based now on a definite radiobiological principle—the oxygen effect. Gray was a product of the Cavendish Laboratory at Cambridge, the arch rival of Berkeley in the early years. The graduate student class at

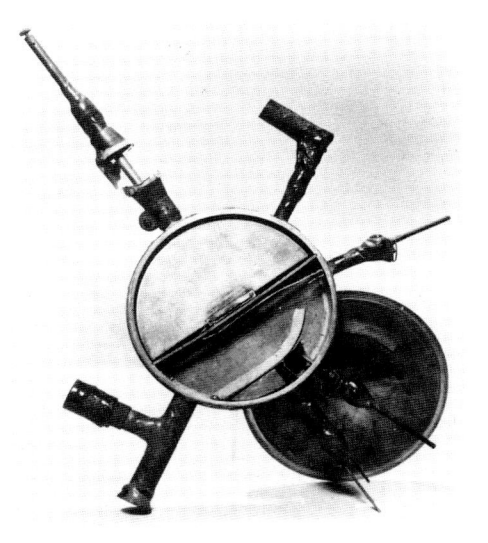

Fig. 5.7 Ernest Rutherford at the Cavendish Laboratory, Cambridge. (Courtesy of the American Institute of Physics).

Fig. 5.8 Ernest Lawrence at about the time he moved to the University of California. He was in many ways the prototype of a new generation of large scale team researchers. (Courtesy of the Lawrence Berkeley Laboratory)

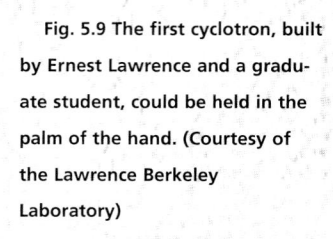

Fig. 5.9 The first cyclotron, built by Ernest Lawrence and a graduate student, could be held in the palm of the hand. (Courtesy of the Lawrence Berkeley Laboratory)

Fig. 5.10 Ernest Lawrence (right) and his second cyclotron. A series of cyclotrons of increasing size were built in the 1930s. They made possible the production of artificial radioactive isotopes, used in radiotherapy and nuclear medicine and now in many branches of medical research. (Courtesy of the Lawrence Berkeley Laboratory)

Fig. 5.11 Sir James Chadwick, who discovered the neutron in 1932.

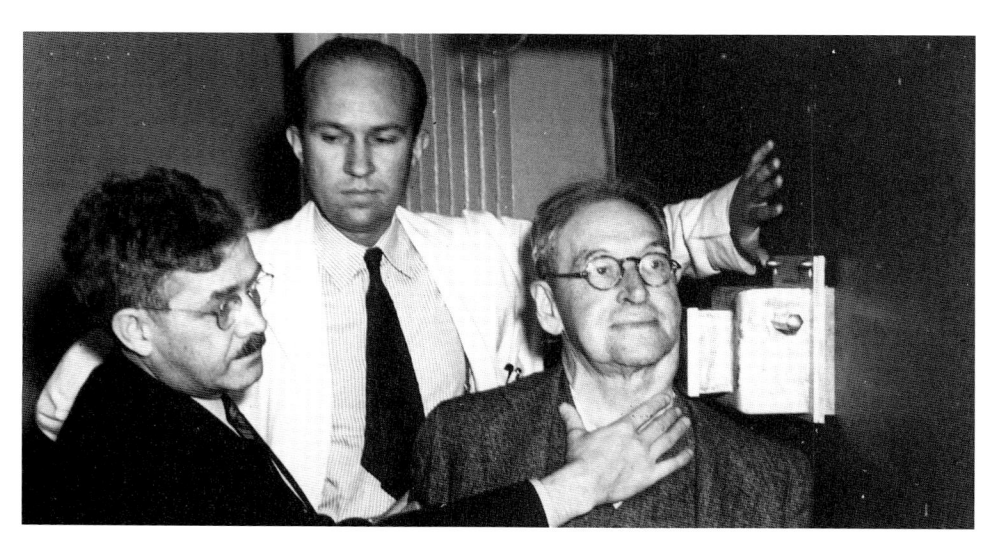

Fig. 5.12 John Lawrence (center), Robert Stone (left) and the first neutron patient treated on the 37-inch cyclotron (1939). (Courtesy of the Lawrence Berkeley Laboratory)

Cambridge in 1934 included an astonishing number of individuals destined to play a major role in medical physics and radiation biology (Fig. 5.13). Of them all, no one was to have more impact than Gray. After developing his theory of the cavity ionization chamber (which has led to the unit of radiation dose being named the *gray* [Gy]), he became a hospital physicist with Mottram at Mount Vernon Hospital in 1932. From his vision and perseverance grew the Medical Research Council cyclotron at Hammersmith Hospital, the first cyclotron dedicated to medical use and the first to be used for a random-

Fig. 5.13 The research group at the Cavendish Laboratory, Cambridge, taken in 1933. (Courtesy of Mrs. Eileen Lea and the Cavendish Laboratory)

ized clinical trial of fast neutrons. But that is another story reserved for the following era.

THE SUPERVOLTAGE ERA: 1950–1970

In Vitro Survival Curves

By the 1950s the lead in radiation research had crossed the Atlantic, largely as a consequence of the Manhattan District Project. The successful development of the atomic bomb, which abruptly ended World War II, led to the establishment of the Atomic Energy Commission (AEC), which made research funds available in the United States in previously unimagined quantities for the development of the peaceful uses of atomic energy ("Atoms for peace!") and the investigation of the biological effects of radiation (Figs. 5.14a and b).

A major breakthrough came from outside the field of radiation oncology when Puck and Marcus at the University of Colorado (T. T. Puck was professor and chairman of the Department of Genetics) developed techniques to grow single human cells in culture, leading to the first radiation survival curve.[27] Many researchers previously had grown mammalian cells in mass culture, but the growth media of the day were not sufficiently sophisticated to allow cells to be grown at low density under conditions where the ability of individual cells to grow into a colony could be observed—the essential prerequisite for the assessment of cell survival following radiation (Fig. 5.15). Puck and his colleagues had the same problem as everyone else, and the (possibly apocryphal) story is told that the breakthrough came when they

Fig. 5.14a The logo of the United States Atomic Energy Commission (AEC). The AEC funded much of the post-World War II research on the effects of radiation.

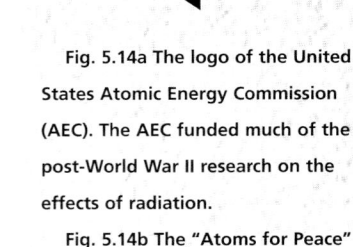

Fig. 5.14b The "Atoms for Peace" logo epitomized the post-World War II spirit to harness the power of the atom for peaceful purposes.

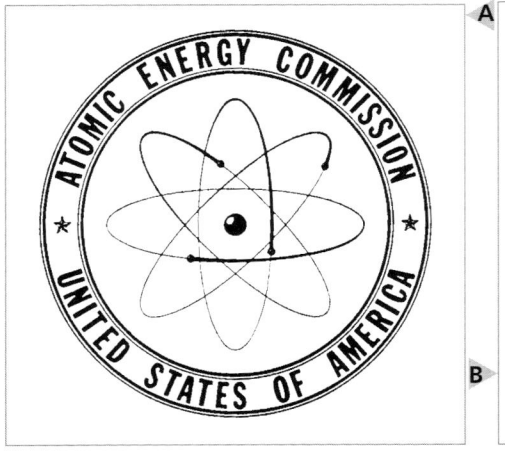

Fig. 5.15 Theodore T. Puck, University of Colorado, developed single cell mammalian cloning techniques, produced the first radiation survival curve for mammalian cells, and opened up the whole field of quantitative biology. (Author's collection)

were visited by Szilard, who said in his broken central European accent, "Don't let them know they are alone." This inspired suggestion led to the idea of feeder cells. The viable cells, whose integrity was to be tested, were mixed with large numbers of radiation sterilized "feeder cells," which could attach and grow for a while, but could never form colonies. Feeder cells allowed mammalian cells to be used for survival experiments by the same techniques that had been used for decades with bacteria.

Split Dose Experiments—Elkind Repair

Cell culture techniques spread like wildfire on both sides of the Atlantic. The split dose experiments of Mortimer M. Elkind had a particularly strong influence on radiation therapy (Fig. 5.16).[28] Elkind was working with a line of cells derived from a Chinese hamster (developed initially because they have half the number of chromosomes as human cells and are therefore easier for cytogenetic experiments). These cells grow rapidly in culture with a cycle of about ten hours and are characterized by having a very large shoulder on their radiation survival curve. It was possible to demonstrate that splitting a radiation dose into two fractions separated by times ranging from one-half hour to several hours led to an increase in survival. This was interpreted in terms of the repair of sublethal damage. The importance of this *in vitro* finding to radiotherapy was evident, since clinical protocols almost always consisted of multiple small doses separated by at least twenty-four hours—long enough for sublethal damage to be repaired. In truth, the experimental observation was not new, since the effect of splitting a dose into two fractions separated by various time intervals had been investigated by John Read before World War II. But Read had used seedlings of *vicia faba*, and his results were not applied to radiotherapy. Elkind's experiments followed on the heels of Puck's development of the survival curve for mammalian cells, and the importance in radiotherapy was recognized immediately.

Radiosensitivity Through the Cell Cycle and Potential Lethal Damage

Len Tolmach spent many years at Washington University in St. Louis. As a result of a falling out with his colleagues in radiation oncology, much of that time was spent in the anatomy department, but in the short space of three years he was involved in three key discoveries in radiation biology (Fig. 5.17).

First, in collaboration with William Powers, head of radiotherapy in St. Louis at the time, he demonstrated the

Fig. 5.16 Mort Elkind as the Failla Lecturer of the Radiation Research Society. Elkind is most well known for his demonstration of sublethal damage repair. (Courtesy of Dr. J. W. Osborne and the archives of the Radiation Research Society)

existence of viable hypoxic cells in mouse tumors which dominated their radiocurability.[29] This is further described later in this chapter.

Tolmach's second discovery was that the radiosensitivity of cells varied with the phase of the cell cycle. With Terasima, a postdoctoral fellow in his laboratory (who subsequently returned to Japan and held several senior positions including that of associate director of the unit at Chiba), Tolmach synchronized HeLa cells by shaking off those in mitosis and showed that their radiosensitivity varied as they moved through the cycle.[30] A more complete study of the "age response function," as it is often known, "including full dose-response curves for the various phases of the cycle" was later published by Warren Sinclair (Fig. 5.18).[31] Similar patterns of the variation of sensitivity through the cell cycle were later demonstrated in organized tissues synchronized by hydroxyurea first by Hall, Brown, and Cavanagh in a plant meristem and later by Withers and colleagues in the mouse intestinal epithelium.[32,33]

Tolmach's third contribution was to coin the term "potentially lethal damage," defined to be that component of radiation damage that can be influenced by the postirradiation conditions. He showed that more cells survived a given dose of X rays if they were kept at a lowered temperature or in a depleted medium for several hours after irradiation.[34] It was left to Jack Little and George Hahn to show that potentially lethal damage could be repaired if cells were kept for a few hours in plateau phase after irradiation and that there is a significantly enhanced cell survival in animal tumors if they are left in place for several hours postirradiation before being excised and assayed for clonogenicity.[35,36] This later work is more frequently quoted because the postirradiation conditions have more direct relevance to radiotherapy, but the concept of potentially lethal damage came from Tolmach. It is of interest to note that Len Tolmach grew up in New York City only a block from Robert Kallman, who coined the term reoxygenation; between them they were responsible for two of the four Rs of radiotherapy.

In Vivo Survival Curves

The next important development came from workers at Westminster Hospital and Hal Gray's British Empire Cancer Campaign (BECC) radiobiology laboratory in London. The collaboration of a pathologist (Harold Hewitt) with a medical physicist (C. W. Wilson) led to the development of the dilution assay technique, making possible the determination of a quantitative cell survival curve for cells of a lymphocytic leukemia.[37] This was the first *in vivo* survival curve. It is true that it related only to a "fluid" tumor, i.e., to leukemia cells in the peritoneal cavity of mice, but the idea was soon extended to solid tumors if they could be dissociated into a single cell suspension. Of more importance in the long run was the arrival in this same

Fig. 5.17 Len Tolmach was involved in three important discoveries in radiation biology: the presence of hypoxic cells in mouse tumors, the variation of radiosensitivity with phase of the cell cycle, and the demonstration of potentially lethal damage. He is shown here with Gloria Li. (Courtesy of Dr. J. W. Osborne and the archives of the Radiation Research Society)

Fig. 5.18 Warren Sinclair published a detailed study of the variation of radiosensitivity through the cell cycle for hamster cells and went on to become president of the National Council on Radiation Protection and Measurements. (Courtesy of Dr. J. W. Osborne and the archives of the Radiation Research Society)

Fig. 5.19 H. Rodney Withers, probably the most creative experimenter in radiation biology, developed an ingenious system to obtain a dose response relationship for reproductive integrity of skin stem cells. Dr. Withers was trained in radiotherapy in Australia, read for his Ph.D. in England, and has subsequently spent most of his professional life in the United States: the radiobiological citizen of the world. (Author's collection)

laboratory at this time of a young radiotherapist from Australia, H. Rodney Withers, who was assigned the Ph.D. project of obtaining a dose response curve for cells of a normal tissue irradiated *in vivo* (Fig. 5.19). He chose to work with mouse skin and developed the ingenious skin colony assay in 1967.[38] At a small radiobiology meeting in London in 1965 he described the mouse skin cells as having "an O.K. D_o of 125 rads"—i.e., the sensitivity of cells irradiated *in situ* in mouse skin was not significantly different from the sensitivity of single cells in a petri dish in the *in vitro* assay. This important result was first publicly reported in the same room at 32 Welbeck St. at the British Institute of Radiology where in 1957 Spear had cast doubt on the usefulness and relevance of Puck's initial work.[39]

It was in the same room, as well, that in 1959 Gray had pointed out that the *in vitro* survival curves of Puck for HeLa cells, with their D_o of "96R" became coincident with the *in vivo* survival curves of Hewitt and Wilson, with their D_o of 140 rads, when the dose was corrected by a factor of 1.45 (because Puck irradiated cells attached to glass) as described by Morkovin and Feldman.[40] In those heady days, people thought that all cell survival curves had the same D_o and the same extrapolation number of two because of the two strands of DNA. This simplistic view did not last long. Incidentally, it was the same Arnold Feldman who changed places with this author in an Oxford-Denver exchange visit so that I could learn the then new technique of mammalian cell culture and continue my transition into radiobiology.

After completing his Ph.D. at the University of London, Rod Withers moved to the United States and developed a whole range of systems which allowed survival curves to be obtained in a variety of normal tissues, including mouse jejunum with Mort Elkind at the National Institutes of Health, testes when he moved to M. D. Anderson, and kidney at Los Angeles.[41] Rodney Withers was the most creative experimentalist in the field. All these assay systems devised by Withers involved the scoring of regenerating clones *in situ* in normal tissues. In fact the very first system involving the regrowth of the clones from a single cell in a normal tissue was devised in Canada by McCullock and Till.[42] They took cells from nucleated bone marrow from donor mice, inoculated them into the tail veins of recipient mice, and scored the colonies that developed in the spleen. At a much later date, Clifton, Jirtle, and Gould developed similar cloning techniques *in vivo* for the hormonally controlled cells of normal thyroid and breast tissue, transplanting irradiated cells of these tissues from donor animals to grow as clones in the fat pods of recipient animals.

Mouse Tumor Hypoxic Cells: Reoxygenation and High Pressure Oxygen

In the early 1950s several groups tested the effect of breathing oxygen, at atmospheric or increased pressure, on the response of tumors to X irradiation.[43] Gray's work attracted particular attention among radiotherapists, due to the suggestion that oxygen might sensitize a tumor to a greater extent than the surrounding normal tissues, i.e., improve the therapeutic ratio. Just fourteen months after the appearance of Gray's paper, I. Churchill-Davidson treated his first patient in high pressure oxygen (HPO).[44]

Gray and his colleagues had pointed out that HPO might sensitize normal tissues to some extent, and the work of E. A. Wright and P. Howard Flanders underlined this danger.[45] Unfortunately, no account was taken of this possibility when randomized clinical trials were performed, with the exception of the small trial by H. A. S. van den Brenk, until the second trial of Henk in 1977.[46,47]

The work of H. B. Hewitt, initially carried out at Westminster Hospital, was of great importance in the development of experimental radiotherapy. Using tumors of spontaneous origin in his own strain of inbred mice, he developed an end-point titration method which provided survival curves for tumor cells *in vivo* (Fig. 5.20). His estimates of D_o (oxic) were in general in agreement with the *in vitro* studies of Puck. Hewitt's technique also made it possible to estimate the state of oxygenation of tumor cells. Leukemic cells infiltrating the liver were found to be entirely "oxic," i.e., gave a D_o characteristic of well-oxygenated cells. The hypoxic base-line could be established by irradiating a dead mouse. The cells of a solid sarcoma were found to be predominately hypoxic.

Powers and Tolmach in St. Louis, using Hewitt's technique, were able to show that a particular tumor, the Gardner lymphosarcoma, was a mixture of oxic and hypoxic cells.[48] They also showed that HPO reduced the hypoxic cell fraction but never abolished it. In fact, every experiment with murine tumors and HPO has indicated the difficulty of fully oxygenating tumors, particularly if they are large. The publication by Powers and Tolmach prompted many investigators to estimate the proportion of hypoxic cells in many different mouse and rat tumors, and it turned out that 10 to 30 percent was a common value.

The experimental work which led to the introduction of HPO in the clinic had all been carried out with single doses. The next vital step was taken in the late 1960s at Stanford University, where Robert Kallman collaborated with a number of Europeans on sabbatical, including van Putten from the Netherlands and Norman Bleehen from London (Fig. 5.21). The critical observation made at about that same time by Herman Suit at the M. D. Anderson Hospital and Hugh Thomlinson at the Hammersmith MRC Unit was that the *proportion* of hypoxic cells in a tumor, which rose to 100 percent immediately after a large dose of X rays, returned to its preirradiation level within a few hours of the radiation exposure.[49] Kallman, and at the same time Thomlinson in London, coined the term *reoxygenation* for this phenomenon (Fig. 5.22). The discovery of reoxygenation by Kallman and van Putten suggested to some radiotherapists that, provided treatment was fractionated, HPO might have no place in radiotherapy. However, the extensive studies of Suit on murine tumors showed that

Fig. 5.20 Harold Hewitt, together with Wilson, developed the dilution assay technique which made possible the first *in vivo* dose response curve. (Courtesy of the archives of the Gray Laboratory)

Fig. 5.21 Bob Kallman coined the term *reoxygenation* and did much to further an understanding of hypoxia in mouse tumors. (Courtesy of the archives of the Gray Laboratory).

Fig. 5.22 Hugh Thomlinson worked with Hal Gray in developing a model for hypoxia in tumors and was responsible for many of the early experiments on reoxygenation at the Hammersmith Hospital. (Courtesy of the archives of the Gray Laboratory)

HPO is generally *more* effective when combined with fractionated radiation than with single doses (Fig. 5.23).[50] The contrasts between this result with HPO and the effect of hypoxic sensitizers are striking; hypoxic sensitizers are *less* effective when given in association with fractionated radiation.

Neutrons and Charged Particles: The Effect of Radiation Quality

As the European countries recovered from the devastation of World War II, they planned their response to the atomic age and the need to understand the effects of ionizing radiations. In the Netherlands the effort was concentrated in a central government funded Radiobiological Research Institute (TNO) at Rijswijk. The group of workers here had a major influence on radiobiological research of relevance to radiotherapy during the 1960s and 1970s, quite disproportionate to the size of their homeland.

In 1955 Eddie Barendsen was a post-doctorate follow in "physics applied to archeology" at Yale University, where he witnessed Ernest Pollard irradiating enzymes with cyclotron-accelerated alpha particles to study inactivation cross sections (Fig. 5.24). Upon his return to the Netherlands he combined this technology with the new Puck culture technique for growing cells of human origin. He first analyzed the relative biological effectiveness (RBE) and oxygen enhancement ratio (OER) of natural alpha particles and later used deuterons and helions from the Hammersmith cyclotron in London. These latter studies were started after he was introduced to Jack Fowler at a meeting in Oxford in 1961 by Hal Gray. Later studies involved neutrons of various energies in Rijswijk as well as in London. While many other investigators filled in the details later, these experiments by Barendsen showed the broad picture of the variation of biological effectiveness with radiation quality and the variation of the importance of the oxygen effect with radiation quality.[51] This set the scene for the exploration of both neutrons and accelerated charged particles for radiotherapy.

Neutrons

This era saw the research efforts that resulted in the use of neutrons in place of X rays for the treatment of cancer. The first use of neutrons at Berkeley in the late 1930s and early 1940s had not been based on any particular radiobiological rationale. The post-World War II

Fig. 5.23 Herman Suit performed many key experiments to demonstrate the role of hypoxia in both mouse and human tumors. He is seen here as president of the Radiation Research Society. (Courtesy of Dr. J. W. Osborne and the archives of the Radiation Research Society)

neutron effort at the Hammersmith Hospital was based squarely on:

(1) The experimental observation that neutrons were less dependent than X rays on molecular oxygen to produce their cytotoxic effect, i.e., OER was smaller for neutrons than for X rays.

(2) The premise that the control of human tumors by X rays was limited by the presence of viable hypoxic cells.

The cyclotron at Hammersmith Hospital was the brainchild of Hal Gray, based on experiments he and his collaborators had performed over the years to elucidate the biological properties of neutrons using a deuteron on beryllium machine that he and John Read had built in the late 1930s for $500 (Fig. 5.25). Following a dispute with the medical director, Gray left Hammersmith to found the BECC Radiobiological Research Unit at Mount Vernon Hospital, which became the Gray Laboratory after his death. The preclinical neutron studies at Hammer-smith were carried out by Tikvah Alper, Shirley Hornsey, Juliana Denekamp, and Stan Field, with Jack Fowler, head of medical physics, coordinating some of the effort (Fig. 5.26).[52] Jack Fowler had been a hospital physicist, first at Newcastle and then at King's College London (Fig. 5.27). He accepted the position of professor of medical physics at the Royal Postgraduate Medical School at Hammersmith, with responsibility for all aspects of a large program in medical physics. During this era, though, he is remembered most for his work on neutrons, using pig skin as a biological endpoint.

Neutron experiments were carried out with a wide range of biological systems, from cells in culture to mouse tumors to mouse normal tissue systems to pig skin. The basic principles of neutron radiobiology became clear. These included the variation of RBE with dose per fraction and the great variation of the RBE between different tissues. This work led to the first clinical trials of neutrons at Hammersmith, led by Catterall.[53] These early trials were greatly hampered by the physical characteristics of this low energy cyclotron-produced beam, poor depth doses, a fixed horizontal beam, and primitive collimating devices. The interpretation of the results of the early trials became highly controversial, but at the time neutrons appeared to yield superior local tumor control and set the stage for a massive research effort involving neutrons in the United States.

The Perceived Problem of Hypoxia

This era also saw the genesis of other strategies to overcome the per-

Fig. 5.24 Eddie Barendsen elucidated the relation between RBE and LET and between OER and LET. His work set the scene for later clinical trials of neutrons. (Author's collection)

Fig. 5.25 The 15 MeV cyclotron at Hammersmith used for the first controlled clinical trials of neutrons. (Courtesy of the Cyclotron Unit, Hammersmith Hospital)

Fig. 5.26 Tikvah Alper led the research team at Hammersmith Hospital during the preclinical experimentation with neutrons. (Courtesy of the archives of Hammersmith Hospital)

ceived problem that the presence of hypoxic cells limited the curability of human tumors by X rays. A common sight in many radiation therapy departments in the 1960s was the hyperbaric oxygen tank. Patients were equilibrated with pure oxygen at a pressure of 2 or 3 atmospheres prior to their treatment with X rays. The first efforts were by Churchill-Davidson at St. Thomas Hospital in London and van den Brenk in Melbourne, Australia, but many efforts were to follow.[54]

These procedures involved heroic efforts. Because of the technical difficulties, treatment protocols were frequently changed to only a few dose fractions. When this was done, HBO treatments were certainly superior to the controls breathing air. The art and science of clinical trials were at a primitive state at this point, and it was a long time before it was appreciated that a fair test of the importance of HBO must include a control arm involving a conventional number of fractions. Dedicated teams on both sides of the Atlantic and in Australia carried out a variety of HBO trials which showed a modest gain for HBO—but not enough to justify the trouble and expense involved in the treatments. Two retrospective observations are in order. First, the gains produced by HBO, while modest, have seldom if ever been achieved by any other technical modification since HBO was abandoned. Second, it is a miracle in hindsight that no patient was incinerated during the treatment of thousands of individuals sealed in high pressure oxygen tanks.

One of the reasons for the demise in the popularity of HBO treatment was the promise that chemical sensitizers would soon be available that would achieve the desired end of sensitizing hypoxic cells without the expense, technical problems, and hazard of high pressure tanks. Led largely by Ged Adams, the 1960s saw the beginning of the effort to produce hypoxic cell radiosensitizers, chemicals that would specifically and differentially sensitize hypoxic cells to killing by X rays while having no effect on normal aerated cells (Fig. 5.28).[55] Some of the principles involved in these oxygen-mimicking compounds were understood, but the compounds available in bacteria were much too toxic for use in animals, much less humans.[56] It was, however, the birth of a field that was to assume great importance in the next era, because it led to the concept of bioreductive drugs and hy-poxic cytotoxins.

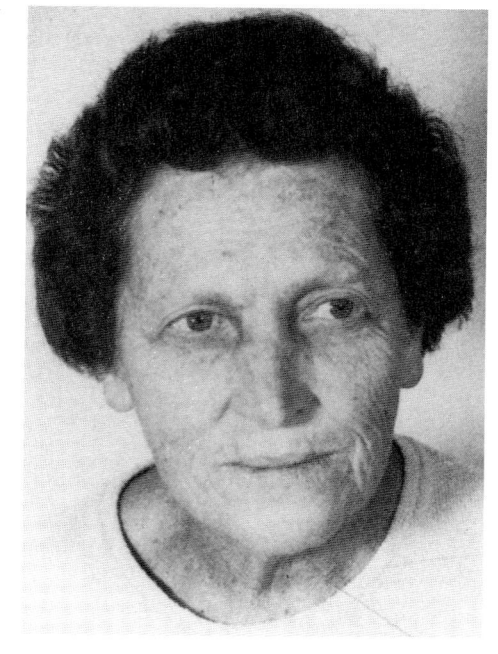

Cell, Tissue, and Tumor Kinetics

The 1950s and 1960s saw the introduction of autoradiography, made possible by the availability of radionuclides as labels, a direct spin-off from the Manhattan District Project of World War II. This resulted in

Fig. 5.27 Professor "Jack" Fowler, the great communicator and teacher of radiation biology. Trained as a physicist, he became director of the Gray Laboratory. Photographed here with Liz Travis, one of many visitors from the United States who spent time at the Gray Lab. (Courtesy of Dr. J. W. Osborne and the archives of the Radiation Research Society)

Fig. 5.28 Professor G. E. (Ged) Adams (right), the "father" of radiosensitizers. Pictured here as the Failla Lecturer of the Radiation Research Society, with G. Whitmore. (Courtesy of Dr. J. W. Osborne and the archives of the Radiation Research Society)

some of the most clever radiobiological research ever performed. During this era cell population kinetics had a major impact on the thinking of radiotherapists, but any direct impact on clinical practice was delayed to a later time.

The availability of radiolabeled precursors of DNA synthesis led to the important work of Howard and Pelc in delineating the phases of the cell cycle (Fig. 5.29).[57] With a light microscope, the only event that can be seen is mitosis itself, in preparation for which cells round up and the chromosomes condense. Using labeled thymidine and autoradiography, they showed that DNA was synthesized only during a discrete period of the cycle, which they called "S". Mitoses and DNA synthesis were separated by the "the first gap in activity" (G_1), while there was a second "gap in activity" (G_2) before the subsequent mitosis. These terms have been retained to the present day and are widely used throughout biology. Alma Howard was a Canadian who spent most of her life in Great Britain, an example of the reverse "brain drain;" Pelc was a refugee from Hitler's Europe who also settled in London.

Labeled precursors of DNA also made possible the analysis of the length of the cell cycle and its constituent parts by means of the "percent labeled mitoses technique," first used by Quastler and Sherman to analyze the mitotic cycle of cells in the mouse intestinal epithelium.[58] Autoradiography was soon applied to transplanted tumors in experimental animals. Mendelsohn showed that only a proportion of cells were in cycle at any given time, which he called the "growth frac- tion."[59] The other important concept developed was that most cells produced by mitosis are lost from the tumor, the so-called "cell loss factor," first estimated for animal tumors by Steel and colleagues.[60] Steel is also credited with inventing the T_{pot}, the potential doubling time.

Following the successful application of the techniques of population kinetics to animal tumors, groups in France and in England performed similar studies on a limited number of cancer patients (Figs. 5.30 and 5.31).[61] This gave new insight into the pattern of growth of human tumors; the discrepancy between the cell cycle of individual cells (a few days) and the gross volume doubling time (months) was accounted for by the fact that only a proportion of tumor cells were in cycle at any given time and, in particular, by the realization that the cell loss factor in human tumors averaged about 70 percent. Comparable studies were not performed in the United States because of the ethical and legal problems inherent in giving patients tritiated thymidine.

As long as cell cycle analysis depended on autoradiography, results could

Fig. 5.29 Alma Howard was Canadian born but worked most of her life in Great Britain, influenced greatly by Hal Gray. Together with Pelc she developed the technique of autoradiography and delineated the phases of the cell cycle. (Courtesy of Dr. J. W. Osborne and the archives of the Radiation Research Society)

Fig. 5.30 Professor Maurice Tubiana is well known for his contributions to clinical radiation oncology as director of the Institute Gustave Roussy in Paris; he also contributed to important laboratory research on cell kinetics. In this photograph he is shown at left with Dr. Norman Bleehen at the eighth International Congress of Radiation Research. (Courtesy of Dr. J. W. Osborne and the archives of the Radiation Research Society)

never be used prospectively to modify the protocol of an individual patient because of the long time (six to eight weeks) required to obtain an image. Rapid techniques to assess cell cycle parameters had to await the development of flow cytometry and appropriate probes not based on radioactive labels.[62] The first example, involving measurement of T_{pot} from a single biopsy specimen following administration of bromodeoxyurine, is described later.

THE MEGAVOLTAGE ERA: 1970–PRESENT
New Fractionation Concepts

The first mammalian cell survival curves of Puck and Marcus (1956) were obtained with HeLa cells, which are characterized by a very small initial shoulder and were fitted to a multi-target model having two components, n and D_o, the extrapolation number and the 37-percent dose slope. In the context of modeling fractionation and dose-rate effects, it soon became apparent that to account for experimental observations, a model involving a nonzero initial slope was essential.[63]

The area received a new momentum in 1976 when Douglas and Fowler adopted the linear-quadratic (LQ) model, originated in the late 1930s by Lea (Fig. 5.32) and Catcheside and derived on biophysical grounds by Kellerer and Rossi.[64,65,66] The important point they made was that the ratio of the constants in the dose response relationship, i.e., alpha/beta, could be deduced from fractionation experiments in a tissue or organism where measurement of surviving fraction could not be determined. In fact, the principle of using isoeffect fractionation data to estimate the survival curve parameters in circumstances where cell survival cannot be measured *per se* had been described fifteen years earlier.[67] However, that was in the context of multi-target survival curve models and experiments involving *vicia faba*.

In the years following the publication of Douglas and Fowler's work, the LQ isoeffect model grew in influence because it predicted that the fractionation sensitivity of a tissue may be classified according to the ratio alpha/beta.[68] Thames's work helped to spread the influence of this model (Fig. 5.33). However, the key step forward came from the recognition by Withers that early responding tissues tended to have large values of alpha/beta (about 8 to 10 Gy), while late responding tissues were characterized by smaller alpha/beta values (2 Gy).[69] Barendsen made a similar suggestion in the same year. This translates into a more curvy dose response relationship for late responding tissues and a greater sensitivity to frac-

Fig. 5.31 Dr. Edmund Malaise (right) is shown here in conversation with Dr. Lester Peters of M. D. Anderson Hospital, Texas. Malaise, of the Institute Gustave Roussy, made many contributions to the study of cell, tissue, and tumor kinetics, and in later years to the development of predictive assays for radiosensitivity. (Courtesy of Dr. J. W. Osborne and the archives of the Radiation Research Society)

function of time reaching 130 cGy/day.[71] By contrast, in the rat spinal cord, proliferation does not occur until much later, and, thereafter, the extra dose required to counter proliferation increases more slowly. In interpreting these data and extrapolating them to the case of a patient undergoing radiotherapy, Fowler made the following important points:

First, increasing overall treatment time spares *early* responding tissues, but has essentially no effect on *late* responding tissues, at least within the normal radiotherapeutic range. This contrast

tionation. This observation explained the fact that fractionation spares late responding tissues more than early responding tissues, an observation which assumed prominence in a large number of studies of modified dose fractionation as well as neutron therapy in the late 1960s and early 1970s.

This observation of Rod Withers, from laboratory animal data, was confirmed first by data from patients obtained in Götenburg (in Strandqvist's former department) by Turesson and Notter.[70] Other workers, especially Thames at Houston and more recently Soren Bentzen in Denmark, have developed methods of extracting alpha/beta and other radiobiological data from animal and clinical data, with widely influential results on new clinical protocols of fractionated radiotherapy.

The other important step forward in the understanding of the radiobiology of normal tissues came from the experimental work of a number of investigators, notably Juliana Denekamp at the Gray Laboratory and Bert Van de Kogel, who was a graduate student with Eddie Barendsen in the Netherlands and later worked at Los Alamos National Laboratory (Fig. 5.34). Working with mouse skin, Denekamp showed that, when irradiated with a fractionated schedule, skin did not begin to proliferate until about twelve days after the start of treatment—but that once it started, the extra dose required to counter proliferation increased rapidly as a sigmoid

Fig. 5.32 Douglas Lea was trained as a physicist at Cambridge but produced some of the earliest radiation survival curves for bacteria. In collaboration with Catcheside, he introduced the linear-quadratic formula to describe dose-response relationships for chromosomal aberrations. (Courtesy of Mrs. Eileen Lea)

◄

Fig. 5.33 Howard Thames is a brilliant mathematician, modeler, and biostatistician at the M. D. Anderson Hospital and collaborated with Rod Withers and Lester Peters to develop models of the response of early and late responding tissues. (Author's collection)

◄

◄

Fig. 5.34 Juliana Denekamp made many contributions to radiation biology in the areas of cell kinetics, tumor biology, hyperthermia, and tumor blood flow. She succeeded Professor Fowler as director of the Gray Laboratory. She is shown here sharing a lighter moment with Eric Hall at the International Conference on Hyperthermia in 1980. (Courtesy of the archives of the Center for Radiological Research, Columbia University, New York)

was a radiobiological explanation of the policy of using a large number of small fractions in six to eight weeks that has been used and advocated by the Houston group, first under Gilbert Fletcher, with a wide following in the United States. It is contrary to the United Kingdom and ex-British-Empire policy of using fewer and larger fractions but in shorter overall times of three to five weeks. It is only in the last decade that the advantages and disadvantages of both approaches, due to the application of flow cytometry to measure proliferation rates (T_{pot}) in tumors, are being clarified.

Second, the biology implies that in resting normal tissues, compensating proliferation does not occur until two to four weeks in human mucosa or skin but then increases rapidly; this is in contradiction to the constant power law time factor in the NSD equation. In the 1980s it was recognized that there was an onset of accelerated regrowth after a lag period in tumors receiving XRT analogous to that in normal tissues, a phenomenon reported twenty years earlier in animal tumors.[72,73]

These new biological concepts, namely the different sensitivity to fractionation of early and late responding tissues, the pattern of proliferative response of steady state normal tissues, and the accelerated regrowth of tumors, provided a convincing rationale for the clinical trials of hyperfractionation and accelerated treatment, which are at the cutting edge of clinical investigations at the present time.

The linear quadratic model has, to a large extent, replaced the NSD concept for the calculation of doses used in different fractionation schemes to arrive at the same biological effect. The basic idea for these calculations came from Thames, but its use was popularized largely by Jack Fowler.[74,75]

Central to the development of a biological rationale for altered fractionation patterns was the group of radiobiologists that flourished around Gilbert Fletcher (Fig. 5.35) at the M. D. Anderson Hospital in Houston. Fletcher himself was a giant in the field of radiation oncology and a master of clinical research, but he also fostered a most important group of laboratory researchers and modelers. First was Herman Suit in the 1960s and 1970s, who was followed as head of experimental radiotherapy by H. Rodney Withers and later by Lester Peters and Howard Thames. The importance of their contributions cannot be overestimated.

Particle Therapy

The years following 1970 saw the interesting development of a major initiative in particle therapy in the United States. The earliest neutron experience of the Medical Research Council at Hammersmith Hospital had produced encouraging results despite the physical limitations of the low energy cyclotron that was available. This stimulated great interest in neutron therapy in the United States, Europe, and Japan.[76] In the United States the first efforts involved three high-energy cyclotrons that had initially been built for physics research but were by then largely obsolete for that purpose and could be adapted for medical use. They were located at College Station, Texas; the Naval Research Laboratory in Washington, D.C.; and the University of Washington in Seattle. A substantial effort was mounted in pre-clinical studies of physics and biology, but the subsequent clinical results never produced results quite as dramatic as those

Fig. 5.35 Gilbert Fletcher was renowned for his contributions to clinical radiation oncology. He also set up and supported a laboratory research group that included over the years Herman Suit, H. Rodney Withers, and Lester Peters. (Courtesy of the M. D. Anderson Hospital, Houston, Texas)

claimed by Hammersmith Hospital. It was recognized from the outset that using treatment machines in physics research installations was suboptimal, to say the least. The 1970s saw the most coordinated effort ever mounted by the radiation oncology community to obtain substantial support for a specific initiative. Many millions of dollars were allocated by the National Cancer Institute to build three dedicated hospital-based cyclotrons to produce neutrons, as well as supporting trials of negative pi mesons at Los Alamos National Laboratory and heavy ions at the Lawrence Berkeley Laboratory.

Three hospital-based neutron facilities were built at the University of Washington in Seattle (Tom Griffin), the University of California at Los Angeles (Robert Parker), and the M. D. Anderson Hospital in Houston (Lester Peters). These facilities joined the Fermilab (Lionel Cohen), a medical neutron facility that operates parasitically from the huge physics research accelerator at Batavia. Again there was a considerable preclinical effort in physics and biology to ensure uniformity of dosimetry and RBE values prior to the mounting of a number of clinical trials involving many disease sites. Machine reliability and patient accrual proved to be a problem everywhere but Seattle, and after fifteen years, most clinical trials showed no advantage for neutrons except possibly for carcinoma of the prostate, salivary gland tumors, and soft-tissue sarcomas.[77]

Meanwhile, Morton Kligerman had obtained funds for a medical pion facility to be appended to the mile-long high-energy proton accelerator at Los Alamos National Laboratory, with the aim of treating cancer patients (Fig. 5.36).[78] The physics and biology research efforts were impressive, though as more data were accumulated, the advantage of pions over X rays looked less obvious. The problems of coordinating clinical trials high on a mesa in New Mexico proved to be insurmountable, and as research funds became more difficult to obtain, the project came to an end without definitive clinical conclusions.

The history of heavy ions is more complex and took longer to unfold. In the race to be the first to accelerate high-energy heavy ions, the Princeton particle accelerator was several months ahead of the Lawrence Berkeley Laboratory in producing a beam of nitrogen ions. The first radiobiological data were published by the Columbia group in 1973. The Princeton particle accelerator was closed down soon thereafter by the Nixon administration, leaving the BEVALAC at Berkeley as the only heavy ion machine in the world (Fig. 5.37). Led by Tobias, the Berkeley group, with help from colleagues across the United States, made a detailed investigation of the biological properties of heavy ions ranging from nitrogen to silicon.[79] The data were of interest to the National Air and Space Administration because astronauts were exposed to high-energy heavy particles in space and of interest to the radiation oncology community as a potential means of treating cancer. In the context

Fig. 5.36 The Clinton P. Anderson Los Alamos Meson Physics Facility. This machine was over half a mile in length and accelerated an intense beam of protons to 800 MeV, which were then used to produce negative pi mesons for high-energy physics research and biomedical applications. It was built by the United States Atomic Energy Commission (now the Department of Energy) at a cost of $57 million. Its enormous size and spectacular location high on a mesa in New Mexico are both evident from this aerial photograph. The first pion treatment for a cancer patient was given in October 1974 by Dr. Morton Kligerman. The facility was closed some ten years later. (Courtesy of Los Alamos Scientific Laboratory, New Mexico)

Fig. 5.37a Schematic of the link-up of the SUPERHILAC and the BEVATRON to produce the BEVALAC. This was the first facility in the world capable of accelerating heavy ions to energies suitable for cancer therapy and was located at the Lawrence Berkeley Laboratory of the University of California. Much laboratory research and patient treatment were conducted at this site.

Fig. 5.37b Aerial photograph showing the layout of the BEVALAC. Heavy ions were produced and initially accelerated in the SUPERHILAC and then piped 500 feet down the hillside and injected into the BEVATRON to be accelerated to high energies. (Courtesy of the Lawrence Berkeley Laboratory, University of California)

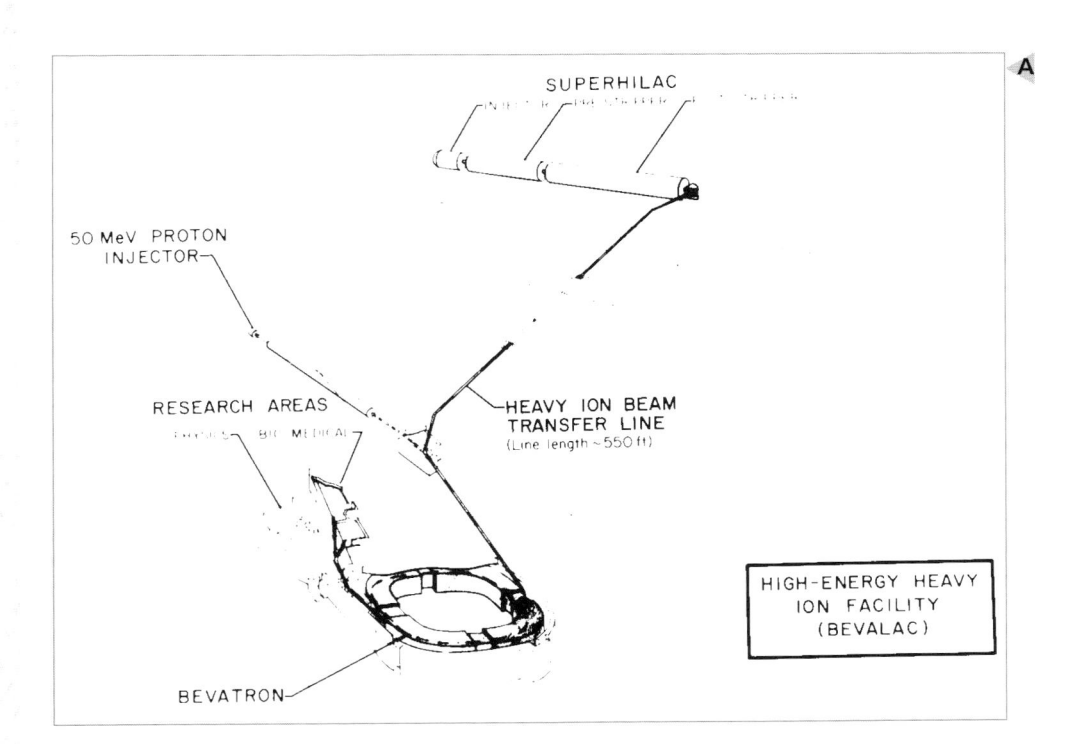

of radiation therapy, beams of heavy charged particles combine the advantages of improved dose distribution, as obtained with protons, with the high-LET advantages of neutrons. It came as something of a surprise (and disappointment) that it was necessary to push to atomic numbers characteristic of argon and silicon before high-LET properties of the beam, spread out to cover a tumor of realistic dimensions, equaled those of neutrons.

Many patients were treated with a variety of charged particles.[80] The sophisticated treatment planning techniques which were developed for the particle therapy program have proved to be of great benefit in external beam radiotherapy with X rays. There have been anecdotal reports of favorable results in patient treatments with heavy ions, but no definite prospective controlled clinical trials were completed before the BEVALAC closed in 1993. It is ironic that there is no heavy ion machine for medical use in the United States at a time when new facilities have been built and are in use in France, Switzerland, Germany, Japan, and Russia.

It was the insights obtained from the difference in the LQ parameter alpha/beta between early and late-responding tissues that eventually clarified high-LET radiotherapy. At high LET it is the alpha component, the initial slope, that is increased with no change in beta. Therefore, most of the sparing effect, which is proportional to alpha/beta, was lost at high LET. The late-responding tissues suffered more damage relative to tumor cell kill, because they had more repair capacity to lose. High-LET treatments are no longer so popular, although the good physical dose distributions of heavy ion beams are a real advantage. These physical advantages can be obtained with proton beams, which are easier and cheaper to generate than the heavier ions of helium, carbon, or argon or than pi meson.

In the long run the least "exotic" heavy particles, protons, have had a more lasting impact on radiation therapy than neutrons, pions, and heavy ions combined. Facilities are in use in the United States, Sweden, and Russia; more than ten thousand cancer patients have been treated, the vast majority of them at the Harvard cyclotron in a team effort led by Herman Suit.[81]

Hypoxic Cell Radiosensitizers and Bioreductive Drugs

The era opened with the knowledge that the potential of a compound as a hypoxic cell radiosensitizer, i.e., to mimic oxygen, depended on its electron affinity.[82] Misonidazole was soon identified at the Gray Laboratory as a potent radiosensitizer of hypoxic cells *in vitro* and in experimental animal tumors and was offered for clinical trials. This compound proved to be much less exciting in the clinic than in the laboratory because the dose that could be administered was limited to suboptimal levels due to the development of neurotoxicity. However, a recent retrospective meta analysis of the twenty or so clinical trials involving misonidazole shows a small but clear improvement in local control when the drug is combined with radiation.[83]

There followed a clever piece of research which led to the synthesis of etanidazole by Stanford Research International (SRI). Bill Lee of SRI and Martin Brown of Stanford University were making and testing analogs of misonidazole when Brown took off for a year's sabbatical in Cambridge, England (Fig. 5.38). By chance, he met Paul Workman there and began doing pharmacokinetics on some of the compounds. As a consequence, they measured the pharmacokinetics of a whole series of compounds with slightly different structures that all had the same radiosensitizing potential *in vitro*.[84] Etanidazole was chosen, which as a hydrophilic compound, does not cross the brain barrier and penetrates poorly into nerve tissue. The limiting normal tissue toxicity is still neurotoxicity but occurs at much higher drug doses than for misonidazole. At the time of writing, the clinical evaluation of this compound as a chemosensitizer and radiosensitizer is still in progress.

However, the direction of developments in this field changed as a consequence of laboratory studies. In the

Fig. 5.38 Dr. J. Martin Brown trained at Oxford and eventually became director of research in radiation oncology at Stanford University, California. He was responsible for major advances in the design and synthesis of hypoxic cell radiosensitizers. (Courtesy of Dr. J. W. Osborne and the archives of the Radiation Research Society)

early 1970s Sutherland showed that metronidazole caused damage in the central regions of spheroids, while Hall and Roizin-Towle showed misonidazole to be preferentially cytotoxic to hypoxic cells.[85,86] Over the past twenty years, it has been slowly realized that it may be much more efficient to *kill* hypoxic cells in a tumor with a hypoxic cytotoxin than to attempt to *radiosensitize* them with a hypoxic cell radiosensitizer. Perhaps this should have been obvious from the start, but it was slow to be appreciated. The upshot is that effective new compounds that operate primarily as hypoxic cell cytotoxins have been synthesized by two groups on opposite sides of the Atlantic, led by Ged Adams and colleagues at the U. K. Medical Research Council and by Martin Brown and colleagues at Stanford. This is a clear change of direction and an exciting development that gives new life to the field.[87] These drugs must still be combined with X rays, however. The radiation kills the aerated cells, and the drug kills the hypoxic cells; hence, the strength of each modality is exploited.

Predictive Assays

Predictive assays have come of age during the modern era. They come in three types: assays for intrinsic radiosensitivity, assays for hypoxia, and assays for proliferative potential.

Assays for inherent radiosensitivity owe much to the work of Malaise and colleagues in Paris and to Gordon Steel and his colleagues in London, who divided tumors into several histological categories and showed that the steepness of the initial slope of the cell survival curve correlates with clinical responsiveness.[88] The first serious attempt to correlate clinical outcome in individual patients with the survival of cells derived from their tumors and grown in a cell adhesive matrix assay was by William Brock, Lester Peters, and colleagues at the M. D. Anderson Hospital in Texas.[89] They focused attention on the SF_2, the surviving fraction at 2 Gy, as a predictor of outcome in head and neck cancer patients receiving postoperative radiotherapy. Although a trend was observed toward higher recurrence rates in patients with more radioresistant tumor cells *in vitro*, this was not statistically significant, possibly reflecting the variability of tumor burden in patients treated postoperatively. This group has also studied the relationship between normal tissue radiosensitivity and tolerance to radiotherapy and has recently demonstrated a significant correlation between late effects in normal tissues and the radiosensitivity of the patient's fibroblasts assayed *in vitro*. The first positive correlation of *in vitro* radiosensitivity and tumor control was published by West and her colleagues at Manchester.[90] In patients with carcinoma of the cervix, the SF_2 was measured by growing cells in soft agar; the probability of local control was poorer in patients for whom the SF_2 was above 0.55.

The need to identify patients whose tumors contain a substantial proportion of hypoxic cells and might, therefore, stand to benefit from hypoxic cell radiosensitizers or neutrons was acknowledged from the outset. A great deal of work was performed in the 1960s to measure oxygen concentration in tumors and to correlate the results with the outcome of radiotherapy for carcinoma of the cervix.[91] A major step forward in recent years was the development of the Eppendorf probe, which allows a rapid profile of pO_2 levels in a tumor to be measured.[92] Hockel and his colleagues in Germany successfully used this improved technology in patients with locally advanced carcinoma of the cervix and found that patients whose tumor exhibited median pO_2 values less than 10 millimeters Hg had a significantly lower recurrence-free survival compared to patients with tumors that showed higher pO_2 values.[93]

At about the same time, Chapman (Fig. 5.39) and his colleagues in Philadelphia successfully linked a radionuclide, iodine-123, to a nitroimidazole, which allows hypoxic regions to be identified by a SPECT scan, since the nitroimidazole breaks down and deposits the radioactive label only in regions of low pO_2.[94] This represents a noninvasive test that can be performed prospectively on individual patients and is an exciting development. Similar labeled compounds have been developed by Koh and colleagues in Seattle.[95]

The cell kinetic studies conducted in the past, based on classical autoradiography techniques, were never suitable as predictive assays in individual patients because of the long time necessary to get a result. The breakthrough came with the development of flow cytometers and the use of nonradioactive dyes as markers. Adrian Begg developed a method to *estimate* the T_{pot} of a tumor within one day by flow cytometry measurements of cells from a single biopsy taken some hours after an injection of bromodeoxyuridine.[96] This work started while he was at the Gray Laboratory in London but reached fruition after he had moved to the Netherlands.

T_{pot} was measured in a group of patients with head and neck cancer in a study by the European Organization for Research and Treatment of Cancer, comparing an accelerated protocol with conventional therapy.[97] If all patients were considered together, there was no significant difference between the conventional and accelerated arms of the study. However, in a small group of patients with fast growing tumors, in which T_{pot} was four days or less, the accelerated protocol resulted in a much improved level of local control. This is the first report in which cell kinetic studies on individual patients were used successfully to select a subgroup of patients who could benefit from a new treatment protocol.

The Gray Laboratory

In terms of research that was to have an impact on clinical radiotherapy, it is impossible to exaggerate the importance of the Gray Laboratory. Founded by Gray as the BECC Radiobiological Research Unit in 1954, it was renamed in his honor at his death. The next director was Oliver Scott, whose ideas influenced so much of the early work on the oxygen effect and whose family had played an important role in founding the lab (Fig. 5.40). When Oliver retired in 1968 due to ill health, Jack Fowler was appointed as his successor. This was a highly imaginative appointment at the time, and he was given a mandate to concentrate the considerable resources of the lab in research in "experimental radiotherapy"—the study of the effects of fractionated radiation on tumors and normal tissues in the mouse. The Gray Laboratory is so important, not only because of key contributions from its well-known staff that included Ged Adams, Jack Boag, Juliana Denekamp (the present director), Harold Hewitt, Barry Michaels, and Adrian Begg, but also from the impressive list of students and visitors who have spent time there and have been inspired to devote their research efforts to radiotherapy; these include Rod Withers, Lester Peters, Liz

Fig. 5.39 Dr. J. Donald Chapman was born and trained in Canada before moving to the Fox Chase Cancer Center in Philadelphia. He made many contributions to research on hypoxic cell radiosensitizers, most notably methods to image hypoxic cells by noninvasive means. (Courtesy of Dr. J. W. Osborne and the archives of the Radiation Research Society)

Fig. 5.40 Sir Oliver Scott was a close friend and collaborator of Hal Gray and became director of the laboratory when Gray died. (Courtesy of Dr. J. W. Osborne and the archives of the Radiation Research Society)

Travis, Janet Rasey, and Jim Fisher. Many of these individuals returned (or emigrated) to the United States after their stay at the Gray Laboratory, bringing their ideas and enthusiasm with them. It is significant and regrettable that there is no comparably large group in the United States, with stable long-term funding, devoted to experimental radiotherapy. The closest equivalent would be the group at the M. D. Anderson Hospital that grew up around Fletcher.

The Dose-Rate Effect and Brachytherapy

The discovery of sublethal radiation damage repair in mammalian cells by Elkind and his group in the late 1950s led to many attempts to model dose response relationships for LDR and also to experiments to determine cell survival curves for protracted LDR irradiation.[98,99,100] These efforts were aimed at providing an experimental basis for brachytherapy. LDRs permit most of the repair that can take place to do so. Because tumor cells have, usually, less repair capability than late-responding tissues, LDR therefore protects the tissues in which late complications occur more than it protects tumors. LDR is therefore good news, from a therapeutic ratio point of view, and we depart from that only with caution.

During the 1960s Paterson and Ellis independently published isoeffect curves for interstitial implants, relating the total dose as a function of implant time that would result in the same level of normal tissue tolerance.[101] In retrospect, these isoeffect curves equate well to radiobiological models based on the linear-quadratic formalism, using values for alpha/beta appropriate for *late* responding tissues. There was a long standing controversy throughout the 1970s and 1980s between those who followed the advice of Paterson and Ellis to correct total dose for overall time and the followers of Pierquin and the Paris School, who maintained that no dose correction was necessary for implant times between three and nine days.[102,103] This implied the absence of a dose-rate effect between 30 and 100 cGy/hour, which was in direct conflict with radiobiological measurements. This controversy was not resolved until the 1990s when a retrospective analysis of the impressive body of data from the Paris group showed that for tumors of the mobile tongue and floor of mouth, both tumor control and the incidence of necrosis depended on dose-rate.[104] For breast cancer, too, local control varied with dose-rate.[105] In both cases, the variation was as predicted from radiobiological data.[106]

The 1980s, too, saw the proliferation of high dose-rate (HDR) afterloaders, supplanting in many institutions the LDR intracavitary techniques that had been used to treat carcinoma of the cervix for three quarters of a century. HDR afterloaders have the advantage of improved radiation protection and allow outpatient treatments, but they also caused great controversy. The outcome of clinical use indicates HDR results that are comparable to LDR, both in terms of local control and the incidence of complications.[107] Radiobiological modeling has been used extensively to suggest HDR protocols in a limited number of fractions (typically five to twelve) that are comparable to LDR.[108]

A further development made possible by the availability of computer-controlled remote afterloaders is pulsed brachytherapy.[109] In this technique, a single radioactive source steps through the catheters of an interstitial implant or through the catheters of an intracavitary applicator, with the dwell time in each source position adjusted to optimize the required dose distribution. This technique buys a number of advantages including only one source to replace, no source preparation, improved dose optimization, better radiation protection, and a constant average dose-rate.

Hyperthermia

The supervoltage era saw the blossoming of hyperthermia as a vigorous and well-funded research area. In various forms the use of hyperthermia as a modality to treat cancer predates the discovery of X rays, but the modern development of hyperthermia based on laboratory research dates from the work of Dewey, Westra, and Eugene Robinson

(Figs. 5.41 and 5.42), who showed that cell survival curves for heat were similar to X rays, with "time at the elevated temperature" replacing absorbed dose.[110] Research into the biological effects of hyperthermia mushroomed, and it soon became evident that the biological properties of hyperthermia made it highly suitable as an anticancer modality. It was discovered that the sensitivity through the cell cycle complemented that for X rays, with radioresistant S-phase cells sensitive to heat.

Hypoxic cells were not resistant to cell killing, as they are to X-ray cytotoxicity, while cells deprived of nutrients or at low pH were found to be highly sensitive.[111] All of these properties favored the killing of cells by heat in large necrotic tumors. There were disadvantages, to be sure, including the phenomenon of thermotolerance, whereby cells became resistant to subsequent heating by a prior heat exposure (Fig. 5.43).[112] In most cases the synthesis of heat shock proteins (also called stress proteins) correlates with the development of thermotolerance. This finding provided the first evidence for the function of these primordial proteins and paved the way for molecular biology in hyperthermia.[113] Thermotolerance makes multiple heat fractions impractical, and it was argued that the best way to handle thermotolerance was to avoid it.[114]

Modest treatments with hyperthermia produced dramatic "cures" in transplantable mouse tumors, perhaps because the encapsulated nature of the tumors and the poor blood supply exaggerated the temperature differential between the tumors and the surrounding normal tissues. Exhaustive studies of normal tissue effects in rodents led to studies in larger animals, including dogs.[115] The stage was set for the trial of hyperthermia in the treatment of human cancer. By the 1990s more than twenty thousand cancer patients had received hyperthermia and several lessons had been learned. First, hyperthermia alone has a very limited place as an anticancer modality, except for palliation.[116] Second, as an adjunct to X rays, hyperthermia was shown to increase the response rate of many types of human cancer, and thousands of patients have been treated, but final proof in the form of prospective randomized controlled clinical trials has been slow in coming.[117] The only phase III trial showing the efficacy of hyperthermia came from the European cooperative group who showed that radiation plus heat doubled the complete response rate and the five-year local control rate of malignant melanoma, compared with radiation alone.[118]

A major and persistent problem was the difficulty of maintaining a uniform high temperature throughout the tumor volume by means of an external heating device, whether ultrasound, microwaves, or a capacitative heating system. As early as the second International Conference

Fig. 5.41 Dr. W. C. "Bill" Dewey made numerous contributions to a range of topics in radiation biology, but in later years became closely identified with cellular research in hyperthermia. (Author's collection)

Fig. 5.42 Eugene Robinson was a pioneer in the cellular studies of hyperthermia. (Author's collection)

Fig. 5.43 Dr. George Hahn devoted much of his career to laboratory studies into the cellular and molecular effects of hyperthermia. (Courtesy of Dr. J. Martin Brown)

on Hyperthermia Oncology in Fort Collins in 1980, it was pointed out that, while the biological properties of heat were very favorable for its use as a modality for the treatment of cancer, the physical principles involved in heating by any of the modalities then in use were not amenable to manipulation to produce controlled uniform volumes of elevated temperature.[119] The slogan, "The biology is with us, but the physics is against us," was controversial and hotly debated in 1980, but the clinical experiences of the next decade proved this prophecy to be more accurate that anyone could have known at the time!

The most promising area of hyperthermia research in the 1990s is the combination of an interstitial implant and hyperthermia also induced by implanted electrodes.[120] It should have come as no surprise that the combination of the best radiation dose distribution with the most uniform heat distribution would result in superior clinical results.

Molecular Techniques in Radiation Biology

Developments and advances in radiobiological research inevitably mirror advances in the broader fields of biology and general science. The past decade or so has witnessed the introduction of the techniques of molecular biology into the field of radiation research. At the time of writing, molecular biology has done little more than influence the outlook and thinking of the radiation biology researcher; there are few examples of impact and none of real importance. Nevertheless, advances in recombinant technology, and particularly in the molecular genetics of cancer, must revolutionize radiation biology as they have changed every other field of biology. For this reason a brief history is included.

Four families of genes have been identified in mammalian cells: repair genes, check point genes, oncogenes, and suppressor genes.

Repair genes involved in the repair of damage by ultraviolet light and by mitomycin-C have been cloned and sequenced.[121] One repair gene in the mammalian cell involved in ionizing radiation repair has also been identified and sequenced, but it may not be a typical example; the radiosensitive mutant for which it corrects was only slightly radiosensitive.[122] By contrast, a number of radiosensitive mutants have been identified in yeast and the genes involved cloned and sequenced.[123] At first it was thought that the genes in yeast were involved solely in the repair of DNA damage by radiation. In fact it turns out that they are also molecular check point genes. Molecular check point genes in general serve the function of ensuring that the initiation of late events are dependent upon the completion of early events. In cells exposed to radiation, the function of the check point gene is to hold the cells in G_2 in order to check for the integrity of their chromosomes before the complex task of mitosis is begun. Cells that are deficient in this gene do not stop in G_2 after radiation but proceed immediately through mitosis, and many more die as a consequence.

In making the jump from laboratory research to the genetics of human disease, the study of individuals that are heterozygous for *atacia telangiectasia* (AT) is a very important consideration. AT heterozygotes comprise only about 1 to 3 percent of the Caucasian population in the United States, but they have a cancer risk two to three times higher than the general population and may account for 9 to 18 percent of breast cancer in young patients.[124] There is also some evidence that AT heterozygotes have a much higher susceptibility to radiation-induced breast cancer than the normal population.

The discovery of the importance of oncogenes in human cancer made it possible to understand why agents as diverse as a retrovirus, ionizing radiation, or chemicals could result in tumors that were indistinguishable one from another.[125] The retrovirus *inserts* a gene into the cell, while radiation or chemicals produce a mutation in a gene that is already present in the cell.

Today well over a hundred oncogenes have been identified in human cancer with more than 80 percent of the transforming genes belonging to the *ras* family.[126] However, activated oncogenes are associated with only 10 to 15 percent of human cancers and tend to be associated with leukemias and lymphomas and less with solid tumors. Oncogenes have been shown to be activated by: (1) a point mutation (for example, *ras*), (2) a deletion (*fos*), (3) a reciprocal translocation (*myc*), or (4) gene amplification (*myc*). Ionizing radiations are not particularly efficient at producing point mutations, but they do cause interstitial deletions and reciprocal translocations with high efficiency. Consequently, in assessing the possible mechanisms by which radiation induces cancer, deletions or translocations would seem to be the most likely candidates.

A number of studies have been performed to investigate the influence of oncogenes on radiosensitivity. In the first such publication, Sklar used NIH3T3 cells, which are immortal, and showed that transfection of *ras* resulted in resistance to radiation, at least at high doses.[127] In a more comprehensive study, McKenna and colleagues used primary rat embryo cells. It was found under these circumstances that transfection of *myc* or *ras* alone had only a modest effect on the radiosensitivity. On the other hand, if *myc* and *ras* were transfected together the cells showed a marked resistance to radiation, particularly in the low dose region at doses of approximately 2 Gy, i.e., at doses comparable to the daily dose used in radiation therapy.[128] This paralleled the earlier work that showed that cooperation between two oncogenes was necessary to produce oncogenic transformation.[129] Although many reports have appeared in the literature showing that transfecting oncogenes into cells *in vitro* can induce resistance, the data are equivocal. There is no real evidence that radioresistance in human tumors is associated with oncogene activation.

Suppressor genes are recessive acting in that both copies must be lost or inactivated for the cell to express the malignant phenotype. Suppressor genes were in fact discovered before oncogenes, at least with cells in culture. Stanbridge and colleagues in the 1970s showed that, if a hybrid was made by fusing a normal human fibroblast to a malignant HeLa cell, the normal cell suppressed the expression of malignancy by the HeLa cell.[130] It was shown further that if, during the repeated subculture of the hybrid cells, chromosome 11 was lost, then the malignant phenotype was restored.[131] From this result, it was inferred that chromosome 11 in the normal human fibroblast contained a gene capable of suppressing the malignant phenotype.

The importance of suppressor genes became evident from the work of Knudsen with retinoblastoma.[132] Retinoblastoma appears in a familial form with high incidence and in a sporadic form at very low incidence. In the familial form, one mutant allele with lost function is inherited from the affected parent. A somatic event during embryogenesis then inactivates the normal allele inherited from the unaffected parent. The probability of this occurring is high and almost all such children are born with bilateral retinoblastoma. On the other hand, in sporadic retinoblastoma *two* somatic mutational events are necessary, the second in a descendant of a cell that suffered the first. This is much less likely to happen, which accounts for the much lower incidence of the sporadic form of retinoblastoma. Knudsen elaborated this two-hit hypothesis in the early 1970s. By the mid-1980s the location of the gene was identified on chromosome 13 and eventually the Rb gene was cloned and sequenced.[133,134] While the Rb gene is associated in 100 percent of cases with retinoblastoma, it is also associated with various other tumors, such as sarcomas, small cell cancer, bladder cancer, and mammary cancer in a relatively smaller proportions of cases.

An obvious mechanism for a suppressor gene to be deleted is by the action of radiation. Since a suppressor gene acts in a recessive way, the deletion would have to occur in both chromosomes of a pair, which of course would be a very low frequency event. In practice, in many cases, it is found that rather than two separate deletions the loss of the pair of suppressor genes occurs by the process of somatic homozygosity.[135] This has been shown to be the mechanism in cases of retinoblastoma, small cell lung cancer, and glioblastoma. The latter is particularly interesting in as much as somatic homozygosity occurs in two different chromosomes for this high grade tumor to be produced. What happens is that one chromosome of a pair is lost, and the remaining chromosome with the deletion is replicated. At the present time there are at least six suppressor genes whose location and function are known. The two most common and most intensively studied are the Rb gene and the p53 gene. Both of these are involved with the arrest of cells in G_1 and in tumor differentiation. It should be noted at this point that double strand breaks in chromosomes have been consistently identified as a type of lesion responsible for most radiation induced cellular responses—including cell lethality and carcinogenesis.

A report in *Science* by the Johns Hopkins group showed that all cases of hereditary nonpolyposis colorectal cancer (HNPCC) showed an alteration on chromosome 2, which may be described as a mutator gene, leading to genomic instability.[136] In a high proportion of the HNPCC tumors, as well as sporadic colon cancer, there was a high incidence of mutations in K, *ras*, p53, and APC genes, though in no instance was the mutation rate 100 percent. The notion that the basic causative event is an alteration in a mutator gene leading to genomic instability may explain what was seen in the past with colon cancer—that there are a whole sequence of changes between the normal epithelium and a metastasizing malignant tumor, including mutations in oncogenes, loss of suppressor genes, and other unidentified chromosomal alterations.[137] The interpretation based on the new data is that genomic instability leads to a cascade of events.

These recent findings may have a dramatic impact on the diagnosis and treatment of cancer because they imply that in many forms of cancer, the disease occurs with significantly high probability in predisposed individuals and that in the near future, these individuals can be identified at birth. This is the challenge that all cancer therapists must face as the next era opens.

▶ ▶ ▶ ▶ ▶ ▶

REFERENCES

Despite the best efforts and intentions, history is highly subjective and selective; while I must bear ultimate responsibility for the content and accuracy of this chapter, I acknowledge with thanks the help of the following friends who read the chapter and made numerous suggestions: Drs. Ged Adams, G. W. Barendsen, J. Martin Brown, Dennis Leeper, Lester Peters, Oliver Scott, and H. Rodney Withers. Without their help, this brief history would be much the poorer. Dr. William Osborne, of Iowa State University and chairman of the History Committee of the Radiation Research Society, provided many of the pictures. Nancy Knight, Ph.D., of the Center for the American History of Radiology provided many more. While the individual pictures are acknowledged, I would not miss this opportunity to thank them for generously giving of their time.

1 Glasser, O., "Röntgentherapie mit Ultraspannungen in Nordamerika," *Strahlentherapie* 60 (1937):557-574.
2 Thames, H.D., and Hendry, J.H. *Fractionation in Radiotherapy.* London: Taylor and Francis, 1987.
3 Kienböck, R., "Zur Pathologie der Hautveränderungen durch Röntgenbestrahlung bei Mensch

und Thier," *Wien. Med. Presse* 42 (1901):873-1040.

4. Albers-Schönberg, G., "Röntgenstrahlenwirkung an den Hoden," *Munch. Med. Wschr.* 43 (1903).
5. Bergonié, J., and Tribondeau, L., "Interpretation de quelques resultats de la radiothérapie et essai de fixation d'une technique rationelle," *Comptes Rendus de l'Academie des Sciences* 143 (1906):983-985.
6. Wintz, H., "Die Einzeitbestrahlung," *Strahlentherapie* 58 (1937):545-551.
7. Regaud, C., and Ferroux, R., "Discordance des éffects de rayons X d'une part dans la peau, d'autre dans le testicule, par le fractionnement de la dose," *Comptes Rendus Societé-Biologique* 97 (1927):431.
8. Coutard, H., "Roentgentherapy of Epitheliomas of the Tonsillar Region, Hypopharynx, and Larynx from 1920 to 1926,". *Am. J. Roent.* 28 (1932):313-331, 343-348.
9. Reisner, A., "Hauterythem und Röntgenstrahlung," *Ergebrisse der Medizinischen Strahlenforschung* 6 (1933):1-60.
10. Duffy, J.; Arnesen, A.N.; and Edward, L.V., "The Rate of Recuperation of Human Skin Following Irradiation," *Radiology* 23 (1934):486-490.
11. MacComb, W.S., and Quimby, E.H., "The Rate of Recovery of Human Skin from the Effects of Hard or Soft Roentgen Rays Or Gamma Rays," *Radiology* 27 (1936):196-204.
12. Strandqvist, M., "Studien über die kumulative Wirkung der Röntgenstrahlen bei Fraktionierung," *Acta Radiologica* (Suppl.) 55 (1944):1-300.
13. Ellis, F., "The Relationship of Biological Effect to Dose-Time-Fractionation Factors in Radiotherapy," in Ebert, M., and Howard, A., eds. *Current Topics in Radiation Research*. Vol. 4. (Amsterdam: North Holland Publishing Co., 1968):359-397.
14. Ellis, F., "Dose Time and Fractionation: A Clinical Hypothesis," *Clinical Radiology* 20 (1969):1-7.
15. Fowler, J.R.; Morgan, R.L.; Silvester, J.A.; et al., "Experiments with Fractionated X-ray Treatment of the Skin of Pigs. I. Fractionation up to 28 Days," *Br. J. Radiol.* 36 (1963):188-196.
16. Turesson, I., and Notter, G., "The Influence of Overall Treatment Time in Radiotherapy on the Acute Reaction: Comparison of the Effects of Daily and Twice-a-Week Fractionation on Human Skin," *Int. J. Rad. Oncol. Biol. Phys.* 10 (1984):607-618.
17. Schwarz, W., *Wein. Klin. Wschr.* 11S (1910):397.
18. Holthusen, H., "Beiträge zur Biologie der Strahlenwirkung," *Pflüger's Arch. Phys.* 187 (1921):1-24.
19. Petry, E., "Zur keuntnis der Bedingungen der biologischen," *Wirkung der Röntgenstrahlen Biochem Zschr.* 135 (1923):353-383.
20. Crabtree, H.G., and Cramer, W., "Action of Radium on Cancer Cells: Some Factors Affecting Susceptibility of Cancer Cells to Radium," *Proc. Roy. Soc.* (London, Biol.) 113 (1933):238.
21. Mottram, J. C., "On the Alteration in the Sensitivity of Cells Towards Radiation Produced by Cold and by Anaerobiosis," *Br. J. Radiol.* 8 (1935):34-39.
22. Mottram, J. C., "On the Skin Reactions to Radium Exposure and Their Avoidance in Therapy: An Experimental Investigation," *Br. J. Radiol.* 29 (1924):174-180.
23. The products of the Read/Gray collaboration published in the *British Journal of Radiology* between 1942 and 1952 include: Gray, L. H., and Read, J., "The Effect of Ionizing Radiations on the Broad Bean Root. Part I. General Notes," *Br. J Radiol.* 15 (1942a): 11-16; Gray, L.H., and Read, J., "The Effect of Ionizing Radiations on the Broad Bean Root. Part II. The Lethal Action of Gamma Radiation," *Br. J. Radiol.* 15 (1942b):39-42; Gray, L.H., and Read, J., "The Effect of Ionizing Radiations on the Broad Bean Root. Part III. The Lethal Action of Neutron Radiation," *Br. J. Radiol.* 15 (1942c):72-76; Gray, L.H., and Read, J., "The Effect of Ionizing Radiations on the Broad Bean Root. Part IV. The Lethal Action of Alpha Radiations," *Br. J Radiol.* 15 (1942d):320-336; Gray, L.H., and Read, J., "The Effect of Ionizing Radiations on the Broad Bean Root. Part VI. The Summation of the Effects of Radiation of Different Ion Density. A) Neutrons and Gamma Radiation, B) Alpha and X-radiation," *Br. J. Radiol.* 17 (1944):271-273; Gray, L.H., and Read, J., "The Effect of Ionizing Radiations on the Broad Bean Root. Part VII. The Inhibition of Mitosis by Alpha Radiation," *Br. J. Radiol.* 23 (1950):300-303; Gray, L.H.; Read, J.; and Poynter, M., "The Effect of Ionizing Radiations on the Broad Bean Root. Part V. The Lethal Action of X-radiation," *Br. J. Radiol.* 16 (1943):125-128; Gray, L.H., and Scholes, M.E., "The Effect of Ionizing Radiations on the Broad Bean Root. Part VIII. Growth Rate Studies and Histological Analysis," *Br. J. Radiol.* 24 (1951):82-92, 176-180, 228-236, 285-291, 348-352; Read, J., "The Effect of Ionizing Radiations on the Broad Bean Root. Part X. The Dependence of the X-ray Sensitivity on Dissolved Oxygen," *Br. J. Radiol.* 25 (1952a):89-99; Read, J., "The Effect of Ionizing Radiations on the Broad Bean Root. Part XI. The Dependence of the Alpha Ray Sensitivity on Dissolved Oxygen," *Br. J. Radiol.* 25 (1952b):651-661; Thoday, J.M., "The Effect of Ionizing Radiations on the Broad Bean Root. Part IX. Chromosome Breakage and the Lethality of Ionizing Radiations to the Root Meristem," *Br. J. Radiol.* 24 (1951):572-622.
24. Rutherford, E., "Anniversary Address 30 November 1927," *Proc. Royal Soc. London* 117A (1928):300-316.
25. Chadwick, J., *Proc. Royal Soc.* 136A (1932):692.
26. Stone, R.S., *Am. J. Roent.* 59 (1948):771.
27. Puck, T.T., and Marcus, P.I., "Actions of X-rays on Mammalian Cells," *J. Exp. Med.* 103 (1956):653-666.
28. Elkind, M.M., and Sutton, H., "X-ray Damage and Recovery in Mammalian Cells in Culture,"

Nature (London) 184 (1959):1293.
29. Powers, W.E., and Tolmach L.J., "A Multicomponent X-ray Survival Curve for Mouse Lymphosarcoma Cells Irradiated *In Vivo*," *Nature* 197 (1963):710-711.
30. Terasima, R., and Tolmach, L.J., "X-ray Sensitivity and DNA Synthesis in Synchronous Populations of HeLa Cells," *Science* 140 (1963):490-492.
31. Sinclair, W.K., and Morton, R.A., "X-ray Sensitivity during the Cell Generation Cycle of Cultured Chinese Hamster Cells," *Rad. Res.* 29 (1966):450-474.
32. Hall, E.J.; Brown, J.M.; and Cavanagh, J., "Radiosensitivity and the Oxygen Effect Measured at Different Phases of the Mitotic Cycle Using Synchronously Dividing Cells of the Root Meristem of *Vicia faba*," *Rad. Res.* 35 (1968):622-634.
33. Withers, H.R.; Mason, K.; Reid, B.O.; et al., "Response of Mouse Intestine to Neutrons and Gamma Rays in Relation to Dose Fractionation and Division Cycle," *Cancer* 34 (1974):39-47.
34. Phillips, R.A., and Tolmach, L.J., "Repair of Potentially Lethal Damage in X-irradiated HeLa Cells," *Rad. Res.* 29 (1966):413-432.
35. Little, J.B., *Rad. Res.* 56 (1973):320-333.
36. Little, J.B.; Hahn, G.M.; Frindel, E.; and Tubiana, M., "Repair of Potentially Lethal Radiation Damage *In Vitro* and *In Vivo*," *Radiology* 106 (1973):689-694.
37. Hewitt, H.B., and Wilson, C.W., "A Survival Curve for Mammalian Leukemia Cells Irradiated *In Vivo*." *Br. J. Cancer* 13 (1959):69-75.
38. Withers, H.R., *Br. J. Radiol.* 40 (1967):187.
39. Spear, F.G., "On Some Biological Effects of Radiation," *Br. J. Radiol.* 31 (1967):114-124.
40. Morkovin, D., and Feldman, A., "End Point of One of the Actions of Radiations on Living Tissue Important in Radiation Therapy and in Acute Radiation Syndrome," *Br. J. Radiol.* 33 (1960):197.
41. Examples of the the varied tissues used in Rod Withers's studies include: Withers, H.R., and Elkind, M.M., "Dose-Survival Characteristics of Epithelial Cells of Mouse Intestinal Mucosa," *Radiology* 91 (1968):998-1000; Withers, H.R., "Regeneration of Intestinal Mucosa after Irradiation," *Cancer* 28 (1971):7581; Withers, H.R.; Hunter, N.; Barkley, H.T.; and Reid, B.O., "Radiation Survival and Regeneration Characteristics of Spermatogenic Stem Cells of Mouse Testis," *Radiation Research* 57 (1974):88-103; Withers, H.R.; Mason, K.A.; and Thames, H.D., "Late Radiation Response of Kidney Assayed by Tubule-Cell Survival," *Br. J. Radiol.* 59 (1986):587-595.
42. McCulloch, E.A., and Till, J.E., *Radiation Research* 16 (1962):822.
43. Following is a listing of groups that, during the mid-1950s, tested the effect of oxygen on the response of tumors to X-irradiation: Hollcroft, J.W.; Lorenz, E.; and Mathews, M., "Factors Modifying Effect of X-irradiation on Regression of Transplanted Lymphosarcoma," *J. Nat. Cancer Inst.* 12 (1952):751-763; Gray, L.H.; Conger, A.D.; Ebert, M.; et al., "The Concentration of Oxygen Dissolved in Tissues at the Time of Irradiation as a Factor in Radiotherapy," *Br. J. Radiol.* 26 (1953):638-648; Dittrich, W., and Uhlmann, G., "Resisten zentwicklung maligner ti, premgegem Röntgenstrahlen," *Naturwissenschaften* (1954):69-70.
44. Churchill-Davidson, I.; Sanger, C.; and Thomlinson, R.H., "High Pressure Oxygen and Radiotherapy," *Lancet* 1 (1955):1091-1095.
45. Wright, E.A., and Flanders, P.H., *Nature* 175 (1955):428-429.
46. van den Brenk, H.A.S., "Hyperbaric Oxygen in Radiation Therapy," *Am. J. Roent.* 102 (1968):8-26.
47. Henk, J.M.; Kunkler, P.B.; and Smith, C.W., "Radiotherapy and Hyperbaric Oxygen in Head and Neck Cancer: Final Report of First Clinical Trial," *Lancet* 2 (1977):101-103.
48. Powers, W.E., and Tolmach, L.J., "A Multicomponent X-ray Survival Curve for Mouse Lymphosarcoma Cells Irradiated *In Vivo*," *Nature* 197 (1963):710-711.
49. Groundbreaking papers concerning the proportion of hypoxic cells at and after irradiation are outlined in the following references: Kallman, R.F., and Bleehen, N.M., "Post-Irradiation Cyclic Radiosensitivity Changes in Tumors and Normal Tissues," in Brown D.G.; Cragle, R. G.; and Noonan, J.R., eds., *Proceedings of the Symposium on Dose Rate in Mammalian Radiobiology, Oak Ridge, TN, 1968.* (Springfield, Va.: CONF-680410, 1968):20.1-20.23; Kallman, R.F.; Jardine, L.J.; and Johnson, C.W., "Effects of Different Schedules of Dose Fractionation on the Oxygenation Status of a Transplantable Mouse Sarcoma," *J. Natl. Cancer. Inst.* 44 (1970):369-377; van Putten, L.M., and Kallman, R.F., "Oxygenation Status of a Transplantable Tumor during Fractionated Radiotherapy," *J. Natl. Cancer Inst.* 40 (1968):441-451; Thomlinson, R.H., "Changes of Oxygenation in Tumors in Relation to Irradiation," *Front. Rad. Ther. Oncol.* 3 (1968):109-121.
50. Highlights of Herman Suit's extensive studies dealing with the relationship between high-pressure oxygen and radiation dose fractionation in murine tumors include: Suit, H. D., and Suchato, C., "Hyperbaric Oxygen and Radiotherapy of a Fibrosarcoma and of a Squamous Cell Carcinoma of C3H Mice," *Radiology* 89 (1967):713-719; Suit, H.D., and Maeda, M., "Hyperbaric Oxygen and Radiobiology of a C3H Mouse Mammary Carcinoma," *J. Nat. Cancer Inst.* 39 (1967):639-652; Suit, H.D.; Howes, A.E.; and Hunter, N., "Dependence of Response of a C3H Mammary Carcinoma to Fractionated Irradiation on Fractionation Number and Intertreatment Interval," *Rad. Res.* 72 (1977):440-454.
51. The following are examples of Barendsen and colleague's research on variations in biological effectiveness due to radiation quality and its dependence on oxygen: Barendsen, G.W., "Impairment of the Proliferative Capacity of Human Cells in Culture by Alpha Particles with Differing Linear Energy Transfer," *Int. J. Rad. Biol.* 8 (1964):453-466; Barendsen, G.W.; Beusker, T.L.J.; Vergroesen, A.J.; and Budke, L., "Effects of Different Ionizing Radiations on Human Cells in

Tissue Culture. II. Biological Experiments," *Rad. Res.* 13 (1960):841-849; Barendsen, G.W., and Walter, H.M.D., "Effects of Different Ionizing Radiations on Human Cells in Tissue Culture. IV. Modification of Radiation Damage," *Rad. Res.* 18 (1963):106-119; Broerse, J.J., and Barendsen, G.W., "Relative Biological Effectiveness of Fast Neutrons for Effects on Normal Tissues," *Curr. Top. Rad. Res. Q.* 8 (1973):305-350.

52. Some of the continuing preclinical neutron studies at Hammersmith carried out after Hal Gray's departure were: Fowler, J.F., and Morgan, R.L., "Pretherapeutic Experiments with the Fast Neutron Beam from the Medical Research Council Cyclotron. VIII. General Review," *Br. J. Radiol.* 36 (1963):115-121; Field, S.B.; Jones, T.; and Thomlinson, R.H., "The Relative Effect of Fast Neutrons and X Rays on Tumor and Normal Tissue in the Rat. II. Fractionation Recovery and Reoxygenation," *Br. J. Radiol.* 41 (1968):597-607; Bewley, D.K., "Radiobiological Research with Fast Neutrons and the Implications for Radiotherapy," *Radiology* 86 (1966):251-257; Bewley, D.K.; Field, S.B.; Morgan, R.L.; et al., "The Response of Pig Skin to Fractionated Treatments with Fast Neutrons and X rays," *Br. J. Radiol.* 40 (1967):765-770.

53. A listing of Catteral's nascent clinical trials of neutrons: Catterall, M., "The Treatment of Advanced Cancer by Fast Neutrons from the Medical Research Council's Cyclotron at Hammersmith Hospital, London," *Eur. J. Cancer* 10 (1974):343-347; Catterall, M., "Results of Neutron Therapy: Differences, Correlations, and Improvements," *Int. J. Rad. Oncol. Biol. Phys.* 8 (1982):2141-2144; Catterall, M.; Sutherland, I.; and Bewley, D.K., "First Results of a Clinical Trial of Fast Neutrons Compared with X or Gamma Rays in Treatment of Advanced Tumors of the Head and Neck," *Br. Med. J.* 2 (1975):653-656; Catterall, M., and Vonberg, D.D., "Treatment of Advanced Tumors of the Head and Neck with Fast Neutrons," *Br. Med. J.* 3 (1974):137-143.

54. Two classic clinical efforts to use high-pressure oxygen in X-ray treatments are: Churchill-Davidson, Sanger, and Thomlinson, "High Pressure Oxygen and Radiotherapy," and van den Brenk, "Hyperbaric Oxygen in Radiation Therapy."

55. Highlights from Ged Adams's pivotal work to produce hypoxic cell radio sensitizers include: Adams, G.E., "Chemical Radiosensitization of Hypoxic Cells," *Br. Med. Bull.* 29 (1973):48-53; Adams, G.E.; Clarke, E.D.; Flockhart, I.R.; et al., "Structure-Activity Relationships in the Development of Hypoxic Cell Radiosensitizers. I. Sensitization Efficiency," *Int. J. Rad. Biol.* 35 (1979):133-150; Adams, G.E., and Dewey, D.L., "Hydrated Electrons and Radiobiological Sensitization," *Biochem. Biophys. Res. Commun.* 12 (1963):473-477.

56. Skarsgard, *Radiation Research* 8 (1969):493-500.

57. Howard, A., and Pelc, S.R., "Synthesis of Deoxyribonucleic Acid in Normal and Irradiated Cells and its Relation to Chromosome Breakage," *Heredity* (Suppl.) 6 (1952):261.

58. Quastler, H., and Sherman, F.G., "Cell Population Kinetics in the Intestinal Epithelium of the Mouse," *Exp. Cell. Res.* 17 (1959):420-438.

59. Mendelsohn, M.L., "The Growth Fraction: A New Concept Applied to Tumors," *Science* 132 (1960):1496.

60. Steel, G.G.; Adams, K.; and Barratt, J.C., "Analysis of the Cell Population Kinetics of Transplanted Tumors of Widely Differing Growth Rate," *Br. J. Cancer* 20 (1966):784-800.

61. Steel, G.G., "Cell Loss as a Factor in the Growth Rate of Human Tumors," *Eur. J. Cancer* 3 (1967):381-387; Tubiana, M., and Malaise, E.B., "Growth Rate and Cell Kinetics in Human Tumors: Some Prognostic and Therapeutic Implications," in Symington, T., and Carter R.L., eds. *Scientific Foundations of Oncology* (Chicago: Year Book, 1976):126-136.

62. Gray, J.W., "Cell-Cycle Analysis of Perturbed Cell Populations: Computer Simulation of Sequential DNA Distributions," *Cell Tissue Kinet.* 9 (1976):499-516; Gray, J.W.; Dolbeare, F.; Pallavicini, M.G.; et al., "Cell Cycle Analysis Using Flow Cytometry," *Int. J. Rad. Biol.* 49 (1986):237-255.

63. The following articles are some of the first to present models involving a nonzero initial dose slope: Barendsen, G.W., "Dose-survival Curves of Human Cells in Tissue Culture Irradiated with Alpha, Beta, 20 kV X and 200 kV X-radiation," *Nature* 193 (1962):1153-1155; Oliver, R., "A Comparison of the Effects of Acute and Protracted Gamma Radiation on the Growth of Seedlings of *Vicia faba* II. Theoretical Calculations," *Int. J. Rad. Biol.* 8 (1964):475-488; Deutreix, J.; Wambersie, A.; and Bounik, C., "Cellular Recovery in Human Skin Reactions. Application to Dose, Fraction Number, Overall Time Relationship in Radiotherapy," *Euro. J. Cancer* 9 (1973):159-167.

64. Douglas, B.G., and Fowler, J.F., "The Effect of Multiple Small Doses of X-rays on Skin Reactions in the Mouse and a Basic Interpretation," *Radiation Research* 66 (1976):401-426.

65. Lea, D.E., and Catcheside, D.C., "The Mechanism of the Induction by Radiation of Chromosome Aberrations in Tradescantia," *J. Genet* 44 (1942):216-245.

66. Kellerer, A.M., and Rossi, H.H., "The Theory of Dual Radiation Action," *Curr. Topics Rad. Res. Q.* 8 (1972):85-158.

67. Hall, E.J., "A Method of Deducing a Dose Response Relationship for Productive Integrity of Cells Exposed to Radiation by Means of Fractionation Experiments," *Br. J. Radiol.* 35 (1962):398.

68. Thames, H.D.; Withers, H.R.; Peters, L.J.; and Fletcher, G.H., "Changes in Early and Late Radiation Responses with Altered Dose Fractionation: Implications for Dose Survival Relationships," *Int. J. Rad. Oncol. Biol. Phys.* 8 (1982):219-226; Barendsen, "Dose-survival Curves of Human Cells."

69. Key articles that detail the differences between alpha/beta ratio values and tissue response times to irradiation include: Withers, H.R.; and Thames, H.D.; and Peters, L.J., "Differences in the Fractionation Response of Acutely and Late-Responding Tissues," in Kaercher K.H.; Kogelnik, H.D.; and Reinartz, G., eds. *Progress in Radio-Oncology.* Vol. II. (New York: Raven Press,

1982):287-296; Withers, H.R.; Thames, H.D.; Peters, L.J.; and Fletcher, G.H., "Keynote Address: Normal Tissue Radioresistance in Clinical Radiotherapy," Chapter 21 in Fletcher, G.H.; Nervi, C.; and Withers, H.R., eds. *Biological Bases and Clinical Implications of Tumor Radioresistance* (New York: Masson Publ., 1983):139-152.

70 Turesson, I., and Notter, G., "The Influence of Overall Treatment Time in Radiotherapy on the Acute Reaction: Comparison of the Effects of Daily and Twice-a-Week Fractionation on Human Skin," *Int. J. Rad. Oncol. Biol. Phys.* 10 (1984):607-618.

71 Denekamp, J.; Ball, M.M.; and Fowler, J.F., "Recovery and Repopulation in Mouse Skin as a Function of Time after X-irradiation," *Radiation Research* 30 (1969):361-370; Denekamp, J., "Changes in the Rate of Repopulation during Multifraction Irradiation of Mouse Skin," *Br. J. Radiol.* 46 (1973):381-387.

72 Withers, H.R.; Taylor, J.M.G.; and Maciejewski, B., "The Hazard of Accelerated Tumor Clonogen Repopulation during Radiotherapy," *Acta Oncol.* 27 (1988):131; Maciejewski, B.; Withers, H.R.; and Taylor, J.M.G., "Dose Fractionation and Regeneration in Radiotherapy for Cancer of the Oral Cavity and Oropharynx. I. Tumor Dose-Reponse and Repopulation," *Int. J. Rad. Oncol. Biol. Phys.* 16 (1989):831.

73 Hermens, A.F., and Barendsen, G.W., "Changes of Cell Proliferation Characteristics in a Rat Rhabdomyosarcoma Before and After X-irradiation," *Eur. J. Cancer* 5 (1969):173.

74 Thames, H.D.; Withers, H.R.; Peters, L.J.; and Fletcher, G.H., "Changes in Early and Late Radiation Responses with Altered Dose Fractionation: Implications for Dose Survival Relationships," *Int. Rad. Oncol. Biol. Phys.* 8 (1982):219-226.

75 Fowler, J.F., "Review Article: The Linear-Quadratic Formula and Progress in Fractionated Radiotherapy," *Br. J. Radiol.* 62 (1989):679-694.

76 The earliest neutron experience of the Medical Research Council at Hammersmith Hospital produced encouraging results, despite the physical limitations of the low energy cyclotron that was available. This stimulated great interest in neutron therapy in the United States, Europe, and Japan. Battermann, J.J., and Mijnheer, B.J., "The Amsterdam Fast Neutron Radiotherapy Project: A Final Report," *Int. J. Rad. Oncol. Biol. Phys.* 12 (1986):2093; Duncan, W.; Arnott, S.J.; Orr, J.A.; et al., "The Edinburgh Experience of Fast Neutron Therapy," *Int. J. Rad. Oncol. Biol. Phys.* 8 (1982):2155; Griffin, T.W.; Davis, R.; and Hendrickson, F.R., "Fast Neutron Radiation Therapy for Unresectable Squamous Cell Carcinomas of the Head and Neck: The Results of a Randomized RTOG Study," *Int. J. Radiat Oncol. Biol. Phys.* 10 (1984):2217.

77 Griffin, T.W.; Pajak, T.F.; Laramore, G.E.; et al., "Neutron vs. Photon Irradiation of Inoperable Salivary Gland Tumors: Results of an RTOG-MRC Cooperative Randomized Study," *Int. J. Rad. Oncol. Biol. Phys.* 15 (1988):1085; Griffin, T.W.; Pajak, T.F.; Mayor, M.H.; et al., "Mixed Neutron/Photon Irradiation of Unresectable Squamous Cell Carcinomas of the Head and Neck: The Final Report of a Randomized Cooperative Trial," *Int. J. Rad. Oncol. Biol. Phys.* 17 (1989):959.

78 Kligerman, M.; Tsuji, H.; Bagshaw, M.; et al., "Current Observation of Pion Radiation Therapy at LAMPF," in Abe, M., and Sakamoto, K., eds. *Treatment of Radioresistant Cancers* (Amsterdam: Elsevier/ North Holland Biomedical Press, 1979):145-157; Bush, S.E.; Smith, A.R.; and Zink, S., "Pion Radiotherapy at LAMPF," *Int. J. Radiation Oncol. Biol. Phys.* 8 (1982):2181-2186.

79 Tobias, C.A.; Blakely, E.A.; Alpen, E.L.; et al., "Molecular and Cellular Radiobiology of Heavy Ions," *Int. J. Rad. Oncol. Biol Phys.* 8 (1982):2109-2120; Phillips, T.L.; Ross, G.Y.; Goldstein, L.S.; et al., "*In Vivo* Radiobiology of Heavy Ions," *Int. J. Rad. Oncol. Biol. Phys.* 8 (1982):2121-2125.

80 Castro, J.R.; Saunders, W.M.; Tobias, C.A.; et al., "Treatment of Cancer with Heavy Charged Particles," *Int. J. Rad. Oncol. Biol. Phys.* 8 (1982):2191-2198.

81 Suit, H.; Goitein, M.; Munzenrider, J.; et al., "Evaluation of the Clinical Applicability of Proton Beams in Definitive Fractionated Radiation Therapy," *Int. J. Rad. Oncol. Biol. Phys.* 8 (1982):2199-2205.

82 Adams and Dewey, "Hydrated Electrons and Radiobiological Sensitization."

83 Overgaard, 1993

84 Brown, J.M. *Modification of Radiosensitivity in Cancer Treatment* (Tokyo: Academic Press, 1984):139-176.

85 Sutherland, R.M., "Selective Chemotherapy of Non-Cycling Cells in an *In Vitro* Tumor Model," *Cancer Res.* 34 (1974):3501.

86 Hall, E.J., and Roizin-Towle, L., "Hypoxic Sensitizers: Radiobiological Studies at the Cellular Level," *Radiology* 117 (1975):453.

87 Jenkins, T.C.; Naylor, M.A.; O'Neil, P.; et al., "Synthesis and Evaluation of 1-(3-(2-haloethylamino)propyl)-2-nitroimidazoles as Pro-Drugs of RSU 1069 and its Analogues, which are Radiosensitizers and Bioreductively Activated Cytotoxins," *Med. Chem.* 33 (1990):2603-2610; Zeman, E.M.; Brown, J.M.; Lemmon, M.J.; et al., "SR 4233: A New Bioreductive Agent with High Selective Toxicity for Hypoxic Mammalian Cells," *Int. J. Rad. Oncol. Biol. Phys.* 12 (1986):1239-1242.

88 Malaise's and Steel's exceptional histological and kinetic studies of tumors were crucial to the development of inherent radiosensitivity assays and are detailed in the following articles: Fertil, B., and Malaise, E.P., "Intrinsic Radiosensitivity of Human Cell Lines is Correlated with Radioresponsiveness of Human Tumors: Analysis of 101 Published Survival Curves," *Int. J. Rad. Oncol. Biol. Phys.* 11 (1985):1699-1707; Deacon, J.; Peckham, M.J.; and Steel, G.G., "The Radioresponsiveness of Human Tumors and the Initial Slope of the Cell-Survival Curve," *Radiother. Oncol.* 2 (1984):217-323; Malaise, E.P.; Fertil, B.; Chavandra, N.; and Guichard, M., "Distribution of Radiation Sensitivities for Human Tumor Cells of Specific Histological Types: Comparison of *In*

Vitro to *In Vivo* Data," *Int. J. Rad. Oncol. Biol. Phys.* 12 (1986):617-624.
89. Brock, W.; Campbell, H.; Goepfert, H.; and Peters, L.J., "Radiosensitivity Testing of Human Tumor Cell Cultures: A Potential Method of Predicting the Response to Radiotherapy," *Cancer Bull.* 39 (1987):98-102.
90. West, C.M.L.; Davidson, S.E.; Hendry, J.H.; and Hunter, R.D., "Prediction of Cervical Carcinoma Response to Radiotherapy," *Lancet* 338 (1991):818.
91. Bergsjo, P.; Christensen, O.J.; and Kolstad, P., "Oxygen Tension in Cancer of the Cervix Following Administration of Vasodilator Drugs during Oxygen Inhalation," *Cancer* 20 (1967):1625-1634.
92. Vaupel, P.; Schlenger, K.; Knoop, C.; and Hockel, M., "Oxygenation of Human Tumors: Evaluation of Tissue Oxygen Distribution in Breast Cancers by Computerized O_2 Tension Measurements," *Cancer Res.* 51 (1991):3316-3322.
93. Hockel, M.; Knoop, C.; Schlenger, K.; et al., "Intratumoral pO_2 Predicts Survival in Advanced Cancer of the Uterine Cervix," *Radiother. Oncol.* (in press).
94. Parliament, M.B.; Chapman, J.D.; Urtasun, R.C.; et al., "Non-Invasive Assessment of Human Tumor Hypoxia with ^{123}I-Iodoazomycin Arabinoside: Preliminary Report of a Clinical Study," *Br. J. Cancer* 65 (1992):90-95.
95. Koh, W.J.; Rasey, J.S.; Evans, M.L.; et al., "Imaging of Hypoxia in Human Tumors with ^{18}F-fluoromisonidazole," *Int. J. Rad. Oncol. Biol. Phys.* 22 (1992):199-212.
96. Begg, A.C.; McNally, N.J.; Shrieve, D.; et al., "A Method to Measure the Duration of DNA Synthesis and the Potential Doubling Time from a Single Sample," *Cytometry* 6 (1985):620-625.
97. Begg, A.C.; Hofland, I.; Van Glabekke, M.; et al., "Predictive Value of Potential Doubling Time for Radiotherapy of Head and Neck Tumor Patients: Results from the EORTC Cooperative Trial 22851," *Sem. Rad. Oncol.* 2 (1992):22-25.
98. Elkind and Sutton, "X-ray Damage and Recovery."
99. Lajtha, L.G., and Oliver, R., "Some Radiobiological Considerations in Radiotherapy," *Br. J. Radiol.* 34 (1961):252-257.
100. Hall, E.J., and Bedford, J.S., "Dose Rate: Its Effect on the Survival of HeLa Cells Irradiated with Gamma Rays," *Rad. Res.* 22 (1964):305-315.
101. Paterson, R. *The Treatment of Malignant Disease by Radiotherapy*. 2nd edition. London: Edward Arnold, 1963; Ellis, F., "Dose, Time and Fractionation in Radiotherapy," in Ebert and Howard, " Current Topics in Radiation Research," pp. 359-397.
102. Pierquin, B.; Chassagne, D.; Baillet, F.; et al., "Clinical Observations on the Time Factor in Interstitial Radiotherapy Using Iridium-192," *Clin. Radiol.* 24 (1973):506-509.
103. Pierquin, B., "Dosimetry: The Relational System," in *Proceedings of a Conference on Afterloading in Radiotherapy* (Rockville, Md.: U.S. Department of Health, Education and Welfare, Publication number (FDA) 72-8024, 1971):204-227 .
104. Mazeron, J.J.; Simon, J.M.; Le Pechoux, C.; et al., "Effect of Dose-Rate on Local Control and Complications in Definitive Irradiation of T1-2 Squamous Cell Carcinomas of Mobile Tongue and Floor of Mouth with Interstitial Iridium-192," *Radiother. Oncol.* 21 (1991):39-47.
105. Mazeron, J.J.; Simon, J.M.; Crook, J.; et al., "Influence of Dose-Rate on Local Control of Breast Carcinoma Treated by External Beam Irradiation Plus Iridium-192 Implant," *Int. J. Rad. Oncol. Biol. Phys.* 21 (1991):1173-1177.
106. Hall, E.J., and Brenner, D.J., "The Dose-Rate Effect in Interstitial Brachytherapy: A Controversy Resolved," *Br. J. Radiol.* 65 (1992):242-247.
107. These studies demonstrate the comparable effectiveness of high-dose rate therapy and low-dose rate therapy: O'Connell, D.; Howard, N.; Joslin, C.A.F.; et al., "A New Remotely Controlled Unit for the Treatment of Uterine Carcinoma," *Lancet* 2 (1965):570-571; Joslin, C.A.F.; Smith, C.W.; and Mallik, A., "The Treatment of Cervix Cancer Using High Activity Cobalt-60 Sources," *Br. J. Radiol.* 45 (1972):257-270; Shigmatsu, Y.; Nishiyama, K.; Masaki, N.; et al., "Treatment of the Uterine Cervix by Remotely Controlled Afterloading Intracavitary Radiotherapy with High Dose Rate: A Comparative Study with a Low Dose Rate System," *Int. J. Rad. Oncol. Biol. Phys.* 9 (1983):351-356; Newman, H.; Jur, B.; James, K.W.; and Smith, C.W., "Treatment of Cancer of the Cervix with a High Dose Rate Afterloading Machine," *Int. J. Rad. Oncol. Biol. Phys.* 9 (1983):931-937; Utley, J.F.; Essen, C.F.; Horn, R.A.; and Moellen, J.H., "High Dose Rate Afterloading Brachytherapy in Carcinoma of the Uterine Cervix," *Int. J. Rad. Oncol. Biol. Phys.* 10 (1984):2259-2263; Thomadsen, B.R.; Shahabi, S.; Stitt, J.A.; et al., "High Dose Rate Intracavitary Brachytherapy for Carcinoma of the Cervix: The Madison System: II. Procedural and Physical Considerations," *Int. J. Rad. Oncol. Biol. Phys.* 24 (1992):349-357; Orton, C.G.; Seyedsadr, M.; and Somnay, A., "Comparison of High and Low Dose Rate Remote Afterloading for Cervix Cancer and the Importance of Fractionation," *Int. J. Rad. Oncol. Biol. Phys.* 21(6) (1991):1425-1434.
108. These studies use radiobiological modeling to compare high-dose rate exposure to low-dose rate: Brenner, D.J., and Hall, E.J., "Fractionated High Dose-Rate Versus Low Dose-Rate Brachytherapy of the Cervix. I. General Considerations Based on Radiobiology," *Br. J. Radiol.* 64 (1991):133-141; Dale, R.G., "The Application of the Linear-Quadratic Dose-Effect Equation to Fractionated and Protracted Radiotherapy," *Br. J. Radiol.* 58 (1985):515-528; Dale, R.G., "The Use of Small Fraction Numbers in High Dose-Rate Gynaecological Afterloading: Some Radiobiological Considerations," *Br. J. Radiol.* 63 (1990):290-294; Fowler, J.F., "Modelling Altered Fractionation Schedules," *Br. J.*

Radiol. (Suppl.) 24 (1992):187-192; Orton, C.G., "Fractionated High Dose Rate Versus Low Dose Rate Cervix Cancer Regimes," *Br. J. Radiol.* 64(768) (1991):1165-1166.

109 Brenner, D.J., and Hall, E.J., "Conditions for the Equivalence of Continuous to Pulsed Low Dose Rate Brachytherapy," *Int. J. Rad. Oncol. Biol. Phys.* 20 (1991):181-190; Fowler, J., and Mount, M., "Pulsed Brachytherapy: The Conditions for No Significant Loss of Therapeutic Ratio Compared with Traditional Low Dose Rate Brachytherapy," *Int. Rad. Oncol. Biol. Phys.* 23 (1991):661-669.

110 The following articles detail nascent laboratory research of hyperthermia: Dewey, W.C.; Hopwood, L.E.; Sapareto, L.A.; and Gerweck, L.E., "Cellular Responses to Combinations of Hyperthermia and Radiation," *Radiology* 123 (1977):463-474; Westra, A., and Dewey, W.C., "Variation in Sensitivity to Heat Shock during the Cell Cycle of Chinese Hamster Cells *In Vitro*," *Int. J. Radiat. Biol.* 19 (1971):467-477; Robinson, J.E., and Wizenburg, M.J., "Thermal Sensitivity and the Effect of Elevated Temperatures on the Radiation Sensitivity of Chinese Hamster Cells," *Acta Radiol.* 13 (1974):241-248.

111 Examples of research that address the efficacy of hyperthermia under varying conditions, *e.g.*, hypoxia, low pH, include: Harisiadis, L.; Hall, E.J.; Kraljevic, U.; et al., "Hyperthermia: Biological Studies at the Cellular Level," *Radiology* 117 (1975):447-452; Gerweck, L.E.; Nygaard, T.G.; and Burlett, M., "Response of Cells to Hyperthermia under Acute and Chronic Hypoxic Conditions," *Cancer Res.* 39 (1979):966-972; Hahn, G.M., "Metabolic Aspects of the Role of Hyperthermia in Mammalian Cell Inactivation and Their Possible Relevance to Cancer Treatment," *Cancer Res.* 34 (1974):3117-3123; Freeman, M.L.; Raaphorst, G.P.; Hopwood, L.E.; et al., "The Effect of pH on Cell Lethality Induced by Hyperthermic Treatment," *Cancer* 45 (1980):61-70.

112 The mechanism and effects of thermotolerance on the use of hyperthermia as a clinical treatement are reported in: Gerner, E.W., and Schneider, M.J., "Induced Thermal Resistance in HeLa Cells," *Nature* (London) 256 (1975):500-502; Gerner, E.W.; Boon, R.; Conner, W.G.; et al., "A Transient Thermotolerant Survival Response Produced by Single Thermal Dose in HeLa Cells," *Cancer Res.* 36 (1976):1035-1040; Leeper, D.B., and Henle, K.J., "Hyperthermia: Effect of Different Temperatures on Normal and Tumor Cells," Wizenberg, M.J., and Robinson, J.E., eds. *International Symposium on Cancer Therapy by Hyperthermia and Radiation*. (Chicago: American College of Radiology, 1975):47-60.

113 Hahn, G.M., and Li, G.C., "Thermotolerance and Heat Shock Proteins in Mammalian Cells," *Rad. Res.* 92 (1982):452-457; Li, G.C., *Int. J. Rad. Oncol. Biol. Phys.* 11 (1985):165.

114 Field, S.B., and Hume, S.P., "Hyperthermia in Animals," in Fielden, E.M.; Fowler, J.F.; Hendry, J.H.; and Scott, D., eds. *Radiation Research*. Vol. 2 (London: Taylor and Francis, 1987):960-965.

115 Gillette, E.L., "Hyperthermic Effects in Animals with Spontaneous Tumors," in Dethlefsen, L.A., and Dewey, W.C., eds. *Third International Symposium: Cancer Therapy by Hyperthermia, Drugs, and Radiation* (Washington: National Cancer Institutes Monographs, 1989):361-364; Dewhirst, M.W.; Sim, D.A.; Sapareto, S; and Connor, W.G., "Importance of Minimum Tumor Temperature in Determining Early and Long-Term Responses of Spontaneous Canine and Feline Tumors to Heat and Radiation," *Cancer Res.* 44 (1984):43-50.

116 Overgaard, J., "The Design of Clinical Trials," in Field, S.B., and Franconi, C., eds. *Hyperthermic Oncology: Physics and Technology of Hyperthermia*. Amsterdam: Martinus Nijhuff Publishers, 1982.

117 The following studies illustrate hyperthermia's efficacy in increasing the response rate of many types of human cancer to radiation treatment: Arcangeli, G.; Cevidalli, A.; Nervi, C.; et al., "Tumor Control and Therapeutic Gain with Different Schedules of Combined Radiotherapy and Local External Hyperthermia in Human Cancer," *Int. J. Rad. Oncol. Biol. Phys.* 9 (1983):1125-1134; Overgaard, J., "Hyperthermia Modification of the Radiation Response in Solid Tumors," in Fletcher, G.; Nervoi, C.; and Withers, H., eds. *Biological Basis and Clinical Implications of Tumor Resistance*. (New York: Masson Publishing, 1983):337-352; Overgaard, J., "Historical Perspectives of Hyperthermia," in Overgaard, J., ed. *Introduction to Hyperthermic Oncology*. Vol. 2. New York: Taylor and Francis, 1984; Overgaard, J., "Rationale and Problems in the Design of Clinical Studies," in Overgaard, J., ed. *Hyperthermic Oncology*. Vol. 2, (London: Taylor and Francis, 1985):325-338; Overgaard, J., and Overgaard, M., "Hyperthermia as an Adjuvant to Radiotherapy in the Treatment of Malignant Melanoma," *Int. J. Hyperther.* 3 (1987):483-501; Oleson, J.R.; Manning, M.R.; Sim, D.A.; et al., "A Review of the University of Arizona Human Clinical Hyperthermia Experience," in Vaeth, J.M., ed. *Frontiers of Radiation and Oncology*. Vol. 18. (Basel: Karger, 1984):136-143; Perez, C.A.; Kuske, R.R.; Emani, B.; and Fineberg, B., "Irradiation Alone or Combined with Hyperthermia in the Treatment of Recurrent Carcinoma of the Breast in the Chest Wall: A Nonrandomized Comparison," *Int. J. Hyperther.* 2 (1986):179-187.

118 Overgaard, J.; Gonzalez-Gonzalez, D.; Arcangeli, G.; et al., "Hyperthermia as an Adjuvant to Radiation Therapy of Recurrent or Metatastic Malignant Melanoma (ESHO 3-85): A Multicenter Randomized Trial," 13th North American Hyperthermia Society/41st Radiation Research Society meeting, Dallas, TX, 1993, Abs. #P-01-17.

119 Hall, E.J., "Hyperthermia; An Overview," in Dethlefsen and Dewey, *Third International Symposium*, pp. 15-16.

120 Coughlin, C.T.; Wong, T.Z.; Strohbehn, J.W.; et al., "Intraoperative Interstitial Microwave-Induced Hyperthermia and Brachytherapy," *Int. J. Radiat. Oncol. Biol. Phys.* 11 (1985):1673-1678; Emami, B.H.; Perez, C.A.; Leybovich, L.; et al., "Interstitial Thermoradiotherapy in Treatment of Malignant Tumors," *Int. J. Hyperther.* 3 (1987):107-118.

121 Rubin, J.S.; Joyner, A.L.; Bernstein, A.; and Whitmore, G.F., "Molecular Identification of a Human

DNA Repair Gene Following DNA-Mediated Gene Transfer," *Nature* 306 (1983):206; van Duin, M.; deWit, J.; Odijk, H.; et al., "Molecular Cloning and Characterization of the Human Excision Repair Gene ERCC-1: cDNA Cloning and Amino Acid Homology with the Yeast DNA Repair Gene," *Cell* 44 (1986):913-923.

122. Thompson, L.H.; Brookman, K.W.; Jones, N.J.; et al., "Molecular Cloning of the Human XRCC1 Gene, Which Corrects Defective DNA Strand Break Repair and Sister Chromatid Exchange," *Mol. Cell Biol.* (1990):6160-6171.

123. Lieberman, H.B.; Hopkins, K.M.; Laverty, M.; and Chu, H.M., "Molecular Cloning and Analysis of *Schizosaccharomyces pombe rad9*, a Gene Involved in DNA Repair and Mutagenesis," *Mol. Gen. Genet.* 232 (1992):367-376; Lieberman, H.B., "Extragenic Suppressors of *Schizosaccharomyces pombe rad9* Mutations Uncouple Radioresistance, Hydroxyurea Sensitivity and Cell Cycle Checkpoint Control," *Genetics* 1993 (in press).

124. Swift, M.; Morrell, D.; Massey, R.B.; and Chase, C.L., "Incidence of Cancer in 161 Families Affected by Ataxia-Telangectasia," *New Eng. J. Med.* 325 (1991):1831-1836.

125. Bishop, J.M., "Cellular Oncogene Retroviruses," *Annu. Rev. Biochem.* 52 (1983):301-354; Bishop, J.M., and Varmus, H.E., "Functions and Origins of Retroviral Transforming Genes," in: Weiss, R.; Teich, N.; Varmus, H.; and Coffin., J., eds. *RNA Tumor Viruses: Molecular Biology of Tumor Viruses*. 2nd ed. (Cold Spring Harbor, N.Y.: Cold Spring Harbor Laboratory, 1984):990-1108.

126. Bos, J.L., "The *ras* Gene Family and Human Carcinogenesis," *Mutat. Res.* 195 (1988):255-271.

127. Sklar, M.D., "The *ras* Oncogenes Increase the Intrinsic Resistance of NIH3T3 Cells to Ionizing Radiation," *Science* 239 (1988):645-647.

128. McKenna, W.G.; Weiss, M.C.; Endlich, B.; et al., "Synergistic Effect of the v-*myc* Oncogene with H-*ras* on Radioresistance," *Cancer Res.* 50 (1990):97-102.

129. Land, H., Parada, L. F., Weinberg, R. A., "Tumorigenic conversion of primary embryo fibroblasts requires at least two cooperating oncogenes," *Nature* 304 (1983):596-602; Hunter, T., "Cooperation Between Oncogenes," *Cell* 64 (1991):249-270.

130. Stanbridge, E.J., "Suppression of Malignancy in Human Cells," *Nature* 260 (1976):17-20.

131. Saxon, P.J.; Srivatsan, E.S.; and Stanbridge, E.J., "Introduction of Normal Human Chromosome 11 via Microcell Transfer Controls Tumorigenic Expression of HeLa Cells," *EMBO J.* 5 (1986):3461-3466.

132. Knudson, A.G., "Mutation and Cancer: Statistical Study of Retinoblastoma," *Proc. Natl. Acad. Sci.* 68 (1971):820-823.

133. Cavenee, W.K.; Hansen, M.F.; Nordenskjold, M.; et al., "Genetic Origin of Mutations Predisposing to Retinoblaststoma," *Science* 228 (1985):501-503.

134. Lee, W.H.; Bookstein, R.; Hong Young, L. H., Shew, E. Y., and Lee, H. P., "Human Retinoblastoma Susceptibility Gene: Cloning Identification and Sequence," *Science* 235 (1987):1394-1399.

135. Cavanee, W.K., "Tumor Progression Stage: Specific Losses of Heterozygosity," *Int. Symp. Princess Takamatsu Cancer Res. Fund.* 20 (1989):33-42.

136. Peltomaki, P.; Aaltonen, L.A.; Sistonen, P.; et al., "Genetic Mapping of a Locus Predisposing to Human Colorectal Cancer," *Science* 260 (1991):810-812; Aaltonen, L.A.; Peltomaki, P.; Leach, F.S.; et al., "Clues to the Pathogenesis of Familial Colorectal Cancer," *Science* 260 (1991):812-816.

137. Vogelstein, B., "A Deadly Inheritance," *Nature* 348 (1990):681.

CHAPTER SIX

Training and Education

Nancy Knight

The extraordinary attention that greeted Röntgen's announcement of the X rays extended well beyond the academic confines of medicine and physics. "With no discovery within my recollection has the immediate and general excitement been so intense...as with this discovery of Professor Roentgen," noted a San Francisco physician in 1897 in the *Journal of the American Medical Association (J.A.M.A.)*.[1] As the early unrestricted and empirical applications of the rays to disease became more structured and as awareness of the potential dangers grew, distinct schools of practice evolved around various specialists in the United States and Europe. What ensued was a virtual apprenticeship system in academic and clinical medicine, where specific and often idiosyncratic approaches to radiation therapy were learned and passed on to physicians in training. The very limited numbers of persons specializing in the field kept this system in place until well into the twentieth century. The advent of standardized radiation measurement, of enhanced efforts at radiation protection, of increasingly stringent requirements for training and certification, and of large controlled studies has carried the profession into modern times.

This brief overview of instruction and education will survey what has formed the backbone of the history of radiation oncology: those mechanisms by which the field has communicated its growing body of knowledge and experience to successive generations of dedicated practitioners.

The Early Years: 1896–1915

The X ray was a particularly democratic phenomenon in its earliest years—seeing through rich and poor equally and appearing to offer career opportunities to an eclectic range of professionals and hobbyists. Edward Trevert, author of the popular how-to manual, *Something About X Rays for Everybody*, noted in June 1896 that "The layman as well as the professional has been experimenting and trying to obtain knowledge of the source, action, and effect of these rays" (Fig. 6.1).[2] By "professional" Trevert did not mean to imply only the licensed medical doctor. His definition included professional physicists, electricians, photographers, glassblowers, and inventors. The true

Fig. 6.1 Numerous early textbooks were available to teach professionals and amateurs how to generate the new rays of Röntgen. (Courtesy of the Center for the American History of Radiology, Reston, Va.)

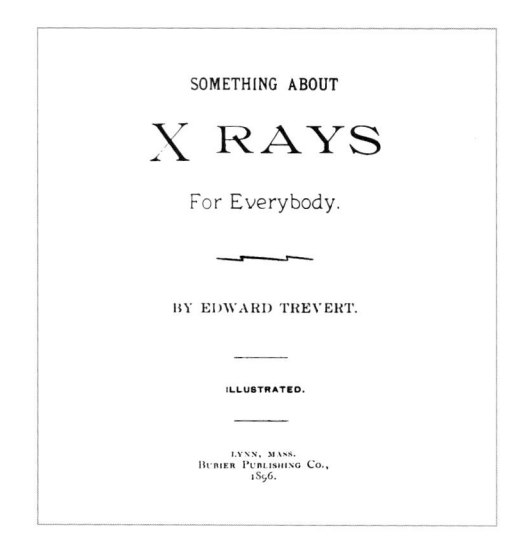

amateur was the hobbyist, interested in seeing the wonders of Röntgen's discovery in his or her own home. Trevert assured these legions of science fans that "with proper care and the necessary apparatus even an amateur may meet with wonderful success" in obtaining radiographs of everything from the family cat to long occluded foreign bodies.[3] Trevert's volume was followed closely by W. Meadowcroft's breezy *The ABC of the X Rays*, a title implying the very simplicity with which the field could be mastered.[4]

Soon a variety of instructional materials were available, many through the mail, to assist fledgling roentgenologists with their work. Röntgen's own *Eine Neue Arte von Strahlen*, reprinted for the third time in the spring of 1896, boasted advertisements for a variety of illustrative materials, including Röntgen's own radiograph of the hand of the anatomist Von Kolliker.[5] The earliest issues of the *Archives of Clinical Skiagraphy*, published later that year, featured similar compendiums of illustrative radiographs and instructional booklets for sale (Fig. 6.2).[6] In New York City, also in 1896, William J. Morton's educational X-ray prints were sold and offered as premiums by E. B. Meyerowitz, an X-ray equipment manufacturer.

But despite a growing medical literature citing the possibility and soon the certainty that the new rays had startling and beneficial effects on a variety of surface and deep-seated diseases, almost all of the earliest instructional materials concentrated exclusively on radiography and diagnosis. The emphasis was on providing examples of anatomical radiographs (on the not unreasonable assumption that many of the recipients had neither studied anatomy nor witnessed traditional postmortem dissections) and descriptions of workable combinations of generators, X-ray tubes, and screens. Instruction in therapy seemed to follow without additional comment. The implication was that one sat the patient down, aimed the X-ray tube with optimistic intent, turned the machine on, waited an unspecified period of time, and hoped for the best.

Throughout this early period tips on shielding, distance, voltage, and tube placement appeared in both medical and popular journals. Publications from the *New York Times* to *Electrical Engineering* to *Century Magazine* issued detailed drawings and instructions.[7] Soon a number of individuals began to specialize in the new rays, while others sought instruction in their use.

Although there were no hard and fast lines of demarcation, medical practice in the United States at the turn of the century was divided. In the east, urban-oriented physicians schooled in mainstream institutions remained closely tied to European traditions of medical education. In the rest of the country, physicians were trained at

Fig. 6.2 Advertisements for equipment, instructional books, and radiographic study series appeared in early issues of the *Archives of Clinical Skiagraphy* in England. The journal was read on both sides of the Atlantic by both medical and nonmedical practitioners. (Courtesy of the Center for the American History of Radiology, Reston, Va.)

Fig. 6.3 Harry Preston Pratt offered a variety of instructional services at his Chicago X-ray laboratory. (Courtesy of the Center for the American History of Radiology, Reston, Va.)

Fig. 6.4 Although M. E. Parberry was trained as an electrician, he offered instruction in X-ray technique and therapy in a fully-equipped lab to interested physicians at the turn of the century. (Courtesy of the Center for the American History of Radiology, Reston, Va.)

Fig. 6.5 Heber Robarts, founder of the American Roentgen Ray Society and its first official organ, the *American X-Ray Journal*, offered to teach physicians how to apply the rays to a range of ailments. (Courtesy of the Center for the American History of Radiology, Reston, Va.)

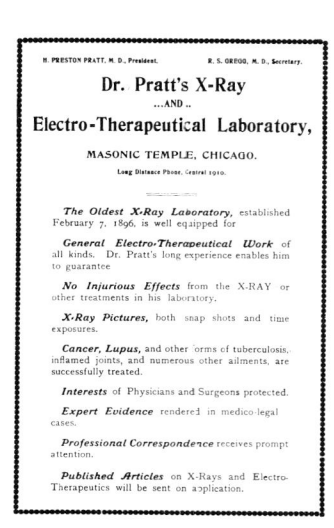

more eclectic schools and received the bulk of their education as apprentices to older physicians.[8] The two groups and their mutual disdain would be reflected in divisions among radiologists for decades.

The X ray was received quite differently by the two groups. In the east radiology developed within teaching hospitals, many of which opened their own X-ray rooms. In the midwest X-ray education appeared simultaneously in a variety of settings: independent X-ray "laboratories," private clinics, osteopathic and homeopathic schools, and in the fraudulent vending of certificates of expertise. Much more attuned to the promising notion that the X rays offered "something for everybody," midwesterners were the first to begin formalized (if eclectic) instruction in radiology and to experiment with possible therapeutic effects.

In the 22 July 1896 issue of the *Electrical Engineer*, Harry Preston Pratt, a Chicago physician, claimed to have treated "an almost hopeless case of consumption with results 'so far but little short of marvelous'."[9] Advertisements for Pratt's "X-Ray and Electrotherapeutic Laboratory" appeared almost immediately, noting that published articles would be sent out for instructional purposes (Fig. 6.3).[10] Wolfram Fuchs, proprietor of the large Chicago X-ray Laboratory, also advertised instruction in the use of the new light.[11] In St. Louis both M. E. Parberry, an electrician, and Heber Robarts, a physician, offered to teach the elements of radiation therapy at their respective offices (Figs. 6.4 and 6.5). Physicians were invited to drop in at Robarts's office between 10:00 and 2:00 to see "patients under treatment, with cancer, lupus, rodent ulcers, neuralgias, and those diseases rebellious to medicine, or inoperable."[12] What these and others offered was specialized, one-on-one instruction by example—short apprenticeships—in the mechanics of the X ray, with additional pointers on successes and failures noted so far.

Émil Grubbé, the putative American pioneer of radiation therapy (see Chapter 1), felt keenly the opprobrium of practice on the fringes of respected medicine. At the 1898 establishment of an X-ray clinic at Hahnemann University Hospital in Chicago, Grubbé (armed with a newly-minted medical degree of his own) was named professor of roentgenology. He found that much of the work pertained to diagno-

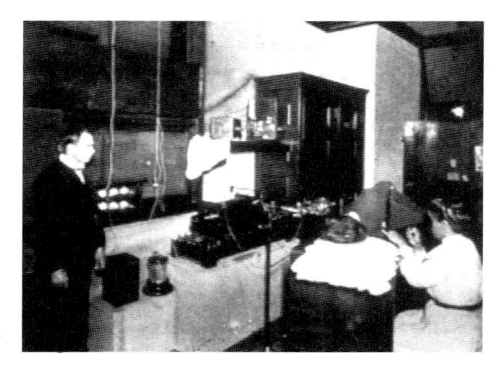

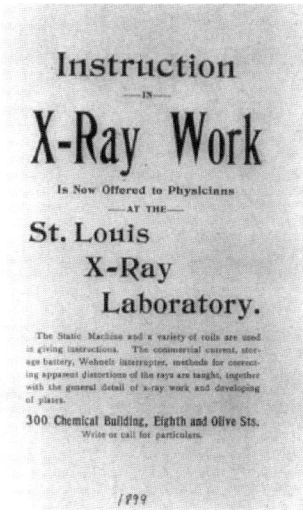

Knight

sis, however. In an effort to elevate instruction in the use of the new rays in therapeutic applications, Grubbé spearheaded the movement to incorporate the Illinois School of Electrotherapeutics, located in the Champlain Building, with a faculty of licensed physicians and the advertised promise of a "handsome engraved certificate" for the three-week course.[13] For physicians in a hurry, the "two weeks' course will make you self-dependent."

The school opened in August of 1899, with several rooms outfitted with equipment for demonstrations in both electrotherapy and radiation treatment (Figs. 6.6 and 6.7). Patients were accepted by referral from staff and other physicians. Grubbé's designation of this venture as a medical school may seem somewhat generous today, but it was probably no better or worse than many similar enterprises claiming to give medical instruction at the turn of the century. In fact, Grubbé was scrupulous in refusing to issue diplomas by mail to students who had never attended classes. To his delight, he found that there were students eager to attend the school:

> ...at last, a turning point had been reached; the medical profession had awakened from its sleep and was really becoming interested in x-ray therapy. I had positive evidence of this awakening for, nearly every day, physicians would call at my office to have me explain this new treatment, and to see with their own eyes just how x-ray applications could be made to various pathological conditions located in, or on, different parts of the human body.[14]

Many of these visitors attended classes at the Illinois School of Electrotherapeutics, which, in its twenty-one years graduated more than 5,000 students. The school closed in 1920 because of what Grubbé called "the disruption of postgraduate medical teaching produced by World War I," a defensive way to rationalize the advent of the more stringent standards for medical education that would close many such schools.[15]

While much of the early instruction in X rays in the east came directly from physicians associated with teaching hospitals, the region was not immune to less orthodox teaching methods. The Electrical Institute of Correspondence Instruction in New York City advertised a mail order course in roentgen ray therapeutics which by 1900 had been "followed successfully by thousands."[16] The Brooklyn Post-Graduate School of Clinical Electrotherapeutics (often referred to simply as the New York School), headed by noted electrotherapist and X-ray author Samuel Monell, advertised a wide curriculum, including the option of education by correspondence using Monell's textbooks (Fig. 6.8).[17]

Doubtless many electricians and photographers used diplomas from these and other short courses to gain entrance into legitimate medical practice. Despite the outpouring of invective against such practitioners by mainstream physicians, it is not at all clear that in their later work they did not perform as well as their more traditionally trained colleagues.[18]

In fact, the best medical training in radiation therapy in these earliest years (and for years to come) was through either one-on-one association or through journals and textbooks. Practitioners, however they were trained, looked to the literature for

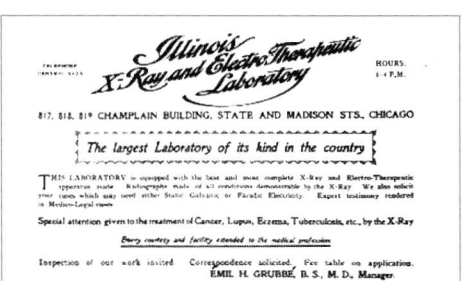

Fig. 6.6 The faculty at Grubbé's Illinois School of Electrotherapeutics was, in the tradition of the midwest, eclectic. Pictured here in 1905 are, from left to right, G. H. Somers, Milton H. Mack, May Cushman Rice, Noble M. Eberhart, Grubbé, Charles S. Neiswanger, W. H. Webster, B. F. Andrews, Mary E. Gardner, Charles H. Treadwell, and H. M. Eisler. (Courtesy of the Center for the American History of Radiology, Reston, Va.)

Fig. 6.7 Grubbé's school grew directly out of his busy laboratory. The two shared equipment and a floor of the Champlain Building. When Grubbé and his colleagues issued common stock in the school in 1908 the dividing line between the lab and school became even more hazy. (Courtesy of the Center for the American History of Radiology, Reston, Va.)

innovations in apparatus, advice on techniques, and reports on therapeutic successes and failures. Medical journals were a primary resource for anyone hoping to remain in the field. The leading American journals of the time, the *Boston Medical and Surgical Journal, American Medicine, Medical News,* and *J.A.M.A.,* supplied a steady stream of reports on therapeutic achievements with the new rays. Most attempted to give enough description and details so that the reader could attempt to replicate the results.

Typical of these were regular reports in *J.A.M.A.* by William Pusey. In May 1900 Pusey noted distance, timing, and voltage for treatment of lupus, eczema, and "for the purposes of epilation." He also listed criteria for determining when a single exposure had "been carried far enough":

1. Appearance of erythema and pigmentation.
2. Blanching of the hair.
3. Loosening of the hair.[19]

However subjective such observational methods of dosimetry may appear today, they were the types of benchmarks by which a generation of radiation therapists honed their practice.

So astonishing were the results reported by some practitioners, and so urgently did some editors feel the need to disseminate this news, that a single isolated case was often hailed as a proven new regimen. A successful treatment of lupus vulgaris was reported by J. T. Knox of Cincinnati in 1901. After detailing his efforts at shielding, multiple treatments, and distance, Knox stated, "Although this is the first case of lupus I have treated by this method, I have no hesitancy in stating that I regard it as an infallible one, if properly applied and continued a sufficient length of time."[20] The interested reader of these reports of idiosyncratic cures must have been further cheered by accompanying photographs of miraculous transformations in patients treated with radiation (Fig. 6.9). The unfortunate fact that many of these photos were clearly airbrushed beyond clinical significance was apparently an acceptable journalistic practice at the time.

Another reason for the careful inventory of treatment variables and outcomes in these early articles was the prevention of so-called "X-ray burns." Although the cause was unknown, an armamentarium of dos and don'ts was built up in an effort to spare both patients and practitioners. E. A. Codman, writing in the *Philadelphia Medical Journal* in 1902, noted that "...therapeutic exposures will continue to be dangerous, and it is therefore important to record the exact condition of the patient's local and constitutional idiosyncrasies, as well as the tube."[21] The idea that masses of unrelated, accretive reports might somehow yield up useful verities was not isolated to roentgenology in these years.

Articles proliferated with detailed case histories of patients, their treatments, and their reactions. Pusey included thirty-six detailed and disparate case histories in a 1902 article.[22] Others reported case after case, usually with few deductions offered about general application. This growing mass of episodic reports of treatment must have seemed baffling to practitioners who wanted to take up the new field. The suggestion was that, with a certain amount of caution and the right apparatus, almost any treatment could be attempted—and the results would be eminently publishable.

Some authors produced tables and

Fig. 6.8 An unfortunate juxtaposition of journal advertising seemed to suggest a potential fallback profession if instruction at Monell's New York School failed to prove profitable. (Courtesy of the Center for the American History of Radiology, Reston, Va.)

Fig. 6.9 Numerous before-and-after photographs of patient treatment appeared in both general medical journals and the fledgling radiology magazines. The hand of the touch-up artist, as in these 1901 photos, often made it difficult to tell exactly how much benefit the X rays had brought to the patient. (Courtesy of the Center for the American History of Radiology, Reston, Va.)

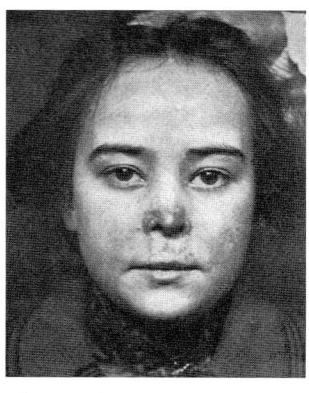

charts designed to assist the beginning radiation therapist with his or her work. Francis Williams, drawing on his groundbreaking work at the Boston Hospital, published a chart of relative distances and intensities for the rays, deriving from these a column of "safety limits" based on skin erythema (Fig. 6.10).[23] The problem with these tabulations and others like them was that, invariably, they omitted one or more of the crucial variables necessary to achieve even vague comparisons between different sets of results.

The advent of journals specializing in radiology and radiation therapy did not go far in giving coherence to the confused mass of individual cases, results, and advice.[24] The *American X-Ray Journal*, founded by Robarts in St. Louis in May 1897, published results of radiation therapy and electrotherapy, as did the early *Transactions* of the American Roentgen Ray Society (1902–1908). Both were widely read within the field but did nothing to provide guidelines aside from description and comparison of cases and methods. *The American Quarterly of Roentgenology*, which began publication in 1906 and became the *American Journal of Roentgenology* in 1913, printed numerous articles on radiation therapy. Novel applications and new diseases treated seem to have earned first publication in these journals, with the emphasis more on possibilities and opportunities than on explanations for results.

"What we need," wrote one roentgenologist, "is a large series of carefully reported cases and lapse of time sufficient to prove that treatment is of lasting success."[25] The author was calling for an end to education by the accretion of empirical evidence and a move to a more scientific basis for the field. A system was needed, and several authors sought to provide it by codifying a number of cases and observations into textbooks.

These early textbooks, ranging from physics and electrical how-to manuals to complete and well-informed volumes, were the mainstay of radiation therapy education in these earliest years. The most widely consulted included Francis Williams's monumental *Roentgen Rays in Medicine and Surgery* (1901), Samuel H. Monell's *System of Instruction in X-Ray Methods* (1902), W. A. Pusey and E. W. Caldwell's *Roentgen Rays in Therapeutics and Diagnosis*, W. H. Rollins's prescient *Notes on X-Light* (1904), and Mihran Kassabian's *Roentgen Rays and Electrotherapeutics* (1907), as well as texts by Carl Beck and George McKee (Fig. 6.11).[24]

Most of these texts were quite lengthy, giving detailed instruction in the physics and production of the rays, with sections on diagnosis and therapy. Many included sections on other light and ray therapies: Grenz rays, N-rays, electrotherapy, actinotherapy, and Finsen ray treatments. Often the radiation therapy sections were longer than their diagnostic counterparts, and most included extensive descriptions of protective measures and elaborate methods for estimating dosage.

Fig. 6.10 Williams, a conscientious investigator and clinician, recorded his own experiences with the rays in an effort to assist others in forming judgments about treatment distances, voltages, and safety. This is typical of early charts made in an effort to standardize and assess treatment success. (Courtesy of the Center for the American History of Radiology, Reston, Va.)

170   Training and Education

Fig. 6.11 Typical of the early textbooks, this volume by William Benham Snow (1903) combined instruction in X rays with a number of different and sometimes exotic therapies. (Courtesy of the Center for the American History of Radiology, Reston, Va.)

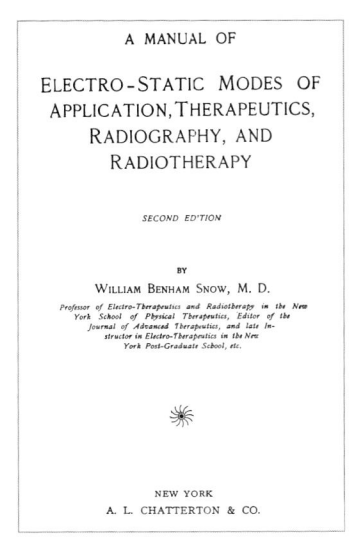

A MANUAL OF
ELECTRO-STATIC MODES OF
APPLICATION, THERAPEUTICS,
RADIOGRAPHY, AND
RADIOTHERAPY

SECOND EDITION

BY

WILLIAM BENHAM SNOW, M. D.

Professor of Electro-Therapeutics and Radiotherapy in the New York School of Physical Therapeutics, Editor of the Journal of Advanced Therapeutics, and late Instructor in Electro-Therapeutics in the New York Post-Graduate School, etc.

NEW YORK
A. L. CHATTERTON & CO.

Kassabian's book, written as the author was already disfigured by exposures to radiation and only three years before his death, was a thorough and reasoned synopsis of all that was known about the X ray in 1907. To be absolutely complete, Kassabian surveyed thirty-one American and European practitioners specializing in roentgenology. One appendix included not only thumbnail descriptions of their preferred apparatus, but also complex charts surveying additional apparatus, dosage, and practical applications.[27] From these charts it was possible to replicate as closely as possible the clinical scene and treatments of the great names in the field, among them Freund in Vienna, Baetjer at Johns Hopkins, Girdwood in Montreal, and Caldwell and Morton in New York. Kassabian's book, like that of Williams, served the confused practitioner in that it assembled many of the separately published case histories, grouping them by disease. But little attempt was made to draw conclusions about similarities or differences among results in these cases. Instead, the books were aimed at "the practical physician," providing the facts—and lots of them—and, as Kassabian proudly noted, "no space has been encumbered with the recital of fanciful theories or those of a controversial nature."[28]

The volume by Pusey and Caldwell was the most widely used by roentgenologists and general practitioners. In his preface to the first edition section on X-ray therapeutics, Caldwell set out his own notion of a textbook:

> I have given as fully as possible the details of my own experience and of the experience of other workers, for it is only by the accumulation of such data that it becomes possible to arrive at a satisfactory estimate of the value of the method.[29]

What followed there and in the second edition of the book were more than 300 of the authors' own cases, with many more descriptions from other physicians. The results from these early aggregate texts may have been the illusion of an education in the field, but two things were needed before true education could begin: first, adequate apparatus and dosimetry devices to ensure comparability of results and, second, an informed rationale for applying the rays in a specific manner to individual diseases.

Comparability would have to wait, but soon volumes devoted exclusively to radiation therapy appeared, and many of their authors had methods which could be interpreted as systems. The 1904 translation of Leopold Freund's now classic *Elements of General Radio-therapy for Practitioners* was widely consulted and cited.[30] Some readers even referred enthusiastically to the "Freund" method, though the elements of that method seem today difficult to tease out from the mass of disparate data, techniques, and observations.[31] Nobel M. Eberhart, a professor at Grubbé's Illinois School of Electrotherapeutics, published *Practical X-Ray Therapy* in 1907.[32] Two years later another Illinois School colleague, Gordon G. Burdick published *X-Ray and High-Frequency in Medicine*.[33] Other specialized books, of varying quality, appeared with regularity.

But at whom were these texts and specialized manuals aimed? There were no formal hospital programs in radiation therapy, and the number of persons in the United States and Canada concentrating exclusively in the field could be counted on two hands. Eberhart, who referred to his work as "a brief and handy working manual," made clear his intended audience. The book, he wrote:

> ...is especially designed to meet the requirements of the busy practitioner who has installed an X-ray outfit in his office and, after having been instructed by the maker of the apparatus in the general management of the same, finds himself confronted by [numerous] questions.[34]

Freund noted in his introduction that he had "purposely presupposed but little knowledge on the part of the reader."[35] Other radiation therapy textbook authors, like one-time electrotherapist William Benham Snow in 1903, remarked on the inexperience of most who intended to use the X ray in therapeutics.[36] In fact, despite the growing mass of clinical reports and evidence, running just beneath the surface of early instructional literature was the notion that radiation therapy offered something for *every doctor*. There was no hint in any of these books that the reader might choose to specialize in general radiology, much less radiation therapy. This information was aimed at a general awareness of the field for those who wished to employ it as an adjunct to broader medical practice.

Teaching hospitals and medical schools offered little more to the interested student than a brief acquaintance with the new rays. Although radiation therapy with both X rays and radium had earned a place in many hospitals prior to World War I (most notably with Pfahler at the Hospital of the University of Pennsylvania, William Morton in New York, Francis Williams in Boston, and Howard Kelly and Frederick Baetjer at Johns Hopkins), there was no systematized course of training.[37] Henry Pancoast recalled that as a young physician at the University of Pennsylvania in 1902 he was asked casually whether he would like to "try the new X-ray work." Training for the new position meant spending eight to ten hours with William Goodspeed in the physics department. "At the expiration of that time," wrote Pancoast, "he told me that was all he knew of the subject and that any further knowledge would have to be gained through experience—and so it had to be."[38]

As time went by, the reluctance to institute formal training stemmed in part from the fact that both radiology and radiation therapy were thought of as mere adjuncts to more established fields within the hospital. One influential German physician was quoted in American publications averring that "radiologists should be trained dermatologists to begin with."[39] At a 1905 meeting of the College of Physicians of Philadelphia devoted entirely to radiation therapy, radiologist George Johnston rose to proclaim that in therapy "the legitimate X-ray man does not consider himself the successor of the surgeon, but is proud to be accorded the position of assistant."[40] Many hospitals consigned all of radiology and its practitioners to the regions of the hospital usually reserved for what were at that time designated as "auxiliary services:" brace-and-limb, medical photography, and physical therapy. Few United States hospitals had separately recognized X-ray departments in these early years; none had departments for radiation therapy. Providing courses of instruction in such circumstances was difficult if not impossible. Interested students could learn by spending more time with the radiation therapists—for all practical purposes apprenticing, just as their non-hospital trained counterparts did. At institutions like the Memorial Hospital in New York, where James Ewing and a group of surgeons and pathologists were beginning pioneering work in radiation and radium therapy, a few students were able to carve out for themselves an education in the field.[41]

In a search for a reliable system and for instruction that was not based solely on a compilation of unrelated experiences, many North American physicians went to Europe to learn about radiation and radium therapy. These seekers were part of a long tradition of medical "grand tourists," who since the 1820s had pursued the education and added cachet offered by European universities and specialists. With the advent of radium and the rise of different schools of medical thought on radiation therapy in Europe, many Americans went for brief periods of study and observation with the Curies, Albers-Schönberg, Freund, Béclère, and other noted professors.[42] Robert Abbé of New York made several trips to Paris to study with Marie Curie, bringing back radium for use in his own practice. Others who went for long or short periods of European study included Pfahler, Morton, and Williams. In 1902 the Johns Hopkins Medical School sent Fred Baetjer on a year-long expedition to study European X-ray and radium treatment of cancer.[43] Those who followed Baetjer in the next ten years would add much to the American store of information on radiation therapy and would begin to formulate ideas about ways in which the field should be organized and taught. It would not be until after the unexpected exposure to European methods occasioned by service in World War I and a radical change in thinking about all types of medical education that United States physicians would look seriously at formalizing education in radiation therapy.

BECOMING A PROFESSION: 1915–1950

Historians often designate the first World War as a dividing line between the old world and the new. In radiation therapy the war years had special meaning. They saw the beginning of the widespread use of Coolidge's improved X-ray tube and resulting improvements in reliability and results. Efforts were underway to achieve international agreement on standards for radiation dosimetry. Physicians who served abroad were made aware of teaching traditions already well formed and active in London, Paris, Vienna, and Stockholm. On their return they found a medical educational system at great pains to remake itself in the wake of damning reports on widespread inefficiency and corruption.[44] The scene was set for reform in medical education, and the right tools finally were available for radiation therapy to become a science. The question was whether or not the field was ready to meet this challenge.

In 1914, as a partial response to the revelations made in the Flexner Report and other studies critical of medical education, the AMA's Council of Medical Education and Hospitals (CMEH) published a list of hospitals with approved internships.[45] This would be followed by a list of hospitals with approved graduate education in medical specialties. This mild attempt to impose standards had almost no effects on radiation therapy. Most of those interested in the field continued to receive training by what amounted to ad hoc preceptorships within the few institutions with radiation therapy facilities. Strong loyalties tended to attach to the individual mentors willing to provide education in the field, a tradition which has lingered in radiation oncology. James Case, in looking back on the history of education in the field, noted the extraordinary influence of teachers like Baetjer and Hickey.[46]

A few hospitals offered more structured education in the field. Memorial Hospital in New York was one of the first institutions to formalize training in the early 1920s when Cornell medical students began rotations through the cancer treatment facilities.[47] Memorial also adopted the European tradition of welcoming professional "visitors," who observed for periods from two days to six months. In 1921 and 1922 more than 300 visiting physicians observed radiation therapy at Memorial. Other institutions, including the Mayo Clinic, the Massachusetts General Hospital, the Hospital of the University of Pennsylvania, and Johns Hopkins, opened radiation therapy sections through which medical students rotated or visited regularly as an adjunct to other service rotations. The short-lived Philadelphia Post-Graduate School of Roentgenology, sponsored in 1915 by the Philadelphia Roentgen Ray Society, may well have been the first school dedicated to the specialized academic training of radiologists but graduated no radiation therapists.[48]

But this was not, it must be emphasized, the formal training of radiation therapists. It was a process by which specialists in other fields could become acquainted with radiation therapy. Even at Memorial, with the first three-year Rockefeller research grants offered in 1926, the fellowship holders were surgeons, gynecologists, and pathologists. The full year they spent studying the applications of X rays, radium, and surgery in the treatment of cancer made them better equipped in their own fields, not as radiation therapists. Clear-eyed analysts could see the possibility of radiation therapy collapsing as a field, as the very act of education parceled out its special skills to other areas of medical practice. As late as 1936 this route was actually advocated in the pages of the *American Journal of Roentgenology:*

> The gynecologist, urologist, or proctologist does not call in a general surgeon to carry out surgical procedures on his patients. He has appropriated surgical procedures for his own use. It seems reasonable that eventually he will do the same thing with radium therapy,...and unbiased opinion will concede that this is not only logical but more efficient.[49]

The radiological literature was filled with articles debating the merits and possibilities for special training in both radiology and radiation therapy, as well as discussions of both undergraduate and postgraduate requirements.[50] Most dealt at length with aspects of roentgenology and fluoroscopy as teaching tools in the general medical curriculum but said little about radiation therapy education.

In some of these articles there is a subtle tension in the question of whether undergraduate medical students should rotate through radiology at all. In 1932 Case wrote:

> ...On the whole they [academic radiologists] have had more interest in instructing postgraduate physicians desiring to devote themselves to the specialty of radiology than in the teaching of undergraduate medical students.[51]

At a time when the field was struggling to establish itself as a separate entity, initiating only those who had already chosen to specialize in the field might have seemed the most prudent course.

Case, like a growing number of others, believed that acquainting the undergraduate with all aspects of radiology was essential both in turning out well-prepared physicians and in gaining recognition for the specialty. Even he acknowledged, however, that the chances for formal undergraduate study of radiation therapy were not good. And, while at least 619 hospitals in the United States offered some form of postgraduate education (internships) in 1932, Case was realistic about the limits of the standard internship and radiation therapy:

> The field of roentgen therapy is so complicated that it is to be doubted whether more can be done in the way of instruction during the intern year than to teach the general principles of treatment, the selection of cases, the realization that it is possible to make accurate dosage, and the limitations of the method.[52]

The Move Toward Certification

Efforts were already underway that would make postgraduate education in radiology, however problematic, a requirement. This process would establish, once and for all, that radiology was a medical specialty, but the status of radiation therapy would remain less clearly defined.

In 1917, as a response to unregulated practice, the American Board of Ophthalmology was formed to set standards and confer specialty certificates. Like radiologists, ophthalmologists struggled in a field in which both physicians and lay persons regularly usurped their area of expertise.

Throughout the 1920s radiologists looked at ways to upgrade their own standards for teaching and practice in an effort to bring greater legitimacy to the specialty. Albert Soiland's efforts to have a section on radiology authorized as part of the AMA in 1923 and his subsequent founding of the American

College of Radiology (ACR) were a part of this larger effort.[53] Others proposed nationally standardized programs of postgraduate education, with three-year courses with projected curricula in which diagnosis would take clear precedence over therapy.[54]

Standardization was at least a more plausible eventuality in radiation therapy than it had been twenty years earlier. With more reliable apparatus able to generate larger and larger voltages, improved radium devices, and more effective radiation protection, the field was achieving remarkable results in a widening range of benign and malignant conditions. At the second International Congress of Radiology in 1928 (where the theme was "Education and Medical Training in Radiology"), the Committee on Units and Measures reached an agreement on an international standard radiation measure, the roentgen. At least in theory, results from all over the world could now be compared and favorable clinical outcomes duplicated. Radiation therapy was ready to move forward as a specialty. The ties between diagnosis and therapy meant that they would move together.

In 1933 representatives from five radiological organizations (ARRS, ACR, American Radium Society, [ARS] the Radiological Society of North America, [RSNA] and the Section on Medicine of the AMA) met in Milwaukee to discuss the formation of a "qualifying board."[55] The American Board of Radiology (ABR) was incorporated in Washington, D. C., in May of 1934. During its first year 404 candidates were approved as certified radiologists. The oral examination (which many of the original candidates bypassed) was directed by an examining board of eminent radiologists, with examiners in physics added later. Certificates were granted in radiology, roentgenology, diagnostic roentgenology, therapeutic radiology, and therapeutic roentgenology—a series of classifications that remains as confusing today as it was in the 1930s.

By 1937 the ABR's Committee on Graduate Radiologic Training recommended a one-year medical internship followed by at least three years of postgraduate work in a recognized institution. The plan called for "Examination in the basic sciences of radiology as well as in the clinical aspects thereof," with no specific mention of the weight to be given to therapeutic radiology in this training.[56] Byrl R. Kirklin outlined a suggested course of instruction to satisfy the new requirements.[57] His three-year residency program would include a year of diagnosis, nine months of basic and technical medicine, six months of electives, and nine months of radiation therapy. Other radiologists proffered different potential curricula for radiology residents.

United States medical schools were not entirely ready to take on the new burden of educating radiologists. In 1937 there were sixty-eight "Class A" medical schools with four-year programs leading to the medical degree. Of these, forty had separately chaired radiology departments. One offered no instruction in radiology, while twenty-seven of the schools included radiology under other department headings: anatomy, pathology, and surgery.[58] Moreover, many of these schools "farmed out" instruction in radiation therapy to private practitioners with hospital ties. This practice was deplored as leading to idiosyncratic and unreliable techniques, but throughout the 1930s even ABR board examiners advertised private preceptorships for interns and residents.[59,60]

While the formation of an effective specialty board brought recognition to radiology in general, it did little to train professional radiation therapists. Of the first seventy-two diplomates certified in radiology, forty-four were surgeons, gynecologists, and dermatologists. Still more were general radiology candidates who failed the diagnostic portion of the exam and were granted instead a certificate for therapeutic radiology.[61] This awarding of certification in therapeutic radiology almost as a consolation prize persisted into the 1950s.

In 1939 there were fewer than fifty physicians devoted to the exclusive practice of radiation therapy in the United States. Of these, a number had immigrated from training centers in Europe.[62] There was not a critical mass of academicians sufficient to realize a broadly-based standardized course of instruction in radiation therapy. Exactly where such training would fit within the larger field of radiology was unclear.

In 1943 George Holmes suggested a simplification of ABR certification to three areas: radiology, diagnostic radiology, and therapeutic radiology.[63] The war years froze changes in the ABR, and Holmes's suggestion was not even printed until 1946 and did not see implementation until 1959. His idea was that this simplification would clear up the confusion of the old categories, at the same time that it would consolidate and give additional weight to the practice of radiation therapy. Holmes knew quite well that he was fanning a fire that had been smoldering since the incorporation of the ABR: the threat of civil war in the ranks of radiologists.

A Separate Specialty?

For years radiation therapists had fought off the claims of other medical specialists to their field. With the establishment of the ABR, radiation therapists at last had bona fide standing within a specialty. They were members of radiological organizations (ACR, ARRS, RSNA, and ARS) that exerted increasing influence on medical and social issues and legislation. Yet the radiation therapists found themselves outnumbered and sometimes outvoted, both in organizational and institutional settings, by their diagnostic colleagues. As early as 1938, Swedish roentgen pioneer Gösta Forssell advocated the separation of therapy from diagnosis:

> A complete divorce of these two branches of radiology offers such great advantages for research and instruction, as well as for the utilization of radiology in practical medicine, that it is fervently to be hoped for as soon as possible at university and other large hospitals.[64]

Douglas Quick, in his Carman Lecture to the ARS in 1947, gave a long and impassioned rationale for the separation of diagnosis and radiation therapy into separate hospital departments, noting the tendency of practitioners to be overshadowed by their diagnostic colleagues. He went on to suggest that radiation therapy and radium therapy be divided services (a situation in place at the Mayo Clinic since the 1920s). "Nothing short of the insistence of the Board [the ABR]," he said, "...will change this indifference to the establishment of adequate training."[65] Quick also pointed out the operational flaw in his plan: the problem of "how the additional trained personnel in therapy is to be supplied." A concerted effort, he believed, including the use of resources like those of the proposed Registry of Radiologic Pathology at the Army Institute of Pathology, would soon yield a functional separate field of radiation therapy.

If many radiation therapists were unresponsive to these suggestions, a number of general radiologists found them appealing. In a starkly honest appraisal of the situation, Thomas Groover, founder of the large Washington-based practice Groover, Christie, and Merritt, had advocated dumping radiation therapy from private practice partnerships. He found radiation therapy not as profitable, and, "from a strictly economic standpoint, our group is overbalanced on the side of therapy. Radiologists [diagnostic] cannot continuously provide out of their earnings the subsidy which efficient radiotherapy of cancer requires."[66] Few hospital radiation departments shared Groover's views, and fewer department heads wanted to see a portion of their staffs and revenues taken away to form new departments.

Among radiation therapists and radiation educators, however, feelings were moving toward a split. Ursus V. Portmann surveyed the current state of radiation therapy instruction in 1950 and found it wanting. He noted the new vistas opened by radioisotopes and treatment units, the poor instruction received by many students, and the loss at which many who wished to specialize in the field found themselves:

> In view of the vast differences in diagnostic and therapeutic radiology it should be obvious that separation of these branches is advisable, and should be initiated in large teaching institutions in this country as soon as possible to improve the status of both.[67]

But many in the field, both diagnosticians and therapists, were threatened at the notion of divided academic programs within a divided field. Portmann had correctly pointed out that change was needed. At mid-century radiation therapy was poised for expansion and ready for systematized educational reform. In contemplating a break from the long partnership with diagnostic radiology, some radiation therapists ignored new partners on the post war horizon: organized medicine and the federal government.

1951–PRESENT: NEW PARTNERSHIPS TOWARD RADIATION ONCOLOGY TRAINING

The academic, organizational, and clinical worlds of radiation therapy were changing rapidly at mid-century. Powerful new weapons against cancer had become available, and their origins in war research put them under the close scrutiny of new government agencies. American medicine was becoming more organized, encouraging standardization of medical education and postgraduate training. In radiology, physicians who had returned from the war with X-ray training swelled the field's diagnostic ranks, while radiation oncology remained what several writers referred to as the "stepchild" of the field.

Those who had sought to sever radiation therapy from diagnostic radiology faced an insurmountable obstacle: there were not enough radiation therapists to see the numbers of patients who sought treatment, much less to form a separate and active specialty. Either massive funding in the form of not entirely welcome government interaction was needed to create, almost overnight, a new field, or radiation therapy would need to remain in concert with the rest of organized radiology in addressing mutual goals.

Arguing against precipitous action, Juan del Regato summarized this dilemma in 1953:

> To call for division of the large departments of radiology into departments of radiodiagnosis and radiotherapy is to ask for delivery without gestation for, at the present time, even cancer hospitals with a high sense of responsibility for the lives of their patients must do without radiotherapy for lack of well qualified aspirants to the positions. To divide departments of radiology and to entrust radiotherapy to self-trained neophytes may prove to be the hardest blow yet given to therapeutic radiology. To stiffen the specialty board examination would result only in penalizing young students for a poor quality of training for which they are not responsible; and to require special certification by a radiotherapy board of examiners offers no better solution.[68]

del Regato called for a versatile, patient-oriented clinical training in special cancer centers. At the time he was working toward developing such a center at the Penrose Cancer Hospital in Colorado Springs.[69] Similar efforts continued at Memorial Hospital in New York and were underway with Gilbert Fletcher at the M. D. Anderson Hospital in Houston and Henry Kaplan at Stanford. At each institution it would take the force of individual will to shape a well-codified teaching program, and the graduates of each program would bear the stamps of these forceful mentors.

Difficulties in reaching agreement on uniform standards in radiation therapy training reflected both the strong spirit of the few teachers specializing in the field and the scattered institutions in which they were taught. As late as 1960 the twenty-nine institutions which claimed to have training programs in "straight" (full-time) radiotherapy listed only twenty-five residents.[70] Most of these hospitals had only one or two radiation therapists on staff with only one resident. In such situations elaborate teaching programs were not possible.

But organized medicine was moving toward systematized accreditation of

The American Club of Therapeutic Radiologists/The American Society for Therapeutic Radiology and Oncology

Founding a specialty medical society can be a delicate undertaking. At mid-century, despite growing numbers of practitioners and an increasingly complex field, those who specialized in radiation therapy in the United States had no society of their own. They were represented in the ranks of the American Radium Society (ARS), but so were the surgeons, gynecologists, and other specialists who believed the use of ionizing radiation should fall with their purview. Radiation therapists were represented in the Radiological Society of North America (RSNA), in the American Roentgen Ray Society (ARRS), and in the American College of Radiology (ACR), but in each they formed only a very small percentage of the total membership.

Juan A. del Regato and other radiation therapists who quietly discussed the possibility of a separate society understood that members of these larger organizations might perceive this as a threat. Several radiation therapists had already proposed splitting off radiation therapy from diagnostic radiology in hospital services, and del Regato and his colleagues did not want to appear to be leading a secession movement. It was necessary to conduct the most subtle of organizational moves and to make it clear in doing so that radiation therapy intended to remain allied with all of radiology.

In 1953 the International Club of Radiotherapists was organized in Copenhagen. del Regato served as secretary of the American wing of the International Club and took the initiative of inviting the other American members (fifteen in all) to a dinner during the annual meeting of the RSNA in 1955. He noted that the RSNA, like the ARRS and the ACR, had turn downed requests to form separate sections for radiation therapy and was suspicious of his request. To contract for a dining room he had to get written permission from the RSNA secretary. Other radiation oncologists were invited as guests of the International Club members. The meeting was a great success, and the group met twice annually thereafter at the RSNA and ARS. Although substantive matters were discussed, the meetings had a social character which none of the larger societies perceived as a threat. del Regato later recalled:

> The gatherings represented two groups, members of the Club and their guests. As expected, the idea arose to create a single American Club to include everyone. I was ready with a one-page Founder's Agreement. In Chicago, in December of 1958, fifty-six of us approved the foundation of the American Club of Therapeutic Radiologists, with provisions to eventually 'formalize the informality.'

Those who know del Regato well suspect that he had been carrying the founders' agreement since the first meeting in 1955. He understood that the club had to develop naturally out of the collegiality of its members, rather than as a reaction against other organizations in radiology. Fifty-four members signed the founders' agreement.

The club was incorporated under the laws of Colorado in 1962, and a crab surrounded by electronic orbits (designed by del Regato) was adopted as its official seal. By 1962 membership had risen to 252, and by majority agreement the name was changed to the American Society of Therapeutic Radiologists. In 1970, with 308 in attendance, the society held its first separate scientific meeting. In 1972 *Cancer* became the official publication of the society, and in 1976 the society sponsored the *International Journal of Radiation Oncology•Biology•Physics* as its official organ. Known in the field as the "red journal," the it became the official and exclusive journal of the society in 1985.

In 1983 the society adopted a new mission statement:

> to advance the practice of radiation oncology by disseminating the results of scientific research, providing opportunities for education and professional development of its members, and promoting a health care environment conducive to optimal patient care.

The society adopted a vision statement which called for the pursuit of excellence in practice and in the multidisciplinary setting of cancer care. To reflect its broader mission the name of the organization was changed to the American Society for Therapeutic Radiology and Oncology (ASTRO).

Today ASTRO is the largest society of radiation oncologists in the world, with 3,157 active members, 573 associate members, 160 corresponding members, 9 affiliates, 1,009 junior members, and 35 corporate members. The annual meeting now includes extensive scientific and commercial exhibits and is attended by thousands of radiation oncologists from the United States and abroad. ASTRO sponsors a number of educational programs and is active in the work of other major radiological societies. Although its beginnings were quiet, the organization now gives a strong voice to the specialty of radiation oncology.

---

teaching programs. Throughout the 1930s and 1940s the AMA had continued to publish its lists of approved specialty boards and internships. In the early 1950s the AMA's CMEH sponsored conference committees in each specialty to review and evaluate graduate medical education. A system of residency review committees (RRCs) was set up, with the RRC for Radiology meeting for the first time in 1953.[71] Representing the ABR on this committee were Kirklin, Quick, and H. Dabney Kerr, while Eugene Pendergrass, Warren Furey, and Edward Leveroos represented the AMA CMEH.[72] The charge of the RRCs was then, as it is now, to review training programs, discuss relevant issues, and decide which programs deserved full accreditation.

Accreditation from the RRC for Radiology became one of the necessary credentials for securing government research grants in cobalt and other new technologies in the 1950s. With the advent of widely funded government grants for graduate medical education in the 1960s, all RRCs were made up of broadened boards under the Liaison Committee for Graduate Medical Education (LCGME), representing five major medical groups. The move toward increasing the numbers of separately accredited programs in radiation

Association of Residents in Radiation Oncology

The Association of Residents in Radiation Oncology (ARRO) was officially formed in October 1983 during the annual meeting of the American Society for Therapeutic Radiology and Oncology (ASTRO). Discussions for organization of such a society had begun the previous year, when an ad hoc committee of chief residents, led by Daniel Flynn, elected Francine Halberg as resident representative to the Residency Review Committee in Radiology. In the fall of 1983 radiation oncology residents across the country elected six officers to form the Executive Committee of ARRO. This committee met in March 1984 and outlined the following goals and objectives: (1) to disseminate information to all radiation oncology residents; (2) to formalize residents' input in professional organizations and committees affecting radiation oncology residents and residency training; and (3) to provide a forum for radiation oncology residents each year in conjunction with the ASTRO meeting.

ARRO Executive Committee members represent resident interests to a number of organizations, including the American College of Radiology, the Society of Chairmen of Academic Radiation Oncology Programs, the Residency Review Committee, the Intersociety Commission radiology summit, and others. ARRO provides its members with information on fellowships, sources of research funding, and elective rotations at other programs. The organization maintains an updated recommended text and journal outline and sponsors a variety of programs for residents during the annual ASTRO meeting. Results of annual ARRO questionnaires and surveys contribute to decisions made by the American Board of Radiology about examinations and are useful for other decision-making bodies in the field.

ARRO provides a strong voice for the residents who will lead the field in the future, at the same time that it provides for them an avenue of participation in the larger events and discussions that affect radiation oncology.

—David H. Hussey, M.D.

---

therapy was aided by the establishment and growth of the American Club of Therapeutic Radiologists (today the American Society for Therapeutic Radiology and Oncology), in the 1950s.

A picture of training in the field during the 1950s and 1960s can be gathered from the residents' program at Penrose. Although each program varied in content, scope, and aim, Penrose's program was among the best and most thorough and is well worth scrutiny today. Candidates in straight radiation therapy stayed in Colorado Springs for two years, followed by a fellowship year at other institutions in the United States or abroad. Residents served three-month rotations in anatomical pathology, surgical oncology, and nuclear medicine, as well as their regular clinical and research work in radiation therapy. del Regato targeted seven areas to be mastered for a successful education in the field:

1. Clinical manifestations of neoplasms and their pathobiology.

2. The accurate diagnosis, differential diagnosis, and staging of the various forms of cancer.

3. Indications for and technical aspects of multimodal management

of a broad spectrum of malignant neoplasms, including the side effects and late morbidity associated with each of the different modalities.

4. Continuous follow-up care of all patients regardless of type of treatment administered or even when therapeutic abstention was decided and including the terminal care of some patients.

5. The discipline of accurate and orderly record keeping and the maintenance of tumor registry systems.

6. Participation in clinical research through intramural as well as cooperative group studies.

7. Educational programs including didactic lecture, conferences, tumor boards, etc.[73]

Between 1949 and 1974 del Regato trained fifty-five full-time radiation therapists at Penrose—more than doubling the number of practitioners who had been in the field at the start of this period. Together with the graduates of similar programs these men and women would form the first generation of radiation oncologists.

Regular surveys revealed that the number of institutions training postgraduates in radiation therapy tripled between 1960 and 1970, reaching 66, and the number of residents in training increased more than six fold, to 150.[74] By 1972 500 residency positions were being offered, but only 244 were filled. There was a clear need for more physicians in the field, a need fueled by the numbers of patients seeking radiation therapy. In 1951, for example, 1,555 new cases were treated at the Memorial Hospital Department of Radiation Therapy, with 798 patients seen for follow-ups and additional therapy.[75] Ten years later these numbers had nearly doubled, but the staff remained the same. At M. D. Anderson, where some planners had worried that the term "cancer center" would alienate both the public and physicians, patients were soon writing in to refer themselves for treatment on the new cobalt machine.[76]

The advantages of such controlled national studies was obvious. In 1963, at the request of Dr. Kenneth Endicott, director of the National Cancer Institute, a Committee for Radiation Therapy Studies (CRTS) was formed to serve as an advisory group to the National Advisory Cancer Subcommittee for Diagnosis and Treatment. Gilbert Fletcher served as the first chair, with del Regato, Milton Friedman, Manuel Garcia, Henry Kaplan, Morton Kligerman, Victor Marcial, Walter Murphy, and James Nickson as members. The CRTS, subsequently the Committee for Radiation Oncology Studies (CROS), provided not only the means for setting national standards of practice in the field, but provided manuals, guidelines, and goals for the field.[77]

In 1968 CRTS launched the Radiation Therapy Oncology Group (RTOG), which received initial funding from the National Cancer Institute in 1971. Responsible for the last twenty-five years for coordinating clinical trials in radiation oncology, the RTOG has had a tremendous impact on training through government grants, through the pool of cases studied, and through the mass of significant clinical results and subsequent effects on technique.

By 1978 107 institutions offered training in straight radiotherapy, with 344 of 595 available fellowship positions filled.[78] By 1982 there was cause for worry in these numbers. Only 89 institutions offered training, with 339 of 501 positions filled.[79] The need for trained radiation therapists had not diminished, but the numbers of institutions willing to train specialists had dropped off. Moreover, the authors noted an increasing difficulty in persuading graduates of American medical schools to choose the field as a specialty. Six years later the numbers of institutions and positions remained virtually unchanged, but the number of residents in training had risen to 517.[80]

In the following year, 1988, Robert G. Parker published an assessment of current training programs and recommendations for needed changes.[81] He noted that, ironically, in the short period of sixteen years the analyses of planners in the field had gone from a perceived shortage of radiation therapists to concerns about an oversupply.[82] While this did not diminish the need for new practitioners, it suggested that new members of the field should be of the highest quality and that areas like research, previously given less attention in training, should receive new emphasis. Parker listed the numbers of grants available for the support of residents and fellows and noted the declining federal dollars available for these positions. For the first time, Parker suggested that "the current requirements for the management of patients in the community appears to be satisfied."[83]

Efforts at improving the quality of training continued. In 1984 the ACR administered its first in-training examination. By 1987 445 examinees took the test at 93 locations.[84] The test covered radiation biology, physics, and clinical practice. Today the examination is required in 91 percent of radiation oncology training programs and provides a benchmark by which cognitive skills in the field can be measured in an effort to enhance both individual and institutional performance. Demographic feedback from the tests can indicate broad areas of shortfalls in clinical experience with specific aspects of the field, and these are translated back into recommendations for training standards.[85]

Over the last twenty-five years, universities across North America have moved to establish separate departments of radiation oncology and sometimes separate institutions. For the most part this has not been a subject of debate, as both diagnostic and therapeutic radiology have each moved forward, represented at the same time by both exclusive and shared organizations. Within radiation oncology and in the ecumenical radiation community as a whole, the quest for quality training to produce able and talented practitioners has produced remarkable achievements. For one hundred years, from the early guidance of mentors in empirical assays to today's carefully planned and supervised courses of study, the goal of quality education in the use of radiation has remained the same.

▶ ▶ ▶ ▶ ▶ ▶

REFERENCES

Dr. David Hussey, who has done extensive research on the history of education in radiation oncology, provided numerous references and resources helpful in the preparation of this chapter.

1. Jones, Philip M., "X-Rays and X-Ray Diagnosis," *J.A.M.A.* 29 (6 Nov. 1897):945.
2. Trevert, Edward. *Something About X Rays for Everybody*. (Lynn, Mass.:Bubier Publishing Company, 1896): preface.
3. Ibid., p. 29.
4. Meadowcroft, William H. *The ABC of the X Rays.* New York: Excelsior Publishing House, 1896.
5. Röntgen, Wilhelm C. *Eine Neue Arte von Strahlen.* Third Printing. (Würzburg: Physik-Med Ges. Würzburg, 1896): inside back cover.
6. *The Archives of Clinical Skiagraphy* (later to become the *British Journal of Radiology*) published numerous advertisements for the earliest manufacturers of X-ray equipment, as well as stunning folio-sized sepia-tone radiographic prints that were widely used as wall art in X-ray laboratories and clinics.
7. See for example "Hidden Solids Revealed," *New York Times* (16 January 1896):9; "Roentgen or X-Ray Photography," *Scientific American* 74 (15 February 1896):103; and Daniels, N.H., "Photographing Through Opaque Bodies," *Electrical World* 27 (7 March 1896):243.
8. For an excellent analysis of the professional history of physicians in the United States and of the issues of geographic differences in approaches to training, see Paul Starr, *The Social Transformation of American Medicine,* New York: Basic Books, 1982.
9. Pratt, Harry Preston, *Electrical Engineer* (22 July 1896):261.
10. Advertisements for Pratt's laboratory can be found in the Grigg, Grubbé, and Manufacturing Collections of the Center for the American History of Radiology, Reston, Va.
11. Fuch's advertisements, photographs of his extensive clinic, and many of his large signature sepia-toned radiographs are on file at the Center for the American History of Radiology, Reston, Va.
12. A series of advertisements from Parberry and Robarts as their separate clinics and interests evolved are on file in the Grigg Collection at the Center for the American History of Radiology, Reston, Va.
13. Grubbé, Émil H. *X-Ray Treatment. Its Origins, Birth, and Early History.* St. Paul: Bruce Publishing

Co., 1949. Additional original advertisements, stock certificates, correspondence, and curriculum information on the Ilinois School are on file in the extensive Grubbé Collection at the Center for the American History of Radiology, Reston, Va.

14 Grubbé, *X-Ray Treatment*, p. 64.
15 Grubbé, *X-Ray Treatment*, p. 75.
16 Advertisements and early examples of the correspondence course from this school can be found at the Center for the American History of Radiology, Reston, Va.
17 Monell's school flourished; advertisements are on file at the Center for the American History of Radiology, Reston, Va. Monell's advertisements also show up in many early manufacturer's catalogues.
18 Feelings ran high; see for example Edwards, G.P., "Some Observations in X-Ray Work," *International Journal of Surgery* (October 1903); and Bevan, Arthur Dean, "The X-Ray as a Therapeutic Agent," *J.A.M.A.* 42 (1904):26-30.
19 Pusey, William A., "Roentgen Rays in Skin Disease," *J.A.M.A.* 34 (1900):1185.
20 Knox, J.T., "Treatment of Lupus Vulgaris with X-Rays, *J.A.M.A.* 34 (1900):1210.
21 Codman, E.A., "X-Ray Burns," *Phil. Med.* J. (29 March 1902).
22 Pusey, William A., "Report of Cases Treated with Roentgen Rays," *J.A.M.A.* (1902):911-919.
23 Williams, Francis H., "X-Ray Therapy," *Medical News* (March 1904).
24 One recent survey of the history of radiological literature is Karim Valji, "Radiologic Publications," in Vol. I, *Diagnosis,* in Gagliardi, Raymond A., ed. *A History of the Radiological Sciences* (Reston, Va.: Radiology Centennial Inc., 1996).
25 Delavan, R., "X-Rays in Laryngeal Cancer," *Medical Record* (1 Nov. 1902).
26 Williams, Francis H. *Roentgen Rays in Medicine and Surgery.* New York: The MacMillan Company, 1901; Monell, Samuel H. *System of Instruction in X-Ray Methods* ; Pusey, William A., and Caldwell, Eugene W. *Roentgen Rays in Therapeutics and Diagnosis.* Philadelphia: W.B. Saunders, 1903, second edition 1904; Rollins, William H. *Notes on X-Light.* Boston: privately published, 1904; and Kassabian, Mihran. *Roentgen Rays and Electrotherapeutics.* Philadelphia: J.B. Lippincott Co., 1907. These were among the most influential of the many X-ray related texts published during the first decade after the discovery.
27 Kassabian, p. 253 through unnumbered appendices.
28 Ibid., p. iii.
29 Pusey and Caldwell, p. 219.
30 Freund, Leopold. *Elements of General Radio-Therapy for Practitioners*. Translated by G.H. Lancashire. New York: Rebman Co., 1904.
31 In another *J.A.M.A.* article by Pusey the author cited his applications of "exposure to X-rays in the manner of Schiff and Freund in Vienna." See Pusey, William A., "Lupus Healed with Roentgen Rays," *J.A.M.A.* 35 (1900):1476-1478.
32 Eberhart, Noble M. *Practical X-Ray Therapy.* Chicago: New Medicine Publishing Co., 1907.
33 Burdick, Gordon G., *X-Ray and High Frequency in Medicine.* Chicago: Physical Therapy Library Publishing Co., 1909.
34 Eberhart, p. 9.
35 Freund, p. v.
36 Snow, William B. *A Manual of Electro-Static Modes of Application, Therapeutics, Radiography, and Radiotherapy* (New York: A.L. Chatterton and Company, 1903):217-219.
37 The interested reader is referred to the two biographical anthologies of Juan del Regato, *Radiological Physicists* (New York: American Institute of Physics, 1985), and *Radiological Oncologists* (Reston, Va.: Radiology Centennial Inc., 1993).
38 Pancoast, Henry E., "Reminiscences of a Radiologist," *Am. J. Roent.* 39 (1938):169-186.
39 Riehl, G., "Zur Röntgen-Therapie," *Wien. Klin. Wschr.* (4 March 1904); translated in 42 *J.A.M.A.* (1903):1113.
40 Johnston, George, "Meeting of the College of Physicians, Phila.," 45 *J.A.M.A.* (1905):899.
41 For an excellent review of Ewing's work see the chapter on his life in del Regato, *Radiological Oncologists,* pp. 65-76.
42 The thumbnail sketches of radiation therapists and physicists appended to del Regato's *Radiological Oncologists* provide numerous examples of Americans who journeyed to Europe in these early years for instruction.
43 Baetjer's planned trip was reported in *J.A.M.A.* 39 (1902): 263.
44 Numerous medical and social historians have written about the effects of the Flexner report on medical reform, about efforts to standardize and regulate drugs, and the reshaping of medical services which followed. For a summary, see Starr, pp. 116-144.
45 AMA, "Provisional List of Hospitals Furnishing Acceptable Internships for Medical Graduates," issued in 1914 as a pamphlet and reprinted in *J.A.M.A.;* followed by "Hospitals Approved for Residencies in Certain Specialties," 1924.
46 Case, James T., "Teaching of Radiology," in Glasser, Otto. *Science of Radiology* (Springfield, Illinois: Charles C. Thomas, 1933):344.
47 The progress of radiation therapy using both X rays and radium can be read in the meticulous *Reports of the Memorial Hospital,* published under the direction of Ewing, Failla, Janeway, Barringer, and others throughout the 1910s, 1920s, and 1930s.
48 Pancoast, p. 183.
49 Grier, G.W., "Qualifications for the Practice of Radium Therapy," *Am. J. Roent.* 36 (1936):455.

50 See for example Hickey, Preston M., "Instruction of Undergraduates in Radiology," *Radiology 2* (1924):62; Hickey, Preston M., "Postgraduate Instruction in Radiology," *Radiology 8* (1927):379; and Case, James T., "Teaching of Radiology to Interns," *Radiology 18* (1932):957.
51 Case, in Glasser, p. 345.
52 Ibid., p. 350.
53 See del Regato's biography of Soiland in *Radiation Oncologists*, pp. 77-86.
54 See for example Desjardins, Arthur U., "Radiologic Education," *J.A.M.A.* 99 (1932), and Case, J.T., "Teaching Radiology," *J.A.M.A.* 99 (1932).
55 Krabbenhoft, Kenneth L., "Certification and Education," in Vol. I, *Diagnosis*, in Gagliardi, Raymond A., ed. *A History of the Radiological Sciences* (Reston, Va.: Radiology Centennial Inc., 1996).
56 For a history of the ABR and the reproduction of some early documents and decisions, see Jenkinson, E. A. *History of the American Board of Radiology, 1934-1964*. Privately published and bound, 1964.
57 Byrl Kirklin, *J.A.M.A.* 109 (1937):633-634.
58 Many such statistics can be found in the yearly roundups of the *Yearbook of Radiology* (Chicago: Yearbook Publishers). These figures are from the 1937 *Yearbook*.
59 See for example Hodges, Fred J., and Pendergrass, Eugene, in *J.A.M.A.* 109 (1937):634-637.
60 Grigg cites these advertisements in his timeline of education. Grigg, E.R.N. *Trail of the Invisible Light* (Springfield, Illinois: Charles C. Thomas, 1965):838.
61 del Regato, Juan A., "The Training of Therapeutic Radiologists," *Radiology* 95 (1970):703.
62 Ibid.
63 Holmes, George, "Therapeutic Radiology," *Radiology* 47 (1946):652-659.
64 Forssell, Gösta, "Role of Radiology in Medicine," *Radiology* 30 (1938):12.
65 Quick, Douglas, "Therapeutic Radiology; Carman Lecture," *Radiology* 50 (1948):283-296.
66 Groover, Thomas, "in," *Radiology* 32 (1939):617-621.
67 Portmann, Ursus V., "Therapeutic Radiology as a Specialty," *Am. J. Roent.* 63 (1950):2.
68 del Regato, Juan A., "Training Centers in Therapeutic Radiology," *Postgraduate Medicine* 14 (1953):161-162.
69 Wilson, J. Frank, and Chabazian, C.M., "Penrose Cancer Hospital, 1949-1974: A Quarter Century of Achievement," *Int. J. Rad. Onc. Biol. Phys.* 15:1475-1483.
70 del Regato, "Training of Therapeutic Radiologists," p. 704.
71 CMEH, "Background and Development of Residency Review and Conference Committees," *J.A.M.A.* 165 (1957):60-64.
72 Pendergrass, Henry P.; Brady, Luther W., Jr.: and Weinlader, James P., "Accreditation of Graduate Medical Education in Radiology: Historical Perspectives," *Investigative Radiology* 21 (1986):812-814.
73 Wilson and Chabazian, p. 1481.
74 del Regato, "Training of Therapeutic Radiologists," p. 704.
75 del Regato, Juan A., "Survey of Training of Therapeutic Radiologists in the United States," *Radiology* 106 (1973):225-226.
76 Statistics for numbers of patients treated can be found in the annual *Memorial Hospital Reports* for these years. The reports were often summarized in the *Year Books of Radiology* as well.
77 del Regato, Juan A., "Report to the Clinical Studies Panel. Suggestions for Incorporating Studies Using Radiation Therapy into the Clinical Studies of the Cancer Chemotherapy National Service Center," *Cancer Chemotherapy Reports* 7 (1960):47-49.
78 For an excellent summary of the efforts of CRTS/CROS and RTOG, see Brady, Luther W., "Gold Medal Address: The Radiation Therapy Oncology Group—1987," *Int. J. Rad. Onc. Biol. Phys.* 15 (1988):537-542.
79 del Regato, Juan A., "The 1978 Survey of Training of Therapeutic Radiologists in the United States," *Int. J. Rad. Onc. Biol. Phys.* 5 (1979):93-97.
80 del Regato, Juan A., and Pittman, Donna D., "The 1982 Status of Training of Therapeutic Radiologsts in the United States," *Int. J. Rad. Onc. Biol. Phys.* 9 (1983):387-392.
81 Cox, James D.; Flynn, Daniel F.; Pittman, Donna D.; Brady, Luther W.; and del Regato, Juan A., "Radiation Oncology: Postgraduate Medical Education in the United States, 1988," *Int. J. Rad. Onc. Biol. Phys.* 16 (1989):1577-1982.
82 Parker, Robert G., "Training in Academic Radiation Oncology in the United States: A Report to the Inter Society Council for Radiation Oncology," *Int. J. Rad. Onc. Biol. Phys.* 18 (1990):1245-1248.
83 For the change in perceived manpower issues in the field, see CRTS, "Crisis in Radiation Therapy Training and Practice. Final Report to the National Cancer Institute from the Subcommittee for Training of the Committee for Radiation Therapy Studies," 1 September 1972. On file at the American College of Radiology; and Davis, L.W.; Cox, James D.; Diamond, J.; et al., "The Manpower Crisis Facing Radiation Oncology," *Int. J. Rad. Onc. Biol. Phys.* 12 (1986):1873-1878.
84 Parker, p. 1248.
84 Wilson, J. Frank, and Diamond, J., "Summary Results of the ACR Experience with an In-Training Examination for Residents in Radiation Oncology," *Int. J. Rad. Onc. Biol. Phys.* 15 (1988):1219-1221.
85 For a general assessment of requirements for residency training in radiation oncology today, see Marks, James E.; Armbruster, Judith S.; Brady, Luther W.; et al., "Special Requirements for Residency Training in Radiation Oncology," *Int. J. Rad. Onc. Biol. Phys.* 24 (1992):815-817.

CHAPTER SEVEN

BRACHYTHERAPY

Basil S. Hilaris, M.D.

In 1896, stimulated in part by Professor Röntgen's laboratory-generated rays, Antoine Henri Becquerel (1852–1908) discovered that uranium salts spontaneously emitted rays of an unknown nature. Marie Curie (1867–1934) and her husband Pierre (1859–1906), fascinated by this discovery, began to study Becquerel's invisible rays.[1] Three years later, the Curies, working with one ton of Austrian pitchblende (uranium ore) at their laboratory at the Municipal School of Physics and Chemistry in the outskirts of Paris, identified two radioactive elements. The first element was named polonium after Marie's native country, and the second element was named radium. The Curies' tedious work with pitchblende is a familiar story, as is that of their triumphant announcement to the French Academy of Science on 26 December 1898. Some lines of this communication are as follows:

> The various reasons we have just enumerated lead us to believe that the new radioactive substance contains a new element to which we propose to give the name of radium. The new radioactive substance certainly contains a very strong proportion of barium; in spite of that its radioactivity is considerable. The radioactivity of radium, therefore, must be enormous.[2]

In 1901 Becquerel carried a small amount of the radium salts in his vest pocket and subsequently found that an ulcer was produced on the skin of his abdomen, directly under the pocket. Pierre Curie was intrigued by this and, as Marie Curie wrote in her thesis, "he applied a weak radium amount upon his arm for ten hours. A redness appeared immediately, and later a wound was caused which took four months to heal. The epidermis was locally destroyed, and formed again slowly and with difficulty, leaving a very marked scar."[3] The Curies loaned a small radium tube to Henri Danlos, a physician at the Saint-Louis Hospital in Paris who irradiated a patient with lupus.[4] In 1903 the Curies and Becquerel jointly were awarded the Nobel Prize in physics for the discovery of radioactivity. The nuclear age had begun.

Alexander Graham Bell, in August 1903, wrote a letter to a Dr. Sowers, who subsequently forwarded it to the editor of *American Medicine*. In his letter Bell suggested the application of "a tiny fragment of radium sealed up in a fine glass tube and inserted into the very heart of the cancer, thus acting directly upon the diseased material."[5] The discovery of the X rays by Wilhelm Röntgen in 1895 and of radium by Marie and Pierre Curie in 1898 led immediately to cancer therapeutic trials. It appears that the first treatments were carried out with skin and breast cancer, but cancer of the cervix was also among the first cancers in which the new radiation was tried. Remarkable early results in skin cancer and lesions in the oral mucosa led to an explosion of interest in applications of natural radiation to a range of diseases.

Soon physicians in many countries in Europe established radium institutes. In 1906 Louis Wickham, a dermatologist, and Paul Degrais, a surgical pathologist, founded the Biological Laboratory of Radium in Paris. Henri Dominici was appointed as the clinical director with Jacques Danne in charge of physics. In 1907 Dominici demonstrated the different biological effects of the various qualities of rays emanated by radium salts and found that the superficial burn caused by a radioactive substance was due to both beta and soft alpha rays. Moreover, he found that these rays could be filtrated by encasing the radium salt in a lead container, permitting the passage of only the "ultra-penetrating" rays with no detrimental effect on healthy tissue. Wickham and Degrais published an English version of their experiences with radium in 1910 (Fig. 7.1).[6]

While crossing a street in Paris on a rainy afternoon in April 1906, Pierre

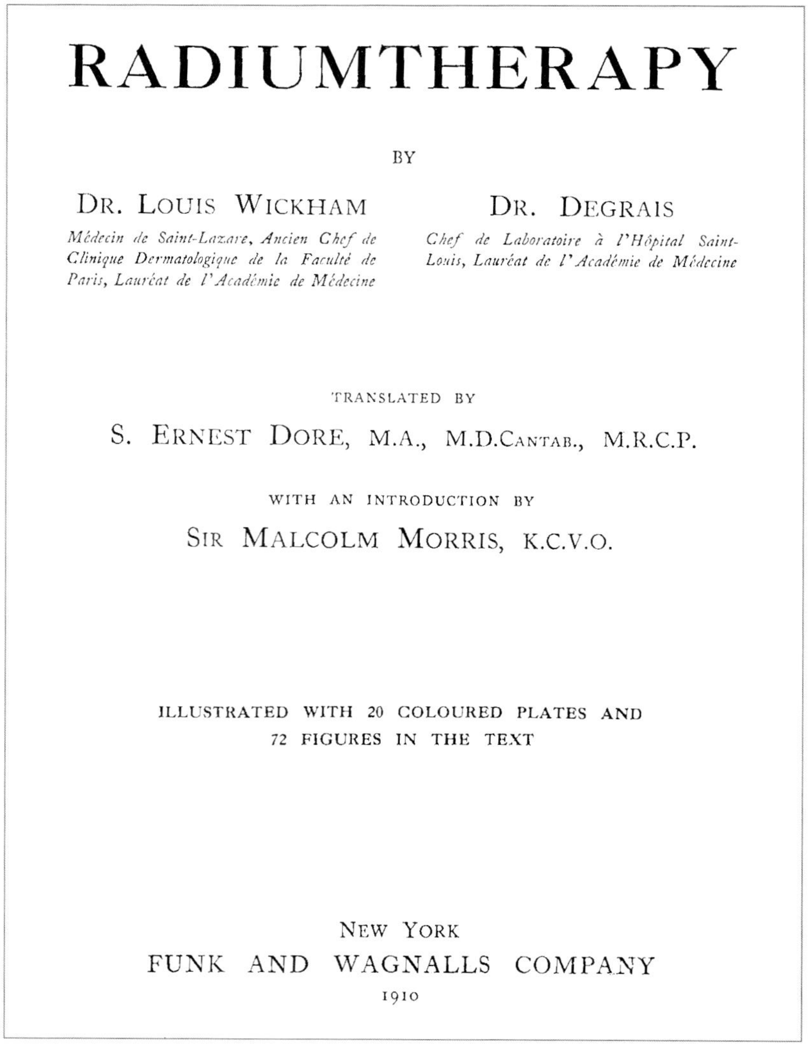
RADIUMTHERAPY

BY

DR. LOUIS WICKHAM
Médecin de Saint-Lazare, Ancien Chef de Clinique Dermatologique de la Faculté de Paris, Lauréat de l'Académie de Médecine

DR. DEGRAIS
Chef de Laboratoire à l'Hôpital Saint-Louis, Lauréat de l'Académie de Médecine

TRANSLATED BY

S. ERNEST DORE, M.A., M.D.CANTAB., M.R.C.P.

WITH AN INTRODUCTION BY

SIR MALCOLM MORRIS, K.C.V.O.

ILLUSTRATED WITH 20 COLOURED PLATES AND 72 FIGURES IN THE TEXT

NEW YORK
FUNK AND WAGNALLS COMPANY
1910

Fig. 7.1 The English version of the Wickham and Degrais text on radiumtherapy, issued in 1910. (Courtesy of the Center for the American History of Radiology, Reston, Va.)

Curie was struck and killed by a heavy wagon drawn by two horses. Marie continued her research, determining the atomic weight of radium in 1907 and subsequently succeeding in isolating the element in a pure state (an operation which was never to be repeated). In 1911 she again received the Nobel Prize, this time in chemistry for the discovery of radium, and was the first person to receive the award twice. Marie Curie died in a sanatorium at Sancellemoz on 4 July 1934 of aplastic pernicious anemia, injured by a lifetime's work with radiation.

In 1912 the renowned Pasteur Institute and the University of Paris founded the Radium Institute devoted to the science of radioactivity. It was divided into the Curie Pavilion, directed by Marie Curie, for research in physics and chemistry, and the Pasteur Pavilion, directed by Claudius Regaud, for research in medicine and biology. "These two institutions, materially independent, were to work in cooperation for the development of the science of radium."[7] In 1910 the Radiumhemmet (literally, "Radium Home") was established in Stockholm with Gösta Forssell (1876–1950) as its first director. Forssell was to become the godfather of a new specialty using radium in the treatment of disease and would name it *brachytherapy* in 1931.[8]

Brachytherapy, the treatment of cancer by radioactive sources placed at a short distance from a tumor, was thus born within a short period after the discovery of radium. However, the interest in radium as a cure for cancer created such a demand that its price soared from an already daunting $10,000 per gram in 1904 to $150,000 in 1918. The cost of science was beginning to increase.

EARLY HISTORY OF RADIUM IN THE UNITED STATES

As Marie Curie wrote in the 1920s,

The Buffalo Society of Natural Science has offered me, as a souvenir, a publication on the development of the radium industry in the United States, accompanied by photographic reproductions of letters which Pierre Curie had received in 1902 from Buffalo, New York. In those letters, engineers wanted to learn the process of radium purifications.[9]

Pierre Curie, after consulting with Marie, had replied most fully (and without reservations as to patents or proprietary claims) to the questions asked by the American engineers.

One of the first published claims to the use of radium was that of Margaret A. Cleaves (1848–1917) of New York.[10] She borrowed two sealed glass tubes of radium, imported from Paris, from Professor Charles Baskerville of the department of chemistry of the University of North Carolina. She treated a patient with advanced carcinoma of the cervix, through the vagina. She also treated a patient with recurrent sarcoma of the cheek and another with a recurrent carcinoma of the scalp. Shortly afterward Robert Abbe, also of New York, used radium imported from Germany to make a vaginal application for carcinoma of the cervix (Fig. 7.2).[11]

In 1903 Francis Williams of Boston, whose book *The Roentgen Rays in Medicine and Surgery* had been received with enthusiasm two years earlier, went to France (Fig. 7.3).[12] Antoine Béclère (1856–1939), who had just translated

Fig. 7.2 Robert Abbe (1851–1928) (Courtesy of the Center for the American History of Radiology, Reston, Va.)

Williams's book into French, interrupted his summer vacation to return to Paris and meet the admired American. Williams obtained firsthand information on radium and current therapeutic trials from Béclère and purchased a few milligrams (mg). Upon his return to Boston, he attempted to compare the effects of roentgen rays and radium in the treatment of tumors. He concluded that radium had definite advantages for intracavitary applications.[13] In collaboration with his brother Charles, Williams investigated the possibilities of the use of radium in the treatment of inflammatory diseases of the eye.[14] He also developed an intraoral instrument for the treatment of hypertrophic tonsils by radium.[15]

Another pioneer of radium, Albert Soiland, who had organized the first department of radiology at the University of Southern California Medical College in 1904, traveled to Europe and purchased a plaque of 5 mg of radium suitable for the treatment of cancer of the skin. He remained a dedicated brachytherapist (Fig. 7.4).[16,17]

The daughter of James Douglas, a Canadian mining engineer and president of the Phelps-Dodge Corporation, was diagnosed with breast cancer in 1907. She was operated on five times in New York but each time developed a local recurrence. Douglas took her to London for treatment with radium, which he obtained privately from Paris at an exorbitant cost. Despite these efforts, his daughter died in the spring of 1910. Soon afterward, Douglas, who was eager to put his money and energy to work in the cancer battle, met with James Ewing, professor of pathology at Cornell University. From a letter written by Ewing in December 1910, it is clear that he discussed the importance of cancer research and was persuaded to donate for this purpose a great deal of money to Memorial Hospital in New York. In addition, because radium was unobtainable in any quantity within the United States, he formed a partnership with Howard Kelly, a professor of gynecology at Johns Hopkins University, and Charles Parsons, director of the United States Bureau of Mines, to found the National Radium Institute to mine carnotite deposits on leased lands in Colorado (Fig. 7.5).[18] From 1913 to 1917 the partnership produced and divided among themselves 8.5 grams of radium. Kelly's share of radium went to his private clinic serving the Johns Hopkins Hospital, Parsons's went ultimately to Harvard University and other institutions around the country, and Douglas's was sent to Memorial Hospital in New York.

Douglas himself experimented with radium in his small laboratory at the rear of his office in the Phelps-Dodge Corporation headquarters. He developed a radium concoction which he drank regularly, applied to his wife's

Fig. 7.3 Francis H. Williams (1852–1936) (Courtesy of the Center for the American History of Radiology, Reston, Va.)

Fig. 7.4 Albert Soiland (1873–1946) (Courtesy of the Center for the American History of Radiology, Reston, Va.)

Fig. 7.5 Transporting hundreds of thousands of pounds of raw ore from the mines to Denver for refining into radium was a mammoth undertaking, even for the powerful partnership that made up the National Radium Institute. Here, ox carts haul raw ore from the mines to waiting railroad cars, 1914. (Courtesy of the Center for the American History of Radiology, Reston, Va.)

feet, and on occasion offered to visitors. Douglas died on 25 June 1918 at the age of eighty-one of pernicious anemia, perhaps related to his radium intake.[19]

Memorial Hospital

William Duane, professor of physics at Harvard, had perfected a radium emanation extraction and purification plant which made possible the production of glass radon seeds for interstitial implantation.[20] Duane agreed to install a model of his plant at Memorial Hospital (Fig. 7.6).[21]

In 1914 a radium department was established at Memorial Hospital, and Henry Janeway (Fig. 7.7), a surgeon with unusual ingenuity, was charged with developing new techniques of radium therapy. The next year Janeway brought in a young engineering student, Gioacchino Failla (Fig. 7.8), as an assistant physicist to take care of the radon plant and study methods of improving radon applications in cancer treatment. Failla learned to operate the plant, but he also learned everything known about radioactivity and its medical uses. Within a short time he became an authority to whom everyone else turned for information. The radium supply increased slowly from 36 mg in 1914 to 4 grams in 1917.[22] It became possible to compress radon into capillary glass tubes. The vault for the storage of the radium had been skillfully connected with a pumping apparatus, so that the daily withdrawal of the radon gas for therapeutic use was relatively safe and rarely interrupted.

The year 1917 was noteworthy in the development of radium therapy at Memorial Hospital.[23] At first the glass containers of radon had been placed in contact with accessible tumors in the same manner as surface applications of radium. Benjamin Barringer, the urologist at Memorial, now introduced them through the cystoscope and laid them directly on the surface of tumors in the bladder. Janeway provided other surgeons with radon sources for the treat-

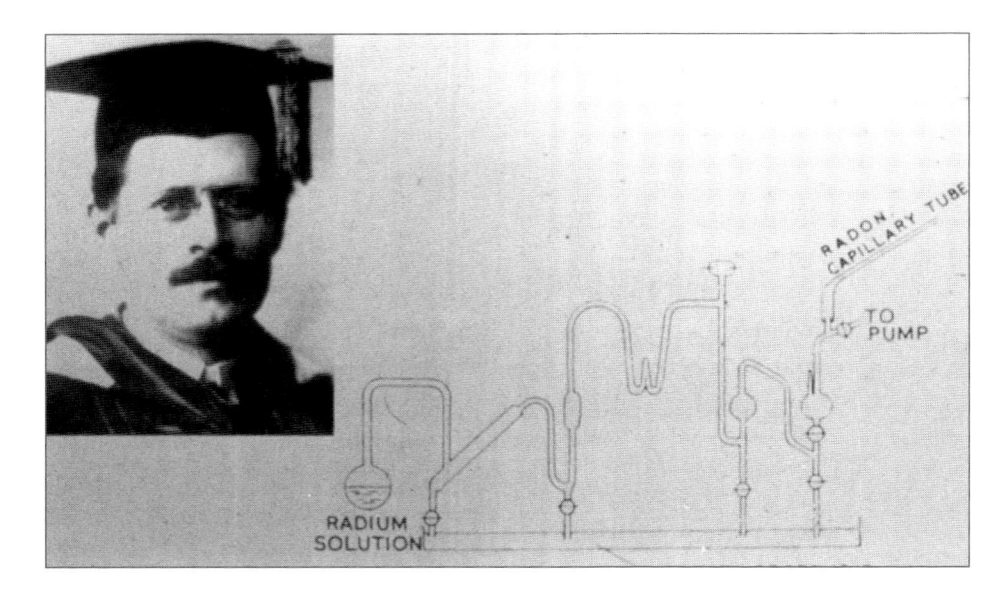

Fig. 7.6 Professor William Duane (1872–1935) and a diagram of his radon extraction plant, designed in 1915. (Courtesy of the Center for the American History of Radiology, Reston, Va.)

Hilaris

ment of malignant tumors of various parts of the body. In the same year Janeway, Barringer, and Failla co-authored a book describing the early results of brachytherapy at Memorial Hospital (Fig. 7.9). The first fifty pages of the book were devoted to a didactic discussion of the physics of radioactivity by Failla. Janeway discussed the principles and methods of the application of radium to cancer and gave details of his experience with cancers of the skin and oral cavity. Barringer gave the results of the irradiation of patients with cancer of the bladder and cancer of the prostate.[24]

At the same time as these events at Memorial, John Jolly (1867–1933) and Walter C. Stevenson (1877–1931) in Dublin were successful in placing the capillary containers into the lumen of ordinary steel needles.[25] This not only provided desirable filtration but, in addition, permitted the interstitial implantation of the radioactive sources into the tumors. Barringer adopted the innovation to facilitate the implantation of radon into carcinomas of the prostate using local anesthesia through the perineum.

In 1918 for the first time small capillary glass tubes containing radium emanation were introduced directly into the tissues through fine trocar needles and left in place—initiating the practice of permanent interstitial radiation. During the same period applicators with dental molding compound were constructed for holding radium in place on the skin and within the oral cavity. Other types of applicators were also constructed to allow the use of similar techniques on the eye and the vocal cords. In the outpatient radium department at Memorial, an average of fifty to sixty patients were seen daily.[26]

In addition to operating the emanation plant and the calibration of its products, Failla took an active part in the various approaches and trials with radioactive sources. He developed a machine shop in the basement of the hospital and took delight in the design and construction of all kinds of accessories. Among many other gadgets, he built a bell-shaped lead container to hold radon so that it could be brought into contact with the cervix. The device was used by Harold Bailey (1878–1929) and William Healy (1876–1954), the institution's gynecologists.

World War I interrupted this work, and the young Italian-born engineer was needed for duty abroad. When he returned after the armistice, he was ready in earnest to develop a research laboratory. In 1919, at the insistence of Janeway, the hospital created a department of physics of which Failla was appointed director. In the same year, Edith Quimby (1891–1982) joined as an assistant to Failla. She later wrote that "I was fortunate enough to apply and be accepted for the position and thus started an association which was terminated only when we

Fig. 7.7 Henry H. Janeway (1873–1921) (Courtesy of the Center for the American History of Radiology, Reston, Va.)

Fig. 7.8 Gioacchino Failla (1891–1961) (Courtesy of the Center for the American History of Radiology, Reston, Va.)

Brachytherapy

Fig. 7.9 Janeway, Barringer, and Failla coauthored a thorough report in 1917, describing the early results of brachytherapy at Memorial and giving detailed accounts of applications tailored to specific conditions. (Courtesy of the Center for the American History of Radiology, Reston, Va.)

both reached retirement age in 1961."[27]

The Treatment of Uterine Cancer by Radium

The oldest claim to the use of radium in cervical cancer in the United States, to our knowledge, is that of Margaret Cleaves, who apparently used a combination of X rays and radium. In a 1903 publication she stated:

> The pelvic case has been under care for three months and has improved to date under the combined influence of the x-ray (internal applications entirely) and ultraviolet light. There is, however, a discharge of blood from the rectum upon defecation and now and then at other times. Because of this and believing that the radium rays would penetrate more deeply than the x-rays, the radium was used. A radium bromide tube was placed in the vagina and allowed to lie on the posterior surface for five minutes, and for an additional five minutes in the anterior wall. A second

RADIUM THERAPY IN CANCER

AT THE MEMORIAL HOSPITAL
NEW YORK

(FIRST REPORT: 1915 - 1916)

BY
HENRY H. JANEWAY, M.D.

WITH THE DISCUSSION OF TREATMENT OF CANCER OF THE
BLADDER AND PROSTATE
By BENJAMIN S. BARRINGER, M. D.

AND AN INTRODUCTION UPON THE PHYSICS OF RADIUM
By GIOACCHINO FAILLA, E. E., A. M.

NEW YORK
PAUL B. HOEBER
1917

five minute application was made a few days later. Five days subsequent to the use of radium, no bleeding, no odor, no discharge, no ulceration and vaginal and cervical mucous membrane are normal in appearance. There has been no bleeding from the rectum since the radium was used.[28]

Howard Kelly was another pioneer in the use of radium for treatment of gynecological cancers (Fig. 7.10). He devised the cystoscope, the Kelly pad, and rectal and vaginal specula among many related items. He began using radium in 1908, and in the beginning treated only inoperable cases and recurrences after surgery. In 1913 he extended the treatment to operable cases, giving prophylactic treatment before hysterectomy.

Despite the expense of acquiring radium and the difficulties of working with it on a daily basis, a number of clinicians by the 1910s were regularly not-

Fig. 7.10 Howard Kelly (1858–1943)

ing its beneficial effects on their patients. Clark began his radium therapy in 1913 and reported good results in 1917. In 1915 Bailey and Healy began systematic radiotherapy of carcinoma of the cervix at Memorial Hospital.[29] Henry Janeway's work there, beginning in 1914, with radium in uterine cancer was most notable. He was the first in the United States to advocate radium as the treatment of choice in cervical carcinoma. He developed the technique of burying radium emanation in the cervix, and, in fact, all his work with radium emanation needles was primarily original. His work on conservative surgery plus radium was recognized early, and his paper, "Treatment of Uterine Carcinoma," remains a classic in the field.

The Founding of the American Radium Society

On 22 June 1916, during the annual meeting of the American Medical Association (AMA), a group of physicians involved in radium therapy met in Detroit and agreed to found a society in which workers in different disciplines would meet and exchange experiences relative to the therapeutic uses of radium. A few months later, on 26 October 1916 in Philadelphia, Henry Schmitz (1871–1939) of Chicago presented the bylaws for a new association, the American Radium Society (ARS), which were approved by the twenty present prospective members. The aims of the society were declared to be the promotion of the scientific study of radium, its physical properties, and its therapeutic applications (Fig. 7.11). William Aikins (1859–1924) of Toronto was elected as the first president. It was decided that the society would meet annually for one day before the AMA meeting, with annual dues set at five dollars. In the course of time, eight of the twenty charter members would become ARS presidents.[31,32] In 1921 the *American Journal of Roentgenology* became the official organ of the ARS, adding *Radium Therapy* to its title in 1923.

The first honorary members were selected in 1921: Marie Curie of Paris, Howard Kelly of Baltimore, and Francis Williams of Boston. The following year, four additional honorary members were chosen: James Ewing of New York, William Duane of Boston, William Coolidge of Schenectady, and Claudius Regaud of Paris. A Janeway lectureship was established in 1933 by the initiative of ARS's seventeenth president, Burton Lee (1874–1933). The first Janeway lecturer was James Ewing, who spoke on early experiences in radium therapy.

Both radium and X rays were increasingly applied indiscriminately by inadequately trained persons, opportunists, and frank charlatans in these years. As early as 1922 an ARS committee was appointed to seek agreement with other radiological societies to establish an accreditation institution for radiology.[33] It was not until 1933, however, that the American Board of Radiology was founded to protect the public from irresponsible practices and to preserve the dignity of qualified professionals.

1920–1940: GOLDEN ERA OF RADIUM

By the early 1920s radium work was done in several institutions in the United States. Marie Curie, in a May 1920 interview with Mrs. Meloney, editor of a New York magazine, stated with some envy that "America has about fifty grams of

Fig. 7.11 Formally organized in 1916, the American Radium Society originally included physicians, researchers, and even manufacturers, in an effort to establish an ecumenical organization. The original minutes of the society, dating to 1916 and including a series of evolving logos, are available at the Center for the American History of Radiology, Reston, Va.

Fig. 7.12 Marie Curie with President Harding in 1921. (Courtesy of the Center for the American History of Radiology, Reston, Va.)

radium, four of these are in Baltimore, six in Denver, seven in New York."[34] Mrs. Meloney, after her return to New York, organized in 1921 the Marie Curie Radium Fund—with Robert Abbe among the active committee members—in order to collect enough money to buy a gram of radium as a gift to the discoverer. In 1925, in an official ceremony at the White House, a gram of radium was symbolically presented to Marie Curie by President Warren Harding (Fig. 7.12). The actual gift gram of radium was deemed too dangerous and valuable for a public ceremony.[35]

During this period, in addition to work by Failla and Quimby at Memorial Hospital, others, including Stenstrom at the New York State Institute for the Study of Malignant Disease (later Roswell Park), Weatherwax at Philadephia General Hospital, and Glasser and Fricke at the Cleveland Clinic, began the development of radiation physics departments that steadily contributed to the science of radiation therapy. The Mayo Clinic was at the time one of the few institutions in the world where the practice of radiology was divided into three sections: diagnostic roentgenology, therapeutic roentgenology, and radium therapy. Henry Bowing (1884–1953) served as chief of radium therapy. It is of interest that Ralston Paterson's first exposure to radium therapy came in 1926 when for a month he rotated through Bowing's department as a fellow at the Mayo Clinic.[36]

Failla and Quimby during this period studied the production of erythema of the skin in laboratory animals and patients and concluded that the erythema could be taken as a biologic indicator with some degree of accuracy. Failla designed and built an apparatus that facilitated uniform segmentation and calibration of the glass radon seeds in a few minutes. This plant was semiautomatic in operation and greatly lessened the exposure of operators to radiation (Fig. 7.13).

In 1920 Janeway published his experience with the interstitial use of radon seeds in the *American Journal of Roentgenology and Radium Therapy*, outlining the indications and advantages of this technique.[37] He recommended the use of radon in combination with external irradiation in lip and intraoral tumors; rectal lesions;

Fig. 7.13 Improved version of the automatic radon plant, installed at Memorial in the 1930s by Failla and Quimby. Note the radium safe incorporated into the unit. The sign "Holmes Electric Protection" refers not to an early radiation safety device but to a burglar alarm. (Courtesy of the Center for the American History of Radiology, Reston, Va.)

Hilaris

cancers of the cervix, prostate, bladder, and breast; in primary and metastatic tumors involving lymph nodes; and in sarcomas of the extremities.

In 1922 Quimby published tables of intensity distributions at various distances from point, linear, circular, square, and rectangular sources.[38] Within a few years, several other physicists published similar dosage tables, notably Paterson and Parker, who gave the doses in terms of roentgens.[39]

The principal objection to the use of glass radon seeds, however, was the lack of filtration and inhomogeneous irradiation of the affected tissues. It is of interest that Janeway, who had suffered for twenty-one years from an adamantinoma of the mandible and had treated himself with glass radon seeds, had extensive necrosis of the soft tissues. Failla verified in laboratory animals the necrotic effect on the tissues adjacent to the source.[40] Quimby made comparative tests of various metals and found that gold tubes were adequately segmented and calibrated.[41]

Several large radon plants were built in the United States, according to a scheme devised by Failla.[42] As a result of the easy availability, radon "gold" seeds became quite popular and were widely adopted and used for permanent implantation.[43]

Some pioneering work performed during the late 1920s and 1930s at the General Radium Service of Memorial Hospital is worthy of special mention (Fig. 7.14). Hayes Martin developed "nerve-injection" for the relief of pain and, together with Edward Ellis of the pathology department, made a "notable contribution to the technical problem of obtaining biopsies through needle puncture by aspiration." Many new surgical procedures were devised to meet the "...peculiar circumstances incident to radium application, which might well be designated as cancer surgery...laryngotomy for the purpose of accurate implantation of radium rather than for removal of a portion of the larynx...intensive radiation of rectal growths followed by a more limited type of surgical removal...."[44]

THE DARK AGE OF BRACHYTHERAPY

Radium, hermetically sealed in tubes, needles, or capsules, allowed the development of well-established techniques that produced satisfactory clinical results. However, since the radium salt was in the form of fine powder, rupture of the sealed container resulted in dispersal of the active material, with disastrous results owing to the long half life of radium and its high radiotoxicity.

Fig. 7.14 Attending staff of the Memorial Hospital, 1938, including some of the most influential figures in the development of radium therapy. Seated, left to right: Norman Treves, George Ernest Binkley, Hayes Elmer Martin, William Bradley Coley, Benjamin Stockwell Barringer, James Ewing, Frank Earl Adair, and Fred Waldorf Stewart. Standing: Ralph Eugene Herendeen, biologist Halsey Joseph Bagg, George Thomas Pack, William Spencer MacComb, Gordon Palmer McNeer, Alfred Franklin Hocker, George Hall Hyslop, Joseph Helms Farrow, Archie Leigh Dean, Gray Huntington Twombly, Lloyd Freeman Craver, superintendent George Holmes, physicist Gioacchino Failla, Edgar Leonard Frazell, and Frank Raymond Smith. (Courtesy of the Center for the American History of Radiology, Reston, Va.)

Moreover, the gamma rays emitted by radium sources were of sufficiently high energy to present a serious problem in personnel exposure.

Despite good clinical results, professional concern regarding the harmful effects of radium exposure, technical difficulties related to source construction and availability, and laborious dose calculations limited the use of brachytherapy to major centers. In the 1940s and 1950s spectacular technical developments in the field of external beam therapy and improvements in surgical techniques resulted in a declining interest in brachytherapy and a marked decline in its use. Many radiation workers felt that these disadvantages of radium warranted the investigation of other radionuclides for clinical use.

This became possible with the development of nuclear reactors. Enrico Fermi (1901–1954) and Leo Szilard (1893–1964) were among the individuals who produced the first self-sustained nuclear chain reaction leading to the release of nuclear energy, at the University of Chicago on 2 December 1942. The design and development of nuclear reactors, although initially devoted to the atomic bomb and nuclear destruction, had other consequences more important for mankind—the production of artificial radionuclides for a myriad of medical applications.

Advances Through Atomic Medicine

Increased production of artificial radionuclides resulted in a variety of commercially available sources for brachytherapy, each with certain advantages and disadvantages. Phosphorus-32 (^{32}P) was the first artificial radionuclide to be produced in the cyclotron for clinical use in 1936. With the advent of nuclear reactors, however, production of artificial radionuclides was accelerated: gold-198 (^{198}Au) was produced in 1947, followed by cobalt-60 (^{60}Co) in 1948; iridium-192 (^{192}Ir) in 1954; yttrium-90 (^{90}Y) in 1956; and cesium-137 (^{137}Cs) in 1957, which in the 1970s replaced radium-226 (^{226}Ra) as the radionuclide of choice for intracavitary therapy.

^{192}Ir, in the form of wires and/or seeds, was introduced in 1955 by Henschke as a substitute for radium needles and/or radon seeds.[45] In the early 1960s, however, the United States Atomic Energy Commission (AEC) imposed several restrictions on its use for permanent implantation, and, as a result, the technique was abandoned following 361 implants performed at Memorial Hospital. ^{192}Ir seeds are commercially available today in groups of twelve, spaced a centimeter apart in a nylon ribbon, and have completely replaced radium needles. The use of ^{192}Ir wires, because of concerns about possible radioactive contamination, was also restricted in the United States; yet it became the standard practice for temporary implantation in Europe.

Low-energy iodine-125 (^{125}I) sources were introduced in 1965 at Memorial Hospital as a substitute for ^{198}Au and ^{222}Ra in permanent implantation.[46] Their soft radiation had the advantage of being well localized and not exposing distant portions of the bone marrow. Localization and dosimetry procedures were developed specifically for these sources and contributed to their increasingly wide utilization for cancer therapy. The first clinical study with ^{125}I was conducted at Memorial Hospital from 1966 through 1967, investigating low energy radionuclides for permanent interstitial implantation and measuring the reduced emission of radiation from their use.[47] A more detailed study followed from 1968 to 1971. A final report was published by the Food and Drug Administration in 1975.[48] The AEC removed the ^{125}I seeds from the investigational procedure list on June 1975.

The availability of new radionuclides made it possible to introduce improved techniques for interstitial, intracavitary, and surface applications and created a renewed interest to brachytherapy.

Afterloading of Radioactive Sources

Afterloading, a household word today among radiation oncologists, was first described in 1903 by H. Stroebel, who afterloaded a radium source with the help of a guide tube implanted into a tumor.[49] Abbe is credited with a similar procedure in a 1910 reference.[50]

The principle was reintroduced and the method refined and developed further by Ulrich Henschke (1914–1980), first at Ohio State University and, after 1955, at Memorial Hospital in New York. Henschke simultaneously studied medicine and physics at the University of Berlin and graduated in both fields with the highest possible marks in each of the twenty-eight required examinations. From 1937 to 1940 he was assistant to Professor W. Friedrich, the most famous of Röntgen's pupils. Henschke's contributions to science were many. His investigations at Friedrich's Institute for Radiation Research at the University of Berlin led to the first accurate measurement of the roentgen unit using a large pressure chamber.[51] At the radiation clinic of the Charity Hospital of the University of Berlin, he pioneered intraoperative radiation therapy of lung and stomach cancers, working together with surgeon F. Sauerbruch, and published on contact X-ray therapy and rotational therapy. Soon afterward he worked at Davos in Switzerland, analyzing the spectrum of the sun rays to determine the cause of skin burning and publishing the biological effects of ultraviolet and infrared radiation. In the early 1950s, after his arrival in the United States and in cooperation with H. Mauch, Henschke developed an artificial leg with a hydraulic system that is still widely used, enabling handicapped people to live more normal lives. "The field of radiation owes a large debt to Henschke for his grand vision. He was a brilliant inventor and innovator, and towards the end of his life a pioneer in bringing effective cancer management to many parts of the third world."[52] Henschke's major contributions to radiotherapy were the development of afterloading in brachytherapy, the introduction of ^{192}Ir sources for temporary interstitial implants, and the design of a simple double-headed cobalt teletherapy machine, appropriately named Janus and well adapted to third world conditions.

Afterloading of the radioactive sources has been responsible for the present renaissance in brachytherapy, providing greater flexibility of radiation source distribution, greater accuracy and control of the procedure, and enhanced personnel protection from radioactivity. The principle of afterloading consists of two steps: the insertion of unloaded tubes or applicators, and the afterloading with radioactive sources.[53] Afterloading techniques are now used throughout the world. Three types of afterloading in brachytherapy should be distinguished:

(1) Operative afterloading, in which radioactive sources are inserted in the operating room after unloaded needles have been placed inside the tumor. This reduces, but does not eliminate, radiation exposure in other hospital areas.

(2) Postoperative afterloading, in which radioisotopes are inserted on the hospital ward hours or days after the implantation of empty needles or tubes. This eliminates all radiation exposure in the operating room, the recovery room, the diagnostic radiology department, and in hallways and elevators of the hospital. However, postoperative afterloading is still burdened with radiation exposure to the nursing staff and anyone who enters the patient's hospital room.

(3) Remote afterloading, in which radioactive sources are inserted in a special sealed room from an outside control, and no one but the patient is exposed to radiation. Remote afterloading is, therefore, clearly the best technique for interstitial and intracavitary brachytherapy in terms of radiation protection.

Remote afterloading

Remote afterloading was developed in the 1960s along two different lines: high-activity and low-activity remote afterloaders. High-activity remote afterloading was favored by the group at Memorial Hospital from as early as 1961. Small cobalt sources of high activity, moving back and forth, were used to simulate sources of different and longer active length. It was concluded in a paper on this subject that "...on the basis of our limited experience with such short treatment times in the last three years, we feel that they may be used with impunity if the total dose is divided into more fractions."[54] A subsequent paper stated that "...Moving source remote afterloaders can be used with all gamma emitting radioisotopes, but ^{137}Cs appears most suitable except in the case of short treatment times, for which ^{60}Co and ^{192}Ir are preferable because of their higher specific activity, which in turn permits smaller sources and applicators."[55,56] The remote afterloader developed at Memorial Hospital in 1964 was commercially marketed by Atomic Energy of Canada under the name Brachytron and was installed in several medical centers, including those at the University of California in San Diego, the University of Southern California in Los Angeles, and the Cancer Institute in Bejing. The first model at Memorial Sloan-Kettering Cancer Center in New York was used in the irradiation of tumors of the vagina, nasopharynx, and mouth. It was also used to treat cancers of the cervix and endometrium. The "hot room" consisted of a treatment room surrounded by a thick wall, with a control room on the other side of the wall. Communications between the treatment room and the control room were by closed circuit TV and a speaker system. A lead safe in the wall held the radiation sources—tiny, stainless steel rods containing radioactive cobalt. The rods were welded to the ends of long cables threaded into flexible plastic tubes, which extended out of the safe into the treatment room. In an actual procedure, to irradiate the vagina for example, a hollow plastic applicator was inserted into the vagina and three plastic tubes were connected to it. The medical personnel left the room. From the control room the three sources of radioactive cobalt were advanced electrically out of the safe, through the plastic tubes, and into the applicator. A sensor inserted in the rectum continuously monitored the treatment to avoid over-irradiation and damage to healthy tissue. Early results were reported in 1974.[57] This remote afterloader remained in use at Memorial Sloan-Kettering from its installation in 1964 until 1979, when it was replaced by a commercial unit (Gamma Med II).

1980 TO THE PRESENT

In the United States the role of brachytherapy in the last fifteen years has expanded dramatically. This success story is driven by extensive technological development, an increasing number of physicians who practice brachytherapy, and by the interest generated in other specialties due to its undisputed effectiveness. This increased role is reflected in the large brachytherapy literature (more than three thousand brachytherapy citations for the past ten years alone!); the formation of the American Endocurietherapy Society (AES), which lists more than four hundred members; and the journal *Endocurietherapy/Hyperthermia Oncology (E.C.H.O.)* which has been published under AES auspices since 1985.[58]

The introduction of computer technology and the resulting ability to produce radiation dose distributions projections individualized for each patient has tremendously increased the accuracy of brachytherapy. Commercially produced high dose-rate (HDR) units, optimization of brachytherapy, three-dimensional brachytherapy planning, and ultrasound-based real-time planning are now becoming available and are being investigated intensively. New radionuclides with lower energy, improved physical dose distribution, and radiobiological effectiveness were introduced in the 1980s: palladium-103 and ^{137}Cs as alternatives to ^{125}I, ytterbium-169 as an alternative to ^{192}Ir, and americium-241 intended to replace ^{137}Cs. Dose rates varying from low dose-rate (LDR) to HDR and/or pulsed (PDR) brachytherapy—which allows a pulsed variable low dose rate—are currently being explored.

The development of integrated brachytherapy units which will combine all steps required for the performance of a procedure are also being actively pursued. Such efforts attempt not only to integrate all the various pieces of equipment but also to integrate the information flow and medical procedures, provide a sterile environment, and coordinate the activities of the surgical facility, fluoroscopy, radiography, treatment planning, and HDR afterloading treatment.[59]

Tumor models developed in the laboratory offer a reliable quantitative approach for the evaluation of various combinations of brachytherapy, chemotherapy, and/or immunotherapy in an attempt to eradicate tumor cells at local and distant sites while decreasing morbidity. The realization of these goals will definitely have an impact on the future utilization of brachytherapy alone or in combination with other cancer modalities as we enter the next century.

▶ ▶ ▶ ▶ ▶ ▶

REFERENCES

1. Becquerel, A.H., "Sur quelque propriete nouvelles des radiations invisibles émisés par divers corps phosphorescents," *CR Acad. Sci.* (Paris) 122 (1896):559.
2. Curie, E. *Madame Curie: A Biography*, V. Sheean, trans. (New York: Pocket Books, 1967):173.
3. Curie, M., "Radioactive Substances," a translation from the French of the classical thesis presented to the Faculty of Sciences in Paris (New York: Philosophical Library, 1961):67.
4. Danlos, H., "Sur l'action physiologique et thérapeutique du radium," *Med. Sci. Pharmacol.* 9 (1904):65-74.
5. Sowers, Z.T., "Clinical Notes and Correspondence," *Am. Med.* 6 (1903):261.
6. Wickham, L., and Degrais, P. *Radium Therapy.* New York: Funk and Wagnalls, 1910.
7. E. Curie, *Madame Curie.*
8. Ibid.
9. Forssell, G., "La lutte sociale contre le cancer," *J. Radiol.* 15 (1931):621-634.
10. Cleaves, M., "Radium (With a Preliminary Note on Radium Rays in the Treatment of Cancer)," *Med. Rec.* 64 (1903):601-606.
11. Abbe, R., "Notes on the Physiologic and Therapeutic Action of Radium," *Wash. Med. Ann.* 2 (1904):363-377.
12. Williams, F.H. *The Roentgen Rays in Medicine and Surgery.* New York: Macmillan, 1901.
13. Williams, F.H., "A Comparison Between the Medical Uses of the X Rays and the Rays from the Salts of Radium," *Boston Med. Surg. J.* 150 (1904):206-209.
14. Williams, F.H., "Notes on Radium: Production of X Rays from the Gamma Rays of Radium; Use of Radium in Some Diseases of the Eye," *Boston Med. Surg. J.* 150 (1904):554-561.
15. Williams, F.H., "Treatment of Hypertrophied Tonsils and Adenoids by Radium: A Preliminary Report," *Boston Med. Surg. J.* 184 (1921):256-257.
16. Soiland, A., "The Treatment of Inoperable Cancer of the Pelvis by Radium," *Am. J. Roent.* 12 (1924):378-379.
17. Soiland, A., "Comments on the Use of Radium for Intraoral Cancer," *Am. J. Roent.* 13 (1925):102.
18. Parsons, C.P.; Moore, P.B.; Liad, S.C.; and Shaefer, O.C. *Extraction and Recovery of Radium, Uranium, and Vanadium from Carnotite. Bulletin 104.* Washington, D.C.: Department of the Interior, Bureau of Mines, 1915.
19. Considine, B. *That Many May Live.* New York: Memorial Center for Cancer and Allied Diseases, 1959.
20. Duane, W., "On the Extraction and Purification of Radium Emanation," *Phys. Rev.* 5 (1915):311-314.
21. del Regato, J.A., "William Duane," *Int. J. Rad. Oncol. Biol.* Phys. 4 (1978):717-729.
22. Janeway, H.H.; Barringer, B.S.; and Failla, G. *Radium Therapy in Cancer at the Memorial Hospital.* New York: Paul B. Hoeber, 1917.
23. *Annals of Memorial Hospital, Radium Department*, 1917 (New York: Memorial Hospital, 1917).
24. Stevenson, W.C., "Preliminary Clinical Report on a New and Economical Method of Radium Therapy by Means of Emanation Needles," *Brit. Med. J.* 2 (1914):9-10.
25. Janeway, Barringer, and Failla, *Radiation Therapy.*
26. Annals of Memorial Hospital, report of the General Radium Department, 1918.
27. Quimby, E., "Medical Radiation Physics in the United States," *Radiology* 78 (1962):518-522.
28. Cleaves, M.A., "Radium (With a Preliminary Note)."
29. Healy, W.P., and Bailey, H., "Cancer of the Uterine Cervix Treated by Irradiation: Method of Treatment and Results in 1,024 Cases," *J.A.M.A.* (1924):1055-1056.
30. Janeway, H.H., "The Treatment of Uterine Cancer by Radium," *Surg. Gyn. Obs.* 29 (1919):242-265.
31. American Radium Society Minutes, 1916– 1935. Center for the American History of Radiology, Reston, Va.
32. del Regato, J.A., "The American Radium Society: Its Diamond Jubilee," *Am. J. Clin. Oncol.* 14 (1991):93-100.
33. del Regato, "American Radium Society."
34. Curie, E., *Madame Curie.*
35. Ibid.

36. del Regato, J.A., "Ralston Paterson," *Int. J. Rad. Oncol. Biol. Phys.* 13 (1987):1081-1091.
37. Janeway, H.H., "The Use of Buried Emanation in the Treatment of Malignant Tumors," (1920):325-327.
38. Quimby, E.H., "The Effect of the Size of Radium Applicators on Skin Doses,"*Am. J. Roent.* 9 (1922):671-683.
39. Paterson, R., and Parker, H.M., "A Dosage System for Gamma-Ray Therapy, Parts I and II," *Br. J. Radiol.* 7 (1934):592-632.
40. Failla, G., "The Development of Filtered Radon Implants," *Am. J. Roent.* 16 (1926):507-525.
41. Quimbly, E.H., "Comparison of Different Metallic Filters Used in Radium Therapy," *Am. J. Roent.* 13 (1925):330-342.
42. Failla, G. US Patents No. 1,553 794 and No. 1,609 614.
43. del Regato, J.A., "Gioacchino Failla," *Int. J. Rad. Oncol. Biol. Phys.* 19 (1990):1609-1630.
44. Quick, D., "Report of the General Radium Service for 1926-1929," Memorial Hospital.
45. Henschke, U.K., "Interstitial Implantation with Radioisotopes," in Hahn, F., ed., *Therapeutic Use of Artificial Radioisotopes* (New York: Wiley, 1956):375-397.
46. Hilaris, B.S.; Henschke, U.K.; and Holt, J.G., "Clinical Experience with Long Half-Life and Low-Energy Encapsulated Radioactive Sources in Cancer Radiation Therapy," *Radiology* 91 (1968):1163-1167.
47. Hilaris, B.S.; Holt, G.J.; and St. Germain, J. PHS grant EC-00113, Bureau of Radiological Health, 1966.
48. Hilaris, B.S.; Holt, G.J.; and St. Germain, J. *The Use of Iodine-125 for Interstitial Implants.* Washington, D.C.: U.S. Department of Health, Education, and Welfare; Public Health Service; Food and Drug Administration, 1975.
49. Stroebel, H., "Vorschlæge zur radiumtherapie," *Deutsche Med. Zeit.* 24 (1903):1145-1146.
50. Abbe, R., "Short Paragraph Describing His Method," *Arch. Roentgenol.* 15 (1910): 74.
51. Friedrick, W.; Schulze, R.; and Henschke, U., "Beitrage zum problem der radiumdosimetrie IV," *Strahlentherapie* 60 (1937):38.
52. von Essen, C.F., "In Memoriam: Ulrich Konrad Henschke," *Int. J. Radiation Oncol. Biol. Phys.* 8 (1982):947-948.
53. Henschke, U.K.; Hilaris, B.S.; and Mahan, G.D., "Afterloading in Interstitial and Intracavitary Radiation Therapy," *Am. J. Roent.* 90 (1963):386-395.
54. Henschke, U.K.; Hilaris, B.S.; and Mahan, G.D., "Remote Afterloading for Intracavitary Radiation Therapy," *Radiology* 83 (1964):344-345.
55. Henschke, U.K.; Hilaris, B.S.; and Mahan, G.D., "Remote Afterloading for Intracavitary Radiation Therapy," in Ariel, Im, ed. *Progress in Clinical Cancer* (New York: Grune and Stratton, 1965):127-136.
56. Henschke, U.K.; Hilaris, B.S.; and Mahan, G.D., "Intracavitary Radiation Therapy of Cancer of the Uterine Cervix by Remote Afterloading with Cycling Sources," *Am. J. Roent.* 96 (1966):45-51.
57. Hilaris, B.S.; Ju, H.; Lewis, J.L.; et al., "Normal and Neoplastic Tissue Effects of High Intensity Intracavitary Irradiation: Cancer of the Corpus Uteri," *Radiology* 110 (1974):459-462.
58. Wilson, J.F., and Hilaris, B.S., "Brachytherapy as a Model for Subspecialization within Radiation Oncology," *Int. J. Rad. Onc. Biol. Physics.* 24 (1992):877-879.
59. Van't Hooft, E., "The concept of an integrated brachytherapy unit," *Proceedings of the Seventh International Brachytherapy Conference.* Baltimore/Washington, 6-8 Sept. 1992.

*P*articipation in World War II increased both the numbers of persons working with radiation and United States involvement with radiation issues. Here the Picker field unit, used in the field for both therapy and diagnosis, was prepared for mass shipment in 1943. (Courtesy of the Center for the American History of Radiology, Reston, Va.)

CHAPTER EIGHT

Intersociety, Government, and Economic Relations

Carl R. Bogardus, Jr., M.D.

▶ ▶ ▶ ▶ ▶ ▶ ▶ ▶ ▶ ▶ ▶ ▶ ▶ ▶

It is impossible to separate entirely the early developments in therapeutic radiology and diagnostic radiology, as they were parallel in their growth and maturity. The two specialties melded together politically and economically for the first seventy-five years of their existence. It is true that there were a few physicians who specialized solely in radiotherapy in the earlier years, but economic changes and interactions with hospitals, fledgling insurance companies, and the federal government were always directed at "radiology" in general.

Many of the early pioneers of radiation therapy were not even radiologists but surgeons or gynecologists who used radiation in the cancer treatment portions of their specialties. Even today membership in the American Radium Society (ARS) is not confined to radiation oncologists but is open as well to members of surgical specialties such as gynecology, otorhinolaryngology, and general surgery.

At the beginning of this century, when medicine was developing its various subspecialty groups, patients were responsible for their own medical payments. There were no health insurance companies as we know them today, no government regulations, and no third party intermediaries. The only political and socioeconomic issues addressed by physicians in these years were directly involved with hospital boards and occasionally with their fledgling specialty societies.

Following the end of World War II there was rapid development and proliferation of new and sophisticated equipment for radiation therapy, starting with cobalt teletherapy units and progressing to modern linear accelerators. This led to a marked increase in the range and availability of applications in radiation therapy and, consequently, increased the demand for physicians specializing in this field. Developments in apparatus, scientific and biological knowledge, and in the specialized training of new physicians in radiation oncology are covered in detail in other sections of this book. Development along these numerous parallel paths led to the growth and eventual independence of radiation oncology as a fully recognized medical specialty.

Despite this modern independence, the economic history of radiation oncology has remained tightly entwined with all of radiology, and the benefits of this relationship have far outweighed the bad points. Radiation oncology, if left alone as a small specialty, would have had little or no political influence on its fate.

THE RELATIONSHIP OF RADIATION ONCOLOGY WITH HOSPITALS

It should be remembered that the hospital was originally planned as an institution for the indigent, often annexed to an almshouse or prison. The doctor expected the hospital to provide space and some of the necessary tools, as well as assistance in the hospital, particularly since the doctor's services were usually rendered without charge. Until the last years of the nineteenth century, persons who were economically competent never thought of entering the hospital as a patient. Later, as these individuals began to patronize hospitals, the institution continued to furnish to the doctor, without charge, a shop in which to work, the tools of his trade, and a corps of professional and technical assistants. This placed the doctor in a very delicate position with respect to control of hospital policies.[1]

Diagnostic X-ray units were constructed within months of Röntgen's announcement and first appeared in private clinics and offices. But hospitals very quickly realized that diagnostic X rays could be a lucrative source of income and began to build specialized X-ray rooms within the hospital. Logically, these should be staffed by the physicians who were becoming expert in the reading of these radiographs, and the specialty of radiology was born of necessity—with the hospitals serving as midwives. This early symbiotic relationship between the new specialty physician and the hospital as owner of the equipment has kept the radiologist in a somewhat vulnerable position, at the same time that it set tones and precedents for increasing technical specialization in other fields. The question of *how* the radiologist would be compensated dominated economic discussions in the field for years to come.

The Intersociety Committee for Radiology reported in 1940:

> ...for several years, radiologists have warned surgeons, internists, obstetricians, urologists, orthopedists, et al., that hospitals would one day dominate all medicine if a few specialties were sacrificed to economic convenience. If hospitals could practice radiology, it has been said they could likewise practice other specialties. If radiology is included as a *hospital benefit* in group hospitalization plans, there is nothing to prevent the inclusion later of other services.[2]

The process of hospital control of medical costs had started slowly. In the 1920s a few hospitals instituted flat rate plans under which patients paid a flat fee covering the cost of medical services and hospital facilities. The first instance of this type of flat rate or early health maintenance organization (HMO) began in Dallas in 1927, when Baylor University Hospital started a group hospitalization and insurance plan. Teachers at Baylor paid 50¢ per month for three weeks of "free of any additional cost" hospital care at Baylor Hospital. By 1929 other employee groups asked to join, and this was the beginning of what eventually would become the Blue Cross Hospital Insurance Plan.[3] Steady income from insurance premiums provided new revenues for Baylor Hospital. Patients received their entire care from the hospital with no additional fees. Early plans like these soon showed that radiology and pathology were potent revenue sources, with income from these areas used to cover losses in other departments.

Most radiology services in the 1930s were billed as hospital services with the income split between the radiologist and the hospital. These percentages varied across the country and from institution to institution. Over the years this ratio gradually stabilized with 40 percent going to the radiologist and 60 percent retained by the hospitals—a split which provided the historical basis for today's Medicare technical/professional/global ratios. Many hospitals attempted to divide the billing process so that radiologists would bill for their own services, but, amazingly enough, radiologists steadfastly held to the percentage split from the hospital-based billing. They reasoned that if radiologists were forced to bill separately, the hospitals would keep a larger percentage of the generated income from the department of radiology.

Other hospitals took a different approach and defined radiology as a simple hospital service (like pharmacy or brace-and-limb) and consequently put the radiologists on salary. Many radiology practices held on to these arrangements with their hospitals for years, until the Health Care and Financing Administration (HCFA) finally mandated that all physicians must bill separately for their professional services in 1983.

In 1939 the American College of Radiology (ACR) recognized the seriousness of hospital employment of the radiologist. The American Hospital Association (AHA) proposed to take over control of radiology, anesthesia, and pathology, on a salaried basis, making these physicians employees of their hospitals. This would be a striking contrast with other branches of medicine, but it was felt that this ownership of physicians' services was necessary to develop sufficient income to cover other costs under hospital insurance plans. The ACR strongly recommended that physicians be separate from any plan for hospital insurance because of the serious risks to their professional integrity. They recommended that medical services be entirely eliminated from any type of hospital insurance plan and that radiologists bill patients directly and assume full responsibility for their financial return, just as did other physicians in private practice.[4]

In June 1939 the American Medical Association (AMA) noted at their House of Delegates meeting that the physician roentgenologist should preferably be one who was a diplomate of the American Board of Radiology (ABR) and stated further that it should not be the policy of the hospital to make a profit from the department of radiology.[5] The AMA noted as well that the hospital was rapidly becoming a dominant factor in the delivery of medical care in the United States. The importance of the traditional family doctor as the main healthcare provider was being obscured in a trend toward the institutionalization of medicine, increased technology, and the specialization of physicians. The hospital was rapidly becoming the central nexus of healthcare in this country. Hospitals began to encroach on private office-based outpatient care by opening pay clinics and outpatient departments.[6] These early incursions into the outpatient field were the precursors to today's extensive ambulatory care facilities, many owned and operated by hospitals.

In 1940 the AMA stated publicly that many patients who were placed in the expensive facilities of hospitals could just as well be cared for in their own homes. The AMA felt that this overutilization stemmed from hospitals encouraging the use of their facilities, from patients demanding admission, and from physicians who found hospital care much more convenient than the inefficient but time honored house call.[7]

In the late 1930s some hospitals began using technicians—often without physician supervision—to operate radiology departments, not only for the taking of diagnostic X rays, but for the rendering of therapeutic X-ray treatments. The Kansas Medical Society noted in 1939 that "all X-ray therapy should be administered under the personal supervision of a doctor of medicine. Therapeutic roentgenology is a specialized and technical branch of the practice of medicine which, in unskilled hands, might be dangerous. While it is recognized that there is a need for skilled technicians as a means of assistance to the physicians, these technicians should work only under the direct supervision of a doctor of medicine."[8]

The Vincent Bill, introduced in the New York legislature in 1939, prohibited the practice of radiology by nonphysicians. It specifically stated, "To prevent terminology controversies, the Bill defines radiology as the diagnosis or treatment of disease by exposure to radium or roentgen rays." The bill explicitly defined radiology and limited its practice to physicians, dentists, and osteopaths.[9]

In the 1930s radiology was practiced in three different types of hospital settings: (1) those occupied exclusively by nonpaying patients, (2) those occupied by a combination of nonpaying and paying patients, and (3) those occupied by paying patients alone. Roentgenologists were paid differently in each setting. They could work for a straight salary in the hospital with nonpaying patients. In the hospitals with a mix of patients, the hospital billed on a fee-for-service basis, and the roentgenologist would receive a discount on the renting of space or facilities but would be expected to treat charity patients without charge. In the full-pay hospitals, the roentgenologist would lease the space and equipment and pay the hospital a monthly rental. The hospital usually billed straight fee-for-service with some split of collected revenue. In the late 1930s the rent charged to the roentgenologist could vary from as little as $300 per month in a small hospital to as much as $3000 per month in a large affluent institution.[10]

The ACR estimated that over 57 percent of hospital radiologists in 1939 practiced on a percentage basis, sharing the gross receipts with the hospital. Approximately 36 percent of radiologists were paid a salary, often based upon the amount of work performed. A small percentage, about 7 percent of radiologists, collected their own fees and had no financial arrangement at all with hospitals.[11]

The November 1939 *Bulletin* of the Intersociety Committee for Radiology quoted Arthur Christie's book on the economic problems of medicine:

> During the years when radiology was being advanced from a physical technique to a clinical specialty, the hospitals also acquired a position of new importance in the delivery of medical services. In 1939, there were over 6,000 registered hospitals in the United States, with a bed capacity of more than a million and a half, and almost all of them had a radiology department. During this development, it was natural that radiology should become a medical science practiced almost 90 percent in the hospital.[12]

In 1940 one-fourth of the cost of medical care in the United States went to pay hospital bills. The total cost of hospitalization in 1940 was $656 million dollars, $300 million of which was covered by federal, state, and local taxes, and $302 million by patient fees and insurance, with the remainder by contributions and endowments.[13]

The years of the second World War postponed much of the conflict between physicians and hospitals, as the nation's entire effort for that five-year period was devoted to operating as efficiently as possible with severely reduced numbers of physicians and technical assistants. Price freezes prevented any major changes in economic policies in the hospital/radiology relationships, just as freezes on the manufacture of X-ray equipment relieved hospitals of any obligation to update radiology facilities.

Following the end of the war and throughout the 1950s, hospitals proliferated and prospered. Many new facilities were built, aided by government funding, and many included magnificent radiology departments. Radiologic procedures increased at an astonishing rate, and radiology flourished as a specialty. In spite of this prosperity, radiation therapy was often relegated to a relatively primitive facility, usually in the basement, primarily using orthovoltage equipment and operating as "a philanthropic sideline of diagnostic radiology."[14]

The advent of cobalt teletherapy units, followed by the application of linear accelerators, brought radiation therapy out of the dark corners and into new dedicated facilities. The relationship with the hospital, however, did not change appreciably, and many of the early radiation therapists remained salaried or working on percentage contracts with the institutions. The hospitals still tightly controlled the tools of the trade of radiology, especially in therapy, since the equipment was so expensive that very few radiation therapists could set up private and independent offices.

At the time of the inception of

Medicare in 1964, a high percentage of both diagnostic and therapeutic radiologists were still working on some type of percentage contract or straight salary from hospitals. Some of these individuals remained on a percentage contract until 1983, when the realization that Medicare would no longer pay for the physician's component of radiological services if it were bundled into the hospital's unified or global billing finally convinced them to do otherwise. Global billing at this time remains one of the options in freestanding centers but is no longer recognized in hospital situations.

THE DEVELOPMENT OF THIRD PARTY PAYERS

Income from Radiologic Practice

The minutes of the early meetings of the ACR and the ARS demonstrate that the problems members faced in the 1930s were similar to the major issues radiologists confront today. By 1937 sixty separate small group hospitalization plans were in operation in the United States. The largest, by far, was the Blue Cross–Blue Shield plan. Of these plans, fifteen provided full reimbursement for X-ray and radium therapy. Following the successful lead of the Blue Cross and Blue Shield plans, commercial insurance companies took up health insurance. The Aetna Insurance Company wrote its first hospital insurance policy in 1936; the Equitable Insurance Company followed in 1939. By the early 1940s Traveler's Insurance, Metropolitan Life, and Prudential Insurance were also in the health insurance business. Health care insurance, covering both physician and hospital costs, became a lucrative market.[15]

In 1940 the ACR noted that twenty-eight state medical societies were seeking authority to start Blue Shield plans called "The Doctor's Plan."[16] The question was, "will radiology be included as a part of the professional benefits offered, or will these services be regarded as part of hospital care and be included in payments made to the hospitals both for radiation therapy and diagnostic radiology?" It was noted that the accepted practice in many of the larger hospitals was to make the services of diagnostic and therapeutic radiology available as part of hospital care. It seems that fault was found in the method of payment for these services through the hospital, with particular opprobrium directed against full-time salaried hospital radiology positions.[17] The 1942 ACR *Annual Report* quoted the 1941 presidential speech of Dr. Henry Walton: "We can all recall instances in which radiologists, who through their own ability, had built up attractive practices in hospitals only to be replaced by the hospital in the later years by younger men who were willing to accept positions under arrangements which permitted the hospital to keep larger incomes from the department receipts."[18] Translation: This usually meant working for a lower salary than the incumbent radiologist was asking.

Dominance and medical leadership in the health provider organizations has historically remained with surgery and internal medicine, with radiology playing a relatively passive role. Even now, we see government health plans driven by primary care physicians. One explanation for the quiet role of radiology may be its consistent place among the highest paid specialties—with infrequent cause to press for special considerations.[19]

Medical Economics reported in 1930 a median gross physician income of $12,000 with a net of $7,147. Surgeons were the only specialists reporting higher gross incomes than radiologists. By 1935, however, *Medical Economics* indicated that radiologists had become the highest paid of the medical specialties, with an average net income of $6,590. Surgery was second with a net of $5,961.[20] This represented a substantial income in the Depression years of increased buying power per dollar.

In 1936 the average net income for all physicians had dropped to $4,143. The average net income for a radiologist on this survey was $9,700 with only 3 percent of radiologists reporting incomes greater than $50,000.[21] By 1939 the average income for radiology was

$16,700 with a net of $9,860, indicating a cost of practice of about 32 percent.[22]

The fee schedules for diagnostic radiology were often described in detail in early ACR bulletins, but the fees for therapeutic radiology have always remained relatively vague—in part because of the wide range in variations in treatment methods among different practitioners and institutions. Occasionally global fees would be mentioned for radium therapy. In 1938 these ranged from $50 to $300 for radium applications. Superficial X-ray treatments averaged about $5 per treatment, and deep X-ray therapy averaged about $10 per treatment, with no other procedures listed. In 1938 the average global fee for a chest X ray in the radiologist's office was approximately $12, with a total of twenty-five diagnostic X-ray procedures comprising the entire fee schedule.[23] The most expensive procedure in diagnostic radiology was a three-film encephalogram at $50.

By 1947 the average physician grossed $17,476 and after expenses netted $9,884 per year. At this level, physicians were in the top 3 percent of earners in the United States. The total gross income for all physicians in 1947 was approximately $2.5 billion dollars, with the single highest gross income reported at $180,000 and only 0.1 percent of physicians admitting to grossing over $100,000 per year. The seven physicians who made up the top income reports in 1947 were headed by a proctologist, and no radiologist made the list. The gross average income of a radiologist was $34,693 with an after expenses net of $20,319 per year. Radiology was reported as the highest average income medical specialty in the United States in 1947, exceeding surgery and obstetrics and gynecology by a significant margin.[24]

A survey done by the ACR in 1952 showed an income range of between $16,000 and $37,000 a year for the chief of radiology position at a university hospital. Surveys have been conducted over a number of years by the Society of Chairman of Academic Radiation Oncology Programs (SCAROP); in 1990 the seventy-fifth percentile of salaries in university hospitals was noted to be $251,000 per year for a chairman. This rose to $350,000 in 1993. Entry level positions into academic radiology in 1993 were reported between $125,000 and $150,000 for the first year.[25] At the same time, entry level into radiology private practice was approximately $225,000 per year, rising to full partnership in the practice within an average of two to three years. Radiology continues to be a high income specialty but is now surpassed by cardiology, thoracic surgery, and in many situations medical oncology. The *AMA News*, 10 January 1994, reported from its Center for Health Policy Research that radiologists netted a median income of $240,000 per year compared to $148,000 per year average for all physicians in 1993.[26]

Billing for Radiological Services

In 1939 Albert Rayle noted in the *Journal of the Medical Association of Georgia* and was quoted in the *Bulletin* of the Intersociety Committee for Radiology that "physicians should devote themselves to the scientific phase of their work and let the economics take care of themselves." He went on to say, "But now the economic phases have been thrust upon us and we have to talk about them, however distasteful it may be. We physicians have nothing to sell but our services. We have spent many years and many thousands of dollars to make those services worth something to the public, and we are entitled to sell them in a fair market without improper restrictions or unfair competition."[27]

The *Bulletin* of the Intersociety Committee for Radiology in October of 1939 quoted a book by Theodore Wiprud, "It is with such items as office procedures, accounting systems, and collection methods that the young, inexperienced doctor is apt to have difficulty. Every doctor must be a bit of a businessman and, rarely having received academic training in these subjects, he should seek assistance in sound business details. A proper accounting system, adequate files and

Table 8.I
A German Relative Value Scale, 1938

Procedure	Professional Charge ($)
Superficial therapy, per area, total course of therapy	2
Dermatological treatments per field, total course	2–6
Deep therapy by an approved roentgenologist	40
Complete treatment for cancer of the uterus	50–60

records, and efficient clerical methods are essential to the conduct of a successful practice."[28,29] As you can see, very little has changed in the last fifty years. The major problems with billing have not changed in nature, despite their growing complexity.

Relative Value Scale

Some of the earliest notations of the cost of medical services are related in *The History of Medicine* by Ralph Major. In this book, Major notes that in the Code of Hammurabi (ca. 1950 B.C.) the fees for surgeons were specified, together with the penalties for surgical failure: "If a physician set a broken bone for a man or cure diseased bowels, the patient shall give five shekels of silver to the physician." But, "if a physician operate on a man for a severe wound and cause the man's death, they shall cut off his fingers."[30]

The *Bulletin* of the Intersociety Committee for Radiology in 1939 published a pioneering German relative value scale (RVS). (These values have been converted into United States dollars and reproduced as Table 8.I.) The plan had one restriction: specialists (roentgenologists) were entitled to administer any roentgen ray therapy in their specialty. Dermatologists, however, were allowed to give only superficial therapy; deep therapy could be administered only by an approved roentgenologist.[31] In order to be paid under the German system, every physician was obliged to give an exact report about each patient and procedure. A report was transmitted to the insurance company and to the medical examining board, while copies were retained by the roentgenologist and sent to the referring physician.[32] Each bill submitted was required to have the following items:

1. Name of patient
2. Statement of disease
3. Type of radiologic treatment or radiation therapy, the total "R dose" recorded
4. The roentgen diagnosis
5. Name and address of the referring physician
6. Calculation of charges divided into expenses and fees

Payment for radiological services was made by a special commission of radiologists, who supervised the qualifications of those desiring to practice radiology and audited all accounts submitted.

In 1937 the ACR and its radiologists struggled to retain global billing with a percentage division to the radiologist rather than splitting the bills into technical and professional components. The hospitals, on the other hand, were the proponents of division into technical and professional components and stated, quite logically, that the use of the operating room by the surgeon allowed the hospital to recover technical costs for the operating suite and its attendant expenses; the surgeon would then bill the patient for his professional services. The radiologist, they argued, should be no different. The hospitals owned the X-ray machines and should be entitled to collect the technical component for the use of the equipment, while the radiologist should bill the patient for professional opinions on interpreting films or expertise in treating malignancies. The ACR was quite adamant that the hospital should not pay radiologists a salary but should divide payments for the procedures with the radiologists. The argument was not fully resolved until HCFA mandated the institution of separate billing for the professional and technical components of all of radiology in 1983.

The total number of radiologists rose from 1,005 in 1931 to 2,191 radiologists in 1938. More than 75 percent of the physicians claiming to be radiologists were certified by the ABR and limited their practice exclusively to radiology. In 1937 the majority of radiologists apparently devoted less than one-third of their practices to therapeutic radiology, and only 1 percent of the radiologists stated they practiced therapeutic radiology 100 percent of the time. Very few of the radiation therapists in those days owned their own equipment; almost all were working in large hospital centers as employees of the institutions.[33]

By the mid-1960s radiation therapists were far more likely to be billing for professional services and acting as medical specialists than their counterparts in diagnostic radiology. A great deal of effort was exercised by the ACR and other specialty societies to convince members to scrap hospital billing contracts and bill for their services like other physicians. Radiation oncology led the way and laid a path for the rest of radiology to follow, a path that that would ultimately save them from becoming hospital employees. By 1980 20 percent of diagnostic radiologists were still working under hospital billing contracts, at a time when almost all radiation oncologists were billing on a fee-for-service basis as individual practitioners.

In the diagnostic radiology volume of this Centennial set, Otha Linton comments that in late 1965 a fierce debate raged between the United States House of Representatives and the Senate regarding the definition of radiology as either a medical specialty or a hospital service. The swing vote in favor of a medical specialty was carried because Senator Russell Long of Louisiana had been convinced that radiology was a bona fide specialty during his mother's radiation treatment for cancer. One must conclude that radiology as a medical specialty was saved by radiation therapy.

With the advent of Medicare in 1964, the federal government wisely decided to contract out the provision and administration of the plan to already established health provider companies. One can be certain that the well-organized health providers did not want a competing government agency, and, at the same time, they realized that lucrative contracts could be achieved by providing services as Medicare intermediaries.

When Medicare was formally enacted in June of 1965, it was assigned to the Social Security Administration (SSA), a branch of the Department of Health and Human Services (HHS). SSA created the Bureau of Health Insurance (BHI), to which it assigned the task of designing and managing Medicare. HCFA was a contrivance of the Carter administration, announced and put into effect by HSS Secretary Joseph A. Califano, Jr., in 1977. This imposed four layers of management above the people in the BHI who had been managing the program and made the administrator of HCFA subject to Senate ratification.

In the late 1960s Medicare was still working with the usual and customary payment system (UCR). Medicare reimbursement to radiology was based on the California Relative Value Schedule (CRVS), which was accepted by the majority of insurance companies across the country. Prior to the advent of Medicare, there was little reason for radiologists to have any real direct relationship with the federal government. The development of RVSs or fee schedules was a very slow process. The very earliest of the radiology fee schedules demonstrated charges for radiation therapy by the disease category or by the treatment or treatment course. These sometimes related to the X-ray energy used but had no consideration for any of the other procedures or thought processes that went into developing and performing the course of treatment.

In August 1957 the ACR *Bulletin* stated that a survey established a 60 percent overhead factor for an office, with 40 percent going to the physician for his professional services, based on

Table 8.II Relative Value Schedule

RADIOTHERAPY: (X-Ray—Radium or Radioactive Isotopes, including provision of Radiation sources) Values for Full Course of Treatment Proven Malignancy or Tumors, Global Fees

Head and Neck		CRVS UNITS
7500	Brain, including pituitary	60.0
7502	Oral Cavity	50.0
7504	Orbit	60.0
7506	Nasopharyngeal	60.0
7510	Larynx	50.0
7514	Thyroid	50.0
Chest		
7522	Mediastinum	60.0
7524	Pleura	60.0
Abdomen		
Gastro-intestinal tract:		
7530	Esophagus	60.0
7532	Stomach	20.0
7534	Small intestine	20.0
7540	Colon	40.0
7542	Rectum	60.0
7544	Anus	60.0
Genito-urinary tract:		
7550	Kidney	50.0
7556	Bladder, cancer, complete course	50.0
7558	Testicle	60.0
7560	Prostate	50.0
7563	Penis	40.0
Gynecological tract:		
7570	Ovaries	60.0
7572	Fallopian tubes	60.0
7574	Uterus corpus	50.0
7576	Cervix, complete course radium and X-ray	70.0
7578	Vagina	70.0
7580	Vulva	70.0
7582	Peritoneum	60.0
Breast:		
7590	Preoperative	40.0
7592	Postoperative	40.0
7594	Primary (radiotherapy only)	50.0
7596	Recurrent, chest wall	20.0
7598	Metastases	20.0
Bone:		
7600	Primary	40.0
7604	Metastatic, one area	20.0
Miscellaneous:		
7610	Leukemia, maximum per annum	60.0
7612	Hodgkins, maximum per annum	60.0
7614	Lymphosarcoma, fibrosarcoma, neurosarcoma, and other soft tissue sarcomas, maximum per annum	60.0

Miscellaneous (cont):		CRVS UNITS
7616	Spinal cord lesions	60.0
7618	Polycythemia vera	20.0
Superficial:		
7620	Skin neoplasm up to three cm. diameter	15.0
7622	Lip	20.0
7624	Lymph node metastatic	20.0
7626	Endocrine system, pituitary, adrenal	20.0
7638	Single high voltage treatment	2.6
7639	Consultation on therapeutic procedures	5.0
Non-Malignant Diseases		
7640	Acne, single treatment	1.5
7642	Adenitis, single treatment	2.0
7644	Angioma	3.0
7646	Arthritis, periarthritis	7.0
7648	Asthma, course	7.0
7650	Bursitis, course	7.0
7652	Carbuncle, per treatment	2.0
7654	Cellulitis, per treatment	2.0
7656	Dermatitis	5.0
7657	Infection, course	5.0
7658	Endometriosis	20.0
7660	Enteritis, regional	20.0
7662	Erysipelas, per treatment	2.0
7664	Fibroids	20.0
7668	Fungus infection, per treatment	1.5
7670	Keratosis, per treatment	2.0
7672	Keloid, per treatment	2.0
7674	Hyperthyroidism	50.0
7676	Menorrhagia, Metrorrhagia	20.0
7678	Lymphoid tissue	7.5
7680	Ovarian dysfunction	10.0
7682	Paronychia, per treatment	1.5
7684	Parotitis, per treatment	2.0
7686	Pruritus, per treatment	2.0
7688	Tinea capitis, full course	10.0
7690	Syringomyelia, full course	40.0
7692	Verucca, per treatment	2.0
7698	Single low voltage treatment	1.5

Prepared by the Committee on Fees of the Commission on Medical Services, California Medical Association, 12 February 1956

the historic average split of the radiology fees between the physician and the hospital. A complete fee schedule, including relative value units and professional and technical components, was included. This was the first widespread publication and distribution of an ACR RVS. The first CRVS was published by the California Medical Association in February of 1956 (Table 8.II). Radiotherapy was covered as a disease-oriented system based upon diag-

Table 8.III
A Reference for California Radiology Relative Values and Professional Components of Radiology Services (1964)

	CRVS Value
Radiation Therapy	
Teleradiotherapy: X-ray–1000 KVP and higher, cobalt per treatment	4
Teleradiotherapy: linear accelerator, betatron, etc., per treatment	5
Teleradiotherapy: X-ray less than 1000 KVP, telecesium	3
Consultation and treatment planning	15
Radium Therapy (or other sealed sources of radio-elements used similarly)	
Consultation, dosage calculation, preparation and supervision of application of radio-element	35
Application only, intracavitary	50
Application only, interstitial	75
Application only, superficial plaque or mold	15
Nuclear Medicine	
Radioisotopes–Therapeutic	
Thyroid cancer	40
Thyroid ablation	35
Polycythemia vera, metastatic cancer of bone	40
Intracavitary or interstitial radioisotope therapy	100
Hyperthyroidism	30

▲ ▲ ▲ ▲ ▲

nosis of disease. This was a global billing system with a conversion factor varying from $5 to $10, with an average of about $8 per CRVS unit. The factor was established by the hospital and the radiologist for the area of practice. The hospital billed the patient or the insurance company and paid the radiologist a set percentage or a salary. The average case treated with "deep X-ray" would net the radiologist $100 to $250.

By 1964 the CRVS had simplified the system by eliminating disease categories and based the new system on treatment modalities, a practice which expanded over the next thirty years (Table 8.III). Almost all diagnostic and therapeutic radiology professional service billing was based on the CRVS. Initially, the CRVS was a global billing system, and the radiologists would negotiate a percentage split with the hospital. Radiation oncology was included under the billing system of radiology; note the small number of procedures in the CRVS section for radiotherapy and the scarcity of reimbursement codes. Initially, HCFA utilized the 1964 CRVS to determine payment for radiation therapy for Medicare reimbursement. The conversion factor remained from $5 to $10 per unit depending upon the part of the country where the practice was located.

The ACR was well aware that the UCR reimbursement system was contributing to wide variations in payment values for similar procedures across the country. This was pointed out to HCFA, but, with little supporting data and the reasoning that "things had always been that way," HCFA saw little reason to change the system. Radiation oncology made up a relatively small percentage of the overall billing for radiology, consequently receiving little attention from HCFA. The ACR continued working with HCFA on a regular basis. The minutes from the various meetings of the ACR Board of Chancellors reflect these interactions but do not truly reflect the "behind the scenes" negotiations carried out by members of the ACR staff on behalf of American radiology, including radiation oncology.

In 1972 HCFA notified the ACR that the UCR billing and payment procedure for radiation oncology was creating an undue amount of wasted administrative time throughout the system because of wide inconsistencies in radiation oncology coding and billing. Many radiation oncologists had drifted away from the CRVS and were literally

Table 8.IV
Summary of Procedures and Values from the ACR Supplement, Number 2, 1975

First *User's Guide for Radiation Oncology*	
Procedure	**Range of Fees 1975 to 1988**
Treatment Planning	$25–$600 Often charged multiple times
Simulation	$25–$100 Usually a one time charge
Dosimetry	$10–$50 Seldom charged
Special Treatment Aids (Blocks)	$5–$100 Seldom charged
Intracavitary Brachytherapy	$50–$1,200 Per application
Interstitial Brachytherapy	$200–$3,000 Per application
Megavoltage Treatment Management	$2–$30 Per treatment
Megavoltage Treatment Management	$50–$300 Per course
Kilovoltage Treatment Management	$1–$5 Per treatment
Kilovoltage Treatment Management	$25–$200 Per course
Special Treatment Procedure	$50–$600 Seldom charged
Port Film Interpretation	$5–$50 Per film, week, course; wide range

making up their own billing systems. HCFA is reported to have stated to the ACR at one point, "for less than 1 percent of Medicare payments, it was spending almost 4 percent of administrative time sorting out the paper work." For this, and other reasons, Medicare notified the ACR that unless radiation oncologists standardized their own nomenclature and billing for Medicare, a standard nomenclature would be devised for them and mandated into use. A preliminary meeting was held in Chicago in the fall of 1973 to begin to study this problem. Representatives from the ACR and the leaders of American radiation oncology were brought together to be apprised of the urgency of arriving at a workable solution.

This task force began gathering billing data from many practitioners across the country and compiled this information in a document published by the ACR entitled, "A Standard Nomenclature for Radiation Therapy, ACR Supplement #2, 1975," the first *User's Guide for Radiation Oncology*. The concept of front end loading—placing the values for radiation therapy in the appropriate context of physician work—was first set out in this publication. Prior to that time, most radiation therapy reimbursement was based upon the delivery of daily treatments, although this actually played a smaller role in terms of physician time and involvement. The designation of treatment planning, simulation, dosimetry, and fabrication of treatment devices as front end services was established, and appropriate values were determined for these procedures. The original 1975 *User's Guide* was supposed to have included an RVS, but this was pulled at the last minute on the recommendation of the ACR counsel as a potential problem with the Federal Trade Commission (FTC). The codes were submitted to the AMA for inclusion in the radiation therapy section of the *Physicians' Current Procedural Terminology* (CPT). It was accepted and published as CPT-IV, and HCFA embraced the system as a workable solution to their needs for standardization.

Table 8.IV is a broad range of the average charges for some of these procedures over the years. These values have been compiled from many payment schedules studied as part of research for the 1988 ACR-RVS Valuation Panels. The wide range of values illustrates the problems caused by the longstanding FTC ban on the publication of RVSs. The ranges should actually start at zero fee since many practices failed to realize that they could charge for items such as treatment planning, simulation, and dosimetry. Diagnostic radiology-based practices with general radiologists performing radiation therapy were more likely to miss these procedures than full-time radiation oncologists. All practices suffered from a wide range of values as there was no baseline from

Table 8.V
National Relative Value Scale for Selected Radiation Oncology Procedures (1995)

CODE	DESCRIPTION	GLOBAL RVU	PROF RVU	TECH RVU
77261	Th Rad Tx Planning, Simple	2.10	2.10	–
77262	Th Rad Tx Planning, Intermed	3.19	3.19	–
77263	Th Rad Tx Planning, Complex	4.74	4.74	–
77280	Th Rad Simulation; Simple	4.59	1.07	3.57
77285	Th Rad Simulation; Intermed	7.23	1.58	5.85
77290	Th Rad Simulation; Complex	8.96	2.37	6.59
77295	Th Rad Simulation; 3-Dem	35.18	6.86	28.32
77300	Rad Dosim Calcul	2.30	0.94	1.36
77305	Teletx, Isod; Simple	2.96	1.07	1.89
77310	Teletx, Isod; Intermed	3.95	1.58	2.37
77315	Teletx, Isod; Complex	5.07	2.37	2.70
77321	Spec Teletherapy Plan, Particles	5.53	1.44	4.09
77326	Brachy Isodose Plan; Simple	3.81	1.41	2.40
77327	Brachyth Isodose Plan; Intermed	5.62	2.10	3.52
77328	Brachyth Isodose Plan; Complex	8.19	3.16	5.03
77331	Spec Dosim (TLD, micro etc.)	1.83	1.32	0.51
77332	Teletherapy Devices; Simple	2.19	0.83	1.36
77333	Teletherapy Devices; Intermed	3.21	1.28	1.93
77334	Teletherapy Devices; Complex	5.15	1.86	3.29
77336	Cont Rad Phys Support	3.02	–	3.06
77370	Special Rad Phys Consult	3.54	–	3.54
77419	Conformal Treatment Management	–	5.44	–
77420	Weekly Trmt Mgmt; Simple	–	2.44	–
77425	Weekly Trmt Mgmt; Intermed	–	3.71	–
77430	Weekly Trmt Mgmt; Complex	–	5.44	–
77431	1/2 Week Trmt Mgmt	–	2.74	–
77432	Stereotactic Trmt Course	–	13.27	–
77470	Spec Tx Proc	14.44	3.16	11.28
77750	Infusion Instill Radioelem Sol	8.29	6.94	1.35
77761	Intracav Brachy Appl; Simple	7.93	5.38	2.55
77762	Intracav Brachy Appl; Intermed	11.75	8.09	3.65
77763	Intracav Brachy Appl; Complex	16.64	12.09	4.55
77776	Interstit Brachy Appl; Simple	9.27	7.06	2.21
77777	Interstit Brachy Appl; Intermed	14.87	10.57	4.30
77778	Interstit Brachy Appl; Complex	21.02	15.82	5.20
77781	Remote Afterload Brachy; Simple	22.91	2.35	20.56
77782	Remote After Brachy; Intermed	24.10	3.54	20.56
77783	Remote After Brachy; Complex	25.38	5.27	20.56
77784	Remote After Brachy; Extended	28.49	7.93	20.56
77789	Surface Appl Brachy	2.04	1.58	0.46
77790	Supervision Handling Load Radioelem	2.09	1.58	0.51

▲▲▲▲▲

which to work.

Over the years the *User's Guide* has been revised several times by the ACR, each time with the blessing of the AMA and with the endorsement of HCFA, thus promoting nationwide standardization of coding (Table 8.V).

Congress mandated that a fee schedule for radiology services go into effect in January 1989, and as a consequence a full RVS for radiology had to be developed. Many ACR groups worked diligently throughout the summer and fall of 1988 developing the appropriate RVS for all of radiology, including diagnostic radiology, therapeutic radiology, and nuclear medicine. The radiology RVS was developed from a vast amount of information gathered from billing systems across the country as well as from HCFA's own internal Part B payment data. As these billing systems were reviewed, it became obvious that the discrepancies that had developed over the years between various localities were more numerous and complex than anyone had suspected.[34] This was the beginning of the standardization of radiology fees nationwide and gradually brought an end to wide disparities in Medicare payments to physicians across the country for radiologic procedures.

The ACR Consensus Panel in 1988

developed global, professional, and technical relative value units for the CPT-IV procedures for diagnostic radiology and therapeutic radiology. The original relative value units were based on radiology procedures, but when all of medicine was mandated to come under a single RVS, the relative value units were adjusted to match the nationwide resource-based relative value scale (RB-RVS). Table 8.V reflects the 1995 relative value units for the more common radiation oncology procedures. This is not a complete listing of all of the present descriptors utilized in radiation oncology.

The average cost for the physician component for a full course of curative radiation therapy in 1964 was $300. In 1967 it would cost an average of $325, and by 1973 this same course of treatment would be valued at about $400. The 1996 Medicare RVS, with a conversion factor of approximately $34.63 per relative value unit, limits the physician professional payments to an average of about $2,100 per case. The multitude of new and sophisticated treatment methods and modalities has contributed to significant increases in the technical costs associated with adequate radiation therapy services. However, there has been very little change in the professional component value in almost thirty years.

The Development of RVS and CPT as Related to Radiation Oncology

Most of the roentgenologists in the early part of this century practiced both diagnostic and therapeutic radiology. Diagnostic roentgenology, however, was the predominant source of income, with therapeutic radiology practiced almost as a sideline. In the 1930s, with the development of supervoltage X-ray units, came the emergence of consistently curative radiation therapy. A few physicians began to split off from diagnostic radiology and practice therapeutic radiology full time.[35]

Following the end of World War II, scientific advancements were everywhere in medicine, and atomic power was the key to high-activity radioactive sources. The advent of the cobalt-60 teletherapy units in the late 1950s heralded a long awaited breakthrough and boosted the practice of therapeutic radiology to its full potential. These units made the widespread use of curative cobalt radiation therapy, with its average energy of 1.2 megavolts (MV), available in most large hospitals nationwide. Megavoltage linear accelerators soon followed and have now become the standard equipment in the majority of treatment centers.[36]

The development of national specialty associations, increased output from radiation therapy residency programs, and better equipment all worked together to bring about a small but growing nationwide corps of full-time therapeutic radiologists. Close communication and cooperation among these newly emerging specialists resulted in a rapid standardization of treatment protocols. With this proliferation of knowledge through scientific meetings and publications, standardization of treatment techniques soon followed.

Unanimity in nomenclature and its associated billing and collecting processes, however, was slow in developing, with the first published nomenclature and RVS to be utilized nationwide being the 1956 CRVS. In 1977 the ACR was placed under a consent order from the FTC stating that all communication relative to the values of procedures performed in radiology would cease and all copies of the CRVS were to be collected and destroyed to prevent possible price fixing and restraint of trade.

The ACR remained under this oppressive regulation for the next ten years. During this period there was little or no communication among physicians regarding the amounts that were being charged for their various services. The insurance carriers knew every physician's fee profile, and Medicare knew all the values. Each carrier made up its own payment scheme to fit these national averages, but physicians were not allowed access to these fee schedules. Medicare was divided into more than two hundred payment localities, each with its own allowables based on the usual and customary charges of the physicians in that area. This lack of uniformity in nomenclature was brought into sharp focus as Medicare began to assume a significant portion of the financial burden of many cancer patients in the early 1970s. It was only after the consent order was lifted by an act of Congress in 1987 that the full extent of this variation became public knowledge.

CPT is a five-digit system for coding procedures and services performed by the physician. It is produced and copyrighted by the AMA and goes through yearly updates with input from all of the medical specialty societies. CPT was not originally intended to include any type of hospital technical coding. The Omnibus Reconciliation Act (OBRA), which went into effect 1 October 1988, directed hospitals to bill all outpatient Medicare charges under Part B and to use CPT as the proper nomenclature for billing the services rendered. CPT was originally designed as a coding and reporting system but gradually, over the years, came to be used as a vehicle for carrying the appropriate nomenclature for the billing of procedures to Medicare and the other insurance carriers. Changes in Medicare have forced the AMA to concede that CPT must also be used for billing of some selected technical services. In 1991 CPT for radiation oncology published an entirely new set of codes which covered "technical only" procedures. CPT codes are now utilized for both professional and technical values. CPT provides a uniform language to accurately describe both hospital and physician services while allowing a framework for charges in freestanding centers. The RVS attaches value modifiers to the CPT code which, with the proper conversion factor and locality modifiers, will relate directly to a dollar reimbursement figure.

As we look back through the ACR *Users Guides,* starting with the standard course of nomenclature in 1975, we see that the concept of simple, intermediate, and complex has been expanded to cover many procedures, such as simulation, treatment management, and isodose plans. Over the ensuing years many of the early codes were collapsed, combined, and eliminated. Newer, more explicit codes were devised to take their place. Today forty-six physician-specific radiation oncology codes exist, and sixteen technical-only codes are used by hospitals or centers for radiation therapy. In addition, radiation oncologists have access to thirty-nine evaluation and management codes that may be used by any physician. Radiation oncologists may also utilize other special codes, such as endoscopic procedures found in other sections of CPT. Relative value numbers exist for all of these codes and are standardized nationwide.[37]

The ACR, working under a direct mandate from Congress and in cooperation with HCFA and the AMA, was the first to develop and implement a full RVS for a medical specialty. This was done at a time when the Board of Chancellors of the ACR predicted an RVS for all United States medicine in the near future. By being allowed to develop the system from the beginning, radiology was given the privilege of designing the nomenclature and the appropriate weight of various procedures based on its own needs and perceptions. Many of the other medical specialties were not allowed this luxury, and now all of medicine is encompassed in a single RVS.

THE RELATIONSHIP OF RADIATION ONCOLOGY WITH THE FEDERAL GOVERNMENT AND HCFA

Mac Cahal, the executive secretary of the ACR in 1937, remarked to the Board of Chancellors, "The public is aware that the present day methods of treatment for cancer have yielded only a moderate degree of success, hence the desire for an active federal program directed toward the eradication of this scourge of mankind. We must be prepared to demonstrate that the private physician is capable and prepared to lead the war against cancer."[38]

He went on to say, "The public health service's war (1920–1937) on tuberculosis, syphilis, gonorrhea, diphtheria, and other diseases will now logically lead to the diagnosis and treatment of cancer as a national medical responsibility. It is essential that the profession let the public know that the private practitioner is fulfilling the duties of prevention and treatment." Medical research and the cure of many common diseases had allowed cancer to progress from sixth place to second place as the medical cause of death in America in 1937, a position it continues to hold.[39]

In August of 1937 President Roosevelt signed a bill appropriating $750,000 for a cancer institute in Washington with an annual appropriation of $700,000 for its operation. The institute was to devote its attention to research only. This was the beginning of what would would become the National Cancer Institute (NCI). The fear at that time was that once the NCI was organized the government would start a series of free clinics across the country for diagnosis and treatment of cancer.[40] Although the federal government did ultimately appropriate funding and assist in the construction of cancer hospitals in various locations, the NCI never became involved in the actual treatment of cancer as a sponsored service of the federal government, remaining instead a research institution.

Cahal went on to state in his 1937 report:

> Private medical practice has everything to offer that could be offered by a bureaucratic government in a war on cancer. Unfortunately, the public does not know that. An intelligent campaign of institutional advertising might help to make them aware of this fact. Unless the public has a course of intelligent education offered to them in a form which they will accept, the hands of Mr. Average Citizen will be roused in thunderous applause for the first senator who proposes to command a large portion of the practice of private radiologists, dig deeply into the pockets of the tax payer, and set up a federal program for the diagnosis and cure of cancer.[41]

This more than fifty years ago and sounds like the "new" health plans of the 1990s.

The *Bulletin* of the Intersociety Committee for Radiology in May of 1939 quoted Albert Rayle's paper in the *Georgia Medical Association Journal*, "We are all interested in what we call 'socialized medicine', which, for our purposes, we shall define as any system whereby either the professional or economic aspects of practice are controlled by laymen. The two go hand-in-hand, for economical control will lead to professional control. It is axiomatic that the man who pays the fiddler eventually will call the tune."[42]

Senator Henry Cabot Lodge of Massachusetts introduced on 19 March 1940 a bill to provide for a federal health insurance program and the furnishing of those medical services and facilities which had become standardized in their nature but which, because of high cost, were seldom used. To quote Senator Lodge,

> I believe it is not disputed that countless instances occur every day in which X-ray examinations are desirable—nay essential—but are not given because of the prohibitive cost. The suffering which could be prevented by prompt X-ray examination is indescribable. The prevention of disease automatically tends to reduce the cost of caring for the disease once it has been allowed to take hold. In the case of X rays, it is not inconceivable that it would become a routine part of every physical examination were it not for the cost.[43]

George Cooper, a radiologist from Virginia and a member of the ACR Board of Chancellors, warned in his address to the Tennessee Radiological Society in 1965 that, "when the new Medicare laws will be enacted, radiology may well be defined as a hospital service and, thereby, fixed forever in that position unless specific changes are enacted." The ACR reacted and proposed a separate fee structure for radiological services and managed to have these placed in the professional service section for physicians in the new Medicare structure. The ACR also developed a nonprofessional component for the hospital with the physician and hospital collecting their separate amounts and billing under the separate Part B and Part A sections of Medicare.

The legislation was eventually signed into law.[44]

In 1984 the federal government was struggling to control health care costs. Awareness of the escalation of the cost of health care began in the early 1970s, shortly after the inception of Medicare. With more accurate tracking of the true cost of healthcare delivery, it was estimated that the national medical bill reached $390 billion in 1984 and made up 11 percent of the gross national product. This amounted to more than $1,500 per United States resident to cover the cost of medical care.

It is easy to look back and identify factors that have caused the cost of healthcare to grow by leaps and bounds. New, innovative, and lifesaving high-technology diagnostic and therapeutic procedures resulting in the cure of potentially fatal diseases have been the most significant factors in the rise in the cost of healthcare. Inflation, however, continues to play an extremely important role and one that is commonly overlooked.[45]

Another significant factor contributing to the escalation of costs was the usual and customary billing scheme, which allowed physicians to continually increase the value of their services. This, however, was surpassed by the ability of hospitals to pass through to the patient all of their related costs of healthcare delivery, significantly contributing to the increase in cost. The RVS and diagnostic related group (DRG) classification have somewhat curbed the rate of price escalation and may eventually completely control this factor. Physicians and hospitals may well end up contributing proportionally more of their potential incomes to the control of healthcare costs than any other facet of society.

The Social Security Amendment of 1965 created a hospital insurance program for the elderly (Medicare, Part A) and a voluntary insurance program to pay for physician services to the elderly (Part B). Expenditures under Medicare increased from $3 billion in its first year to $33 billion in 1982 and exceeded $800 billion in 1993. Everyone benefited: patients, physicians, and hospitals. Medicare reimbursed not only the minimum treatment required for a patient but whatever the physicians deemed worthwhile, including the higher number of diagnostic tests and new modalities of therapy. Other third party payers also paid hospitals either their costs plus all applicable passthroughs and overhead or their flat rate charges. This open-ended method of paying hospitals began to change in 1982 when Congress passed the Tax Equity and Fiscal Responsibility Act (TEFRA).

In 1987 the DRG system was completely in place, and hospitals were paid by Medicare under the new prospective fixed payment system based on 467 DRGs. Instead of paying hospitals for total cost for services, Medicare allotted a predetermined amount based on the DRG under which each patient was admitted and treated, but outpatient payment continued to be partially based on cost passthrough.

In an attempt to further control the costs of hospital care under the DRG system, peer review organizations (PROs) were set up in each state by the federal government to monitor hospital usage under Medicare, with the prime purpose to discourage hospital admissions and excessive surgery. The PROs resembled the professional standards review organizations (PSROs) created in the early 1970s but differed with the additional authority of numerical targets which the hospitals had to meet.

MALPRACTICE AND RADIATION ONCOLOGY

In its earliest applications, the use of radiation to treat cancer was not without risk to patients and practitioners. Litigation over "X-ray burns" occurred as early as 1896, less than a year after Röntgen's discovery. One of the earliest references to the rising malpractice costs for radiation oncology occurred in the minutes of the ARS in 1936. The malpractice insurance premiums for radiotherapists in 1936 were higher than those paid by other physicians, and premiums were based mainly on the cost of litigation and to a lesser extent on judgments against physicians engaged in radiotherapy. In the late 1930s the ARS began to formulate and sanction plans to make the practice of radiation therapy less hazardous. A committee was formed to designate minimum standard parameters for radiation therapy.[46]

This committee was to designate a maximum radiation dosage consistent with the type of disease for which X-radiation was used, namely carcinoma. This should be the highest dosage consistent with safety and at the same time adequate to influence the disease. The radiation dosage should be qualified at 200 kilovolts, 1/2 millimeter aluminum, and 1 millimeter of copper filtration with a skin portal of 10 x 10 centimeters. Tolerances should be established for dry exposed surfaces as well as for moist and unexposed surfaces. The dosage as well as the time interval should be specified and whether the radiation was given daily, every other day, or biweekly. There was no need at that time to include what was referred to as "super high voltage" dosages, as these modalities were too few in number and had been insufficiently studied.[47] Despite the best efforts of the ARS, very few standards were ever developed or codified. It has only been in the last few years that true standards for radiation oncology have been seriously considered, developed, and published.

REGULATIONS OF RADIUM, OTHER RADIOACTIVE MATERIALS, AND THE NUCLEAR REGULATORY COMMISSION

The control of the distribution and use of radium over the years has always been a "hot" topic. Shortly after the discovery of radium, its clinical usefulness was demonstrated. The effects that this easily portable form of radiation had upon surface tumors was, indeed, miraculous. And as with all miracles came the desire for control and profit, as well as doing good for mankind—the old story of "doing well while doing good."

The rarity of radium and its tedious extraction process soon established it as an extremely expensive commodity that was bought, traded, and sold. Its inherent radioactive dangers were probably the only reason that it did not become a medium of exchange in the early years of this century. In its earliest medical applications, the control and safety of radium were relatively lax. There were no state or federal regulating bodies, and only the professional medical societies were interested in maintaining some level of radiation safety, both for the user and for the patient. Volumes of information were written from 1920 through the end of the 1930s discussing—but not agreeing upon—proper usage, distribution, and control of radium.

At a meeting of the ARS in 1922, it was recommended that radium, being used for the proper treatment of certain diseases, should have its purchase price deductible as a business expense from income tax. A petition was made to the Treasury Department that radium should be treated as a current expense of a physician's practice if the substance were rented or as a capital expense if it were actually owned, allowing proper deductions for "precarious obsolescence."[48]

At the eleventh annual meeting of the ARS in 1926, the prime concern was "to overcome the pernicious and promiscuous use of radium by anyone and everyone who could procure it through mail order business, rental companies, and so forth." They noted that public education would be extremely important so that the public would have some opinion as to who should be allowed to treat them with an element such as radium, which was "known to be dangerous even in experienced hands." Standards were to be developed by the ARS and coordinated with the American Society of Gynecologists and Obstetricians and the American Dermatological Society and promoted nationwide.[49]

In 1931 the ARS continued to follow up on the inappropriate use of radium. Many so-called "cancer clinics" in the United States were using near-scientific measures, others were run on a completely fraudulent basis. A minimum standard established by the ARS was that a community cancer clinic should have available at least 200 milligrams of radium and no less than a 200,000 volt X-ray apparatus.[50]

In 1934 a suggestion was made for a bill to be submitted to Congress to permit the president of the United States to accept $10 million worth of radium in partial payment of war debts due from Belgium. The ARS wanted to send representatives of their executive committee to the congressional hearings to express their concerns that government ownership and distribution of this quantity of radium would amount to state control of the practice of radium therapy. The bill, however, did not pass, and the specter of a government-controlled medical specialty was not raised again until after World War II.[51]

The *Journal of the American Medical Association* in May of 1939 presented a resolution through the Illinois delegation that was both praised and contested by many physicians and radiologists from coast to coast. The resolution condemned the rental of radium under certain conditions, namely, prescribing and directing the use of radium in the case of a patient whom the prescriber had not personally examined. The journal reported:

> Such a rigid ruling was thought to work a great hardship upon thousands of members of the American Medical Association and cause great suffering to thousands of citizens of the United States who were remotely situated from sources of radium. Many patients would be denied the use of radium, if they were compelled to make long trips to a source of radium. Many patients were stated to have such great confidence in their physicians that they distinctly preferred that he personally administer such radium treatments rather than travel great distances for a radiologist to administer the treatments. The AMA pointed out that the pioneers in radium therapy, and the greatest advocates of radium, were not radiologists, but were leading gynecologists and dermatologists and many of these specialists were well-qualified to make applications of radium. It was felt that radium could be utilized, and diseases could be satisfactorily treated, with great benefit to the patient by the average physician under proper supervision without a radiologist personally examining the patient.[52]

Because of the great expense of radium it was not practical for the average small town physician to own radium, especially since the average physician saw so few cases in which it was needed; therefore, rental was the only solution to the problem. The same reasoning was utilized by the radiologists in countering this position, stating that the physician who did not have a broad experience with radium was the very person who should not be administering this extremely dangerous modality.[53]

The AMA finally passed a compromise resolution stating that it would be considered ethical for a member of the AMA to personally describe a patient and submit this description to an experienced radiologist, who would then prepare a suitable radium applicator and course of treatment for the case. The case would then be treated by the original physician without the supervision of the radiologist. Opposition was raised by radiology to this compromise resolution, and the debate continued. The AMA House of Delegates finally passed a resolution stating "a physician prescribing for a patient without examining said patient is performing an unethical act." This resolution, however, did not preclude the rental of a definite amount of radium and specific applicator by the physician who had examined the case. The resolution did not specifically state that this physician must be a radiologist and, therefore, the controversy continued unabated.[54]

The problems over the control of radium were never fully resolved as radium, being a natural element, never fell under the control of any regulating body. Some states had relatively weak laws regarding the storage, use, and disposal of radium, but even these laws were lax and seldom enforced. The only truly effective weapon against the indiscriminate use of radium came through the insurance carriers who eventually refused to pay for the application of radioactive materials unless performed by a physician, and eventually the payment schedules were limited to those physicians with expertise in the field of radium therapy. Radium remained an expensive material, even after the appearance of cesium and cobalt, a factor which, along with its inherent hazards, contributed significantly to its ultimate obsolescence.

In the 1930s exotic new radioisotopes that potentially could be used in brachytherapy were predicted, but only small quantities could be produced. During and following World War II, however, the development of nuclear reactors made the production of radioactive materials commonplace and eventually doomed the use of radium as a brachytherapy source. The Atomic Energy Commission (AEC) effectively ended the controversy by licensing the use of the newly created man-made radioelements, eventually making radium obsolete. The development of encapsulated sources of cesium and cobalt began to displace radium as brachytherapy sources. These isotopes were then followed by even safer and more easily produced iridium and a succession of other radionuclides. All of the man-made radionuclides initially fell under the control of the AEC, today known as the Nuclear Regulatory Commission (NRC). Many states have assumed the regulatory activities previously held by the NRC. Under all of these circumstances, the man-made radionuclides are carefully regulated, but in many states the little radium remaining is outside the full regulation of the law.

Licensing by the NRC, and more recently by states, has produced very tight regulations as to the availability of radionuclides and the approval of physicians who may use them. So far, organized radiology has resisted attempted incursions by the NRC into licensure and control of electronic radiation producing devices (linear accelerators). Cobalt teletherapy units have always fallen within the purview of the NRC. Licensing fees continued to increase in cost, and cobalt units, with their lower effective energy, drifted into obsolescence, leading to a decline in the number of cobalt teletherapy units in the United States today.

Regulation of all medical devices is becoming more stringent. Linear accelerators and even the software that drives them must have Food and Drug Administration approval before they are licensed for human use. Safety remains a primary concern of the physicians who use this equipment, and the government remains eager to step in and develop the appropriate layers of regulators and regulations.

THE AMERICAN RADIUM SOCIETY AND RADIATION ONCOLOGY

The earliest record of the beginning of the organization of radiation therapy comes from the minutes of the first ARS meeting. The organizational meeting of the ARS was held in Detroit on 22 June 1916. It was voted to organize and establish a permanent society, dedicated to the study and advancement of radium as a curative agent against many forms of disease. Among the organizers of this first meeting were R. H. Boggs of Pittsburgh, vice president; W. H. B. Aikins, of Toronto, Canada, president; H. K. Pancoast of Philadelphia, corresponding secretary; and R. E. Loucks, of Detroit, Michigan, recording secretary and treasurer. The first formal meeting was convened in Philadelphia on 26 October 1916. One of the first duties, proposed by Albert Soiland of California, was that the annual dues of active and associate members should be $5, payable in advance.[55]

Soiland believed it might be of value to the society to publish the proceedings of the ARS meetings in a respected medical journal There was ill-concealed antagonism between the physicians doing radium work and X-ray work, and he believed that if such an arrangement could be made, they might strike a better note of harmony between the two agents which were so nearly allied. Pancoast noted that for the first time the *American Journal of Roentgenology* (*A.J.R.*) had seen fit to publish papers on radium therapy. The minutes reflect the fact that the *A.J.R*.was recommended, but not formally adopted, as the official publication source for the ARS.[56]

In June 1917 the second meeting was held, and scientific papers were presented that covered the established value of radium as a therapeutic agent in dermatology and gynecology.[57] Over the next few years the meetings consisted of scientific sessions covering the effect of radium and X rays on tissues as well as some of the physics aspects of radiation dosimetry.

In 1919 C. Everett Field of New York City first suggested that a committee should be appointed to prepare the ARS to become a section of the AMA. In 1922 the ARS made application to the American Congress of Physicians and Surgeons to become affiliated with that organization and to meet with them every three years. A committee was appointed by the president of the ARS to confer with the American Roentgen Ray Society (ARRS) and the Radiological Society for North America (RSNA) for the purpose of considering the formation of a national board of examiners to pass upon the qualifications of physician candidates in roentgen diagnosis, roentgen therapy, and radium therapy, and for certificates of efficiency for technicians working under the direction of physicians. This movement would eventually lead to the establishment of the ABR.[58]

At the 1920 meeting in New Orleans, Pancoast noted a letter from the editor of the *A.J.R*, offering to become the official journal for the ARS. The *A.J.R* was finally adopted as the official organ of publication at the annual meeting in 1921. Later the name of the journal reflected this change as it became the *American Journal of Roentgenology and Radium Therapy*. Annual ARS dues were raised to $12, and this amount included a yearly subscription to the journal.[59]

The minutes of the ARS in 1926 reflected more than $4,000 in the bank. There was a note in the minutes thanking the treasurer, who was complimented on getting these funds out of a failing bank just an hour before it closed its doors.[60]

The minutes of the eighteenth annual session of the ARS, held at the Palmer House in Chicago in September 1933, reflect that the ARS, the RSNA and the ACR met jointly as the first American Congress of Radiology. This constituted the first—and last—conjoint scientific meeting of all the radiological societies in America.[61]

In 1941 the ACR changed its methods of organization and invited the ARRS, the ARS, and the RSNA, to elect one member each as chancellors with full powers on the ACR board. This was an important step forward for the field, as radiation therapy was now represented in the ACR by a specific individual from the ARS. Dr. O'Brien of Boston was nominated as the first chancellor from the ARS to the ACR.[62] In 1956 it was recommended that any appointee to the ACR from the ARS should be an individual active in therapeutic radiology procedures in order to better represent the interests of the ARS.[63]

The relationships among the various radiological societies have been, for the most part, peaceful. As one would expect of a large family, there have been squabbles, but overall, the various subspecialty societies of radiology have, by their unity, contributed significantly to the advancement of radiology as a parent specialty.[64]

INTERRELATIONSHIPS OF RADIATION ONCOLOGY WITH THE AMERICAN MEDICAL ASSOCIATION

The early records of the Board of Chancellors of the ACR reflects frequent interactions with the AMA on a variety of issues. The level of intensity of many of these meetings has reflected the changing leadership of the two organizations. It has ranged over the years from cordial to hostile depending on the leaders and the issues.

The AMA maintains the dominant role in the political and economic future of all of medicine, and radiology is still considered by many individuals in the AMA as a hospital-based specialty. This perception is gradually changing as the membership and leadership of the AMA become more sensitive to the unique issues confronting radiology as a specialty. Radiation oncology continues to be associated with radiology in the eyes of the AMA and is seldom singled out for special consideration. The AMA remains responsive to the needs of the radiation oncology community when specific issues are brought before its governing bodies. The issue of self-referral and ownership of freestanding centers is one prime example.

THE RELATIONSHIP OF THE AMERICAN COLLEGE OF RADIOLOGY AND RADIATION ONCOLOGY

The ACR was originally organized in 1923 as a purely honorific society. It existed this way for more than a decade, but eventually the leadership realized that socioeconomic concerns, even in those early days, required a broader and more politically active organization. In 1935 the ACR began to look toward developing an organization that could tackle the socioeconomic problems of radiology. W. Edward Chamberlain, chairman of radiology at Temple University, became chairman of the new Board of Chancellors in 1935. He was given a clear mandate to reorganize the ACR as he saw fit. Ties were established to the other fledgling radiological societies, the ARS, RSNA, and ARRS. Liaisons were established between these societies and the new ACR which resulted in significant dialogue representing organized radiology.

The ACR historically has related well with the various subspecialty organizations of radiology. With its primarily socioeconomic focus, the ACR developed close but not competitive interrelationships with the early scientific radiological organizations. The membership of the ACR was devised to allow representation from these radiological societies. Over the years, as other societies have gained in membership and consequently political importance, they have been allocated representation on the Board of Chancellors. Today radiation oncology is represented by two chancellors, one nominated by the American Society for Therapeutic Radiology and Oncology (ASTRO) and the other from the ARS. This representation gives a strong intersociety voice to radiation oncology.

The Intersociety Committee for Radiology was proposed by the ACR in 1936 and established in 1937. It was founded to study matters of economics in the practice of radiology and to improve the relations of radiologists with organized medicine. Its function was to collect facts and information, to make this information available to members and local groups of radiologists, and to offer advice and counsel when requested.[65] The ARRS, RSNA, and the ARS each had three delegates. This was a time during which it became clear that specialty unity was necessary to counter state legislation and respond to an increasing number of federal constraints on radiological practice. Today's Intersociety Commission and its sponsored summit meetings represent thirty-eight radiological societies comprising all of the societies in radiology.

Representation of the various subspecialty societies of radiology was through appointment to the ACR Board of Chancellors by selected individuals. These individuals, chosen as representative from the various specialty societies as new chancellors, came on to the board when old ones retired. The Board of Chancellors in the 1940s and 1950s was a self-perpetuating group and, although they served the needs of American radiologists appropriately, were still viewed as a closed circle club. It was only after increasing complaints from members and from participating societies that more democratic representation from the various radiological societies began to appear on the Board of Chancellors.[66]

One of the earliest functions of the ACR was to develop a structure that would promote the formation of local and state chapters for participation in the national organization. The ACR was made up of individuals from around the country with keen political interests in advancing the cause of radiology. Many cities and states had unofficial gatherings of radiologists for social as well as scientific purposes. In 1941 the ACR instituted local and state chapter representation through the establishment of a council with elected members. The bylaws were appropriately modified, and the council began to bring the needs and problems of all radiologists to the attention of the Board of Chancellors for resolution.[67] In 1955 thirty-four of these councilors had their first separate formal meeting. In 1959 the ACR adopted a formal state chapter system and won from the Board of Chancellors permission for trial ACR chapters in states whose radiologists wished to organize them. The chapters were established as divisions of the ACR and were bound by its rules and regulations and dedicated to implementing college policies within the respective states. The chapters included both diagnostic and therapeutic radiologists. Chapter officers were designated as officers of the ACR, and the bylaws of the state chapters were patterned after those of the ACR. This resulted in the revitalization of the council of the ACR and greatly strengthened the college, as it was now better able to relate to the concerns of individual radiologists

Radiation oncology was sporadically represented within the chapters, but there were many areas of the country where diagnostic radiology was overwhelmingly dominant and radiation oncologists were unable to achieve the desired level of representation at the governing level of the college. This disparity became more obvious, and in 1978 the bylaws were modified and a nationwide chapter of the ACR known as the Council of Affiliated Regional Radiation Oncology Societies (CARROS) was formed. This allowed radiation oncologists to have better representation in the ACR at the level of the council as well as an alternate route to the rank of fellow (FACR).

The 1959 minutes of the Board of the Chancellors of the ACR, as reported to the Board of the ARS, reflected a budget to finance the rapidly expanding activities of the college, but the budget was so large that its adoption would have produced a deficit of $35,000. It was felt that such spending would completely exhaust the funds of the ACR in approximately two years.[68] In 1960 a resolution was introduced by the chancellor from the ARS recommending that the practice of adopting a budget calling for deficit spending be stopped and suggested that a reserve fund equal to one year's expenses be set up. This resolution was most unpopular and was defeated unanimously.[69] It is interesting to note that twenty-five years later the ACR did adopt a policy against deficit budgeting and established a reserve fund equal to at least one year's full operating budget.

In his 1923 presidential address to the ACR, Russell Carman noted:

> Deep in our thoughts, I feel that all of us cherish a dream of a day when the radiologic workers of America shall be gathered in one mighty union, with permanent headquarters, functioning capably through stable bureaus, possessing a great library, conducting a journal with the newest and best in radiologic literature, advancing education, inspiring its members to constant self-improvement, and leading them to greater achievements in glory.[70]

Even from the earliest time, the dream of a single unified radiological organization has been put forward only to be diluted by a continuing proliferation of small specialty and subspecialty societies and interest groups. Considering the number of avenues of interest and self-interest that have been available over the years, it is a tribute to its strength and necessity that the ACR has survived as the primary socioeconomic organization for American radiology.

In 1965 the ACR *Annual Report* contained numerous items of interest to radiation oncology. The previously existing Committee on Cancer became the Commission on Cancer, a status which it maintains today. The first annual Regionalization Conference to discuss cancer staging and end results reporting was held between the American College of Surgeons and the ACR. Cooperative guidelines were drawn up and minimal standards for radiation therapy and major cancer treatment centers and tumor clinics were established. Guidelines for determining the standards of radiotherapy in approved cancer centers were presented. The concentration of patients with certain types of tumors into designated treatment centers was extremely important to assure the success of such clinical research.[71]

In 1965 the American Cancer Society asked the College of Physicians to make a general statement supporting the Surgeon General's report on cigarette smoking. However, the ACR felt that this was not within their province and such a statement was not recommended by the Committee on Cancer. A discussion on the importance of mammography was stimulated by the comment that a study was conducted in New York City in 1964 to evaluate mammographic procedures as diagnostic aids for early detection of breast cancer. Many members of the committee thought that "this might possibly be of some assistance but that the value might be overly estimated."[72]

In 1965 the Commission on Cancer put forth a new set of standards for radiotherapy in approved cancer centers. These stated that such a center must have a full-time radiotherapist in charge of the radiotherapy section of the facility. It noted that radiotherapy required, in addition to conventional low and medium voltage equipment of modern design, the availability of some adequate device operating in the megavoltage range. It also noted, however, that in "other centers," at least medium voltage equipment of modern design should be available and that megavoltage equipment was usually not recommended for other than major cancer management centers.[73]

THE RELATIONSHIP OF THE AMERICAN SOCIETY FOR THERAPEUTIC RADIOLOGY AND ONCOLOGY WITH RADIATION ONCOLOGY

The most important intersociety relationship that radiation oncology has today is with ASTRO. ASTRO remains the focus of our scientific endeavors and has become the largest scientific organization in the world representing radiation oncology. Radiation oncology prior to the mid-1950s was represented almost entirely by the ACR and the ARS. In 1955, with less than a hundred therapeutic radiologists practicing in the United States, it was felt that an organization should be formed to look toward the economic and educational goals of radiation therapy. The American members of the International Club of Radiotherapists initiated informal periodic gatherings to which other radiotherapists were invited. The first of these meetings took place on 5 December 1955 at Barney's Market House on Randolph Street in Chicago. Subsequently, luncheon or dinner meetings took place during the annual conventions of the ARS or the RSNA.[74]

Juan A. del Regato suggested that the group "formalize their informality" and founded the American Club of Therapeutic Radiologists on 18 November 1958, with 54 initial members. By 1966 membership had risen to 254, and the name of the organization was changed to the American Society of Therapeutic Radiologists. In November of 1970 this fledgling society had its first separate scientific meeting at the Mountain Shadows Lodge in Scottsdale, Arizona, with an initial registration of 308 members. In 1983 the name of the society was changed to the American Society for Therapeutic Radiology and Oncology, and a new mission statement was adopted. Today ASTRO is the largest society of radiation oncologists in the world with almost 3,000 active members. The development of ASTRO and the radiation oncology community has proceeded as a one-to-one relationship. As radiation oncology grew and became more organized, ASTRO grew and followed the success of the specialty.[75]

THE RELATIONSHIP OF THE AMERICAN BOARD OF RADIOLOGY WITH RADIATION ONCOLOGY

The ABR has played a key role in the life of American radiology and the specialty of radiation oncology. In his 1964 book, *A History of the American Board of Radiology*, Edward L. Jenkinson noted:[76]

> With increasing specialization in medicine, as the nineteenth century gave way to the twentieth, there sprang up across America innumerable groups of 'specialists,' looking to improve the quality of practice in their respective fields. The American Roentgen Ray Society was organized in 1900, the Radiological Society of North America in 1915, and the American Radium Society in 1916. Just what constituted a 'specialist' was, however, open to a variety of interpretations. Any Doctor of Medicine was entitled to a listing in the Directory of the American Medical Association as specializing in the field in which he considered himself best qualified. In other words, he was the judge of his own qualifications.
>
> With the stage thus set, the medical profession as a whole had considered there should be at least minimal standards in preparation for their practice of any medical specialty. It was felt that unless some method of control was established, each of the states might well enact laws prescribing requirements in various specialties with the result, there would be a large number of separate state boards of examiners for each specialty. The practical solution would be for each specialty to set a specific examining body in place and begin certification of positions wanting to practice that specialty.[77,78]

The American Ophthalmological Society organized the first specialty board, the American Board of Ophthalmology, in 1916 and incorporated it in 1917.

Despite earlier recommendations to form a radiology examining board, it was not until 1932 that five radiologic societies—the ARRS, the RSNA, the ARS, the ACR, and the Section on Radiology of the AMA—appointed a joint committee of three representatives from each organization to investigate the feasibility of establishing a qualifying board in this field.[79] The following men were appointed by the five organizations:

W. Edward Chamberlain, M.D.
Willis F. Manges, M.D.
Arthur C. Christie, M.D.
John W. Pierson, M.D.
Edwin C. Ernst, M.D.

LeRoy Sante, M.D.
Lester Hollander, M.D.
Henry Schmitz, M.D.
George W. Holmes, M.D.
Albert Soiland, M.D.
Edward L. Jenkinson, M.D.
Walter W. Wasson, M.D.
Lyell C. Kinney, M.D.
Rollin H. Stevens, M.D.
B. R. Kirklin, M.D.

Among these names were some of the early giants of radiology and radiation therapy. Radiation therapy was considered an important part of the specialty board of radiology even at the time of its inception.

The board was incorporated in Washington, D.C., 31 January 1934, by Arthur Christie, Paul B. Cromelin, and Bolitha J. Laws for the avowed purpose of encouraging the study and promoting and regulating the practice of radiology (including diagnostic and therapeutic applications of radiant energy including roentgen rays and radium).

The ABR was to determine the competence of specialists in radiology and to examine and test the qualifications of voluntary candidates. The board would grant and issue certificates in the field of radiology. They would prepare and furnish lists of those who were certified by this board and vowed to protect the public against irresponsible and unqualified practitioners who professed to be specialists in radiology.

A board of trustees was established with fifteen members, three of whom were appointed by each of the founding societies. It is important for the members of the radiation oncology community to remember that radiation therapy was included from the inception of the ABR and continues to play a key role. The ABR in 1934 began examining in diagnostic radiology, general radiology (a combination of diagnostic and therapeutic radiology), and radiotherapy.[80]

The ABR held the first examination for certification in radiology in Cleveland, Ohio, in June 1934. Candidates were divided into three classes: Class A, outstanding radiologists of long experience (who for all practical purposes were certified without examination at this initial meeting); Class B, radiologists of less experience; and Class C, young radiologists who had recently completed training.

At the time of the initial examination in June of 1934, sixteen people were certified by examination and 149 were certified without examination. The number of applicants applying for the Class A examination continued to progressively decrease, and by 1940 it was voted there should be no further Class A or Class B candidates. However, as late as 1980 there was a Class A certification, but the board closed this route forever in 1981. By 1940 there were eighty-eight approved radiological centers training one or more radiologists.[81] In 1993 there were eighty-three approved residencies in therapeutic radiology.

Initially, the examination was divided into five types of certification:

1. Diagnostic roentgenology (diagnostic radiology only)

2. Roentgenology (usually referred to as general radiology with the capability of doing both therapy and diagnosis)

3. Therapeutic roentgenology

4. Therapeutic radiology (the difference between 3 and 4 is not entirely clear)

5. Radium therapy (open to those individuals using only radium and not wishing to be examined in any of the other sections of roentgenology, either diagnostic or therapeutic).

In 1959 these were reduced to three divisions: diagnostic roentgenology, radiology (both specialties), and therapeutic radiology (radiation therapy only). Eventually the section on radiology was dropped so that the board now examines only in diagnostic roentgenology and therapeutic radiology. A section covering the examination on radiological physics was added in 1947.

RADIATION ONCOLOGY AND JOINT VENTURES

The conservative approach of President Ronald Reagan regarding health policy matters during the 1980s emphasized competition, rather than regulation, as the mechanism for balancing healthcare spending. For-profit hospitals were converted, formed, or built throughout the country, and many of these hospital chains invested in outpatient facilities. Physicians, always quick to recognize an opportunity, noted that joint-venture outpatient facilities provided the opportunity for a more convenient way of caring for patients as well as the promise of profit from both professional and technical revenues that such centers were bound to generate. Healthcare spending increased during this time by approximately 10 percent per year, and entrepreneurs quickly seized upon the opportunity. Joint ventures, limited partnerships, and creatively financed outpatient radiation oncology facilities flourished nationwide. Hospital planning requirements and oppressive federal regulations for the procurement of expensive high-technology equipment had stymied small community hospitals from installing much-needed new linear accelerators. Private freestanding outpatient facilities were mostly immune from these regulations and as such provided the opportunity for community hospitals to improve medical care as well as an opportunity for physicians to profit from ownership of these facilities.

The growth of outpatient radiation oncology centers began slowly in the early 1980s and reached a peak by the end of the decade. In some instances, a radiology group or a group of radiation oncologists would raise most of the capital to build these facilities. In other situations, referring physicians, such as medical oncologists and surgeons, would also be significant partners or even sole owners in these joint ventures. Hospitals also invested in these centers hoping to share the profit and retain the radiation oncologist as part of the institutional staff. The dominant pattern, however, was the physician joint venture in which physicians were both the primary source of referrals as well as owners of the centers, a practice by no means limited to radiation oncology.

Questions of ethics arose: for instance, if a physician investor's return from a radiation oncology center was directly tied to the volume of referral, did this not color his or her medical judgment? This was seen as a violation of antitrust law and a violation of the physician/patient relationship in its most traditional form. In 1984 ACR president Jerome Wiot cautioned against abandoning the ethical standards of radiology by supporting the concept of imaging or radiation therapy centers owned partially or totally by physicians other than radiologists.[82]

The ACR Council adopted a guideline which asserted "dividends or profits related to such investments should be commensurate with the individual's investment." In 1985 the council went on to amend and strengthen its statement to "recognize and acknowledge the potential for abuse by self-referral in imaging and/or radiation therapy centers owned either in whole or in part by referring physicians."[83]

By 1988 the ACR Council began to condemn the practice of ownership in these centers stating that, "the practice of self-referral of patients for diagnostic or therapeutic medical procedures may not be in the best interest of the patient. Accordingly, referring physicians should not have a direct or indirect financial interest in diagnostic or therapeutic facilities to which they refer patients."[84]

Also at this time, the United States Office of the Inspector General (OIG) was directed to develop guidelines for acceptable business practices by healthcare providers. This eventually culminated in 1991 when Inspector General Richard Kusserow issued a management advisory and guidelines on what the Medicare program would consider acceptable business practices. These guidelines were expressed as "safe harbors," defining accepted practices and leaving any other arrangement in some jeopardy of possible prosecution.[85]

Documentation of the abuse of referrals to these centers and the increase in charges and procedures performed brought this matter fully into the spotlight of public attention.

Various reports had indicated a marked increase in the utilization of diagnostic procedures in physician-owned imaging centers. Two studies of joint ventures in Florida demonstrated the extent of referring physician investment and concluded that such enterprises markedly elevated the cost of health care.[86] Moreover, there appeared to be adverse effects on patient access to care. Mitchell and Sunshine concluded that self-referral also substantially increased the use of services and costs.[87]

Throughout most of this controversy, the AMA held to the position that the physician/owners of a medical facility had the obligation to advise patients of their ownership interest, but beyond this point, would not condemn the practice. In the late fall of 1991 the entire issue of self-referral reached a pinnacle of political and ethical jousting. A group of radiation oncologists in Florida aggressively challenged self-referral by rallying against a series of joint venture facilities planned by a for-profit group primarily controlled by medical oncologists who would be the owners as well as the prime source of referral to these centers. This group of radiation oncologists found themselves squarely positioned against the Florida Medical Association and indirectly against the position of the AMA. The ACR backed the radiation oncologists, and in early 1993 the Florida legislature passed a very broad resolution condemning the ownership of radiation oncology facilities by referring physicians. Many other states, led by Illinois and California, entered their own versions of strong state laws prohibiting physician ownership in freestanding facilities, including reenactment of strong certificate of need regulations.

During this time, the AMA vacillated between various stances, eventually stating "physician investment in healthcare facilities can provide important benefit for patient care. However, when physicians refer patients to facilities in which they have an ownership interest, a potential conflict of interest exists. In general, physicians should not refer patients to a healthcare facility outside their office practice at which they do not directly provide care or services when they have an investment interest in the facility."[88]

The final result for the radiation oncologists who had invested in or owned these centers was reorganization of the financial structure of the center, which resulted in ownership outside the field of medicine or ownership by the radiation oncologist alone. On the positive side, outpatient centers proved that they could provide radiation oncology services at a much more cost effective rate than comparable hospital facilities. As a consequence, many of these centers are now well poised to move into the next decade of managed care competition. A single center can compete effectively in the new marketplace, and a coalesced group of practices or centers is able to provide broad coverage to a wide area of a community and, thus, effectively compete under the new managed healthcare plans. These freestanding centers are able to effectively compete against the more cost burdened hospital-based systems under the new managed healthcare plans.

Radiation oncology remains an independent medical specialty in the fast changing environment of medicine. The identity of radiation oncology as a clinical specialty has served the specialty well over the years and will continue to do so in the years to come. The lessons of the past are well taught, and any specialty is advised to heed the messages. Radiation oncology is a small specialty, but now, as throughout the last century, its strength lies in unity.

REFERENCES

1. "Bulletin of the Intersociety Committee for Radiology," *Am. J. Roent.* (May 1940):774.
2. "Bulletin of the Intersociety Committee for Radiology," *Am. J. Roent.* (February 1940):288.
3. *Financial World Magazine* (23 November 1993):50.
4. "Bulletin of the Intersociety Committee for Radiology," *Am. J. Roent.* (January 1939).
5. *Proceedings of the AMA House of Delegates Meeting*, June 1939.
6. Ibid.
7. "Bulletin of the Intersociety Committee for Radiology," *Am. J. Roent.* (March 1940):442.
8. "Bulletin of the Intersociety Committee for Radiology," *Am. J. Roent.* (June 1939).
9. Ibid.
10. "Bulletin of the Intersociety Committee for Radiology," *Am. J. Roent.* (November 1939).
11. Ibid.
12. Christy, A.C. *Economic Problems of Medicine*. New York: MacMillan, 1935.
13. "Bulletin of the Intersociety Committee for Radiology," *Am. J. Roent.* (March 1940):442.
14. Travis, Jack, M.D., Topeka, Kansas, personal communication.
15. *Financial World Magazine* (23 November 1993):50.
16. Nyberg, C.E., "The Economics of the Practice of Radiology," May 1947.
17. Ibid.
18. ACR *Annual Report*, 1942.
19. ACR Communication M-86, 1 June 1939.
20. ACR *Annual Report*, 1942.
21. Nyberg, "The Economics of the Practice of Radiology."
22. ACR *Annual Report*, 1942.
23. Christy, *Economic Problems of Medicine*.
24. Richardson, W.A., "Physician's Income: A Portfolio of Facts Derived from the Sixth Medical Economics Survey," *Medical Economics* (1947).
25. Society of Chairman of Academic Radiation Oncology Programs (SCAROP), Survey 1993.
26. *American Medical Association News* (10 January 1994):1.
27. "Bulletin of the Intersociety Committee for Radiology," *Am. J. Roent.* (May 1939):860-861.
28. "Bulletin of the Intersociety Committee for Radiology," *Am. J. Roent.* (October 1939):510-511.
29. Wiprud, T. *The Business Side of Medical Practice*. Philadelphia: W.B. Saunders, 1937.
30. Major, R. *The History of Medicine*. Springfield, Ill.: Charles C. Thomas, 1954.
31. "Bulletin of the Intersociety Committee for Radiology," *Am. J. Roent.* (March 1939):460-465.
32. Ibid.
33. "Bulletin of the Intersociety Committee for Radiology," *Am. J. Roent.* (October 1938).
34. *A User's Guide For The Radiation Oncology-Related CPT Codes—1993*. 4th edition. Reston, Va.: American College of Radiology.
35. Ibid.
36. Ibid.
37. Ibid.
38. Personal Communication, Mac Cahal, in letters to the Chancellors and committee members of the ACR, 1937. Center for the American History of Radiology, Reston, Va.
39. Personal Correspondence from Mac Cahal to the Board of Chancellors, 19 August 1937. Center for the American History of Radiology, Reston, Va.
40. Ibid.
41. Ibid.
42. "Bulletin of the Intersociety Committee for Radiology," *Am. J. Roent.* (May 1939):860-861.
43. "Bulletin of the Intersociety Committee for Radiology," *Am. J. Roent.* (October 1940):744.
44. George Cooper, Address to the Tennessee Radiological Society, October 1965.
45. In looking at the economics of past medical practice it is often helpful (though sometimes deceptive) to put historical dollars into contemporary figures. The following table gives a synopsis in five-year increments dramatically illustrating the effects of inflation, and places the past costs in the perspective of today's values.

    DOLLARS REQUIRED TO EQUAL THE PURCHASING POWER OF $1000 IN 1993

    | 1915 - $69.93 | 1935 - $94.94 | 1955 - $185.72 | 1975 - $372.85 |
    | 1920 - $138.88 | 1940 - $97.02 | 1960 - $205.13 | 1980 - $571.03 |
    | 1925 - $121.27 | 1945 - $124.74 | 1965 - $218.30 | 1985 - $745.67 |
    | 1930 - $115.73 | 1950 - $167.02 | 1970 - $268.88 | 1990 - $905.75 |

    [Prepared by Jonathan Sunshine, Ph.D., ACR Director of Research]

46. Proceedings of the annual meeting of the American Radium Society, 1936, Kansas City, Missouri. Typescript and original, Center for the American History of Radiology, Reston, Va.
47. Ibid.
48. Minutes of the 7th annual meeting of the American Radium Society, May, 1922, St. Louis, Missouri. Typescript and original, Center for the American History of Radiology, Reston, Va.

49 Minutes of the 11th annual meeting of the American Radium Society, 15 April 1926. Typescript and original, Center for the American History of Radiology, Reston, Va.
50 Minutes of the 16th annual meeting of the American Radium Society, 1931. Typescript and original, Center for the American History of Radiology, Reston, Va.
51 Personal correspondence to the ARS President, American Radium Society Notes From the File, 1934. Center for the American History of Radiology, Reston, Va.
52 *J.A.M.A.* 112 (1939):2166-2167.
53 Ibid.
54 *Radiology* 31 (1939):234-236.
55 Minutes of the 1st annual meeting of the American Radium Society, 1916. Typescript and original, Center for the American History of Radiology, Reston, Va.
56 Ibid. Pancoast was incorrect in his statements about the *AJR*. In its earlier incarnation as the *American Quarterly of Roentgenology* the journal published regularly on radiation therapy.
57 Minutes of the 2nd annual meeting of the American Radium Society, June, 1917. Typescript and original, Center for the American History of Radiology, Reston, Va.
58 Minutes of the 7th annual meeting of the American Radium Society, May, 1922, St. Louis, Missouri. Typescript and original, Center for the American History of Radiology, Reston, Va.
59 Ibid.
60 Minutes of the American Radium Society, 11th annual meeting, 15 April 1926. Typescript and original, Center for the American History of Radiology, Reston, Va.
61 Minutes of the 18th annual meeting of the American Radium Society, 27 September 1933, Chicago, Illinois. Typescript and original, Center for the American History of Radiology, Reston, Va.
62 Minutes of the 26th annual meeting of the American Radium Society, 2 June 1941, Cleveland, Ohio. Typescript and original, Center for the American History of Radiology, Reston, Va.
63 Proceedings of 38th annual meeting of the American Radium Society, 9 April 1956, Houston, Texas. Typescript and original, Center for the American History of Radiology, Reston, Va.
64 ACR, Presidential Address, Carl R. Bogardus, Jr., M.D., Scottsdale, Arizona, September 1992.
65 "Bulletin of the Intersociety Committee for Radiology," *Am. J. Roent.* (October 1938):631.
66 Minutes of the 42nd annual meeting of the American Radium Society, 18 March 1960, San Juan, Puerto Rico. Typescript and original, Center for the American History of Radiology, Reston, Va.
67 Minutes of the 26th annual meeting of the American Radium Society, 2 June 1941, Cleveland, Ohio. Typescript and original, Center for the American History of Radiology, Reston, Va.
68 Minutes of the 41st annual meeting of the American Radium Society, March 1959. Typescript and original, Center for the American History of Radiology, Reston, Va.
69 Minutes of the 42nd annual meeting of the American Radium Society, 18 March 1960, San Juan, Puerto Rico. Typescript and original, Center for the American History of Radiology, Reston, Va.
70 Minutes of the annual meeting of the ACR, 1923. Center for the American History of Radiology, Reston, Va.
72 1965 Annual Report of the American College of Radiology, as it appeared in the ACR *Bulletin*.
72 Ibid.
73 Ibid.
74 del Regato, J.A., historical introduction to *The American Society for Therapeutic Radiology and Oncology Membership Directory,* 1993.
75 Ibid.
76 Jenkinson, E.L. *History of the American Board of Radiology: 1934–1964.* Minocqua, Wisconsin.
77 Ibid.
78 Ibid.
79 Minutes of the 7th annual meeting of the American Radium Society, May, 1922, St. Louis, Missouri. Typescript and original, Center for the American History of Radiology, Reston, Va.
80 Jenkinson, *History of the American Board of Radiology*.
81 "Bulletin of the Intersociety Committee for Radiology," *Am. J. Roent.* (March 1940):442.
82 Wiot, J.F., "ACR Presidential Address," ACR *Bulletin* (October 1984).
83 ACR 1985 Council Statement on Joint Venture, ACR *Bulletin* (October 1985).
84 ACR Council Substitute Resolution 39, September 1988, quoted in the ACR *Bulletin* (October 1988).
85 Medicare and State Healthcare Programs: Fraud and Abuse, OIG Anti-Kickback Provisions, 56fr35952.
86 Mitchell, J.M., and Scott, E. "Joint Ventures Among Health Care Providers In Florida, V2 Contract Report For Florida Healthcare Cost Containment Board," September 1991.
87 Mitchell, J.M., and Sunshine, J.H., "Consequences of Physician Ownership of Healthcare Facilities—Joint Ventures in Radiation Therapy," *New Eng. J. Med.* 327 (1992):1497–1501.
88 AMA Council on Ethical and Judicial Affairs, Report C, December 1991.

The first radiograph by an American to be published abroad was this "American Frog" of Elizabeth Fleischmann Ascheim. Ascheim, of San Francisco, was self taught and used the rays in both therapy and diagnosis until her death from radiation injuries in 1905. (Courtesy of the Center for the American History of Radiology, Reston, Va.)

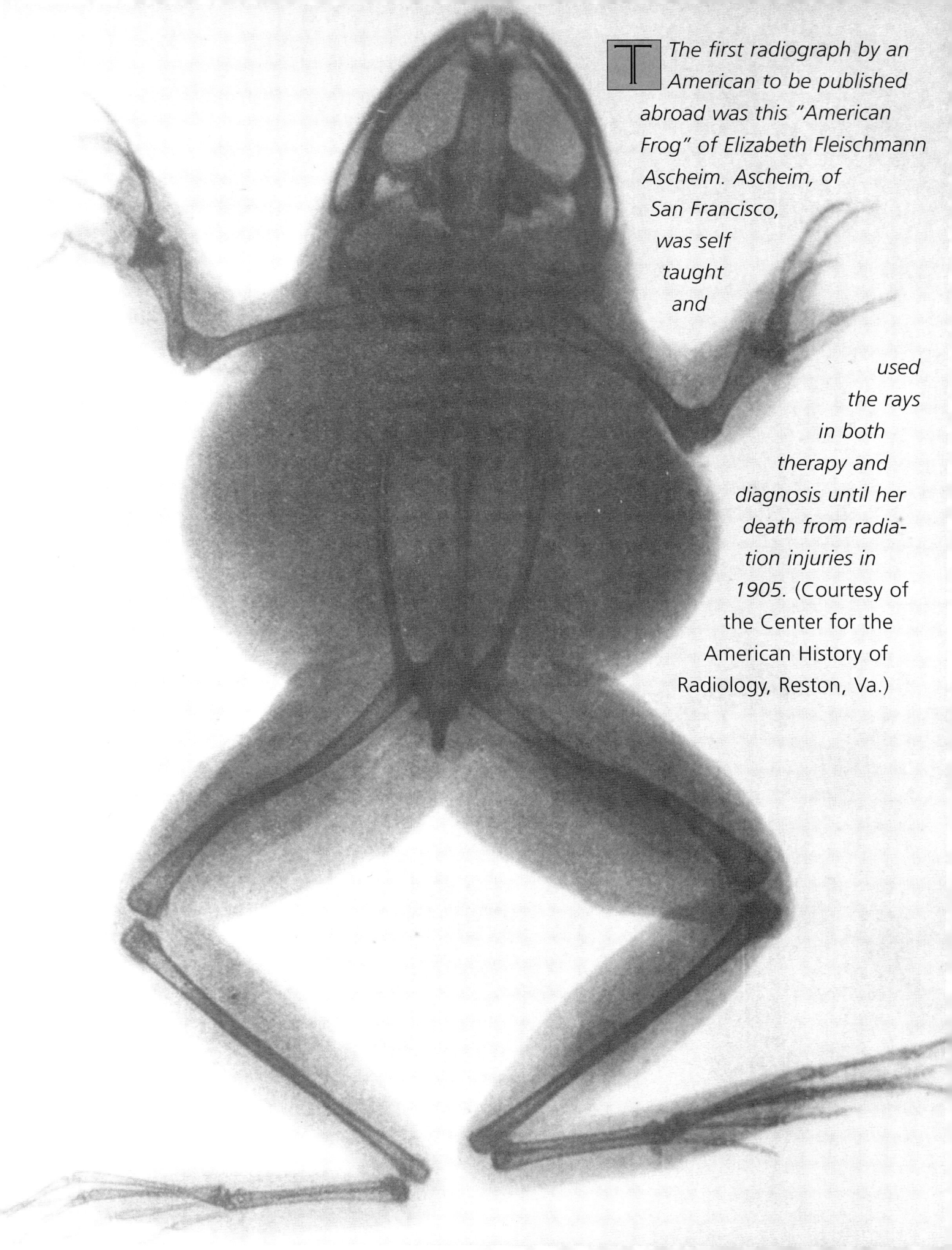

CHAPTER NINE

Women in Radiation Oncology and Radiation Physics

Kate Knepper and Sarah S. Donaldson, M.D.

The written history of radiation oncology for the most part has overlooked the accomplishments of women in the field; their battles for acceptance also have been ignored. In some instances this has been because women, for whatever reason, traditionally have not attained the career highwater marks that make their male counterparts emerge as historical figures: departmental and institutional directorship, national office in medical organizations, or first authorship on numerous publications. This chapter looks at the many women who practiced radiation oncology and radiation physics over the last century, often against great odds and at some personal cost, and whose stories illuminate areas of our history previously unexplored.

The study of women in medicine allows us to examine women's entry into the professional fields. Medicine was one of the first modern professions opened to women, possibly owing to the belief in women's "natural" healing powers. There is an extensive literature on the history of women in medicine, the great bulk of it concentrated from the nineteenth century on. For the most part these studies do not focus on medical specialists, and one finds a few extraordinary women cited repeatedly as examples. The literature which does focus on one specialty within the broad field of women in medicine tends to examine typically "feminine" specialties such as obstetrics, pediatrics, psychiatry, and public health.[1,2,3,4] Several writers have focused on women as medical reformers in the nineteenth century, an activity supported by male physicians who believed that women's moral vision could improve the national health.[5]

Critics of female physicians claimed that medicine might "defeminize" women and destroy their modesty. These critics also asserted that most women were routinely incapacitated by menstruation, adding that intellectual activity further destroyed women's health. Women were quick to point out that no one feared that nursing would defeminize women and insisted that female patients needed female physicians for the sake of modesty. Physicians supportive of women countered the biological argument with research; they found that women who engaged in intellectual pursuits enjoyed excellent health and that most women did not suffer from debilitating menstrual problems.[6]

MARIE CURIE (1867-1934)

Undoubtedly, the most famous woman to influence the field of radiation therapy was Marie Curie. Her discovery of radium created a new discipline of radiation therapy. In the latter years of her career, much of her support came from the appreciation of the applications of her discoveries to the treatment of cancer. She faced many of the same issues that confronted women physicians, such as discrimination and problems balancing work with family.

Marie Sklodowska was born in Poland in 1867, the daughter of a physics and mathematics professor. She worked for several years as a governess supporting her sister Bronya's study of medicine at the Sorbonne; at twenty-four Marie went to Paris to study physics and mathematics at the Sorbonne.[15] Three years into her studies, she met Pierre Curie while requesting space to work in his laboratory, and a little over a year later they married.[16]

Marie Curie began her work on radioactivity when she chose to study the newly discovered Becquerel rays for her doctoral thesis. She discovered that not only was uranium radioactive, but thorium was as well. This led her to hypothesize that radioactivity was not a property of a single element. She also speculated that pitchblende, an ore of uranium oxide, must contain an undiscovered element, because uranium alone could not account for all the radioactivity present.[17] At this point, Pierre gave up his own research and joined her work. Together they discovered polonium and radium. In 1903 Marie and Pierre Curie shared the Nobel Prize in physics with Henri Becquerel for their work in radioactivity. Thereafter, Pierre and Marie continued their work to prove the existence of radium by preparing it as a pure salt, a process which took several years.

(Courtesy of the Center for the American History of Radiology, Reston, Va.)

Their marriage was an unusual collaboration. Through Pierre's academic position, Marie Curie was able to gain access to laboratory facilities from which she otherwise would have been barred. She was careful to take credit for those discoveries which were hers alone, although she was not always believed. The Curies did not challenge every blatant insult to her abilities; for example, when the Royal Institution of London invited Pierre to speak in 1903, Marie sat in the audience as he described their joint project. Neither were concerned with the outward trappings of scientific success, but they did care about laboratory facilities and support so that they could continue their research.[18,19,20]

During these years, Pierre supported her as she attempted to balance science and family. While they did assume that she would be responsible for domestic affairs, they made it possible for her to focus on science as well. They lived in austere surroundings to minimize housework and kept few social ties. When their daughters, Irène and Eve, were born, they hired a nurse. In addition, Pierre's father moved in with them and helped with child care. In a letter to a friend, Marie discussed the issue of balancing family and work:

> It became a serious problem how to take care of our little Irène and of our home without giving up my scientific work. Such a renunciation would have been very painful to me, and my husband would not even think of it; he used to say that he had got a wife made expressly for him to share all his preoccupations. Neither of us would contemplate abandoning what was so precious to us both.[21]

Tragically, Pierre was killed in a street accident in 1906, leaving Marie a widow and single parent at age thirty-eight. Her position was changed from female collaborator to independent scientist. While at first the Sorbonne resisted, under pressure they appointed her as an assistant professor—the first woman to hold such a position there. Years later she attributed her appointment to the emotion surrounding Pierre's death. Without this appointment she would not have been able to continue her research. In the following years she concentrated on proving that polonium and radium were truly elements; she was able to produce salts of radium and polonium and prepare radium as a metal. She received a second Nobel Prize in 1911, this one in chemistry, for this work.[22,23,24]

While Curie received many honors in her life and was the first person, man or woman, to receive two Nobel Prizes, she also had to contend with the perception that she was only a collaborator who had no original ideas of her own. In 1911 a scandal was spread by the wife of a fellow physicist, Paul Langevin, that Marie Curie was having an affair with Langevin. Copies of a letter said to be written by Marie to Paul were leaked to the Paris tabloids.[25] Some critics leapt on these stories, claiming that they proved that a woman could excel in science only when "working under guidance and inspiration of a profoundly imaginative man." They believed her work was not original but was guided first by Pierre Curie and later by Langevin.[26]

Much of Marie Curie's funding and support came from those who hoped her work could be applied for medical purposes, such as cancer therapy. However, medical applications also created a demand for radium which caused its already steep prices to skyrocket out of her reach. She had always refused to patent radium or make any monetary gain from its discovery. In 1920 her Institut du Radium had only one gram of radium and was unable to purchase more for research. An American journalist, Missy Meloney, set up a tour to raise money from American women to buy Curie an additional gram of radium.[27] While Harvard and Yale quibbled about whether she was important enough to warrant an honorary degree, huge crowds came out to welcome her at every turn.[28]

Marie Curie's public image was that of a healer, despite the fact that radium therapy was only a consequence of her work and not the focus of her research. In some ways she found more acceptance in the medical community than in the scientific community. The Academie des Sciences refused to admit her when she applied for membership, while the Academie de Medicine spontaneously appointed her as an independent member, although she had not even applied. While Curie herself was not involved in the medical applications of her work, her discoveries had a tremendous impact on the field of radiation therapy. Without her work, radium might not have been discovered for many years and radiation therapy would have lost one of its most potent weapons. Perhaps more important to this chapter, Curie served as a role model to countless young women contemplating science as a profession and remains today the most famous and easily identified of women scientists in history.

But male physicians had more practical reasons to oppose women's entry into medicine. They feared there was already an oversupply of physicians and that women's entry might damage medicine's struggle for higher status.[7] One way of keeping women out of the field was to deny them training. Before the 1860s recognized medical schools rejected women, although a few schools of homeopathy did accept them.[8] The only place that a woman might receive a decent medical education was in Europe. In 1858 the first women's medical school was founded: the Boston Female Medical College. In 1869 the University of Michigan became the first coeducational medical school. In the next few decades more women's schools were founded, and a few more men's medical schools became coeducational, often encouraged by large donations from wealthy women. The number of female physicians rose throughout the nineteenth century, so that by 1900 women made up 5.6 percent of all physicians.[9] The future looked promising for female physicians—in reality, they were to suffer a backlash from the medical establishment during the early decades of the twentieth century.

Some of the literature on the development of radiology has mentioned the accomplishments of a few individual women, without commenting on the general status of women within the field.[10] Only one book on the history of radiology provided any attention to women as a group and did so in a demeaning and frivolous photograph and caption.[11] In this chapter we discuss the experiences of female physicians and scientists in the "nonfeminine" medical specialty of radiology, with attention to those who practiced radiation oncology and radiation physics during the first half of the twentieth century. The focus has been limited to those women who practiced radiation oncology or its related fields in North America and to the extraordinary women from abroad whose contributions greatly influenced the field in the United States. We investigate the issues pioneering women faced, record their accomplishments, and document their struggles.

THE RADIATION EXPERIMENTERS (1895–1905)

During the late nineteenth and early twentieth century, the medical community changed its perception of illness as an imbalance of the entire body to the more modern view of illness as specific diseases with specific cures.[12] This was largely due to the development of new theories and technologies which were effective at curing certain diseases. The germ theory helped explain and prevent typhoid fever, tuberculosis, and influenza. Vaccines were used to prevent rabies and smallpox, while antitoxins treated diphtheria and tetanus.[13] Microscopes, stethoscopes, and ophthalmoscopes aided in diagnosis.[14]

There was great enthusiasm about the benefits of science and the fruits of experimentation. The medical community and the public were optimistic that disease was no longer a mysterious terror, and that its causes and mechanisms could be understood. Physicians were confident that they had found the method by which all future discoveries would be made. The X ray seemed the very embodiment of all that was hopeful in new technology and invention. Physicians quickly appropriated X rays for their own uses and were eager to experiment with—and speculate about—their possibilities. Radiation therapy held great (if unexplained) promise for those who took the time to learn to use the equipment. The field was new and wide open: any physician could gain expertise quickly through experimentation and reading reports in journals.

At this time the number of women physicians in the United States was growing, and female physicians had won some well-publicized battles against discrimination. In 1893 the Johns Hopkins Hospital was the first prestigious institution to open its doors to women medical students along with its first entering class. However, many schools still would not admit women. Women who did attend coeducational schools were often treated poorly.[29] Ida May Wilson was not allowed to observe "private" operations on men as a medical student in the

1890s and was sexually harassed by a professor at the Ohio Medical University. She then established a private practice only to find that patients were slow to visit a female doctor. Only the encouragement of her brother, also a physician, stopped her from quitting and becoming a nurse.[30] However, Wilson stayed in medicine as a general practitioner and was one of the earliest physicians to experiment with the use of X rays. Like many practitioners then and for years to come, she incorporated X rays into her general office practice. (As late as 1910 only 27 percent of the members of the American Roentgen Ray Society [ARRS] could report that they practiced radiology exclusively.[31]) In 1902 Wilson experimented with X rays in the treatment of three patients suffering with cancer. She believed she had successfully cured two of the patients with skin epitheliomas but that the therapy had failed to affect the third, who had a large cancer which had already spread throughout the arm and axilla.[32] In 1903 she published a paper on her experiments using X rays for the purpose of cosmetic electrolysis, and concluded that "the X-ray method is certainly the ideal way, as it is painless, leaves no scars, and is not tedious, either to the patient or the physician."[33] Wilson did not publish further articles on radiology; perhaps because the radiology literature was more and more dominated by those who practiced and researched exclusively in the field.

Private schools were created to teach the new techniques of radiology to physicians (as well as to photographers and electricians who sneaked in the open back door of medicine to apply their knowledge of X rays). One woman, May Cushman Rice, became a faculty member at the Illinois School of Electro-Therapeutics. Founded in 1899, the school taught the uses of both electricity and X rays.[34] One of the school's advertisements claimed that "a two week course will make you self-dependent."[35] While the male professors taught courses on X-ray diagnosis, radiotherapy, and radiography, Rice's subject area was electrolysis. She may have been involved in the treatment of cancer as well; an article she wrote describing a case of uterine cancer does not make it clear whether she or the school's head, Emil Grubbé, gave the X-ray treatments.[36] However, she presented a number of papers to radiological societies on the therapeutic uses of X rays.[37,38] While she was clearly active in the field, her area of specialization was not prestigious.

Compared to other areas of medicine, radiation therapy offered opportunities for female physicians. There was no need to be associated with institutions such as hospitals, schools, or medical associations to be a radiation therapist; all one needed was the equipment and a desire for experimentation. While women were not welcomed with open arms by the medical community, they were able to secure a place in radiation therapy from the very beginning. Many of the names of these women have been lost to us; they practiced radiation therapy as assistants or as office nurses in the shadow of male physicians. They appear now only as grainy figures in the backgrounds of photographs from the period.

ONLY THE LONELY (1905–1920)

Few women physicians were active in radiation therapy in the period between 1905 and 1920. The number of women physicians in general was declining due to the closure of women's schools and the tight admissions quotas of coeducational medical schools. Only two women who practiced radiation therapy in respected institutions are easily identified, and these two practiced only for a short time. Two other women active in radiation therapy were found, but they worked in areas that were considered unimportant and marginal. The attractiveness of the field of radiation therapy to female physicians may have decreased due to the social reform movement of the 1910s and to the lack of technological progress within radiation therapy. The emerging understanding of the dangers of radiation may have discouraged women in the field as well.

When Johns Hopkins School of Medicine began to admit women in

Margaret Cleaves (1848-1917)

Margaret Abigail Cleaves emerges as the most important woman in radiation oncology in these earliest years. She was the first female physician to be involved in radium therapy and the first physician in the world to use radium in gynecology. She was one of only about twenty physicians in 1903 to gain access to radium and use it for clinical purposes.[39,40] Her story demonstrates the appeal of radiation therapy to female physicians, perhaps because the field was more open to women than better-established (and entrenched) medical specialties.

Cleaves received her M.D. from Iowa State University in 1873 and specialized in psychiatry. Her career in psychiatry looked promising, as she rose to the level of physician-in-charge of the Women's Department at the Harrisburg State Hospital for the Insane in Pennsylvania in 1880.[41] While there she started a new program of treatment and speculated on the causes of insanity in women. At the time the prevailing belief was that gynecological disorders precipitated insanity in women. Female physicians, it was reasoned, should treat the gynecological disorders of these patients in order to improve their mental health. By 1882 Cleaves concluded that diseased reproductive organs led to illness in only a few (mainly venereal) cases; instead, she suggested that female mental illness may have been related to "the endless monotony of the lives of the majority of women" and "too frequent child bearing." One of her biographers, Constance McGovern, suggested that by questioning the gynecological etiology of mental illness, she undermined the need for her position as a female physician.[42]

Cleaves experienced resentment from her colleagues; they did not approve of her autonomy over the female wards. In each of her annual reports to the Harrisburg Board of Trustees, she made reference to this problem. Ultimately in 1883 she resigned her position to go into the private practice of psychiatry. The hospital trustees then changed the administrative structure of the hospital, citing "friction between the superintendent and the women physicians." The subsequent female physicians in charge of the Women's Department of the hospital were relieved of all autonomy in their positions. While practicing psychiatry, once again Dr. Cleaves's role was questioned; when she asked to attend the 1881 meeting of the American Psychiatric Association, the request caused such consternation that she was not allowed to be seated as a member, although fellow male colleagues from her hospital were accorded this privilege.[43]

Twenty years later Cleaves may have been attracted to the field of radiation therapy because it was a new discipline; there were no established institutions to stand in the way of hypothesizing and experimenting. In the early years of radium therapy, anyone who gained access to radium soon became an expert in the field; this was exactly what Cleaves did. In the 17 October 1903 issue of the *Medical Record*, Cleaves related how she inserted radium into the uterus of a woman with cervical cancer.[44] Since 1901 she had been eager to experiment with clinical applications of radium, but was unable to do so until 1903, when Charles Baskerville, director of the

LIGHT ENERGY

Its Physics, Physiological Action and Therapeutic Applications

By
MARGARET A. CLEAVES, M.D.

Fellow of the New York Academy of Medicine; Fellow of the American Electro-Therapeutic Association; Member of the New York County Medical Society; Fellow of the Société Française d'Électrothérapie; Fellow of the American Electro-Chemical Society; Member of the Society of American Authors; Member of the New York Electrical Society; Professor of Light Energy in the New York School of Physical Therapeutics; Late Instructor in Electro-Therapeutics in the New York Post-Graduate Medical School

WITH NUMEROUS ILLUSTRATIONS IN THE TEXT AND A FRONTISPIECE IN COLORS

"But if darkness, light and sight be separate and independent one of the other, then if you remove light and darkness, there is nothing left but void space."—*Buddhistic Sutra*.

NEW YORK
REBMAN COMPANY
1123 WEST 23D STREET, COR. 5TH AVE.

LONDON AGENTS
REBMAN LIMITED
129 SHAFTESBURY AVENUE, W.C.

1904

(Courtesy of the Center for the American History of Radiology, Reston, Va.)

chemistry laboratory at the University of North Carolina, loaned her a tube of radium. The cervical cancer of a patient "had been declared inoperable by the best surgical talent," and internal X-ray treatments had improved it only somewhat. Cleaves believed that rays from radium could penetrate more deeply than X rays and so inserted the tube of radium into the patient's uterus for a total of fifteen minutes over two days. However, she had to return the radium to the chemistry professor, and no more applications were possible. A week later the patient was feeling well, and her condition was considered to be good. Cleaves concluded that since there were no guidelines for the use of radium in therapy, she would err on the side of short exposures with least danger to the patient; she felt her report "foreshadow[ed] an important place in medicine for radium."[45]

Margaret Cleaves wrote one of the first textbooks on radiation and its clinical applications, *Light Energy: Its Physics, Physiological Action, and Therapeutic Applications* (1904). She published regularly, writing a few articles every year about radiation, light therapy, and other uses of electricity in medicine. She maintained an interest in other research topics and wrote popular articles on the prevention of disease and the need for sex education for working-class women.[46,47]

Cleaves was the founder and chief of the Electro-Therapeutic Clinic Laboratory and Dispensary of New York City and a professor of phototherapy at William Snow's New York School of Physical Therapeutics. She served as the American editor of the British journal *Medical Electrology and Radiology* from 1903 to 1904.

Cleaves's views on female physicians' acceptance within medicine are unclear. In her autobiography she stated that "men have always accorded me my place in the profession. I have not had to ask for recognition even."[48] On the other hand, her experiences with the American Psychiatric Association and the Harrisburg Hospital suggest that she did indeed experience some conflict. Cleaves was only one of a long line of female physicians who denied any personal discrimination, despite experiences which would suggest otherwise. However, she did reject what she saw as "the exploitation of women physicians as a separate and distinct labor from the rest of the profession."[49]

Cleaves's autobiography reveals that she was lonely and depressed much of the time. She believed herself to suffer from neurasthenia, a disease of nervous exhaustion in which she was quite interested. Today, she might have been diagnosed as depressive, suffering from radiation sickness or anemia, or having any of number of other diseases which were not well recognized at the time. The typical treatment for neurasthenia was a rest cure; she resisted this, preferring to remain active and care for her patients. Looking back on her life, however, she felt that her work had not been successful: "Scientific experimental work was begun which I never had strength to finish. Others had taken up the same line of work and brought it to a satisfactory conclusion, mine remains only a bit of wreckage on life's tempestuous sea."[50]

Cleaves blamed her failures on her illness, as well as a lack of social support: "all through life there has been stress and strain with no one to look after my needs." She had few friends and her only source of regular social contact was her physician. She described her patients as her family and friends, and declared that "science is my mistress." She felt that she could not relate to other women as equals because too often she felt professionally responsible for them. Being a physician was her primary identity, however, one of her fantasies was to "play I am a woman not a doctor."[51]

Societal expectations at that time placed a heavy burden on any woman who shunned the typical feminine life for a profession. Other women physicians found support and intimacy through marriage, relationships with other women, or adopting children: "for some women who chose to remain single, the decision exacted its price in loneliness."[52] It is likely that some of the strain that Margaret Cleaves experienced was due not only to her illness, but to the lonely life she led and to the difficulty of defying society's pressures. Yet she still made great contributions as a scientist, physician, and author in the new field of radiation therapy.

1893, many female physicians considered this to be a great stride forward—that coeducation was preferable to separate schools for women.[53] By 1910 most of the women's schools had already closed or merged with men's schools. In 1910 the Flexner report on medical education was published, a scathing account of the problems that existed in many schools. It documented that women's medical schools were usually underfinanced and understaffed; subsequently, all but one of the women's schools closed. But the coeducational schools did not take up the slack and admit more women; instead, coeducational schools severely limited the number of female students through quotas and other discriminatory practices.[54] This served to restrict the number of new female physicians during this era.

Leda J. Stacy was one of only two women practicing radiation therapy during this era who was able to practice in a "reputable" institution (Fig. 9.1). Born in 1883, she graduated from Rush Medical College in 1905. Only four teaching hospitals at that time would even accept female interns. The Children's Hospital in San Francisco did accept her and, thereafter, in 1908 she moved on to a practice at the Mayo Clinic. She began her career as an anesthesiologist, then became an assistant in internal medicine. As with many women, her career was enhanced by the effects of World War I. Many patients who had previously received medical treatment in Europe came to the United States for treatment. Some European women requested female physicians, so in 1915, a special department of gynecology was formed at the Mayo Clinic in 1915 to meet their requests. As soon as Stacy was given an assignment in the newly founded department of gynecology, the Mayo brothers decided their clinic should adopt the use of radium therapy. Stacy was then sent to various medical centers, including Johns Hopkins Hospital and Memorial Hospital in New York to learn its use.[55]

Stacy was in charge of radium therapy at the Mayo Clinic from 1915 to 1919 and published a number of papers on the treatment of gynecological cancers.

In 1919 the roentgenology section of the Mayo Clinic was divided into a diagnostic section and a therapy section, which included both X ray and radium treatment. However, soon the radium therapy was removed from Stacy's direction, and she returned to work in internal medicine.[56] It is not known whether this was by her choice, or whether her superiors felt that one of the numerous returning veteran physicians should handle the increasingly important specialty of radium therapy. Her transfer coincided with the end of World War I and perhaps with the end of the demand for female physicians to treat European patients with gynecological cancers. Stacy remained at the Mayo Clinic until 1935. At that time she moved to a private gynecology practice in White Plains, New York.[57] She worked there until her retirement at age eighty-three in 1966 and died in 1973.[58]

Annabelle Davenport was the other woman who practiced in a well-known institution (Fig. 9.2); she was the first radiologist at the University of California at San Francisco.[59] She taught a course in roentgenology in 1912 and 1913, which covered both diagnosis and therapy.[60] She held the post of clinical instructor but never rose to the level of professor.

Women in many fields have often found that they gain the most acceptance in areas which are not prestigious, pay less, or are otherwise marginalized.

Fig. 9.1 Leda J. Stacy (Author's collection)

Fig. 9.2 Annabelle Davenport (Author's collection) ▶

Between 1910 and 1920 X-ray therapy was overshadowed by the success of radium treatment, and both were overshadowed by the preeminence of diagnostic radiology. James Ewing of Memorial Hospital recollected that at that time "roentgenologists who engaged in therapy were looked upon with suspicion....there was little concept of the possibilities of roentgen therapy."[61]

One woman, Mary Elizabeth Hanks, spoke out against the neglect of X-ray therapy. A gynecologist who had studied X-ray therapy in the late 1910s, she decried what she felt was an excessive emphasis on diagnosis: "The whole medical fraternity in its engrossed, eager study of fascinating diagnostic problems, has seemed to lose sight of the prime object of all medical science, the cure of disease." She felt too many roentgenologists did not care about research in therapy and only practiced therapy to bring in money.[62] In an article titled "The Value of the Roentgen Ray in the Treatment of Uterine Fibroids," she extolled the benefits of radiation therapy over surgery: no mortality, no anesthetic necessary, no hospital stay. Yet, because so many doctors were ignorant to the advantages of radiation therapy, "hundreds of women continue to form the same procession to the operating table, just as ten years ago." Hanks believed that if physicians took proper care in technique, radiation therapy held far more promise than surgery in the treatment of fibroids. She blamed her fellow physicians for not taking advantage of the opportunities presented by radiation therapy.[63] Little heed was taken of her opinions until the 1920s, when Coolidge tubes and improved apparatus enabled more powerful treatments.

Mary Arnold Snow was active in an even less reputable field: electrotherapy. Along with her famous husband, William Benham Snow, she coedited the *American Journal of Electrotherapeutics and Radiology* in the late 1910s. She was primarily interested in the clinical applications of mechanical vibrations but published on a number of applications of X ray, including inflammation and the treatment of uterine hemorrhages and fibroids.[64,65] During this time most radiologists were attempting to disassociate themselves from electrotherapists, a group at odds with the American Medical Association (AMA), the ARRS, and most of established medicine. Radiation therapy limited itself to X rays and radium; electrotherapists used electricity, mechanical vibrations, ultraviolet light, sunlight, and a host of other agents. The medical establishment believed electrotherapy's techniques had little medical value and labeled electrotherapists as quacks. The historian E. R. N. Grigg described Dr. Snow and her husband as two of the few ethical and reputable physicians involved in the field.[66] While they may have maintained their individual reputations, their field was not highly regarded.

There are a number of reasons why there were so few women in radiation therapy during this period. Physicians of both sexes had abandoned the field, as it seemed to have lost some of its initial promise. Radium was in short supply, and the difficult-to-operate X-ray apparatus seemed better suited to diagnosis than therapy. Certainly both men and women had doubts about the future of the field.

In addition, fewer women were graduating from medical school, and most of these were faced with the peculiar dilemmas presented by the social medicine of the Progressive era. Although some, like Dr. Mary Putnam Jacobi, believed that women were the equal of men in medical practice and abilities, more popular was the belief that women were better suited to the "feminine" specialties: obstetrics, gynecology, child welfare, nutrition, and hygiene.[67] The Progressive era was characterized by reform—new movements attempted to solve problems connected to housing, sanitation, education, poverty, and health. Female physicians found they could have a place within the movement by concentrating on public health issues—child welfare, nutrition, hygiene —which coincided neatly with the designated feminine specialties.[68] Radiation therapy had little to do with social reform and, with its flying sparks and heavy machines, was not especially feminine. Thus, women physicians may have been more eager to enter specialties where they could join in the social reform and where they were welcomed.

As other fields in medicine welcomed women, radiation therapy seemed to have lost much of its promise. Even though radiation therapy had not yet erected professional barriers to women, it is little wonder that few women chose to enter the field at this time. As for those who did, it must have been a lonely experience.

Opportunities Open (1920–1938)

The numbers of women involved in radiation therapy increased dramatically in the period between 1920 and 1938 as more women were attracted to the field. Some women in this era owned their own equipment, became heads of departments, or ran their own schools. Most of these women had early opportunities in the 1910s upon which they were able to capitalize. Some women were able to advance their careers through the support of male mentors, by filling in for colleagues during World War I, or by establishing their own institutions.

Radiation therapy was more attractive to women in this era for a number of reasons. The effectiveness of the therapy itself had increased. Higher voltage machines and more reliable Coolidge tubes enabled X rays to penetrate further and more predictably into the tissues of the body, raising the chances of curing malignant disease. New treatment techniques were developed using radium. Rather than just placing the radium in contact with the skin or in a body cavity, "seeds" of radon gas were inserted directly into diseased tissue. Radiation therapy was no longer limited to skin conditions and a few restricted types of cancer.[69]

In addition, as the social reform movement of the Progressive era ended, nonfeminine specialties like radiation therapy were becoming increasingly attractive to women. These women believed that the male establishment was more accepting of them and that discrimination was lessening. While hindsight shows this to be overly optimistic, there seems to have been less emphasis on their identities as female physicians. Rather than entering medicine with a spiritual mission to endow it with feminine morals, they could simply be doctors.

Women in this era still faced formidable barriers. A survey of medical school deans showed that in the schools which admitted women there were often policies designed to severely limit the number of female medical students.[70] The number of female medical students accepted stayed at a steady 5 percent until World War II. More students than ever were applying to medical school, allowing the schools to become more selective.

A further barrier to advancement was the lack of postgraduate training for women. Many internships were closed to women: in 1926, 527 hospitals restricted their positions to men only, while only 127 would hire women. Of those 127 hospitals, many would consider a woman only when no qualified man was available.[71] Internships were not yet mandatory to enter the field of radiology; experience could also be gained through preceptorship to an experienced physician, short postgraduate courses, or on-the-job training. Nevertheless, internships were a mark of achievement. In the 1930s, after the newly-formed specialty board, the

Fig. 9.3 Monica Donovan (Author's collection)

▶

American Board of Radiology (ABR), set training requirements for certification, the need for internships became much greater. The increasing need for an internship placed another burden on women: even if they were accepted, this meant delaying full earning capacity for a year or more. Women were less likely to have outside financial support; few families were willing to give their daughters funds for a nontraditional career. Less part-time work was available for women, and when it was available it was at lower wages. In this way, the emphasis on specialization and postgraduate training hurt female physicians.[72]

Internships did open up for women in the latter years of World War I, due mainly to the shortage of men. Monica Donovan, for example, graduated from medical school at Stanford at an opportune time—1917 (Fig. 9.3). She was taken on as an intern in the radiology department that year, and made an associate two years later. She went on to be quite successful; she became an instructor at Stanford Medical School and in the 1930s was the head of the radiology department at St. Mary's Hospital in San Francisco.[73,74]

Jobs as well as internships opened for women as men left for war. Elsie Fox already had training in radiology when she was able to take a male physician's place as a radiologist at the City Hospital on Welfare Island, New York (Fig. 9.4). However, her male colleague returned a year and a half later from his military service, and she was obliged to leave. With her previous experience she was able to find another position as a radiologist for the Bronx Hospital in New York.[75] Gisela Von Poswick had a similar opportunity. She graduated from the Women's Medical College of Pennsylvania in 1911, but her first experience in radiology was not until 1916, when she was able to gain a position as a radiologist at Hahnemann Hospital in Scranton, Pennsylvania. The position lasted until 1918 and the return of veteran doctors; in later years, she would develop her own private practice.[76]

Cassie Belle Rose also had the advantage of beginning her career during World War I and went on to advance farther than any other woman in radiation therapy at the time (Fig. 9.5). She graduated from Rush Medical College in 1914, was an intern at Mary Thompson Hospital, and then returned to Rush to work in the radiology department in 1916.[77] In 1922 she became the head of the department of roentgenology at Presbyterian Hospital in Chicago and served as an associate clinical professor on the faculty of Rush Medical College. After twenty years at Rush she moved to

▶

Fig. 9.4 Elsie Fox (Author's collection)

Colorado and became the radiologist at the Porter and the Boulder Sanitaria in 1936.[78,79] She worked there until her death in 1942.

A number of women found experience through unconventional paths. During World War I the Army refused to commission women as medical officers. Some women found their way around this by setting up their own hospitals in Europe: the American Women's Hospitals. Barbara Hunt organized a field hospital in France and treated several cases of cancer during the war. Previously she had not decided on a specialty, but her wartime experiences sparked her interest in radiotherapy.[80]

Elsie Fox also found training in a nontraditional way. She worked as a volunteer part-time technician and assistant in the Radiotherapy Clinic of the Mount Sinai Hospital Laboratory from 1915 to 1917. The experience she gained through this volunteer work may have helped her gain a paid position as an assistant radiologist in the X-ray Therapeutic Clinic at the New York Postgraduate School and Hospital.[81] From there, she was ready to take advantage of opportunity when World War I began.

A number of the women in this era had the support of men, often more established in the field, who were able to aid them with their careers. Zoe Allison Johnston (Fig. 9.6) had the support of a physician father. After graduating from the Women's Medical College of Pennsylvania in 1911, she practiced medicine with her father for three years, and then went into partnership with a pioneer radiologist, Russell Boggs, until his death in 1922. She continued on in a series of partnerships with male physicians until retirement in 1949. Her skills as a radiologist were gained through her work with Boggs rather than in formal training.[81] Johnston was a well-respected physician, especially noteworthy for her leadership in numerous organizations: she was president of the Pittsburgh Roentgen Ray Society, the Pennsylvania Radiological Society, the American Medical Women's Association, and the Allegheny County Medical Society. In addition, in 1935 she was honored as the first woman elected to be president of the American Radium Society (ARS). Barbara Hunt (Fig. 9.7) also enjoyed a close relationship with her father. He was himself a physician, and was eager to have her follow in his footsteps. When she began to practice medicine in 1912, she assisted him in surgical operations until she left to organize hospitals in France. Thus, she was able to gain experience while supporting herself through his practice.[82]

Several of the women radiation therapists found freedom by practicing

Fig. 9.5 Cassie Belle Rose (Author's collection)

Fig. 9.6 Zoe Allison Johnston (Author's collection)

Fig. 9.7 Barbara Hunt (Author's collection)

by themselves and setting up their own institutions. Hunt converted part of her home into a private hospital for cancer patients when she was beginning to specialize in radiotherapy.[83] Von Poswick had her own practice and was the first woman physician to have her own X-ray equipment (Fig. 9.8).[84] Elsie Fox was the owner and director of the Harvey X-ray and Analytical Laboratories in New York City beginning in 1918, and was the owner, instructor, and director of the affiliated Harvey School from 1939 until her death in 1945.[85]

Fig. 9.8 Gisela von Poswick (Author's collection)

One woman chose not to create an institution but to recreate herself in order to practice in the field. Alberta Lucille Hart graduated with the highest honors from the University of Oregon Medical School in 1917 and was the only woman in her class (Fig. 9.9a). After medical school, she decided to adopt a male identity, and changed her name to Alan Lucill Hart (Fig. 9.9b). By living as a male doctor, Hart enjoyed a power and freedom denied to many women and went on to specialize in radiology and to write a popular book on X-ray therapy. In 1925 Alan Hart took a wife, and after many years of practice in radiology, died in 1962. It is hard to say whether Hart's lifestyle helped or hindered; while there may have been less discrimination on some counts, on at least one occasion Hart was recognized by a past acquaintance and hounded out of town by the local physicians.[86,87,88]

Creating one's own institution is not easy in any specialty, but particularly difficult in radiology. Many historians of medicine have used radiology as the primary example of the hospital-based specialty: "radiology grew directly out of hospital needs and largely in hospital contexts" and "the hospital became the setting for training and a good proportion of radiological practice."[89] Working on their own may have been somewhat unorthodox, but it gave these women more control and a chance to escape from some of the hostilities seen in male-controlled institutions. A few women seem to have thrived in these institutions. Rieva Rosh (Fig. 9.10) worked at Bellevue Hospital as an associate visiting radiation therapist in the 1920s and later as an assistant in surgery at the New York University Medical School. She was interested in bone tumors, published widely in the field, and was an influential teacher.

The appeal of radiation therapy and its increasingly effective technology found more women entering the field in this period. Professionalization of the field sometimes obstructed the entrance and advancement of women; but many found other ways to gain experience and positions. Their methods proved

Fig. 9.9a Alberta Lucille Hart (Author's collection)

Fig. 9.9b Alan Lucill Hart (Author's collection)

successful as they found satisfying and innovative work in radiation therapy.

THE SEARCH FOR AUTONOMY (1938–1950)

From the late 1930s to the 1950s the field of radiation therapy was changing; fueled by war-inspired research, new equipment such as supervoltage machines and the electron beam were introduced. Women during this transitional age began their careers in the so-called Golden Age of radiology, the period between the two world wars, in which many developments occurred in X rays and radium. During this time many women became deeply involved with the research that was changing the field so rapidly. Opportunities were opening for women, but they still faced discrimination in residency training programs, in publication of scientific work, and in academic advancement. Perhaps the most important issue for many women was autonomy; many radiotherapists, both men and women, felt they did not have sufficient control over the management of their patients.

The number of women in medical school stayed at a steady 5 percent throughout this period, except during World War II, when declining male enrollments forced the admission of more women.[90] Once in medical school, women faced additional challenges: Anna Hamann, for example had the privilege of being taught physics by Professor Röntgen himself in Germany. However, Röntgen was not fond of medical students, and even less fond of women medical students. When she was late for class one day, he failed her in the course. She was reinstated only after her father, also a physics professor and respected colleague of Röntgen, intervened on her behalf.[91]

Fig. 9.10 Rieva Rosh, center front, in the fall of 1942, in a photograph taken by Dr. Louis Raider at a staff outing at Dr. Rosh's home in Connecticut. (Author's collection)

By the late 1930s postgraduate training was required for those who wanted to enter the field of radiology. In 1933 the five national societies of radiology gathered to form the ABR, a national board that would "protect the public from irresponsible practices and preserve professional dignity" by certifying qualified radiologists. In addition to passing the examination, candidates also had to have had a one-year internship and a three-year course of study in radiology. Outstanding radiologists were "grandfathered in" without examination for the first few years, and then the process was discontinued.

At the beginning of World War II, of the 712 hospitals which offered internships, only 105 accepted applications from women.[92] Given that an internship was necessary for board certification, this created an even greater institutional barrier for women to overcome. Two of the women of this era managed to escape the American internship. Anna Hamann was already a well-established physician when she immigrated from Germany, and Vera Peters substituted personal training in Canada in place of official postgraduate training.[93]

Even those hospitals that claimed to accept women for an internship year did not always encourage them to enter. Ruth Guttmann, for example, applied for a residency at Memorial Hospital in New York in 1940. The director of the hospital, James Ewing, advised that rather than start a residency, she would be better off to "settle down in the country." She was crushed, but persevered. Ewing agreed to accept her on a six-month probation. At the end of the probationary period, he agreed that she could stay for another six months; she remained at Memorial Hospital for nine additional years, working as a member of the staff when her residency was complete.[94]

Florence Chu chose radiology as a specialty because its fixed hours allowed time for family (unlike surgery or medicine). She then chose therapy over diagnosis because she enjoyed the patient contact.[95] Selma Hyman also looked for a specialty that would permit family time and first chose pathology. She was not accepted into a pathology residency and felt that it may have been because she was a woman. However, she did accept a radiology residency at Brooklyn Jewish Hospital.[96] Esther Marting wanted to combine her love of sculpture with medicine by becoming a plastic surgeon but found that the field was "entirely closed to women." Instead, she entered radiology.[97] Ruth Guttmann chose radiology for a pragmatic reason—it was the only residency open when she applied to Memorial Hospital. She specialized in therapy for the same reason as Chu—she enjoyed the patient contact.

World War II has been acknowledged as a time of opportunities for many women in medicine due to the shortage of men. Even Harvard Medical School, one of the last bastions of male-only medical education, finally became coeducational. Across the country, the numbers of female medical students rose, and many internships were opened to women.[98] The women discussed in this section were already established in their careers by the time of World War II and were not able to benefit from increased medical school admissions or opening of internships.

For two of the women of this era, the political situation before World War II provided urgent reasons to leave their native Germany and come to America. Anna Hamann came from Germany, and Ruth Guttmann was born in Germany and was beginning her internship there when her father's anti-Nazi politics made it necessary for her to leave Germany and move to the United States.[99]

All of these women were involved in research, exploring the possibilities of new technologies and reexamining past assumptions. For some, recognition for their work trailed their research by decades, often because of the hostility of their peers and the rejection of their ideas by radiology journals and meeting program chairs.

Not all women experienced hostility from their peers. Selma Hyman always felt welcomed and accepted by her colleagues. She relayed, however, that she had never been ambitious for status. During her stay at the University of

Anna Hamann (1894-1969)

Anna Hamann was born near Hamburg, Germany, and studied medicine at the University of Munich, where her physics professor was Wilhelm Röntgen. Her doctoral thesis, presented in 1924, was on radiotherapy, which she subsequently practiced in hospital settings in Germany. On attending the 1937 International Congress of Radiology in Chicago, she made arrangements to come to the United States to work as a radiation therapist at the University of Chicago. Her father was a socialist and also was at odds with the Nazi party. It was rumored that Anna had helped Jews leave the country and had been forced to emigrate when suspicion fell upon her.[103]

(Courtesy of the Center for the American History of Radiology, Reston, Va.)

Once in the United States, however, she faced suspicion that she was a German spy. This was in part because of her frequent trips to Germany to visit her ailing mother. More important, she was viewed with suspicion by those in Enrico Fermi's lab, in which key research on the atomic bomb was being conducted. The Federal Bureau of Investigation kept Anna under surveillance, although she was never questioned. Workers at the University of Chicago were warned "not to get too close" to her.[104] One of Hamann's colleagues believed that her German background "denied our [the University of Chicago's radiology] department any but a marginal connection with the Manhattan Project."[105]

Anna Hamann was aware of the distance kept by her fellow physicians, as demonstrated by a story she later told. She once fell into a ditch dug for a sewer line and broke her arm. She jokingly noted that she "must get out in a hurry or some of my professional colleagues from the University of Chicago might start throwing in the dirt."[106] One of her acquaintances believed that she was distrusted because she was German, but it is also possible that her forthright manner did not sit well with her peers.[107] A further difficulty for Hamann was that she had lost her fingerprints from her prior exposure to radium, and this presented a problem for security at the University of Chicago. Her habit of wearing tiny pearl-buttoned kid gloves at all times (even in the clinic) added to her reputation as an eccentric. Denied advancement at Chicago, she went on to head up her own department at the Evanston Hospital.

Fig. 9.11. Ruth Guttmann (Author's collection) ▶

Oregon Medical School, she never had to compete with a man to head the department of radiology because she never wanted to become the chair. She wanted to be involved in the clinical side of medicine, not in climbing the academic hierarchy.[100] This may have removed many potential sources of tension for her.

Three women from this era did become heads of their departments; Ruth Guttman (Fig. 9.11) was the director of the Department of Radiotherapy at the Francis Delafield Hospital at Columbia University from 1955 to 1976, and Florence Chu was the chair of the Department of Radiation Therapy at Memorial Hospital from 1976 to 1984.[101,102] Anna Hamann, on the other hand, was never promoted beyond associate professor during her long career. She did become the director of the Department of Radiation Therapy at Evanston Hospital; when younger men were promoted to head the radiation therapy departments of several hospitals associated with Northwestern University, she felt passed over and commented that they were too young for the jobs.[108,109]

One of the greatest issues for many of these women was that of professional autonomy. For example, when Hamann was at the University of Chicago during the 1940s, she was frustrated by the lack of admitting privileges. Furthermore, younger physicians, even residents in other fields, could discontinue the treatment of her radiotherapy patients. Hamann left the University of Chicago for the Evanston Hospital in 1948. During her negotiations with the director of the Evanston Hospital, she specifically requested written approval for "some sort of security for the position of a hospital radiotherapist as to its independence on changes in the administrative and professional staff."[110] The director was able to assure her that her position as director of the radiation therapy department would be permanent and independent; she accepted his offer and left the University of Chicago for the more secure and independent position at Evanston Hospital.[111]

Other women had similar experiences. According to one of her colleagues, Ruth Guttmann felt that she was nothing more than a technician at Memorial Hospital and did not have control over her work.[112] Guttmann stayed on staff at Memorial until 1950; Florence Chu began working there in 1949, and her opinion was that it was hard for the radiation therapy department to gain independence because Memorial Hospital was a surgically-oriented institution. As the field of radiation therapy became more complex, she felt that the situation improved.[113] Guttmann and Hamann found similar solutions to their similar problems—they left, finding independence in other institutions. Radiation therapists in general had difficulty obtaining referrals from surgeons, but the problem seems to have been enhanced for the female radiation therapists.[114]

Women in private practice had more control over their own practices. For example, when Selma Hyman was frustrated by the lengthy delay and complex bureaucracy surrounding the purchase of a supervoltage machine in her department at the University of Oregon Medical School, she left and joined her husband's private practice of radiotherapy. Together, this husband and wife team bought and ran the first private practice cobalt unit on the West Coast of the United States (Fig. 9.12). But not all women who worked in hospitals felt their autonomy was limited. Vera Peters felt that the surgeons at the Toronto General Hospital were happy to give up cancer patients to the radiation therapists and did not attempt to interfere with their treatment.[115]

As radiation therapy became a more accepted profession, women had to cross even more barriers to advance in the field. There were fewer ways to gain practical experience outside the required internship, and the possibility for doing research became limited to academic institutions. However, within academia female physicians found little autonomy, and a number turned to the more flexible positions offered in private practice. Those who persevered in academic centers found that their work was not always taken seriously. While women were making impressive contributions in radiation therapy, their accomplishments were not always recognized.

RECOGNITION AT LAST (1950–1960)

While the number of women in the field of radiation therapy did not increase during these years, these women are distinguished from those in earlier eras by the fact that they rose to greater heights in their careers and received more of the recognition due to them. Many of these successful women had been able to find supportive work environments and felt that they had not suffered for their gender. Nevertheless, discrimination had not disappeared from some of these women's lives.

During and immediately after World War II, the enrollment of female medical students doubled from 5.3 percent of all graduating students in 1941 to 12.1 percent in 1949. In the postwar era, when male entrants were once again graduating, the percentage of women dropped back to prewar levels.[116] Since the women of this era graduated from medical school during the 1940s and 1950s, one might expect to see this surge in enrollment demonstrated in the number of women active in the field, but this was not the case. Norah Tapley and Patricia Tretter were among the few women physicians in radiation therapy who graduated from coeducational schools during the peak years of 1948, 1949, and 1950. Radiation therapy did *not* experience an influx of women due to World War II.

The women graduating during the peak years of 1948 to 1950 may have found themselves directed into traditionally feminine fields, such as pediatrics or psychiatry. After the war, there was a strong cultural backlash against women who worked. Many women physicians were removed from hospital positions to make room for male physicians returning from war. The "cult of domesticity" demanded that women return to their traditional place at home with family after their brief experience in the working world during World War II. Although the number of employed women continued to rise, new female workers tended to be older women who had raised their children and were entering lower-level jobs in clerical service work.[117] Faced with a society that disapproved of women entering professional fields, female medical graduates may have found it easier to specialize in a feminine subspecialty.

Fig. 9.12 Selma and Milton Hyman (Author's collection)

Fig. 9.13 Lillian Fuller (Author's collection)

▶

A number of the women interviewed for this chapter had strong family support in entering medicine. Eleanor Montague was the child of Italian immigrants, and her father worked as a coal miner, along with other jobs, during the Depression. Her family saw education as the only path for advancement and felt that medicine would be a good career.[118] Lillian Fuller had an interest in commercial art, but decided that the field was very limited. Her secondary interest was medicine, which her parents supported (Fig. 9.13).[119] During medical school Fuller contributed cartoons to the school paper. A professor of radiology asked her to draw illustrations for a cancer detection brochure, and he influenced her decision to enter radiation therapy. Patricia Tretter had initially wanted to be a nurse, until the sixth grade, when her father pointed out that she might prefer giving orders rather than taking them; at that point, she changed her career goal to that of physician.[120]

Eleanor Montague chose radiology because its hours allowed time for family, and radiation therapy because she felt that seeing patients was more exciting than reading films and performing fluoroscopy in the dark.[121] She worked in a radiology department prior to entering medical school and initially had some interest in diagnostic radiology but was only able to find a position in radiation therapy. Norah Tapley had an early interest in diagnostic radiology, but changed to therapy after training under Morton Kligerman. Initially, she had difficulty with the depressing aspects of therapy; on more than one occasion she was found weeping in the corner of Kligerman's office, unable to complete her assignments because she could not bear to see terminally ill patients. However, she was determined to prove to Kligerman that she could do the job. In fact, she did such an outstanding job that he later recruited her to a faculty position at Columbia University.[122]

A number of the women of this era had the same mentor in Morton Kligerman. He helped both Lillian Fuller and Eleanor Montague find positions with Gilbert Fletcher at the M. D. Anderson Hospital. In 1954, when Kligerman was in the midst of recruiting Norah Tapley to a faculty position at Columbia, evidence of salary inequity and discrimination against women became apparent. The hospital intended to pay Tapley considerably less than what they had paid the male who previously had filled the position to which she was being recruited. Kligerman objected and threatened to resign as director of the radiation therapy teaching service if the hospital did not pay Tapley a salary equal to that offered to males. Eventually the hospital relented, and she joined the radiation therapy department at a standard salary level.[123] Many women at the time must not have been so fortunate to have such a determined and powerful mentor.

Three women of this era, Fuller, Montague, and Tapley, worked at the M. D. Anderson Hospital in the radiation therapy department chaired by Gilbert Fletcher (Fig. 9.14). Fletcher was a strong believer in the merits of women physicians; his own wife was a pediatric ophthalmologist. His department had equal pay for men and women for equal levels of advancement. It does not come as a surprise that the women who advanced the farthest came from an unusually supportive environ-

Fig. 9.14 Gilbert Fletcher with three strong women in the department of radiation therapy at the M. D. Anderson Hospital: Lillian Fuller, Norah Tapley, and Eleanor Montague. (Author's collection)

ment, with the encouragement of established colleagues.

In contrast to the prior era, women of this period were promoted within their departments. A number of women became directors or chairs of their departments. Norah Tapley was the director of the radiation therapy department at Presbyterian Hospital in New York in the late 1950s. Martha Southard (Fig. 9.15) was cochair of the Department of Radiation Therapy and Nuclear Medicine at Thomas Jefferson University Hospital. In 1979 her colleagues honored her by presenting her portrait to the university; she was the first woman in the history of the school to receive this honor. She was often considered the alter ego of Simon Kramer, chair of the Department of Radiation Therapy. She took responsibility for both clinical practice and residency training when he was away. She specialized in the treatment of gynecological malignancies and lymphomas. She died from cancer in 1979.[124, 125, 126] Lillian Fuller was the deputy chair of the Department of Clinical Radiotherapy at the M. D. Anderson Hospital from 1986 to 1990.[127] Compared to the earlier experience of Anna Hamann, who retired after a long and productive career while still at the level of associate professor, women in this later era fared much better.

While some women in this period found their achievements ignored or recognition delayed, many received due recognition for their research and clinical skills. Norah Tapley contributed definitive work on the physics and clinical use of electron beam therapy; her book, *Clinical Applications of the Electron Beam*, was the standard in the field (Fig. 9.16). She also did pioneering work in the use of radiation therapy in the treatment of infants with retinoblastoma. She was well recognized by her colleagues: she was a fellow in the American College of Radiology (ACR), an editor of the *International Journal of Radiation Oncology, Biology, and Physics*, a trustee of the ABR, and president-elect of the ARS at the time of her death.[133,134]

Eleanor Montague specialized in the treatment of women with breast cancer; she "pioneered many techniques and approaches to the treatment of breast cancer which are standard in the treatment of breast cancer today" (Fig. 9.17).[135] She received the gold medal of the American Society of Therapeutic and Radiation Oncology (ASTRO), the gold medal of the Radiological Society of North America (RSNA), the Janeway Lecturer of the ARS, and many other honors. She was also active as a member of the board of directors of ASTRO and on many committees

Fig. 9.15 Martha Southard (Author's collection)

Fig. 9.16 Norah duVernet Tapley (Author's collection)

Fig. 9.17 Eleanor Montague (Author's collection)

Fig. 9.18. Joyce Kline Puletti (Author's collection)

of the ACR, the ARS, and the American Cancer Society.[136,137] In 1993 she was inducted into the Texas Women's Hall of Fame.[138]

Other women in the field also have received distinctions and honors for their work. Lillian Fuller, a fellow of the ACR, served on numerous national and international committees concerned with the treatment of Hodgkin's disease. She was the first radiotherapist to organize and chair a radiotherapy section in an oncology cooperative group. Her major focus was in malignant lymphoma. Martha Southard specialized in the treatment of gynecological cancer and lymphomas and became the president of the Keystone Area Society of Radiation Oncologists. Joyce Kline Puletti also specialized in gynecological cancer and was awarded fellowship in the ACR (Fig. 9.18).[139] She was a president of the Wisconsin Society of Radiation Oncologists and was the associate director and clinical director of the division of radiation oncology at the University of Wisconsin Hospital and Clinics.[140] Florence Chu (Fig. 9.19) was acknowledged as a major contributor and received the prestigious Marie Curie Award from the American Association for Women Radiologists (AAWR) in 1993.[141]

Women of this era felt they had chosen their careers well; most believed that women physicians had to work harder than men but that overall they did not suffer for being female. They were able to advance further in their careers and received more honors and recognition than women in prior eras. However, their numbers were still small; they could not depend on other women radiation therapists for support or as role models because there were so few of them. The

Vera Peters (1911-1994)

Vera Peters is acknowledged as a pioneer in research on Hodgkin's disease and breast cancer. Her decision to choose radiology and then radiation therapy was influenced by the inspirational and brilliant Gordon Richards, who treated her mother for breast cancer when Peters was a medical student. When she graduated from medical school, Richards became her personal mentor, directing her postgraduate training.[128]

In 1940 she was the first to examine the use of radiation in the treatment of patients with Hodgkin's disease, then regarded as uniformly fatal. In 1950 she published a seminal paper demonstrating otherwise: patients with pathologically proven Hodgkin's disease treated with radiation therapy had five- and ten-year survivals of 51 percent and 35 percent.[129] She had difficulty getting the paper published. The *Canadian Medical Association Journal* rejected it, claiming its tables were too intricate. She believed that the initial rejection of this paper and the disbelief of the data were related to the fact that she was a woman.[130] Only people who worked closely with Peters believed that her research was valid. In the late 1950s the medical establishment began to take an interest in Hodgkin's disease, and soon others were demonstrating what Peters had already proven. Twenty-nine years after her 1950 paper, the American Society of Therapeutic Radiology (later ASTRO) awarded her its highest honor, the gold medal. In doing so it was acknowledged that: "This [the 1950 paper] was a major paper in the history of the profession's approach and understanding of this disease, although it was some years before many members of the profession accepted the viewpoint demonstrated by the data analyzed and reported by Dr. Peters...."[131] Even after her research was validated, she was not always accepted by her colleagues. Other radiation therapists endeavored to keep her off committees, scoffed at her work, and told her to specialize in a different area, such as breast cancer research. They seemed to be jealous of her success; she felt sorry for them, and tried to ignore such "little things." Her conclusion was that none of the poor treatment she received mattered as long as she knew that she was doing the right thing.

(Courtesy of the Center for the American History of Radiology, Reston, Va.)

Fig. 9.19 Florence Chu (Author's collection)

numbers of women in radiation therapy would not increase until the 1970s.

The numbers of women physicists and biologists involved with radiation therapy increased during this period as well, including the work of Edith Quimby (see sidebar on Quimby in Chapter 4). Henrietta Corrigan was involved in the measurement of radiation dosage, the effects of radiation on the brain, and the application of radioisotopes. She was an assistant professor of radiology at Wayne State University.[142] Lucille Ann DuSault was unusual in having only a bachelor's degree, but was able to contribute a great deal to the area of time/dosage research.[143] Elizabeth Focht was a world authority on the effects of radiation on the eye and was a member of the staff at the New York Hospital.[144]

Anna Goldfeder (Fig. 9.20), a radiobiologist who was neither a physicist nor a physician, made many contributions to cancer research in more than sixty years of work. Born in Poland in 1897, she came to the United States in 1931. She developed methods for culturing human breast cancer cells, bred a strain of mice with no naturally occurring tumors, researched the effectiveness of different radiation dosages, and demonstrated the necessity of radiation shielding during treatment. She was the director of cancer and radiological research at the New York City Hospital Department, and did research at Harvard, Columbia, and New York University. She was extremely dedicated and continued to do research in her laboratory in the Delafield Hospital for years after the building had officially closed. She died in 1993; in an autobiographical essay, she stated that she had no regrets about choosing cancer research despite the complexity of the field, for its very intricacy had been her reason for selecting it.[145, 146, 147]

DEFINING ISSUES FOR WOMEN IN RADIATION ONCOLOGY

Family Decisions

Women physicians in every era have had to consider whether or not to have families of their own. Medicine is a demanding profession, particularly for women, and balancing the demands of both family and career is delicate. When female physicians have chosen both, many found that their male colleagues believed them no longer capable of high quality work.

Fig. 9.20 Anna Goldfeder (Author's collection)

Most of the earliest women radiation oncologists did not marry or have children. Margaret Cleaves is one of the few women who made her views known on the subject, describing her patients as her family and declaring that "science is my mistress."[148] We know little about the women of these early years. Leda Stacy never married; Mary Arnold Snow married and collaborated with her husband in the editing and publishing of their journals. We have no personal information on Annabelle Davenport or Mary Hanks.

Among the women who practiced during the following era, 1920–1938, the group was split between those with families and those without. Elizabeth Newcomer married a fellow radiologist; they met in medical school and later practiced together. They specialized in different areas, he in diagnosis, she in radiation therapy, which may have reduced competition between them. They had one son.[149] Zoe Allison Johnston married an attorney, and they adopted a child.[150] Barbara Hunt never married; she believed "she could never have been a good wife to any man." This did not stop her from raising children—she adopted four and was pleased to see one son follow her to become a physician.[151]

Many of the women radiation therapists in this era did not marry or have children: they include Monica Donovan, Elsie Fox, Gisela Von Poswick, and Cassie Belle Rose. It is interesting to note that the women who did have families were in private practice. The women whose careers advanced further academically or by running institutions did not have families. Single women no doubt had more time to devote to their careers, while married women may have found less acceptance because of their marital status.

During the next period, 1938–1950, much more information was available on how women handled families as well as careers. Those women who had private practices felt that they had an advantage in balancing work and family. Since they set their own hours, they could adjust their schedules to family demands. Selma Hyman felt she was capable as both mother and physician when she practiced with her husband as her partner, and it was easy to be relieved at work if a family emergency called.[152]

Esther Marting took the combination of work and family one step further; she established her radiation therapy practice in the basement of her home. This included several examination rooms and two kilovoltage machines. Her husband had his neurology office on the first floor, while the family lived on the second floor.[153] This integration of professional and personal lives was not disruptive. "For us as children, it was never as though our parents were away at work—just downstairs...if we needed to see mother, more often than not we'd just wander into the examination room after a discreet knock, somewhat to the bemusement of the patients," her daughter wrote.[154] Marting initially had an academic position, which carried with it little flexibility. She was the assistant director of the Chicago Tumor Institute from 1943 to 1946. During this time, while her husband was away at war, she balanced a demanding career and raised her children alone. In 1944, with one infant daughter, Suzannah, and pregnant with her second daughter, Priscilla, she worried about being dismissed from her position because of the pregnancy. She successfully concealed her figure by wearing tight girdles and bulky laboratory gowns. When a colleague at the hospital was told that "the baby has come," he replied, "I didn't know Suzannah [her older child] had been away." Marting kept up a full-time career throughout but did admit, "it isn't easy to juggle family and home with work."[155] Her experiences in an academic practice may have prompted her decision to enter private practice, where she could create a more flexible work environment for herself (Fig. 9.21).

Some women with families stayed in the academic arena. Florence Chu felt that it had been necessary to divide herself between her work and her family, and that her family had suffered as a result. She had a housekeeper to help take care of the children, and her husband shared the responsibilities. On the other hand, Vera Peters felt that bal-

Fig. 9.21 Esther Marting (Author's collection)

ancing work and family had been easy with a supportive husband and live-in help. Not all women had children, of course. Ruth Guttmann was married and did not have children. Anna Hamann never married— she was "married to her work" and thought of herself as a physician before she thought of herself as a woman, according to one of her friends.[156]

In the next era (1950–1960) there was a nearly equal division between women who had children and those who did not. Norah Tapley never married but dedicated herself to her work. Lillian Fuller was married during her career in radiotherapy. Although she had no children, she believed that balancing the responsibilities of a physician, a wife, and a mother was not always easy for her colleagues who had done so.[157] Patricia Tretter found balancing work and family to be easy, as long as she had good live-in help. When the help was not reliable, "then it got rough," but for the most part she felt little difficulty.[158]

Eleanor Montague had two children before her residency and two more later in her career. After the birth of her children she spent only a month or two at home before returning to work. She also had occasional problems arranging for child care; on one occasion she had to take a few weeks off from her residency to find someone to replace a previous helper. Her supervisors allowed her the time off but warned her that it should not happen often. When one of her children needed extra care for several years, Montague moved from full-time work at the M. D. Anderson Hospital to the less pressured Methodist Hospital for a part-time position. She felt that the work was less interesting, but it allowed more time at home. After a few years her father moved in with the family and helped with child care, freeing her to return to a full-time practice of radiation therapy.[159] The authors cannot recall ever hearing a similar story from a male physician.

A gradual evolution seems to have taken place in the way female radiation therapists were able to juggle families and careers. The earliest women did not marry or have children; in the next group, those who had families generally had private practices, while those without families advanced farther within the field. During the more recent era women have found that they could balance a family with an academic career, but that it was difficult. The academic arena was not noted for being accommodating, and women without reliable child care or with children needing extra attention experienced additional burdens. Private practice appeared to allow a flexibility which eased the strain of balancing motherhood with medicine.

Discrimination

Many women in medicine have faced discrimination at some point in their careers. In the early part of the century this was quite blatant. As the years went on formal barriers gave way to more subtle forms of bias. Women were hired but at lower rates of pay and were less likely to be promoted than their male colleagues. Many women were criticized for choosing demanding careers instead of becoming full-time homemakers.

It is difficult to speculate about the experiences of the earliest women in the field. For instance, we do not know whether Leda Stacy wanted to give up her work on radium therapy at the end of World War I or whether she was removed against her wishes. Thanks to personal interviews and better information, we know much more about women's experiences during the next two eras. We have seen that women faced discrimination in many different ways, from helpful suggestions that they should settle elsewhere to the offer of longer hours for less money. Many women physicians believed they had been expected to work harder than men. Esther Marting specifically counseled a group of female premedical students, "You must work harder than your male counterparts, be so good that you are acceptable when the opportunity comes."[160] The female radiation therapists also felt that women were not as likely as men to be promoted to higher positions.

Although many of the women interviewed for this chapter told stories about specific instances of discrimination to others, most asserted that they themselves never had faced insurmountable problems being a woman in medicine. For example, despite Ruth Guttmann's experiences when applying for a residency, she did not feel that she had suffered as a woman.[161] Lillian Fuller felt that she never had any problems as a female physician with any of her colleagues or patients but acknowledged that women in medicine are not as likely to be promoted to higher positions.[162]

This pattern of recognizing societal discrimination while denying personal discrimination has been studied by some psychologists. One explanation is cognitive bias: it is difficult for one person to feel victimized without access to aggregate data about her colleagues. Many people, especially women, also do not want to be labeled as victims and wish to avoid labeling an individual person as a villain. Thus they can acknowledge that society as a whole does not treat a specific group fairly, while believing that they have never suffered personally as members of that group.[163]

Some women allude briefly to possible discrimination, then brush it aside by saying that the people involved were not important. Both Vera Peters and Eleanor Montague coped with criticism or personal attacks by ignoring them or treating them as irrelevant. One woman, Florence Chu, even felt that she may have been treated better than her male colleagues because she was a woman. Selma Hyman also may have benefited by being a woman; she felt that she was chosen for a position as an assistant director of the tumor clinic at Michael Reese Hospital in Chicago because the director felt that she was "safer," and less of a threat than a male rival.[164]

Some women may have felt reluctant to talk about discrimination because they did not want their stories to be published. Many of the participants in these stories are still living; often, there is no hard evidence of discrimination, only allegations and rumors. For instance, some men are more willing to talk about discrimination than their female colleagues; yet they qualified their stories as only rumors, not fit to print. From the stories told by both the men and the women, there is no doubt that women radiotherapists in every era have had to fight discrimination.

APPEAL OF RADIATION THERAPY TO WOMEN

In order to understand whether radiation therapy was more attractive to women than other medical specialties, and whether radiation therapy was preferred over diagnosis, we obtained certification data from the ABR. Since the ABR was not formed until 1933, only post-1934 data are available. The sample size is so small for the pre-1965 data that it is not always included. The system of classification varied over the years: therapy consisted of categories of both therapeutic radiology and radiation oncology, while the diagnostic/general radiology category covered roentgenology, general radiology, diagnostic roentgenology, diagnostic radiology, and nuclear medicine. Figure 9.22 demonstrates the number of men and

Fig. 9.22 The numbers of physicians certified in radiation therapy from 1934 to 1990. (Author's collection)

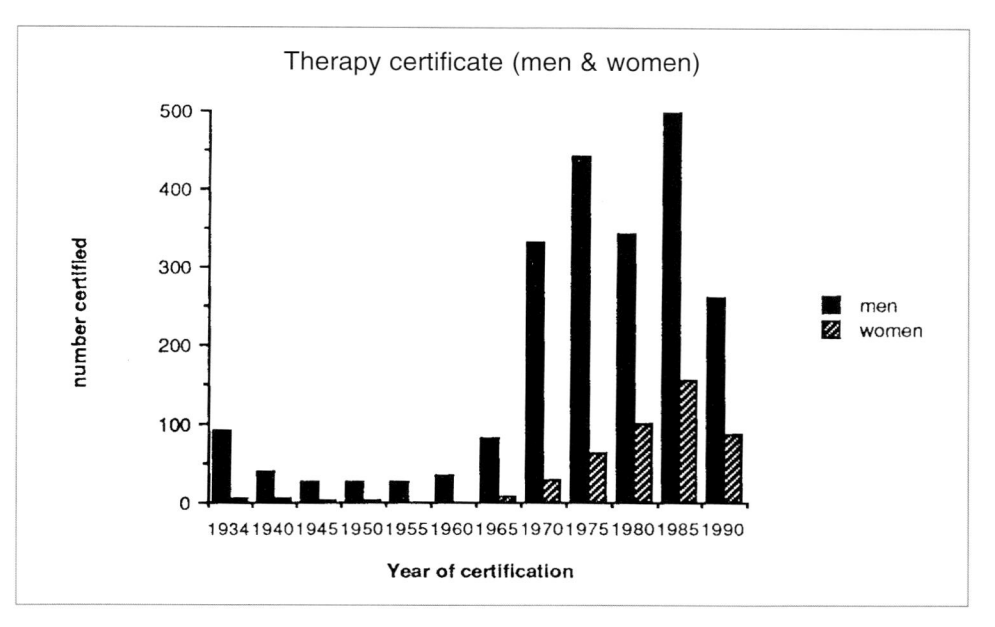

women becoming certified in radiation therapy during the last fifty years.

In order to examine whether radiation therapy was more or less attractive to women than other medical fields, the percent of female radiation therapy diplomates was compared to the percent of female medical school graduates from the preceding five years (Fig. 9.23). Thus, the percent of female diplomates from 1970 to 1974 was compared to the percent of female graduates from 1965 to 1969. The time lag was created because it typically takes four to five years of internship plus residency training before a candidate can become certified in radiology. Thus, the women in those two groups would have graduated from medical school at approximately the same time.

From 1965 on we can see that radiation therapy consistently was more attractive to women than other fields, as demonstrated by the higher percentage of women entering radiation therapy as compared to women entering medicine in general. One of radiology's main attractions for many of the women interviewed was its regular hours, which made it easier to balance practice and family. The women also cited intellectual interest, previous exposure to the field, or inspirational mentors as reasons for their choice.

Radiology does not carry a particu-

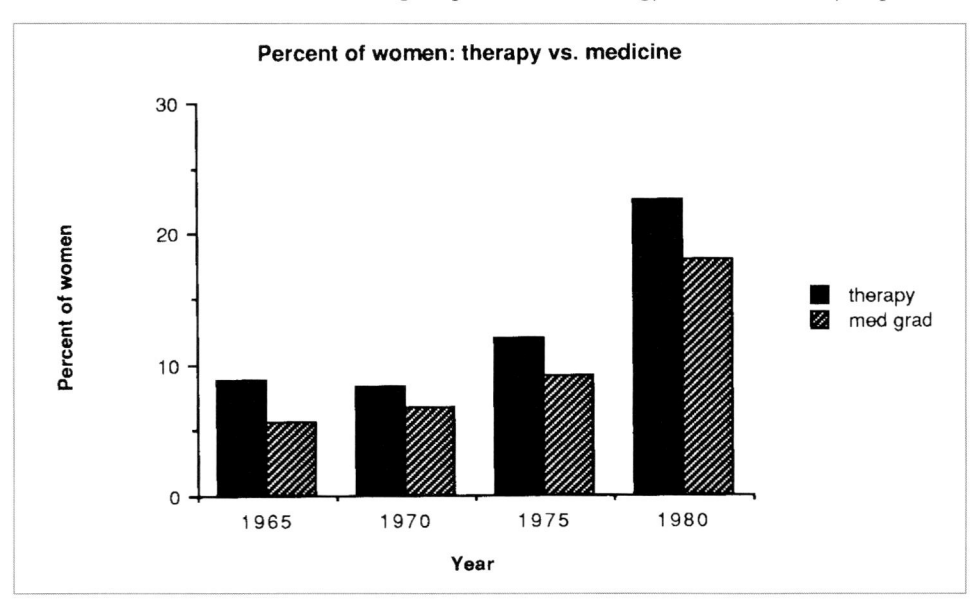

Fig. 9.23 The percent of females in radiation therapy as compared to female medical school graduates. (Author's collection)

Knepper • Donaldson

larly feminine image as has pediatrics, or a masculine image, as has surgery. It carries an emphasis on physics, which traditionally has been an area largely of male expertise. However, all physicians must have a firm science background, including physics, chemistry, and biology; thus, female physicians may not have been deterred by the need for proficiency in physics.

The percentage of women certified in radiation therapy was compared to the percent certified in diagnostic or general radiology (Fig. 9.24). Most radiology residencies through the 1960s focused largely on teaching diagnostic radiology, so many physicians never received much exposure to the field of therapeutic radiology. Therefore, it is surprising to see that therapy has been a more popular choice than diagnosis or general radiology for female radiologists. Many women mention intellectual interest and patient contact as reasons for choosing therapy over diagnosis.

For many years, therapy did not carry as much prestige as diagnosis or other areas of medicine; thus women may have found it easier to secure a residency position in this area. Both Esther Marting and Selma Hyman believed that discrimination may have denied them their desired residencies, in plastic surgery and pathology, and thus accepted positions in radiology. When she began work in radiation therapy in the 1940s, Selma Hyman thought that the field had a poor reputation and that surgeons were reluctant to refer patients to radiotherapists.[165]

The amount of respect given to the field could vary by institution; when Eleanor Montague worked at the M. D. Anderson Hospital, radiation oncologists were treated well and were included in the multidisciplinary team that made treatment and management recommendations for the patient. However, when she switched to working at the Baylor College of Medicine in 1969, she found that the radiation oncologist was called only after the surgeon decided that he was finished with the patient. "Radiation therapy was not only in the basement physically, but also in the basement psychologically."[166] Today, all respected institutions accept the multidisciplinary approach to treating cancer, and radiation therapy is an honored speciality.

QUALITIES

These pioneering female contributors have emerged as leaders and have helped build the discipline of radiation oncology. Although we acknowledge only a few in this chapter, we recognize there were many research scientists, clinician-healers, and educators. Some of these women have been lost to history, their contributions have been overshadowed by the difficult times and the

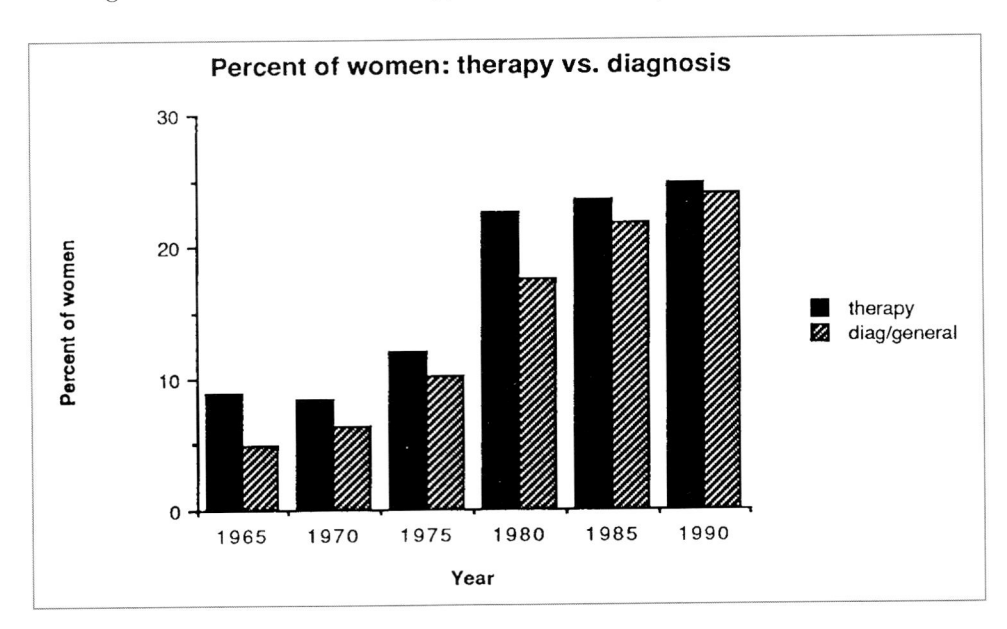

Fig. 9.24 The percent of women certified in radiation therapy as compared to diagnostic and general radiology. (Author's collection)

largely nonsupportive society in which they practiced. In the early days it was not acceptable for a woman to be educated, let alone study science. Medical schools had strict admission policies and tight quotas regarding female applicants. As a specialty radiation oncology was never thought of as a particularly feminine specialty. Not surprisingly, most of the leading women in radiation oncology served as role models to others behind them. All these women share qualities which most certainly contributed to their success.

The women we credit today as the contributing pioneers, who chose a male-dominated profession, succeeded largely because they were confident in their rights and abilities. They held the important qualities which helped shape their success. They were curious and adventuresome in exploring, probing, and testing unknowns; resourceful in securing necessary support; flexible in finding novel solutions; persistent in overcoming barriers; caring in serving others; and optimistic in their endeavors. These women served to inspire all those who have followed in their paths.

▶ ▶ ▶ ▶ ▶ ▶

REFERENCES

1. Wertz, R., and Wertz, D. *Lying In: A History of Childbirth in America.* New York: Free Press, 1977.
2. Horn, M. *Before It's Too Late.* Philadelphia: Temple University Press, 1989.
3. McGovern, C., "Doctors or Ladies? Women Physicians in Psychiatric Institutions, 1872–1900," *Bulletin of the History of Medicine* 55 (1981).
4. Leach, W. *True Love and Perfect Union: The Feminist Reform of Sex and Society.* New York: Basic Books, 1980.
5. Morantz-Sanchez, R. *Sympathy and Science.* New York: Oxford University Press, 1985.
6. Ibid.
7. Walsh, M.R. *Doctors Needed: No Women Need Apply.* New Haven: Yale University Press, 1977.
8. Harris, B. *Beyond Her Sphere.* Westport, Conn.: Greenwood Press, 1978.
9. Walsh, *Doctors Needed.*
10. Brecher, R., and Brecher, E. *The Rays: A History of Radiology in the United States and Canada.* Baltimore: Williams and Wilkins Co., 1969.
11. Grigg, E.R.N. *The Trail of the Invisible Light.* Springfield, Ill.: Charles Thomas Co., 1965.
12. Pellegrino, E., "Sococultural Impact of Modern Therapeutics," in Rosenberg, C., ed. *The Therapeutic Revolution.* Philadelphia: University of Pennsylvania Press, 1979.
13. Ludmerer, K. *Learning to Heal* (New York: Basic Books, 1985):77-79.
14. Rosenberg, C. *The Care of Strangers* (New York: Basic Books, 1987):153.
15. Hellman, S., "Curies, Cure and Culture," *Perspectives in Biology and Medicine* 36 (1) (1992):40-41.
16. Pycior, H., "Marie Curie's Anti-Natural Path," in Abir-Am and Outram, *Uneasy Careers and Intimate Lives.* New Brunswick: Rutgers University Press, 1987.
17. Pflaum, R. *Grand Obsession* (New York: Doubleday, 1989):62-66.
18. Ibid.
19. Curie, E. *Madame Curie.* V. Sheean, trans. New York: Pocket Books, 1937.
20. Reid, R. *Marie Curie.* A Mentor Book. New American Library, 1975.
21. Pycior, "Marie Curie's Anti-Natural Path."
22. Ibid.
23. Pflaum, *Grand Obsession.*
24. Curie, *Madame Curie.*
25. Reid, *Marie Curie.*
26. Pycior, "Marie Curie's Anti-Natural Path."
27. Giroud, F. *Marie Curie: A Life.* L. Davis, trans. New York: Holmes and Meier, 1986.
28. Pycior, "Marie Curie's Anti-Natural Path."
29. Walsh, *Doctors Needed.*
30. Gabel, J., "Medical Education in the 1890s: An Ohio Woman's Memories," *Ohio History* (Spring 1978):53-66.
31. Brecher and Brecher, *The Rays.*
32. Wilson, I.M., "Cancer Treated by X ray," *The Woman's Medical Journal* (October 1902):277-279
33. Wilson, I.M., "Hypertrichosis Treated with the X rays," *American Electro-Therapeutic and X-Ray Era* (July 1903):273.
34. Grubbe, E.H. *X-Ray Treatment: Its Origin, Birth, and Early History* (St. Paul: Bruce Publishing Company, 1949):73-74.
35. *American Electro-Therapeutic and X-Ray Era* (December 1901):15.
36. Rice, M.C., "A Case of Carcinoma," *American Electro-Therapeutic and X-Ray Era* (September

1903):345-347.
37. Society minutes, *Archives of Electrology and Radiology* (January 1904):20.
38. Society minutes, *Archives of Electrology and Radiology* (August (1904):290.
39. Brecher and Brecher, *The Rays*.
40. Davis, K., "The History of Radium," *Radiology* (1924):9.
41. Kelly, H.A., and Burrage, W.L. *Dictionary of American Medical Biography* (Boston: Milford House, 1928/1971):232.
42. McGovern, "Doctors or Ladies?"
43. Ibid.
44. Cleaves, M., "Radium: With a Preliminary Note on Radium Rays in the Treatment of Cancer," *Medical Record* (17 Oct. 1903):603-605.
45. Ibid.
46. Cleaves, M., "Prevention of Disease," *The Ladies' World* (April, May, and June 1910).
47. Cleaves, M., "Education in Sexual Hygiene for Young Working Women," *Charity* 2 (1906).
48. Cleaves, M. *Autobiography of a Neurasthene*. Boston: Gorham Press, 1910.
49. Cleaves, M., *Woman's Medical Journal* (1909).
50. Cleaves, *Autobiography*.
51. Ibid.
52. Morantz-Sanchez, R., "The Many Faces of Intimacy," in Abir-Am, P., and Outram, D., eds. *Uneasy Careers and Intimate Lives,* p. 55.
53. Ibid., p. 187.
54. Ludmerer, K. *Learning to Heal* (New York: Basic Books, 1985):248.
55. Applegate, N., "Dr. Stacy, Retiring, Is Honored at Tea," *Reporter Dispatch* 17 November 1966.
56. Clapesattle, H. *The Doctors Mayo* (Minneapolis: University of Minnesota Press, 1941):604.
57. Applegate, "Dr. Stacy."
58. del Regato, Juan, M.D. Telephone interview. 20 January 1993.
59. Garland, L.H. *A Brief History of Early Radiology in California*. Unpublished, Sept. 1955.
60. *University of California Medical School Announcement*. Berkeley: University of California Press, 1914.
61. Brecher and Brecher, *The Rays*.
62. Hanks, M.E., "The Value of the Roentgen Ray in Uterine Fibroids," *Journal of Roentgenology* 2 (1919):177-189.
63. Walsh, *Doctors Needed*.
64. Snow, M.A., "Physical Therapy in Orthopedic Conditions," *Physical Therapeutics* (1931):65-79.
65. Snow, M.A., "X-Ray Therapy of Uterine Hemorrhage and Fibroids," *Physical Therapeutics* (1926):375-390.
66. Grigg, *Trail of the Invisible Light*.
67. Antler, J., "The Educated Woman and Professionalization," dissertation, State University of New York, Stony Brook, 1977. Ann Arbor: UMI, 1981.
68. Morantz-Sanchez, R. *Sympathy and Science*.
69. Brecher and Brecher, *The Rays*.
70. Morantz-Sanchez, R. *Sympathy and Science*.
71. Van Hoosen, B., "Internships for Women," *Medical Women's Journal* (March 1926):65-66.
72. Morantz-Sanchez, R. *Sympathy and Science*.
73. Moffit, F., "In Memoriam: Monica Donovan," *Bulletin of the San Francisco County Medical Society* (March 1949):22, 57.
74. Garland, *A Brief History*.
75. "In Memory: Elsie Fox," *Medical Woman's Journal* (May 1945):53, 60.
76. "Gisela Von Poswick," *Am. J. Roent.* (February 1941):277-278.
77. *Chicago Medical Blue Book*. 45th edition. (Chicago: Eugene Clark Publisher, 1935):309.
78. "C.B. Rose Resigns After 20 Years in X-ray Department," *Presbyterian Hospital Bulletin*, 1936.
79. "Cassie Belle Rose Thatcher," *Am. J. Roent.* 47 (1942):930.
80. Kobler, J., "Island Doctor," *McCall's* (January 1954):76.
81. "In Memory: Elsie Fox."
82. Osmond, L., "In Memoriam: Zoe Allison Johnston," *Radiology* 77(1961):999.
83. Kobler, "Island Doctor."
84. del Regato, J., "The American Board of Radiology: Its 50th Anniversary," *Am. J. Roent.* 144 (1985):198-199.
85. "In Memory: Elsie Fox."
86. Katz, J. *Gay American History* (New York: Thomas Crowell Co., 1976):258-279.
87. Katz, J. *Gay/Lesbian Almanac* (New York: Harper and Row, 1983):516-522.
88. Hart, A. L., entry in *National Cyclopedia of American Biography* (New York: James T. White and Co., 1960):604-605.
89. Rosenberg, *The Care of Strangers*.
90. Walsh, *Doctors Needed*.
91. Willy, R., "In Memoriam, Anna Hamann, M.D.," *Radiology* 96 (1970):457.
92. Walsh, *Doctors Needed*.
93. Peters, Vera, M.D. Telephone interview. 9 February 1993.
94. Guttmann, Ruth, M.D. Telephone interview. 15 January 1993.

95. Chu, Florence, M.D. Telephone interview. 16 December 1992.
96. Hyman, Selma, M.D. Telephone interview. 26 February 1993.
97. Fabing, S., "Esther Marting, M.D.: A Personal Remembrance." Unpublished, 25 November 1985.
98. Walsh, *Doctors Needed*.
99. Guttmann, telephone interview.
100. Hyman, telephone interview.
101. Guttmann, Ruth, M.D. Curriculum vita.
102. Chu, telephone interview.
103. Jacobsen, Joie. Telephone interview. 13 January 1993.
104. del Regato, telephone interview.
105. del Regato, J., "Hermann Holthusen," *International Journal of Radiation Oncology, Biology, and Physics* 14 (6)(1988):1278.
106. Moss, William, M.D. Letter to Melvin Griem, M.D. 7 January 1993.
107. Moss, William, M.D. Telephone interview January, 1993.
108. Moss, letter to Griem.
109. Hamann, Anna., M.D. Curriculum Vitae.
110. Hamann, Anna, M.D. Letter to Roger DeBusk, M.D. 15 December 1947.
111. DeBusk, Roger, M.D. Letter to Anna Hamann, M.D. 5 January 1948.
112. del Regato, telephone interview.
113. Chu, telephone interview.
114. Hyman, telephone interview.
115. Peters, telephone interview.
116. Walsh, *Doctors Needed*.
117. Chafe, W. *The Paradox of Change* (Oxford: Oxford University Press, 1991):186-192.
118. Montague, Eleanor, M.D. Telephone interview. 11 December 1992.
119. Fuller, Lillian, M.D. Telephone interview, 11 December 1992.
120. Tretter, Patricia, M.D. Telephone interview. 3 March 1993.
121. Montague, telephone interview.
122. Kligerman, Morton, M.D. Telephone interview. 4 February 1993.
123. Ibid.
124. Fletcher, G., and Brady, L., "In Memoriam: Norah duVernet Tapley," *International Journal of Radiation Oncology, Biology, and Physics* 7 (1981):1131.
125. Asbell, S., "In Memoriam: Martha Southard, M.D.," *Radiology* 137 (1980):263.
126. Dobelbower, Ralph, M.D. Letter to Sarah Donaldson, M.D. 4 May 1993.
127. Fuller, Lillian, M.D., curriculum vitae, 1993.
128. Peters, telephone interview.
129. Peters, M.V., "A Study of Survivals in Hodgkin's Disease Treated Radiologically," *Am. J. Roent.*.63 (1950):299-311.
130. Peters, telephone interview.
131. Rubin, P., "Gold Medal Awards of the American Society of Therapeutic Radiologists," *International Journal of Radiation Oncology, Biology, and Physics* 6 (1980):911.
132. Peters, telephone interview.
133. Fletcher and Brady, "In Memoriam."
134. Fletcher, G., "In Memoriam: Norah duVernet Tapley, M.D.," *Radiology* 141 (1981):855.
135. Levitt, S., "Gold Medal Recipient: Eleanor Montague," *Proceedings of the 34th Annual ASTRO Meeting* (1992):15.
136. Ibid.
137. Montague, Eleanor, M.D. Curriculum vitae, 1993.
138. "Shop Talk," *Oncology Times* May (1993):34.
139. Crummy, A., and Javid, M., "In Memoriam: Joyce Kline Puletti," *Radiology* 155 (1985):840.
140. Kline, Joyce Claire, M.D. Curriculum Vitae, 1984.
141. Chu, Florence, "Curie and Distinguished Resident Awards," *AAWR Focus* (Fall 1993): 9.
142. Eyler, W., "In Memoriam: Henrietta Corrigan," *Radiology* 105 (1972):468.
143. Eyler, W., "In Memoriam: Lucille Ann Du Sault," *Radiology* (1977):837.
144. Evans, J., "Elizabeth Florence Focht," *Radiology* 95 (1970):706.
145. Lambert, B., "Dr. Anna Goldfeder Dies at 95," *New York Times* 18 Feb. 1993.
146. Spiegelman, S., "A Tribute to Anna Goldfeder," *Annuals of the New York Academy of Science* 397 (1982):ix-xii
147. Goldfeder, A., "An Overview of Fifty Years in Cancer Research," *Cancer Research* 36 (1976):1-9.
148. *Cleaves, Woman's Medical Journal*.
149. Elizabeth Newcomer, M.D., *Medical Women's Journal* (June 1945):51.
150. Osmond, "In Memoriam."
151. Kobler, "Island Doctor."
152. Hyman, telephone interview.
153. Gibbs, Tilly. Letter to Sarah Donaldson, M.D., 11 February 1993.
154. Fabing, "Esther Marting."
155. Hamric, J., "Lady Doctor in the House," *Cincinnati Times-Star* 4 September 1957.
156. Jacobson, telephone interview.
157. Fuller, Lillian M.D. Letter to Sarah Donaldson, M.D., 12 May 1993.

158 Tretter, telephone interview.
159 Montague, telephone interview.
160 Weichart, Kathryn Ann. "Esther Marting, M.D." Personal papers of Tilly Gibbs.
161 Guttmann, telephone interview.
162 Fuller, telephone interview.
163 Crosby, F., "The Denial of Personal Discrimination," *American Behavioral Scientist* 27 (1984):371-386.
164. Hyman, Selma, M.D. Letter to Sarah Donaldson, M.D., and Kate Knepper, 11 February 1993.
165 Hyman, telephone interview.
166 Montague, telephone interview.

CHAPTER TEN

African American Radiation Oncologists

Carl M. Mansfield, M.D., Sc.D.

▶ ▶ ▶ ▶ ▶ ▶ ▶ ▶ ▶ ▶ ▶ ▶ ▶ ▶

As recently as ten years ago, few persons of color—from any of the multitude of nonwhite ethnicities that make up modern America—were admitted to white training programs in radiology. Of those rare individuals completing such programs, few went on to acceptance at major academic institutions or hospitals. The medical schools of Howard and Meharry universities served as the training institutions for the largest percentage of African American doctors. But there were noted exceptions. In the 1930s and 1940s Memorial Sloan-Kettering, the Universities of Chicago, Pennsylvania and Illinois, the Rush Medical Center, Temple University, and the Pennsylvania and Graduate Hospitals of Philadelphia were frequently listed by African American physicians as places where they had received specialty training. It was a challenge to find the stories of many of these individuals after their training, because African Americans in many states were not permitted to join most radiologic societies. Even those who "beat the system" to attend prestigious white training programs were denied the opportunity to present papers, publish findings, or hold national office.

After having reviewed a number of publications, I had the opportunity to read an unpublished paper by Dr. William E. Allen on the accomplishments of African American radiologists, in which he stated that "from the relatively scarce source of material available" he would "attempt to provide some information on this overlooked and frequently ignored segment of American Radiology."[1] Dr. Allen's paper, titled "History of Black Radiology," was written in the mid-1970s. He noted that "little is

Note: When asked to write a section on the history of African Americans in radiation oncology, I knew the task would be difficult. In fact, of all the historically disenfranchised groups, African Americans have been the most thoroughly excluded from all levels of academic medicine. I was fully aware that the American "system" would not have facilitated accomplishments by African Americans and that there would have been little interest in recording the accomplishments of the few that managed to beat such a system. Mindful of this problem of lack of material, I met with members of the radiology section of the National Medical Association at their 1993 national meeting to ask whether a chapter should be written. Their response was a strong and overwhelming expression that their story should be told, even if it took a great effort to produce a few pages! Their story is not just about accomplishments in the field of radiation oncology and radiology. It is a story that is interwoven into the very fabric of this country's racial attitude. The African American story is about what happened to a race, what happened to a people. They believe that their professional story, regardless of how abbreviated, illustrates their past treatment by the dominant medical culture and illuminates this culture's treatment of their race in general.

known of the history of Black Radiologists, yet they have played a significant role in the development of Radiology in this country, and have made many important contributions to the specialty." He went on to say, "Much of their meaningful work has gone unrecorded and unpublished. In other instances their names have been intentionally or unintentionally omitted from the published reports of investigations or research in which they actively participated." I will draw some of my material from this document. In 1965 Dr. E. R. N. Grigg, a Czechoslovakian-born radiologist, wrote a history of radiology. He inserted a section on "Negroes in American radiology." He recognized that he was doing a courageous thing and stated, "It is (considered) 'politically unwise' to single out the Negroes among American radiologists, but such information is nowhere available in print."[2]

In writing about the history of radiation oncology in the United States, it is necessary to start with the general history of radiology, because of their common origin. Only in the last twenty to twenty-five years has there been a separation and a clearer distinction between the two specialties. Nor can we separate the history of the African American radiation oncologist from the larger history of African American physicians. A short review will show the environment in which African American radiation oncologists developed.

African American Physicians—An Historical Note

In 1944 Bousfield divided the history of African American physicians into three periods. The first was the slavery period, from 1619 to 1865; the second from 1866 to 1930; and the third extended to the time of his publication, 1931 to 1944.[3] During the first period there were a few African American freedmen and slaves who practiced medicine. In 1729 Lieutenant-Governor Gooch of Virginia wrote, "I met a Negro...who performed many wonderful cures of disease." In 1740 the *Pennsylvania Gazette* described a runaway slave named Simon who was able "to bleed and draw teeth."[3] Another slave, James Derham, in 1783, at the age of twenty-one, was called the most distinguished physician in New Orleans. Dr. Benjamin Rush met Derham in 1788 and is quoted as having said, "I expected to have suggested some new medicines to him, but he suggested many more to me." In 1792 a runaway slave named Caesar published an article in the *Massachusetts Gazette* on symptomatology and treatment of poison. The *City Gazette Daily Advertiser* of Charleston, South Carolina, in 1797, advertised to find a runaway slave who "passes for a doctor among people of his color and it is supposed practices in that capacity about town."[4]

A medical education for African Americans was quite difficult to obtain. For example, Dr. Martin R. Delany was born in 1812 and educated in Pennsylvania, where he received "good schooling." He did an apprenticeship with a leading Pittsburgh physician and two additional apprenticeships (at least one being required for medical school admission in that day). He was rejected by all four medical schools to which he applied: University of Pennsylvania, Jefferson Medical College, and two medical schools in New York. Later he was accepted to Harvard Medical School in the class of 1850.[5]

Bousfield believed that the period between 1865 and 1930 was the most difficult for African American physicians. This period marked the beginning of the systematic exclusion of the African American from mainstream medicine and medical societies. The treatment of African Americans in white medical schools was typified by the experience of Dr. Francis Mossell. Dr. Mossell was accepted to the University of Pennsylvania in 1876. He was asked to sit behind a screen in the classroom but refused to do so.[6]

One of many instances of exclusion is exemplified by the struggle of African American physicians in Washington, D.C., to gain admission to local societies. To understand the following narrative, it is important to know that the Howard University and Medical School were established between 1866 and 1868 in response to poor living conditions in the Freedmen's Bureau contraband camp that had been established for escaped slaves. The camp itself became Freedmen's Hospital. Having established a medical school and hospital with the help of white physicians, some of whom remained on the staff, the African American physicians applied to the County Medical Society of the District of Columbia for admission. This was denied. The American Medical Association (AMA) turned down an appeal. The action also kept these physicians from membership in the AMA. At the time of the publication of Dr. Bousfield's paper in 1945, there were still no African American members of the District of Columbia Medical Society.

Because of these membership restrictions, in 1895 African American physicians organized the National Medical Association (NMA). Within this association, the radiology section was started in 1949. Subsequently the radiation oncology portion of the section was begun.

African American physicians were forced to organize their own hospitals, because most institutions refused to admit "colored" patients. The following quote shows how these conditions served as a driving force for the building of separate hospitals: "many of the existing white hospitals refused Negro patients entirely. When they were accepted, they were housed in the basement." In some instances there was only a sheet separating these patients from the furnace. Other hospitals would only accept those specific cases needed for resident training.

For the third period Bousfield noted that there were ten African American physicians graduating per year from the seventy-four white medical schools.[7] As late as 1942, *at least* sixty-four—or 86 percent—of the white medical schools did not accept any African Americans. The remainder accepted one per year or one every four years.

African American women suffered greatly under these circumstances, yet a few were able to excel. For example, Harriet A. Rice, M.D., was decorated by France in 1919 "for her devotion and ability...during the war."[8] From its beginning, Howard University Medical School was also dedicated to the right of women to receive medical education. However, there were obstacles to this approach. For example, in 1877, on a motion by the delegation of the Jefferson Medical College, the Howard delegation was not seated at the AMA convention, because at Howard men and women were taught such delicate subjects as anatomy in the same classes.[9]

It is hard to describe or explain the unique situation in which African Americans have found themselves: a thorough and universal exclusion from all mainstream aspects of American life. This exclusion in medicine in general and in radiation oncology specifically gives a special urgency to finding and telling the stories of those who managed, against the odds, to train, to practice, and to contribute to the field.

I had originally intended to include other minorities in this chapter—the many races and ethnic backgrounds which make up the modern medical setting. However, I soon discovered that few people wanted either themselves or their own ethnic group to be called minorities. The fear of being so designated is understandable when one considers the historical risk of being "nonmajority" in the United States. This is especially understandable when minority status is equated with being black, thus putting minorities at risk for ill treatment and exclusion. I did not speak to every group, but I did speak to enough to realize that it would be unwise to determine in these pages who is or is not a minority.

America has never been kind to its minorities. Therefore, many groups can claim initial or periodic ill treatment. Possibly for this reason, many from these groups have tended to be unsympathetic to the situation of African Americans. They will often cite their own economic rise from poverty to middle class in one to two generations. However, the ill treatment of most other minorities has been short lived, usually spanning only one to two generations. Few minorities have had to endure thorough, pervasive, universal exclusion for more than three hundred years. This exclusion of the African American has resulted in a subculture which by its very difference has produced even more intense efforts to further exclude.

A TIMELINE HISTORY OF AFRICAN AMERICAN RADIATION ONCOLOGISTS

Even in the face of such adverse circumstances, the African American radiation oncologist was able to endure and, in many instances, to achieve. They have contributed articles to peer review journals and, in recent years, have become chairpersons and directors of radiation oncology departments in academic institutions. Their history is, of necessity, episodic; a relatively small number of individuals, often working in isolation, precludes a seamless narrative. What follows are highlights of individual lives and achievements which mark the contributions of African Americans in the first century of radiation oncology.

1899—Marcus Wheatland (Fig. 10.1), who graduated from Howard University in 1895, became the first known African American radiologist. He practiced radiology in Rhode Island and became a charter member of the New England Roentgen Ray Society in 1896. The amount of radiation oncology that he practiced is, of course, unknown. However, his membership in the American Electrotherapeutic Association would indicate that, like many early practitioners in the field, his radiologic practice was both diagnostic and therapeutic.[10]

1918—I. Greely Brown, who practiced in Elizabeth, New Jersey, published an article titled "Treatment of Fibroid Tumors of the Uterus with Radium, " in the *Journal of the National Medical Association (J.N.M.A.).*[11]

1921—F. W. Willis, of Chicago, Illinois, discussed the use of pre- and postoperative irradiation in an article titled "Carcinoma of the Breast: Preoperative and Postoperative X-Ray Treatment," in the *J.N.M.A.*[12]

1923—William S. Bainbridge mentioned X-ray treatment in his article "Multiplex Pathology and the Cancer Problem," in the *J.N.M.A.*[13]

1926—James L. Martin (Fig. 10.2) received formal training in radiation oncology under George Pfahler. He was accepted as a fellow to the University of Pennsylvania in 1921 and studied in the Graduate School from 1921 to 1923. He stayed on as clinical assistant to Dr. Pfahler and as a staff member of the University of Pennsylvania Graduate School of Medicine. In 1926 he published an article with Pfahler in the *American Journal of Roentgenology* on an experimental study regarding the combined effects of roentgen and ultraviolet rays on carcinoma of the skin.[14] He also became a diplomate of the American Board of Radiology (ABR) in 1938.[15]

1927—H. M. Green published "Cancer: A Brief Study in, with Special Reference to Its Surgical Treatment," in the *J.N.M.A.*, in which he, as a surgeon, discussed radiation therapy.[16]

1929—B. Price Hurst published "X-ray Treatment of Hyperthyroidism, " in the *J.N.M.A.* He was a member of the department of radiology at Freedman's Hospital, Washington, D.C.[17]

1929—Columbus Harrison's "Cancer" mentioned irradiation treatment in an excellent review of the history and treatment of cancer in the *J.N.M.A.*[18]

In 1932 there were fewer than fifteen African American radiologists in the United States, and none were

Fig. 10.1 Marcus F. Wheatland (1868–1934) (Courtesy of the Center for the American History of Radiology, Reston, Va.)

Fig. 10.2 James L. Martin (1882–1974) (Courtesy of the Center for the American History of Radiology, Reston, Va.)

Fig. 10.3 William E. Allen, Jr. (1903–1981) (Courtesy of the Center for the American History of Radiology, Reston, Va.)

accepted as members of the American College of Radiology (ACR).[19]

1930s—William E. Allen (Fig. 10.3) must be listed as the outstanding African American radiation oncologist. He graduated from the Howard University School of Medicine in 1930 and did an internship and residency at City Hospital Number 2 in St. Louis. In 1933, disturbed by the lack of training opportunities for African Americans in radiologic technology, he started the first school for their training. In the late 1930s he established one of the first approved radiology residencies for African Americans. In 1935 he was listed by the Council on Medical Education and Hospitals of the AMA as a physician specializing in radiology.[20] These lists preceded the organization of the ABR in 1934. When Dr. Allen took and passed his ABR examination in 1935 he had to ride the freight elevator to his radiotherapy exams at the Jefferson Hotel in St. Louis.[21] He was the first African American to join the ACR in 1940, became a fellow of the college in 1945, and received the gold medal in 1974 (Fig. 10.4). In 1949 he was a founder of the radiology section of the NMA. He was professor of radiology at St. Louis and Washington universities. He volunteered for the armed services and reached the rank of Lt. Colonel in the United States Army during World War II.[22] He was a consultant to the secretary of the army, and served on multiple committees of the ACR. By 1960 he had published more than thirty scientific articles in *Radiology* and *Radiation Oncology*, the earliest of which were "Radiation Therapy in Carcinoma of the Cervix" in 1935 and "Advanced X-ray Therapy" in 1937.[23]

1933—Ulysses G. Dailey was a surgeon who contributed to oncology when he wrote extensively on the treatment of cancer. In 1933 he published an article titled "Treatment of Cancer of the Breast," (*J.N.M.A.*) in which he mentioned irradiation.[24]

1934—J. R. Cuff, who published extensively on the pathology of tumors, was a pathologist at Meharry Medical College who contributed to oncology. He discussed the use of X-rays and radium in his article "Subperiosteal Osteogenic Sarcoma," (*J.N.M.A.*).[25]

In 1934 a few nonblack institutions had at least partially trained some African American physicians in roentgenology. Among these were the Sinai, New York; Columbia University, New York; Massachusetts General, Boston; Rush Medical School, Chicago; Cook County Hospital, Chicago; and Bellevue Hospital, New York.[26] However, it is necessary to recognize the tremendous task of preparing and training African American physicians that fell to two medical schools: Howard University Medical School in Washington, D.C., and Meharry Medical College in Nashville, Tennessee.

Fig. 10.4 Allen received the gold medal of the American College of Radiology in 1974. (Courtesy of the Center for the American History of Radiology, Reston, Va.)

It can be said, with no great exaggeration, that African Americans as a group owe their very health and survival to the men and women who were teachers at and graduates from these institutions. Other than the few African American radiologists–radiation oncologists trained at a few nonblack institutions, the burden of training physicians in this specialty field fell to the African American hospitals, whose staffs were routinely excluded from academic hospitals and other large, well-funded institutions. This made the task of training specialists in all areas especially difficult. This exclusion from mainstream medicine left an entire race at risk for poor health care.

1934—Ulysses G. Dailey published an "Outline of the Present Status of the Surgical Treatment of Hyperthyroidism," (*J.N.M.A.*).[27]

1936—W. B. Stephens presented a description of the technique of X-ray therapy in his article, "The Present Status of the Diagnosis and Treatment of Carcinoma of the Uterus" (*J.N.M.A.*).[28]

1938—Charles H. Kelley, Jr. (Fig. 10.5) taught and practiced radiation therapy at Howard University from this year until his death in 1956. He became a diplomate of the ABR, 1939.[29]

1939—John W. Lawlah (Fig. 10.6) followed Dr. Kelly in teaching and practicing radiation therapy at Howard University. He became a diplomate of the ABR, 1939.[30]

1940—Russell Minton (Fig. 10.7) published "The Possibilities of X-Ray Therapy in the Treatment of Cancer" (*J.N.M.A.*).[31]

1940—W. E. Allen and H. E. Thornell published "A Radiologist's View on the Treatment of Carcinoma of the Breast" (*J.N.M.A.*).[32]

1940—J. R. Cuff published "The Mammary Gland in 702 Autopsy and 9220 Surgical Specimens" (*J.N.M.A.*), and discussed pre- and postoperative X-ray therapy.[33]

1941—William H. Cargill published "Carcinoma of the Cervix: Diagnosis and Treatment" (*J.N.M.A.*).[34]

1942—Lawrence D. Scott (Fig. 10.8) was coauthor on "Observations on the Results of Combined Fever and X-ray Therapy in the Treatment of

Fig. 10.5 Charles Henry Kelley, Jr. (1901–1956) (Courtesy of Alan E. Oestreich, M.D.)

Fig. 10.6 John W. Lawlah (1904–1976) (Courtesy of the Center for the American History of Radiology, Reston, Va.)

Fig. 10.7 Russel F. Minton (b. 1900) (Courtesy of Alan E. Oestreich, M.D.)

Fig. 10.8 Lawrence D. Scott (1907–1980) (Courtesy of Alan E. Oestreich, M.D.)

Malignancy" (*J. Southern Med. Assn.*) and later "Summary of the Results of Combined Fever and X-ray Therapy in the Treatment of Hopeless Malignancies" (1944, *J.N.M.A.*), and "The Histologic and Clinical Response of Human Cancer to Irradiation" (1946, *J.N.M.A.*).[35]

1945—G. Jack Tarleton (Fig. 10.9) received his M.D. from Meharry Medical College. Other achievements include ABR, 1949; fellow, ACR, 1974; chairman, Meharry Department of Radiology, 1949; professor, radiology at Meharry Medical College, 1952. In 1953 he published "Radiosensitivity of Free Sarcoma: 37 Cells Grown in Peritoneal Exudate of the Mouse *In Vivo*," *Federation Proceedings*; in 1954, "Effects of Fractionated Total Body Irradiation on Sarcoma Cells in the Peritoneal Exudate of the Mouse," *Federation Proceedings*; and in 1955 "*In Vitro* Sensitivity to X-rays of Free Tumor Cells in the Peritoneal Exudate of the Mouse," *Federation Proceedings*, as well as other articles in the *Proceedings of the Society for Experimental Biology and Medicine* (1953), *Federation Proceedings* (1956), and *J.N.M.A.* (1958).[36]

1950—Charles W. Thompson published "Carcinoma of the Lip" (*J.N.M.A.*).[37]

During the period between 1950 and 1970 there was gradual improvement in the acceptance of African Americans into radiology and the radiation field. This was during a time of intense civil rights activity by civilian and governmental groups, and it became more difficult to practice policies of exclusion. However, the improvement was due mostly to the fact that more hospital internships and residency positions were available than graduates from medical schools to fill them. This is especially true in radiology and radiation oncology. Whatever the reason, it was during this time that a few African Americans were given the opportunity to demonstrate that their abilities were equal to those of anyone else. The competency of the African American physician was recog-

Fig. 10.9 Gadson Jack Tarleton, Jr. (b. 1920), shown here instructing Joseph D. Hopkins, M.D. (Courtesy of Alan E. Oestreich, M.D.)

Fig. 10.10 Leslie L. Alexander (b. 1917) (Courtesy of the Center for the American History of Radiology, Reston, Va.)

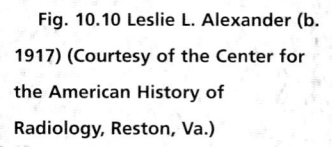

Fig. 10.11 Harold Perry (b. 1924) (Courtesy of Alan E. Oestreich, M.D.)

abstracts.[38] In 1970 Dr. Alexander became the first African American to chair the department of radiology at Down State Medical School, thus becoming the first professor and the first African American to head a department in a white medical school. From 1980 to 1985, he was a member of the Board of Chancellors of the ACR.

1957—Harold Perry (Fig. 10.11) received his M.D. from Howard University in 1948. He became a captain in the United States Air Force, 1953; Diplomate of the ABR, 1955; associate professor, University of Cincinnati College of Medicine, 1963; clinical professor of radiation oncology, Wayne State University, 1982. Today he is chairman of the department of radiation oncology at Sinai Hospital, a position he has held since 1981. He has been president of the Michigan Radiological Society (1977–1978). He has researched three-dimensional treatment planning and electron beam therapy, and has published many articles and abstracts.[39]

1962—Carl M. Mansfield (Fig. 10.12) received his M.D. from Howard University in 1956 and his radiology certification in 1962, and became a fellow of the ABR in 1975. He passed the board examination in nuclear medicine in 1972 and became a fellow of the American College of Nuclear Medicine in 1990. He began to practice radiation oncology in 1964 and in 1974 became

nized and contributed to a decrease in exclusions. They were accepted to radiological societies and began to serve on various committees and commissions within those societies. Thus, the number of African American radiation oncologists has increased from just a few to more than forty, as illustrated in Table 10.I. Many have presented excellent papers and published in peer-reviewed literature.

1956—Leslie L. Alexander (Fig. 10.10) received his M.D. from Howard University, 1952; became a professor at Downstate Medical Center, 1969; director of radiation oncology, Northshore University Hospital, 1970. He has published more than 160 papers and

Fig. 10.12 Carl M. Mansfield (b. 1928) (Courtesy of the Center for the American History of Radiology, Reston, Va.)

the first African American professor at Jefferson Medical College. During this time he did the original research on continuous monitoring of the temperature patterns of breast cancer and the use of nuclear medicine scanning in determining prognosis of liver metastases. In 1976 he became chairman of the department of radiation oncology and therefore the first African American to chair a department at the University of Kansas. While there, he pioneered and was the leading advocate for the use of the perioperative implantation of iridium-192 for the treatment of early breast cancer. He did clinical research with Dr. Carol Fabian on the use of radiation in the treatment of advanced Hodgkin's Disease. He established the first school of radiation oncology technology and the first radiation oncology residency in the state of Kansas. In 1983 he became the first African American to chair a department at the Jefferson Medical College when he was appointed chairman of the department of radiation oncology and nuclear medicine. In 1989 he became president of the American Radium Society, thus becoming the first African American to head a major national radiological society. Today he is the associate director, Division of Cancer Treatment, at the National Institutes of Health in Bethesda, Maryland[40].

Table 10.I

African American Radiation Oncologists*

Olubunmi K. Abayomi	James Fred Littles	George A. Alexander
Carl M. Mansfield	Leslie L. Alexander	William Mansfield
Jean-Philippe Austin	John B. McDay	Darryl R. Barton
Gwendolyn H. Parker	Barbara Binkley	Harold Perry
Raleigh J. Bouldware	Lori Pierce	Jo Ann Collier-Manning
Erich G. Randolph	Mark Cooper	Pamela Randolph
Henry E. Cotman	Mack Roach	William F. Demas
Jerrold P. Saxton	Jackie Dunmore-Griffith	Troy Scroggins
Karen Godette	Joseph Simpson	Alfred L. Goldson
Glenda R. Smith	Michele Y. Halyard	John M. Smyles
Bernard Harris	Sharon A. Spencer	William B. Jackson
Oscar E. Streeter	Maria Jacobs	Doris Taylor
Max S. Laguerre	Oneita Taylor	Jerome Landry
Brian W. Weaver	Judith Lightsey	Jennifer Webb-Holt
Brian Fuller		

*There is a risk in listing names, because there are certainly some of whom I am unaware or have forgotten. For this I humbly apologize. I mean only for the list to show how there has been a wonderful improvement in the opportunities for training and employment in the field of radiation oncology. (This list is courtesy of Dr. Erich Randolph.)

1977—Alfred L. Goldson (Fig. 10.13) received his M.D. from Howard University in 1972 and has served as chairman of the department of radiation therapy, Howard University School of Medicine, 1979; and professor of radiation therapy, 1984. He has performed pioneering research in intraoperative irradiation and has worked in Tanzania and Liberia on cancer programs.[41]

1978—Joseph R. Simpson received his Ph.D. from the University of Chicago in 1967 and his M.D. from Harvard Medical School in 1973, passing the medical boards in internal medicine in 1977 and radiology in 1978. Currently he is an associate professor of radiation oncology at Mallinckrodt Institute of Radiology, Washington University School of Medicine.[42]

1987—Arthur T. Porter received his M.D. from Cambridge University in 1978 and has served as chairman of the department of radiation oncology at Wayne State University School of Medicine since 1991. He has done much of the pioneer work on strontium-89 for treatment of bone metastases.[43]

During the period between 1971 and the 1990s there has been an increase in the number of radiologists and radiation oncologists trained in virtually all of the academic centers and teaching hospitals. More attention is paid to qualifications than to ethnic background. This does not mean that all of the barriers have disappeared; it is only that the barriers are less high and less apparent. I believe that Dr. William E. Allen was correct when he wrote in the 1970s, "The future of Black radiologists is inextricably tied to the future of Civil Rights in this country. Progress in Civil Rights will mean progress for Black radiologists both in number and quality."[44] I believe also that the future of African Americans is bound to the fact that they have demonstrated and continue to demonstrate equal competence. The future is promising.

It is also possible for many African Americans to tell of their personal experience in breaking through what was a seemingly impregnable wall to attain success in the field of radiation oncology. Many who tell the stories of their success in this effort note that they did not do it alone. Often it was the support and acceptance of non-African Americans that made a high level of achievement possible. In this short historical review, many names have emerged from personal talks and from the literature and clearly are representative of many others about whom I did not hear. One early contributor was George Pfahler. Some of the most recent that I have known are Juan del Regato, Simon Kramer, Philip Hodes, Luther Brady, John Curry, Paul Fullagar, and many others.

Fig. 10.13 Alfred L. Goldson (b. 1946) (Courtesy of Alan E. Oestreich, M.D.)

Fig. 10.14 Alan E. Oestreich (b. 1942) (Courtesy of Alan E. Oestreich, M.D.)

Fig. 10.15 Ebrahim Ashayeri (b. 1945) (Courtesy of Alan E. Oestreich, M.D.)

Fig. 10.16 Ulrich Henschke (1914–1980) (Courtesy of the Center for the American History of Radiology, Reston, Va.)

There have been individuals of many ethnic backgrounds who gave of their time and talent to work at predominantly black institutions or to work in African American organizations and societies: Drs. Alan Oestreich (Fig. 10.14), Ebrahim Ashayeri (Fig. 10.15), and Ulrich Henschke (Fig. 10.16). Alan Oestreich has worked diligently in the radiology section of the NMA and presently serves as chairman of that section. He has recently published a centennial history of African Americans in radiology.[45] Dr. Ashayeri has been a member of the department of radiotherapy at Howard University for more than fifteen years. I must make special mention of Ulrich Henschke who devoted so much time and effort as chairman of the department of radiation oncology at Howard University. He trained a number of African American radiation oncologists. At the time of his fatal accident in 1980, he was in Africa working to develop radiation oncology in Tanzania.

This chapter was written for and is dedicated to those many African American physicians who were able to achieve in such adverse circumstances. They were able to train and be trained so that they could bring to their people the high levels of specialty health care available to other Americans. There are probably many accomplishments that I have not included, but there was no intent to be exclusive. This has been an attempt to give examples of the experiences of a people.

▶ ▶ ▶ ▶ ▶ ▶

REFERENCES

I wish to thank Dr. Harold Perry for his assistance in providing names of African American radiation oncologists.

1. Allen, W.E., "History of Black Radiologists." Unpublished paper. In the Grigg Collection, Center for the American History of Radiology, Reston, Va.
2. Grigg, E.R.N. *The Trail of the Invisible Light* (Springfield, Ill.: Charles Thomas, 1956):580-583.
3. Bousfield, M.O., "An Account of Physicians of Color in the United States," *Bulletin of the History of Medicine* 17 (1945):61-84.
4. Morais, H.M.M., "The History of the Negro in Medicine," in *The International Library of Negro Life and History* (New York: Publishers Company, Inc., 1967): 7.
5. Curtis, J.L. *Blacks, Medical Schools, and Societies*. Ann Arbor: University of Michigan Press, 1971.
6. Ibid.
7. Bousfield, "An Account."
8. Ibid.
9. Curtis, *Blacks, Medical Schools, and Societies*.
10. Allen, "History"; and Grigg, *Trail*.

11 Brown, G.I., "The Treatment of Fibroid Tumor of the Uterus with Radium," *J.N.M.A.* 10 (1918):110-112.
12 Willis, F.W., "Carcinoma of the Breast: Preoperative and Postoperative X-ray Treatment," *J.N.M.A.* 13 (1921):241-242.
13 Bainbridge, W.S., "Multiple Pathology and the Cancer Problem," *J.N.M.A.* 15 (1923):16-20.
14 Pfahler, G.E.; Klauder, J.V.; and Martin, J.L., "Experimental Studies on Combined Effects on Roentgen Rays and Ultraviolet Rays," *Am. J. Roent.* 16 (1926)150-154.
15 Holton, J.B., "History of the Radiological Section of the National Medical Association," presented at the annual meeting of the National Medical Association for the fifth W.E. Allen Lecture, 1982.
16 Green, H.M., "Cancer: A Brief Study in, with Special Reference to its Surgical Treatment," *J.N.M.A.* 19 (1927):117-119.
17 Hurst, B.P., "X-ray Treatment of Hyperthyroidism," *J.N.M.A.* 21 (1929):100-102.
18 Harrison, C., "Cancer," *J.N.M.A.* 21 (1929):107-110.
19 Holton, "History."
20 Grigg, *Trail*.
21 Holton, "History."
22 Curtis, *Blacks, Medical Schools, and Societies*.
23 Allen W.E., "Radiation Therapy in Carcinoma of the Cervix," *J.N.M.A.* 27 (1935):1-3; and "Advanced X-ray Therapy," *J.N.M.A.* 29 (1937):98-102.
24 Dailey, U.G., "Treatment of Cancer of the Breast," *J.N.M.A.* 25 (1933):87-89.
25 Cuff, J.R., "Subperiosteal Osteogenic Sarcoma," *J.N.M.A.* 26 (1934):102-110.
26 Holton, "History."
27 Dailey, U.G., "Outline of the Present Status of the Surgical Treatment of Hyperthyroidism," *J.N.M.A.* 26 (1934):119-120.
28 Stephens, W.B., "The Present Status of the Diagnosis and Treatment of Carcinoma of the Uterus," *J.N.M.A.* 28 (1936):115-118.
29 Holton, "History."
30 Ibid.
31 Minton, R.F., "The Possibilities of X-ray Therapy in the Treatment of Cancer," *J.N.M.A.* 32 (1940):162-168.
32 Allen, W.E., and Thornell, H.E., "A Radiologist's View on the Treatment of Carcinoma of the Breast," *J.N.M.A.* 32 (1940):204-206.
33 Cuff, J.R., and Quinland, W.S., "The Mammary Gland in 702 Autopsy and 9220 Surgical Specimens," *J.N.M.A.* 1940; 32:231-234.
34 Cargill, W.H., "Carcinoma of the Cervix: Diagnosis and Treatment," *J.N.M.A.* 33 (1941):61-63.
35 Shoulders, H.S.; Turner, E.L.; and Scott, L.D., "Observation on the Results of Combined Fever and X-ray Therapy in the Treatment of Malignancy," *J. Southern Med. Assoc.* 35 (1942):966-70; Scott, L.D., "Summary of the Results of Combined Fever and X-ray Therapy in the Treatment of Hopeless Malignancies," *J.N.M.A.* 34 (1944):184-186; and Quinland, W.S., and Scott, L.D., "The Histologic and Clinical Response of Human Cancer to Irradiation: Citation of Five Cases," *J.N.M.A.* 38 (1946):171-178.
36 Among Dr. Tarleton's publications are: Goldie, H.; Tarleton, G.J.; and Hahn, P.F., "Effect of Pretreatment with Cysteine on Survival of Mice Exposed to External and Internal Irradiation," *Proceed. Soc. Experi. Biol. Med.* 77 (1951); Tarleton, G.J.; Goldie, H.; Jones, A.M.; and Moore, G.A., "Radiosensitivity of Free Sarcoma: 37 Cells Grown in Peritoneal Exudate of the Mouse *In Vivo*," *Fed. Proceed.* 12 (1953):404; Goldie, H.; Tarleton, G.J.; Jeffries, B.; and Hahn, P.F., "Effect of Repeated Doses of External and Internal Radiation of the Structure of the Spleen," *Proceed. Soc. Experi. Biol. Med.* 82 (1953); Tarleton, G.J.; Goldie, H.; Brantley, M.; and Simpson, M., "Effects of Fractionated Total Body Irradiation on Sarcoma Cells in the Peritoneal Exudate of the Mouse," *Fed. Proceed.* 13 (1954); Tarleton, G.J.; Goldie, H.; and Royal, W., "*In Vitro* Sensitivity to X-rays of Free Tumor Cells in the Peritoneal Exudate of the Mouse," *Fed. Proceed.* 14 (1955):479; Goldie, H.; Tarleton, G.J.; Gordon, J.G.; and Geiger, G.L., "Attenuation by Histamine of Lethal Irradiation Effect in Mice and Its Potentiation by Benadryl," *Fed. Proceed.* 15 (1956):589; and Tarleton, G.J., "Some Human Effects of Radiation Exposure," *J.N.M.A.* 50 (1958):442-448.
37 Thompson, C.W., "Carcinoma of the Lip," *J.N.M.A.* 42 (1950):152-158.
38 Among Dr. Alexander's many publications are Alexander, L.L.; Causing, J.; Schwinger, H.N.; and Li, M.C., "Bronchogenic Carcinoma - A Comparative Study of the Palliative Effects of Radiation Therapy, Radiation Therapy Plus Nitrogen Mustard, and Radiation Therapy Plus Amethopterin and Actinomycin D in Combination," *Am. J. Roent.*.87 (1962):375-384; Alexander, L.L., and Beneventano, T.C., "Treatment of Antral Carcinoma with Radioactive Iridium," *J.N.M.A.* 56 (1964):139-142; Friedman, A.B.; Benninghoff, D.L.; Alexander, L.L.; and Aron, B.S., "Total Abdominal Irradiation Using Cobalt-60 Moving Strip Technique," *Am. J. Roent.* 108 (1970):172-177; Alexander, L.L.; Li, M.C.; Cordice, J.W.V.; et al., "Anaplastic Teratoid Carcinoma of the Mediastinum, Pericardium and Lung: An Example of Successful Cooperative Management," *J.N.M.A.* 62 (1970):327-330; Alexander, L.L.; Medina, A.; Benninghoff, D.L.; et al., "Radium Management of Tumors of the Penis," *N.Y. State J. Med.* 71 (1971):1946-1950; Alexander, L.L., and Atkins, N.M.L., "Burkitt's Tumor," *J.N.M.A.* 65 (1973):386-390.
39 Among Dr. Perry's publications are Perry, H., and Chu, F.C.H., "Value of Radiotherapy in the

Management of Soft Tissue Sarcomas," *Acta Un. Int. Cancer* 16 (1960):1452-1453; Perry, H., and Chu, F.C.H., "Radiation Therapy in the Palliative Management of Soft Tissue Sarcoma," *Cancer* 15 (1962):179-183; Perry, H.; Tsein, K.D.; Nickson, J.J.; and Laughlin, J.S., "Treatment Planning in the Therapeutic Application of High Energy Electrons to Head and Neck Cases," *Am. J. Roent.* 88 (1962):250-261; Berry, H.K.; Saenger, E.L.; Perry, H.; et al., "Deoxycytidine in Urine of Humans after Whole Body Irradiation," *Science* 142 (1963):396-397; and Mantel, J.; Perry, H.; and Weikam, J.J., "Automatic Variation of Field Size and Dose Rate in Rotation Therapy," *Int. J. Rad. Onc. Biol. Phys.* 2 (1977):697-704.

40   Among Dr. Mansfield's publications are: Mansfield, C.M.; Kramer, S.; Southard, M.E.; and Mandell, G., "Prognosis in Patients with Metastatic Liver Disease Diagnosed by Liver Scan," *Radiology* 91 (1968):673-678; Mansfield, C.M.; Galkin, B.; Chow, M.; and Suntharalingam, N., "Three-Dimensional Dose Distribution in Cobalt-60 Therapy of the Head and Neck," *Radiology* 93 (1969):401-404; Mansfield, C.M.; Curley, R.F.; Wallace, J.D.; et al., "A Comparison of the Temperature Curves Recorded over the Normal and Abnormal Breasts," *Radiology* 94 (1970):697; Mansfield, C.M., and Suntharalingam, N., "Dose Distribution for Cobalt-60 Tangential Irradiation of the Breast and Chest Wall," *Acta. Radiol.* 2 (1973):40-46; Mansfield, C.M.; Suntharalingam, M.; and Chow, M., "Experimental Verification of a Method for Varying the Effective Angle of Wedge Filters," *Am. J. Roent.* 120 (1974):699-702; Mansfield, C.M., and Jewell, W.R., "Intraoperative Interstitial Implantation of Iridium-192 in the Breast," *Radiology* 150 (1984):600; and Mansfield, C.M.; Fabian, C.; Jones, S.; et al., "Comparison of Lymphangiography and Computed Tomography Scanning in Evaluating Abdominal Disease in Stages III and IV Hodgkin's Disease: A Southwest Oncology Group Study," *Cancer* 66 (1990):2295-2299.

41   Among Dr. Goldson's publications are: Ellis, F., and Goldson, A.L., "Once Weekly Treatment," *Int. J. Rad. Onc. Biol. Phys.* 2 (1977):537-548; Goldson, A.L., "Preliminary Clinical Experience with Intraoperative Radiotherapy," *J.N.M.A.* 70 (1978):493-495; Goldson, A.L.; Ashayeri, E.; Espinoza, M.C.; et al., "Single High Dose Intraoperative Electrons for Advanced Stage Pancreatic Cancer: Phase I Pilot Study," *Int. J. Rad. Onc. Biol. Phys.* 7 (1981):869-874; Goldson, A.L.; Smyles, J.M.; Ashayeri, E.; et al., "Simultaneous Intraoperative Radiation Therapy and Intraoperative Interstitial Hyperthermia for Unresectable Adenocarcinoma of the Pancreas," *Endocurie/Hypertherm. Oncol.* 3 (1987):201-208; and Goldson, A.L., "Intraoperative Radiotherapy for Pancreatic Cancer– Requiem or Revival? (Editorial)" *Int. J. Rad. Onc. Biol. Phys.* 21 (1991):1389.

42   Among Dr. Simpson's publications are: Simpson, J.R.; Nagle, W.A.; Bick, M.D.; and Belli, J.A., "Molecular Nature of Mammalian Cell DNA in Alkaline Sucrose Gradients," *Proc. Nat. Acad. Sci.* 70 (1973):3660-3664; Simpson, J.R.; Perez, C.A.; Phillips, T.L.; et al., "Large Fraction Radiotherapy Plus Misonidazole in the Treatment of Advanced Lung Cancer," *Int. J. Rad. Onc. Bio. Phys.* 8 (1982):303-308; Simpson, J.R.; Francis, M.E.; Perez-Tamayo, R.; et al., "Palliative Radiotherapy for Inoperable Carcinoma of the Lung: Final Report of an RTOG Multi-Institutional Trial," *Int. J. Rad. Onc. Biol. Phys.* 11 (1985):751-758; Simpson, J.R.; Bauer, M.; Wasserman, T.H.; et al., "Large Fraction Irradiation with or without Misonidazole in Advanced Non-Oat Cell Carcinoma of the Lung: A Phase III Randomized Trial of the RTOG," *Int. J. Rad. Oncol. Biol. Phys.* 13 (1986):861-867; Simpson, J.R.; Purdy, J.A.; Manolis, J.W.; et al., "Three-Dimensional Treatment Planning Considerations for Prostate Cancer," *Int. J. Radia. Oncol. Biol. Phys.* 21(1991):243-252.

43   Amon Dr. Porter's publications are: Porter, A.T., "Syndrome of Inappropriate Antidiuretic Hormone Secretion during cis-Dichlorodiamine Platinum Therapy in Ovarian Carcinoma," *Gyn. Oncol.* 21 (1985):103-105; Porter, A.T., "Is Radiotherapy of Use in Stage-C Prostate Cancer?" *Prog. Clin. Biol. Res.* 305 (1989):211-218; Porter, A.T., and Mertens, W., "Strontium-89 in the Treatment of Metastatic Prostate Cancer," *Can. J. Oncol.* 1 (1991):1-11; Porter, A.T., "Strontium-89 in Metastatic Cancer," *Nucl. Gen. Bull.* 14.2(1992):67-70; and Porter, A.T.; McEwan, A.J.B.; Power, J.E.; et al., "Results of a Randomized Phase-III Trial to Evaluate the Efficacy of Strontium-89 Adjuvant to Local Field External Beam Irradiation in the Management of Endocrine Resistant Metastatic Prostate Cancer," *Int. J. Rad. Oncol. Biol. Phys.* 25 (1993):805-813.

44   Allen, "History."

45   Oestreich, A.E. *A Centennial History of African Americans in Radiology* (Takoma Park, Md.: Section on Radiology of the NMA: 1996)

The deaths of young women who painted dials with radium made the public aware of radiation dangers in the 1920s and 1930s. (Courtesy of Kathren, R.L., Radioactivity in the Environment: Sources, Distribution, and Surveillance. New York: Harwood Academic Press, 1984, with permission)

CHAPTER ELEVEN

Radiomedical Fraud and Popular Perceptions of Radiation

Roger M. Macklis, M.D.

▶ ▶ ▶ ▶ ▶ ▶ ▶ ▶ ▶ ▶ ▶ ▶

In 1918 the famous physician Sir William Osler searched for the perfect metaphor to describe the mystery, vitality, and interest of the human condition: "May not man be the RADIUM of the universe? For us, he is a very potent creature, full of interest, the spark whose mundane story we are only beginning to unravel."[1] Today, it is rare to hear a good word in the popular press about ionizing radiation—much less to see it used as a symbol for all that is good and engaging in humankind.

How did things get so bad? From the vantage point of the 1990s it is easy to imagine that radioactivity and radiation medicine have always endured negative public images based on deep-seated fears of technology and associations with cataclysmic wars and reckless, man made catastrophes. Yet, this gloomy picture is far from accurate. For more than a quarter of a century, radioactivity and nuclear science enjoyed the same sort of breathless reportage and media lionization now reserved for fields like astrophysics and molecular biology.

As physics historian Spencer Weart noted in his book *Nuclear Fear*, radioactivity seemed at first to offer the possibility of a shining utopian society basking in the limitless power, everlasting warmth, and inexhaustible raw materials generated by a ceaseless stream of silent radioactive emissions.[2] Almost overnight, flickering skeletal "cathodographic" images produced by evacuated Crookes tubes truly transformed the art of medical diagnosis. Marie Curie's discovery of radium several years later appeared to some well-known biologists and theologians to provide not only limitless medical promise but a good candidate for the elemental essence of life.[3] Only when the news reports and images of nuclear devastation and radiation-associated deaths at Hiroshima and Nagasaki began to filter back to Western society did the public finally begin to lose faith in this unbottled genie. This chapter will trace the history of the popular conceptions of radioactivity and radiation medicine during the first half of this century and will consider some of the social forces that contributed to the current negative public perception of radiation.

The earliest popular descriptions of the series of experiments conducted by Röntgen in the fall of 1895 described his discovery of "a new kind of light"

WHAT WE MAY EXPECT.
"Say, Dick, lend me that five dollar bill."
"Haven't any, old man."
"Yes, you have; this photograph I just took of you shows it in your left-hand waistcoat pocket."

Fig.11.1 Only weeks after the announcement of Röntgen's discovery the X ray had entered public awareness in cartoons, songs, and poems. This cartoon from the 26 March 1896 issue of the British magazine *Life* reflects the widespread notion that the new rays might become a part of everyday life. (Courtesy of the Center for the American History of Radiology, Reston, Va.)

capable of penetrating wood and flesh.[4] Within a week of Röntgen's first publication on 28 December 1895, newspapers like the *Vienna Presse* were profiling the discovery on their front pages.[5] The *New York Times* picked up the story on 16 January 1896, noting that, "Men of science in this city are awaiting with the utmost impatience the arrival of European technical journals which will give them the full particulars of Professor Routgen's [sic] great discovery."[6] By February famous American inventors, including Thomas Edison, were working to exploit the phenomenon of roentgen rays for industrial use and practical technologic advances. By 25 March Edison had contracted to have the Thomas A. Edison X-ray Kit mass-produced for home use.[7] Although his interest in X-ray technology continued for several years, he was never able to make good on his publicly stated goal of being the first to produce a cathodographic image of the working human brain. Ultimately he turned away from X-ray research, in part because of the clear dangers to his assistants, but in large part because it soon became clear that the X ray would not become a fixture in every home medicine closet.

The first field to wholeheartedly embrace the new radiographic technology was not medicine but photography. For both professionals and hobbyists the prospect of producing "skiagraphic" images of hidden or concealed objects seemed tantalizing and of eminently commercial potential. Writing cautiously in the journal *Photography*, an editor noted that:

> If that which is stated to be true is so, it will be possible for a person fully clothed to "sit" for a photographer who will produce from him a photograph showing his skeleton only...who knows but that the latest fashionable craze will yet be skeleton photographs of yourself and friends—hardly attractive to present ideas, but we may get used to it in time.[8]

X-actly So!

The Roentgen Rays, the Roentgen Rays,
What is this craze?
The town's ablaze
With the new phase
Of X-ray's ways.

I'm full of daze,
Shock and amaze;
For nowadays
I hear they'll gaze
Thro' cloak and gown—and even stays,
These naughty, naughty Roentgen Rays.

Fig. 11.2 The X ray made its appearance in poems and songs in early 1896—part of the "X-ray mania" that followed Röntgen's announcement. (Courtesy of the Center for the American History of Radiology, Reston, Va.)

General interest magazines in these last years of the nineteenth century alternated between astonishment and whimsy when reporting on the developing field of radiography. Cartoons and poems published in *Life* magazine in early 1896 suggested that the new photography might be useful for detecting money in the pockets of miserly friends or for documenting the cold steel hearts of fickle girlfriends (Figs. 11.1 and 11.2).[9] A London firm advertised X-ray-proof underclothing for modest women afraid of peeping Toms, and theologians pondered the moral implications of a power capable of seeing the previously unseen.[10] In a move described by the *New York Times* as "fascinating and coquettish," a bold New York society lady allowed herself to be X-rayed without her customary corset to produce a cathodograph of her "well-developed ribs."[11] An amateur photography magazine noted the hysteria building around Röntgen's discovery and sought to dispel the concerns of some readers that "armed with this light and a camera, a man might go around town taking snap shots of other people's brains and stealing their ideas."[12]

Attempting to establish some kind of credible context for the new phenomenon, *Century Magazine* published "Symposium on the Roentgen Rays" in May of 1896.[13] This multi-author article attempted to trace the history of Röntgen's discovery from the early work of Crookes and Faraday and included the latest investigations by influential scientists such as Thomas Edison and William James Morton. Such sober assessments did little to stem the rising tide of "X-ray mania." Less than six months after the initial discovery and characterization of X rays by Röntgen, popular magazines were speculating wildly on the role of this new physical phenomenon in the fields of medical diagnosis, jurisprudence, energy production, metallurgy, instrument testing, and entertainment.[14,15]

The second phase of the public's love affair with radioactivity began in February of 1896 in Antoine-Henri Becquerel's laboratory at Paris's École Polytechnique. Becquerel had heard about Röntgen's new rays and wondered whether some naturally luminescent materials might give off similar rays after exposure to sunlight.[16] In testing a series of such elements, he noted that uranium ore appeared capable of exposing wrapped photographic plates even before the ore was primed by exposure to the sun. He suggested that uranium and other "phosphorescent bodies" were emitting a new kind of radiation with characteristics somewhat similar to the artificial roentgen rays.[17] Two and a half years later, Marie and Pierre Curie identified a new element in uranium-rich pitchblende ore that appeared to be far more radioactive than uranium itself. They named this element radium, and by 1901 they had partially purified enough of the substance to begin loaning small samples to other interested scientists. Becquerel borrowed a small tube of this partially purified radium and—despite Pierre Curie's warnings—carried it around in his waistcoat pocket for several hours. The subsequent erythematous skin reaction on his abdomen appeared to be identical to the roentgen ray burns that had already been noted by physicians and amateur radiographers, and the link between the biological effects of external X rays from cathode tubes and internally generated gamma rays from naturally radioactive substances became evident, if not immediately explicable.[18]

The discovery of radium marked a major change in the way that the public perceived the potential societal impact of radiation and radioactivity. Whereas the discovery of X rays was lauded as a major advance in technology, the discovery of radium appeared to be a natural gift from God, the revelation of a primordial secret of the ages, perhaps even the holy spark of life itself. Semiscientific accounts in the lay press and in popular science journals dwelled on the fascinating powers of this new element, including its ability to induce radioactivity in some nonradioactive elements left nearby and its mysterious ability to melt its own weight of ice in an

Fig. 11.3 Radium was soon found to have beneficial effects on both surface and deep-seated cancers and increasingly became identified as a radiant hope—perhaps the source of prolonged or even eternal life—for sufferers with a wide range of maladies. This depiction appeared in the *Los Angeles Examiner* on 28 January 1914.

hour. Journalists vied with one another to create ever more powerful images conveying the power locked within this extremely rare and gently glowing substance. Pierre Curie himself calculated that there might be as much as 50 billion calories in a gram of radium (an overestimate), and popular accounts of the element's characteristics held that a single gram could raise "10,000 tonnes a mile high" or provide light equal to "several million candles."[19,20]

Moreover, these new radioactive elements did not simply decay at a fixed rate. Careful measurements revealed that the rate of radioactive emissions slowed as the elements decayed, a phenomenon dubbed the half-life effect. To many it seemed as if the element itself was attempting to conserve energy in order to continue forever. Amateur inventors hailed the discovery of radium as the first step in the development of perpetual motion machines and envisioned radium-powered ships, planes, and rockets. A farmer proposed mixing radium with chicken-feed so that hens would lay hard-boiled eggs. This blend of high technology and natural vitality created a very potent image that was to capture the public's imagination and enthusiasm for decades (Fig. 11.3).[21]

During the early 1900s the popular press reported advances in understanding radiation and radioactivity alongside their features on such other technological marvels as electricity, internal combustion engines, and Marconi's radio waves. This era saw the rise of the professional, full-time academic scientist as well as the professional inventor—a class of scientific entrepreneurs exemplified by Edison and his invention factory, an applied research institute in New Jersey employing fifty engineers and investigators. Many other well-known inventors became infatuated with early medicinal radioactivity, including Edison's chief scientist William Hammer, physicist Ernest Rutherford, and even Alexander Graham Bell, who wrote in 1903 that "[t]here is no reason why a tiny fragment of radium sealed up in a fine glass tube should not be inserted into the very heart of a cancer...."[22]

Weart has traced the emergence of yet another group of individuals whose efforts were to add greatly to the mass appeal of radioactivity.[23] This was the group of professional science commentators whom he has dubbed "missionaries for science," journalists with some degree of technical training who served as the unofficial translators of scientific discoveries into image-laden common language ready for popular consumption. By the 1890s these commentators reported regularly on technology and its meaning for the American future and were spurred on by the sustained and broad-based public interest in exhibition projects such as the electrically-powered Columbian Exposition of 1893 in Chicago, called the "White City of the Future." New high-resolution telescopes detected lines on the planet Mars, leading the French journalist and encyclopedist of astronomy Camille Flammarion to report in 1892 that Mars appeared to have a colossal system of canals and other evidence of an ancient, technologically advanced super-race. This level of technology appeared unachievable on earth as long as man was dependent on dangerous, pollution-producing sources of energy such as coal. Thus both the United States and Europe snapped to attention when reports of a new, nearly inexhaustible

Fig. 11.4 Radium was widely heralded as a potential boon not only to medicine but to technology, transportation, agriculture, and industry—a miracle indeed. (Reproduced from the Hammer Collection, National Museum of American History, Smithsonian Institution)

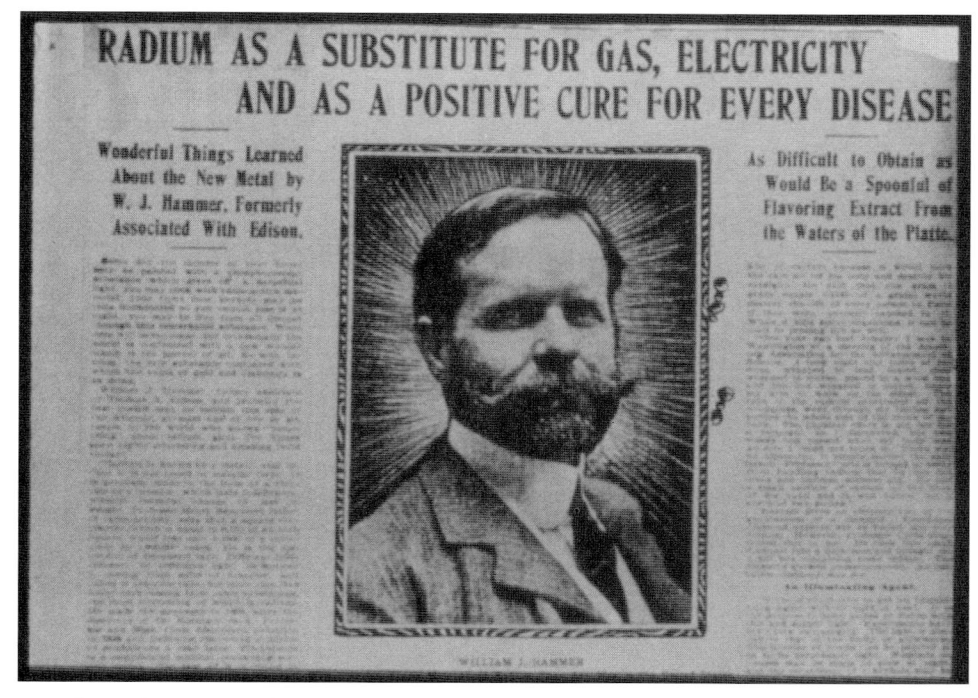

supply of seemingly pollution-free nuclear energy first surfaced in the technical journals at the turn of the century.

From both a medical and a technologic point of view, radium was viewed by some as being the next essential step in the development of a new, more egalitarian society. Journalists like Waldemar Kaempffert, science editor for the New York Times, saw the new science of "atomic" energy as eventually transforming society, allowing for the creation of complete new cities and towns with plentiful, clean energy, healthy occupants, and radioactive rocket-based transportation systems (Fig. 11.4). Moreover, the nuclear "transmutation" of elements such as thorium into radium, first elucidated by Rutherford and Soddy in 1901, seemed to recall medieval quests for effective methods of alchemy and elemental transport, subjects that had never lost their allure for the popular technology audience. As Soddy wrote in 1908 in a popular book titled The Interpretation of Radium, "A race which could transmute matter would have little need to earn its bread by the sweat of its brow. Such a race could transform the desert continent, thaw the frozen poles, and make the whole world one smiling Garden of Eden."[24]

Perhaps the most interesting of the early popular images associated with radiation and radioactivity is the set of associations surrounding the use of radium as a kind of physiological catalyst, a vital force related to the primordial spark of life.[25] Some historians have suggested that the association of radioactivity and life-force may date back to an early mythic consciousness about the centrality of the sun and radiant energy in all vital processes.[26] These rays of life embodied the popular notion of a sort of pure energy soul that could invest matter with animate properties. Displays of electricity had fascinated the public in the late nineteenth century, and the idea that the dead could be brought back to life with the proper application of electrical current was commonplace in popular technical literature and science fiction. The growing public awareness of radiation tapped directly into this set of mythic images but with two important distinctions: First, radioactivity appeared to be a much more powerful and mysterious force than electricity, and its effects on human tissues appeared much more difficult to predict. Second, unlike electricity, the occult power in radium was apparently innate, inexhaustible, and infinitely divisible. Even a small speck of

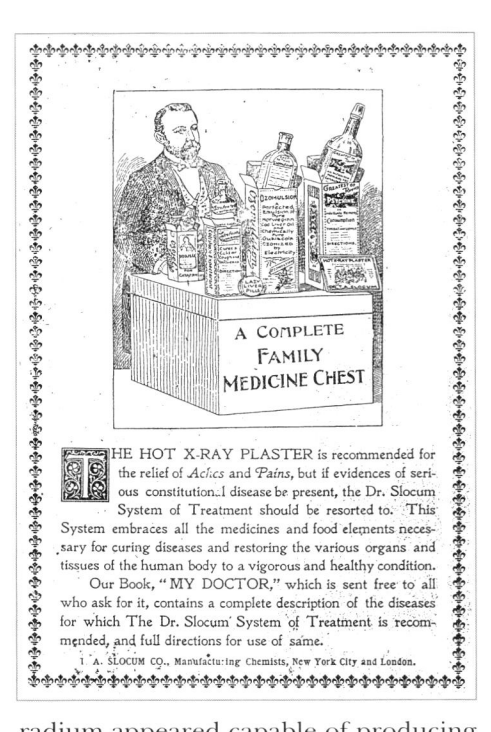

Until 1910 radium was too scarce and expensive to incorporate into such mass-market products.[27] However, when large deposits of uranium were mined in Colorado and Canada, the price dropped from over $500,000 a gram to a relatively conceivable $120,000 a gram, and radium came into wide use in cancer hospitals and dermatologic clinics. At the same time, technical advances in X-ray equipment, especially the invention of the hot cathode tube in 1913 by William Coolidge of the General Electric Company, allowed radiology departments to proliferate and expand their services.[28] Full-time radiologists began to open combined diagnostic and therapeutic practices in many major hospital centers, and academic departments of radiology formed with active teaching, publication, and research programs.

By the beginning of World War I radioactivity and radiologic health sciences were firmly established in medical practice. Marie Curie and a number of other prominent scientists and physicians soon recognized the importance of X-ray images in diagnosing the sort of trauma and foreign body injuries prevalent in wartime, and Curie herself headed up the effort to establish a mobile X-ray service for the French army.

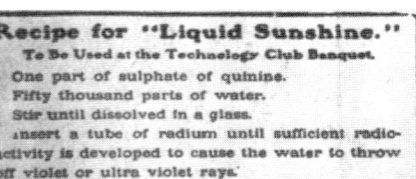

radium appeared capable of producing dramatic effects. This latter observation was of great interest to the burgeoning patent medicine industry, because it suggested the possibility that radioactive compounds in tiny quantities might serve as useful tonics or restorative ingredients when incorporated into salves, pills, and potions (Fig. 11.5 and 11.6).

Fig. 11.5 The notion of marketing the radiant qualities of X rays led some patent medicine manufacturers after 1896 to dub their medicines "X-ray potions." Slocum's 1898 plaster system contained no radiation, but played off the public's interest in the new rays. (Courtesy of the Center for the American History of Radiology, Reston, Va.)

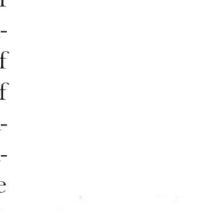

Fig. 11.6 The advent of radium made it possible to put radioactivity into patent medicines and nostrums. In 1904 a group of scientists at the Massachusetts Institute of Technology took advantage of the "solubility" of radium emanations, mixing cocktails made of quinine and water, stirred "with a tube of radium until sufficient radioactivity is developed to cause the water to throw off violet or ultra violet rays." The lights were dimmed, and the scientists socialized in the glow of their "liquid sunshine." (Reproduced from the Hammer Collection, National Museum of American History, Smithsonian Institution).

Fig. 11.7 The inhalation of radium emanation and the ingestion of irradiated water were advertised as beneficial to everything from arteriosclerosis to nervousness. Curiously, while radium implantation and X-ray therapy were making inroads in cancer treatments, radium emanation generally was not directed at malignancies. This 1907 apparatus for inhaling radium emanation could be used in the doctor's office or sent home for self-administration. (Courtesy of the Center for the American History of Radiology, Reston, Va.)

▶

These responsibilities occupied most of her time during the war years, and dramatic pictures of her war service and the utility of field radiography added to the growing luster associated with the benefits radiation promised.

The end of World War I allowed scientists throughout Europe and the United States to refocus their efforts on humanitarian and social concerns. In the field of radiation research, this meant that many scientists and physicians who in wartime had devoted themselves to perfecting devices for X-ray diagnosis of war injuries could now return to their benchtop investigations of the fundamental effects of radioactivity on living organisms. Many of the early radiomedical investigators were physician-tinkerers who became obsessed by the study of these strange, invisible rays that seemed to have such magical effects on living organisms. Marie Curie, returning to her work at the Curie Institute, set as one of the institute's postwar priorities the investigation of the effects of minute, "nontoxic" quantities of radium on cells, animals, and humans. In Great Britain this field was designated "mild radium therapy" to differentiate it from the better known and—by then—well-accepted role of much larger, destructive doses of radium in the treatment of cancer and other neoplasms.[29]

The history of the mild radium therapy movement can be traced back to the homeopathic and physical medicine schools of the nineteenth century. These schools taught that most true healing processes were natural and that tiny (usually infinitesimal) quantities of naturally occurring "stimulatory" compounds (which in large quantities would be toxic) could cure most maladies. Many of these groups also believed in the legendary healing powers of the great European hot springs like Brambach in Germany, Joachimstal in Czechoslovakia, Ischia in Italy, and Sail-les-Bains in France. Devotees established sanitoria on the grounds of many of these hot springs, claiming that the waters were useful in the treatment of various "metabolic" diseases such as melancholy, rheumatism, cretinism, and impotence.

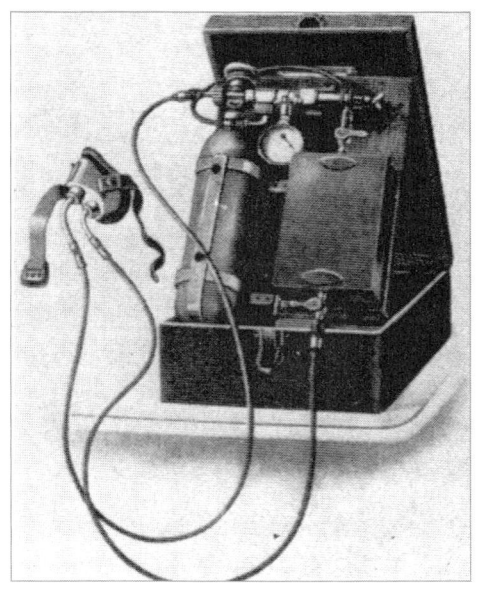

These healing waters posed a mystery, however. They appeared to lose their powers after just a few days when they were removed from the springs and bottled. The great German chemist Justus von Liebig attempted to analyze the waters from the Gastein springs, eventually ascribing their power to a dissolved gas with mysterious and transient electrical effects. In 1903 the discovery was made that the apparent pharmacologic agent dissolved in these waters was radon (radium emanation). After Rutherford's Nobel Prize-winning investigations of alpha particle emissions from radium and radon, the transient healing effects of the hot springs were ascribed to the energetic and short-range alpha particles ejected from the radon nucleus during its short (three-day half life) decay period.[30,31] Radioactivity became identified with the legend of healing waters—in fact, it became the validating and scientific explanation for hydropathic effects—and soon hot springs across the United States and Europe advertised the radiant qualities of their waters (Figs. 11.7 and 11.8).

Although public interest in radioactivity and its healing effects remained high throughout the 1920s, most of the ongoing medical investigations focused on the penetrating beta and gamma emissions of radium and its decay products. The high-energy alpha particles

RADIUM THERAPY

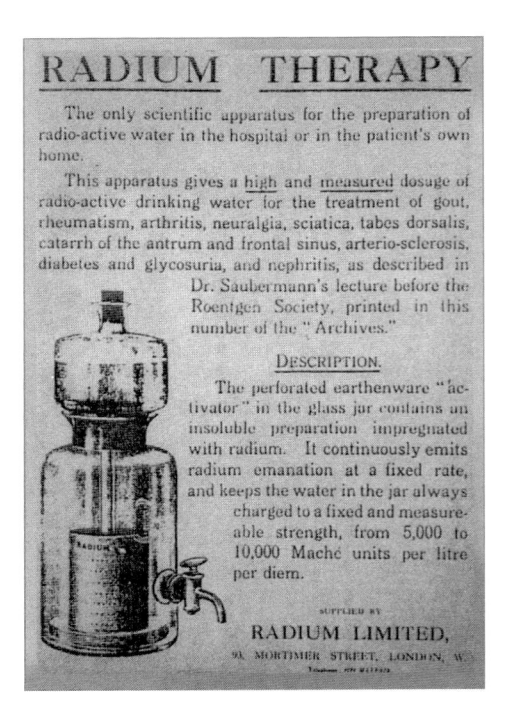

The only scientific apparatus for the preparation of radio-active water in the hospital or in the patient's own home.

This apparatus gives a high and measured dosage of radio-active drinking water for the treatment of gout, rheumatism, arthritis, neuralgia, sciatica, tabes dorsalis, catarrh of the antrum and frontal sinus, arterio-sclerosis, diabetes and glycosuria, and nephritis, as described in Dr. Saubermann's lecture before the Roentgen Society, printed in this number of the "Archives."

DESCRIPTION.

The perforated earthenware "activator" in the glass jar contains an insoluble preparation impregnated with radium. It continuously emits radium emanation at a fixed rate, and keeps the water in the jar always charged to a fixed and measureable strength, from 5,000 to 10,000 Maché units per litre per diem.

SUPPLIED BY
RADIUM LIMITED,
93, MORTIMER STREET, LONDON, W.

ejected at several points in the radium decay scheme were largely ignored or filtered out due to their low penetration and high potential for skin burns. However, these objections did not apply to the use of tiny quantities of radium prepared for oral ingestion, and alpha particles were considered to be the most active class of decay emissions in radiopharmaceuticals prepared for mild radium therapy.

For many devotees the beneficial effects of radioactive water appeared to be the result of energy transfer processes between the alpha particles and metabolically depleted cells.[32] "Could radium," they wondered, "be serving as a kind of metabolic battery charger?" Professor Jacques Loeb's publication in 1912 of *The Mechanistic Conception of Life* reinforced the reductionist philosophy equating complex metabolic processes with sequences of simple energy transfer events. What physiological mechanisms controlled these processes? The answer, it appeared, would be found in the workings of the newly discovered endocrinologic system.

Hence, there was a curious convergence of interest in the newly discovered details of both the endocrinologic system and the radioactive decay systems in the early 1920s. In 1921 Frederick Soddy received the Nobel Prize in chemistry for his work on radioisotopes. In that same year, Banting and Best isolated insulin. In 1923 Banting and Macleod were awarded the Nobel Prize in physiology for their insulin research. It was not clear at the time, however, how the endocrine system actually produced its effects. What was missing, it appeared, was a fundamental biophysical energy transduction mechanism that could account for the apparent energy transfer processes mediated by the endocrine hormones.

To some investigators radioactivity seemed to be the spark required to ignite these major physiologic processes. The German physiologist George Wendt, in his address to the thirteenth International Congress of Physiologists, reported that human leukocytes exposed to low-level radium radiation began migrating toward the radium source and that moribund vitamin-starved rats could be temporarily rejuvenated by exposure to radium.[33] Radium, with its energy-charged alpha particle emissions, truly seemed to be the basis of the fountain of youth legends.

Packaging this mini-miracle for sale seemed an obvious and potentially lucrative next step for those on the fringe of legitimate medicine. Because radium was considered a natural element rather than a drug, it was available over-the-counter in a variety of preparations (in microscopic amounts) and was not regulated by the Food and Drug Administration (FDA), an agency with very limited authority at that time. Radioactive candies, liniments, potions, and creams were widely available by 1915, and pharmacopoeias from the time list dozens of "mildly" radioactive medicinal preparations. Though initially the mild radium therapy movement appears to have been largely confined to Europe, American interest in the medicinal, "catalytic" properties of radium and its decay products surged after double Nobel laureate Marie Curie's whistle-stop railway tour across the United States in 1921. Like the stuff of homeopathic legends, radium appeared to be a substance with two distinct modes of medical efficacy: in large

Fig. 11.8 "Radium drinkers" like this model from the 1920s, usually made of radium-exposed clays and ceramics, were popular in both the United States and Europe. Many can be found today (by using an r-meter) on the shelves of antique and junk shops. (Courtesy of the Center for the American History of Radiology, Reston, Va.)

Fig. 11.9 This radon toothpaste (1920) included a complicated explanation for the benefits of radiation-controlled plaque. Most buyers, however, were attracted by the notion that their teeth would soon be bright enough to glow in the dark. (Courtesy of the Center for the American History of Radiology, Reston, Va.)

quantities, it was destructive and could be used to kill tumors; and in tiny quantities, it appeared to be restorative and could be used to recharge the endocrine system. As one American physician who had worked at the Curie Institute noted, "Having had the luck of studying radium with its discoverer, I wish to say that radium possibilities are innumerable and in the very near future all the progressive physicians will use it, as radium therapy will be one of the most powerful weapons for the elimination of disease and the prolongation of human life."[34]

The proliferation of radium-containing nostrums and home-use products testifies to the powerful public image conveyed by radiation science. Besides radioactive candies and patent medicines produced by companies hoping to cash in on the radium fad, other fraudulent medical companies quickly began producing a full line of radioactive belts, harnesses, and patches designed to rejuvenate aching muscles and flaccid organs. Radioactive toothpaste, hair tonic, contraceptive jelly, eyeglasses, and hearing aids (said to contain a magic ingredient called "hearium") were all widely available in the postwar years (Fig. 11.9).

In 1915 Dr. Sabin von Schocky invented a luminous paint activated by radium-226 and zinc sulphide. Called "Undark," this paint cast an eerie, shimmering glow in the dark.[35] Although Schocky promoted its decorative and artistic uses for the design of self-illuminating toys, costumes, and special effects, the principal commercial use of the material involved the illumination of indicators and dials for watches and machinery. In 1920 more than four million radium-illuminated watches, doll eyes, fishing tools, buttons, and gunsights were produced in the United States. Schocky's U.S. Radium Company employed more than 250 workers during its peak years, many of whom were young women paid by the piece to highlight the numbers on watch faces with the radium paint. Many of these dial painters subsequently were to develop necroses of the jaw from "tipping" the highly radioactive brushes with their tongues so as to be able to paint finer lines on the dials.[36] For the first time reporters began to question self-proclaimed "radioactivity experts" about potential toxicity resulting from exposure to radium and its byproducts.

By the mid-1920s it had become clear that radioactivity and X irradiation were far from harmless. In 1904 Clarence Dally, Thomas Edison's thirty-nine-year-old assistant who was intimately involved in X-ray testing, died of extensive radionecrosis and radiation-induced cancer. Before he died Dally underwent sequential and disfiguring amputations of first his left hand, then most of his right hand, then his left forearm, and finally his right forequarter. Shaken by the experience, Edison announced publicly that he was "through with electricity as a therapeutic agent."[37] By 1908 members of the American Roentgen Ray Society had reported dozens of cases of severe radiation burns to both patients and doctors. Despite the development of rudimentary shielding devices and basic radiation protection principles, radiation-related deaths continued in the medical profession. The great success of diagnostic radiography during wartime muted the calls for greater regulatory oversight and health protection

Fig. 11.10 Eben M. Byers, millionaire and Radithor user, whose death in 1932 would end the era of radioactive patent medicines. (Author's collection)

in radiology, but newspaper articles documenting the dangers of radiation and radioactivity began appearing with increasing frequency after 1920. Nevertheless, during the early 1920s the public fascination with radium seems to have continued unabated, and one gets the impression that most enthusiasts believed that radioactivity was only harmful in large quantities. When radium dial painters and radium chemists from the luminous paint industry began dying of osteonecrosis, many physicians initially attributed the poisoning to chemical irritant effects or impurities within the radioactive paint. However, by the 1930s it was becoming increasingly clear that chronic exposure to even low doses of radiation could prove lethal.[38]

The industrial fatalities unquestionably associated with radium purification and product preparation during the early 1930s did not provoke a major public outcry, probably because the public's attention was focused on the hardships of the Great Depression. The entire radium industry never represented more than a fraction of a percent of total American business, and the total number of deaths associated with the industry was small compared to the numbers killed annually in other industrial accidents associated with mining and heavy industry. Most patients injured by radiation experienced these injuries while being treated for potentially lethal diseases and were expected to experience serious side-effects related to radical treatment. Physicians suffering from radiation exposure often tended to minimize the extent of their injuries, perhaps in efforts to minimize patient fears and concerns.

THE BYERS AFFAIR: A LANDMARK TRAGEDY IN RADIO-ENTREPRENEURSHIP

The general public's attention was finally focused on the dangers of radium and radioactivity by the widely-publicized case of Eben M. Byers, a millionaire sportsman, industrialist, socialite, and man-about-town (Fig. 11.10).[39,40] Byers died early on the morning of Wednesday, 31 March 1932, the victim of a mysterious syndrome that for eighteen months had ravaged his body, corroding his skeletal system until one by one his bones began to splinter and break. Byers had been a broad-chested, multi-talented athlete and sportsman, an expert trapshooter, and the United States Amateur Golf Champion in 1907. As chairman of the A. M. Byers Iron Foundry, he had been the personification of the roaring twenties, a millionaire socialite and steel tycoon who had clambered into the upper reaches of New York's high society and continued to lead a life of privilege even after the market crash of 1929. Well-respected as a prudent captain of industry, Byers was also known as a rakish romancer, and this lifelong bachelor was no stranger to rumors of wild living and trans-Atlantic trysts. At his death Byers's once robust body weighed just 92 pounds. His handsome face had been horribly disfigured by a series of last-ditch surgical operations that removed most of his jaw and part of his skull in an attempt to stop the relentless bone destruction. His bone marrow and kidneys were rapidly failing, but, despite a brain abscess that had left him nearly mute, he remained lucid almost to the end. He died at 7:30 a.m. at Doctors' Hospital in New York

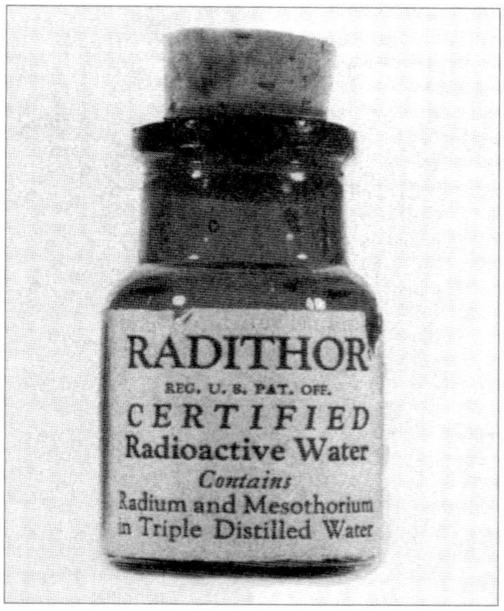

Fig. 11.11 Original Radithor sample bottles. Each bottle cost $1.00 and contained about 16.5 milliliters of fluid. (Author's collection)

City, and his physicians immediately notified the New York Medical Examiner's office.

By the following afternoon the authorities had begun a criminal poisoning investigation and were preparing the body for an extensive forensic autopsy by the Chief New York Medical Examiner, Dr. Charles Norris. The next day, the *New York Times* announced the preliminary results in a front-page headline: "EBEN M. BYERS DIES OF RADIUM POISONING." Byers, it was soon revealed, had for about four years been consuming large quantities of Radithor, a popular and expensive mixture of radium-226 and radium-228 marketed as a health elixir and endocrinologic stimulant, touted as ameliorating the effects of over 150 "glandular" diseases including rheumatism, hypertension, obesity, neurasthenia, impotence, and baldness (Figs. 11.11 and 11.12). Like many of the radioactive nostrums available during this time, Radithor was completely unregulated and was available over the counter and through the mail.

RADITHOR was the crowning entrepreneurial achievement of "Doctor" William J. A. Bailey, dean of the radioactive patent medicine industry that seemed to thrive along with the American economy in the 1920s. Bailey and his career are representative of a group of pop-technology inventor/entrepreneurs who were associated with medicinal radioactivity during its heyday. Born in Boston in 1884, he was a quick-witted, fast-talking Harvard dropout from the class of 1907 (Fig. 11.13). After trying his hand at running an import-export business and promoting himself (unsuccessfully) for the post of trade representative to China under Teddy Roosevelt, Bailey decided to make his fortune as an inventor and high-technology entrepreneur. His attempts in a number of industries were unsuccessful.

Between 1915 and 1921 Bailey appears to have been involved with a number of manufacturing and pharmaceutical scams, including a patent medicine company advertising "restoratives" for impotent men. This era appears to have been the beginning of Bailey's fascination with endocrinologic stimulants and aphrodisiacs, an interest he was later to combine with his enthusiasm for radioactivity and its putative physiologic restorative powers. It is not known whether William Bailey actually encountered Marie Curie during her triumphant 1921 American tour, but it is clear that beginning in the early 1920s he became enraptured with the study of radioactivity and its effects on life. He produced a trans-

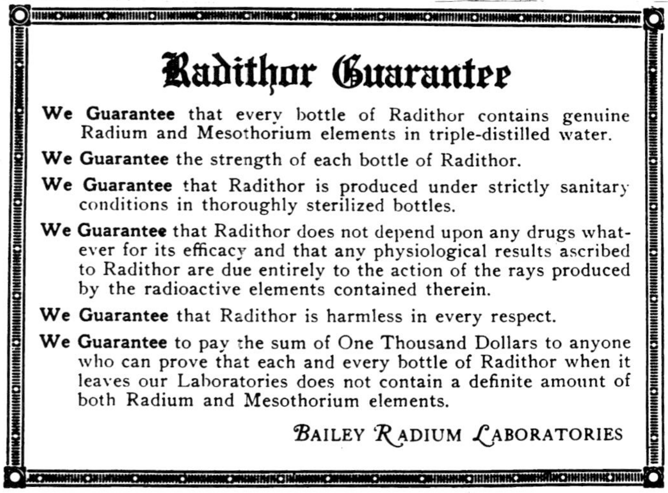

Fig. 11.12 The Radithor guarantee, which eventually resulted in prosecution by the Federal Trade Commission for false advertising. Ironically, the falsity in the advertising was not the list of contents claims, but the assertion that the product was harmless. (Author's collection)

Radithor Guarantee

- **We Guarantee** that every bottle of Radithor contains genuine Radium and Mesothorium elements in triple-distilled water.
- **We Guarantee** the strength of each bottle of Radithor.
- **We Guarantee** that Radithor is produced under strictly sanitary conditions in thoroughly sterilized bottles.
- **We Guarantee** that Radithor does not depend upon any drugs whatever for its efficacy and that any physiological results ascribed to Radithor are due entirely to the action of the rays produced by the radioactive elements contained therein.
- **We Guarantee** that Radithor is harmless in every respect.
- **We Guarantee** to pay the sum of One Thousand Dollars to anyone who can prove that each and every bottle of Radithor when it leaves our Laboratories does not contain a definite amount of both Radium and Mesothorium elements.

Bailey Radium Laboratories

Fig. 11.13 William J. A. Bailey, who, as president of Bailey Radium Laboratories, produced Radithor and other radioactive patent medicines. (Author's collection)

lation of Curie's 1910 classic, *Traité de Radioactivite*. Fascinated by the possibility that trace amounts of radioactive materials might serve to reverse physiologic imbalances and health problems, he incorporated a company in New York called Associated Radium Chemists, and from there put out a line of radioactive patent medicines including Dax for coughs, Clax for influenza, and Arium for run-down metabolisms. Eventually, the operation was closed and the nostrums were confiscated by the Department of Agriculture on grounds of fraudulent advertising. This temporary setback did not appear to deter Bailey, who soon was involved in the formation of two new companies in New York City. The Thorone Company (Thorium Hormones) was focused on the production of radioactive glandular stimulants and aphrodisiacs containing radium and thorium. The American Endocrine Laboratory, located at 113 W. 42nd St. in New York City, produced radioactive medical rejuvenation devices such as the "Radioendocrinator," a gold-plated radium-containing harness with a "gamma ray focusing window" that could be worn either around the neck to rejuvenate the thyroid, around the trunk to irradiate the adrenals or ovaries or, for enervated males, under the scrotum in a special jockstrap adaptor that was provided with the apparatus. The device came in an embossed, velvet-lined case and was approximately the size of a slim deck of cards. It sold first for $1000, then $500, then finally $150 as the rather specialized market became saturated. Even at these exorbitant prices, radioactive nostrums found ready buyers during the energy-crazed 1920s. Dozens of affluent and highly-placed citizens, including Mayor James J. Walker of New York City, later confessed to using radioactive rejuvenator devices and products to treat a variety of ailments during this period.

Bailey and the other radio-entrepreneurs routinely sought opportunities to present their theories at legitimate scientific meetings in hopes of applying the patina of professional endorsement to their unorthodox treatment philosophies. In a public relations coup, Bailey managed to secure an invitation to speak at the medicinal products session of the American Chemical Society meeting in Washington, D.C., on 25 April 1924. Bailey's talk focused on the use of radium in the cure of the hopelessly insane and in the reversal of the aging process. "We have cornered aberration, disease, old age, and, in fact, life and death themselves in the endocrines!" Bailey thundered. "In and around these glands must center all future efforts for human regeneration."[41] The next day, the *New York Times* excerpted the talk in a lengthy and complimentary article about the new science of "radio endocrinology" (Fig. 11.14).

In 1925 Bailey moved to East Orange, New Jersey (no doubt, in part to partake in the nearby aura of Edison's old invention center) and began to market his promotional masterpiece, Radithor. The radioactive tonic was an immediate commercial success, and orders poured in from around the world. Over 400,000 half-ounce bottles were sold between 1925 and 1930. Radithor was advertised as "the climax of thirty years of research and the fullest achievement in internal radioactive treatment," with efficacy against over 150 diseases. Like many other patent medicines of the day, the RADITHOR promotional pamphlets (always shipped separately as "comple-

Fig. 11.14 Newspaper coverage of Bailey's talks on radium was generally complimentary—and optimistic. (Author's collection) ▶

mentary monographs" so that the outrageous claims in the pamphlets could not be construed as warranties of product efficacy) were filled with testimonial letters from satisfied patients and physicians, urging skeptical doctors to "put one or two of your most obstinate cases on radium water so that you may observe its action in your own practice."

By 1930, when Eben Byers first began to experience unusual aches and pains, some of the mania for the expensive radioactive nostrums had begun to subside. Byers began to lose weight and complained of headaches and toothaches. A radiologist in New York, Dr. Joseph Steiner, looked at Byers's radiographs and noticed some similarities between the developing bony lesions in Byers's mandible and those described in the recently deceased radium dial painters. Frederick B. Flinn, the prominent radium expert from the department of industrial medicine at Columbia University, was called in as a consultant and confirmed Steiner's suspicions: Byers's body was slowly decomposing, the result of massive radium intoxication from the Radithor.

A Federal Trade Commission (FTC) attorney sent to Byers's Long Island mansion to take the deposition later described the scene in vivid detail:

> A more gruesome experience in a more gorgeous setting would be hard to imagine. We went to Southampton where Byers had a magnificent home. There we discovered him in a condition which beggars description. Young in years and mentally alert, he could hardly speak. His head was swathed in bandages. He had undergone two successive jaw operations and his whole upper jaw, excepting two front teeth, and most of his lower jaw had been removed. All the remaining bone tissue of his body was slowly disintegrating, and holes were actually forming in his skull.

This lurid description was reprinted in *Time* magazine and in various sporting journals and society papers, and the Byers case rose to international prominence. With his death in 1932 the FTC reopened its investigation and the FDA began campaigning for broader powers. Physicians issued stern warnings on the dangers of radioactivity, in one case even holding up a victim's bones to a Geiger counter on a radio broadcast to demonstrate "the deadly sound of radium."

Medical societies took the welcome opportunity to denounce all patent medicine sales, and calls for radium control laws were voiced throughout America and Europe. The forerunners of the current regulations restricting the sales of radiopharmaceuticals to authorized users actually date specifically to the Byers affair. With the institution of these regulations the radioactive patent medicine industry collapsed, and the era of mild radium therapy came to

The World Famous
FREE ENTERPRISE RADON HEALTH MINE

Box 67 — Boulder, Montana — 59632

"The Unmedical Approach to ARTHRITIS"

Open Since 1952

Fig. 11.15 The public is not unanimous in fear of all things radioactive. As late as the 1980s promoters advertised the health benefits of descending into abandoned radium mines to breathe the emanations. (Courtesy of the Center for the American History of Radiology, Reston, Va.)

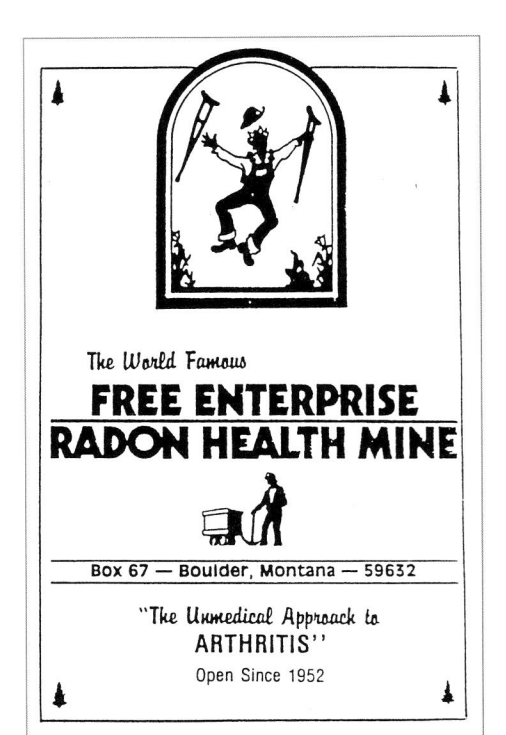

an end. By the late 1930s the public was growing increasingly wary of radiation, and even some oncology experts were becoming skeptical that radiation treatments could significantly impact on the cancer problem.

POSTWAR DISENCHANTMENT

The final phase of public disenchantment with radiation and radioactivity began with the first newsreels showing the catastrophic effects of the atomic bombs dropped on Japan. An uneasy public began to hear more and more about long-term consequences of contact with radioactivity and nuclear energy, and radiation-related concepts such as "fallout," "carcinogenesis," and "genetic mutations" became familiar terms. Highly publicized fears of East-West nuclear exchanges and nuclear power-plant accidents spilled over into public concerns about even low levels of exposure to radiation. The evolution of this broad-based postwar public phobia has been traced in detail in a number of recent publications and will not be repeated here.[42,43]

However, it seems important to note that, even in the 1990s the health effects of chronic exposure to low levels of radiation are not as well established as the public might imagine (Fig. 11.15). Although most industrialized societies continue to reject the use of ionizing radiation for virtually all nonmedical uses (and insist on tight constraints on its use even within the clinical sphere), most objective analyses suggest that public perceptions of ionizing radiation as an environmental menace are somewhat overblown compared to the levels of danger associated with other environmental and behavioral risks in an industrialized society. Nevertheless, this image of radiation as a force for death and destruction, a genie unleashed by an unwitting and arrogant technologic aristocracy, has continued to exert a powerful influence on society's view of radiation and radioactivity.

CONCLUSION

Are there any lessons to be learned from the saga of the rise and fall of the public's interest in and infatuation with radioactivity? Several come to mind. First, it is clear that the discovery of radioactivity coincided with the culmination of a long period of popular fascination with electromagnetic beams and invisible but powerful physical forces.[44] The bioeffects of these mysterious forces had been studied since before the time of Paracelsus and appealed especially to the enlightened turn-of-the-century environment that celebrated the elucidation and demystification of "vital" living processes. It seemed that radioactivity might indeed be the missing link that connected chemistry and physiology. Contemporary discoveries in the fields of biochemistry, enzymology, evolutionary biology, and endocrinology had strengthened the reductionist trend toward thinking of living organisms as simply machines made of common elements. Radioactivity seemed a perfect natural "fuel" to explain the life force that flowed through these elemental machines. In addition, although the influence of church theocracy had waned considerably by the late nineteenth century, people still clung to the notion that there must be some powerful, mystical force that imbued living creatures with the power to move and function. Radioactivity appeared to be a good God-given candidate for that function, and appealed with particular urgency to those seeking to reconcile the new science with religion. The mysterious ability of radiation to at least temporarily (and in many cases permanently) alleviate many types of skin diseases, inflammatory conditions, and cancers seemed clearly to identify radiation as a healing force. Only later, as the chronic effects of radiation overexposure became apparent, did the public image become more negative and skeptical.[45]

Historian James Harvey Young has commented on the gradual decline during the twentieth century of what he calls the "Doctrine of Progress," the popular conception that technologic advances will inevitably lead to social improvements and the betterment of mankind.[46] Public infatuation with Curie, Edison, Lindbergh, and the other great heroes of the age of technology gradually gave way to skepticism and disillusionment with science and progress—a jaded anti-technical point-of-view only ameliorated in the last ten years by the proliferation of the personal computer and the so-called Information Revolution. Young quotes a character in dramatist Maxwell Anderson's 1935 play *Winterset* who says, "What faith men will then have, when they have lost their certainty of salvation through laboratory work, I don't know." The rise and fall of the public's enthusiasm for radiation science mirrors a broader skepticism and disillusionment with scientific progress of all types. Radiation and radioactivity seemed to be the archetypical example of the public's greatest fear: a powerful force unleashed by arrogant and naive technocrats who lacked the moral maturity to understand the disastrous social consequences of their scientific investigations. The ultimate development of the atomic bomb completed this apparent re-enactment of the story of Dr. Frankenstein's monster, underlining the parallels between the history of radiation science and the famous fictional tale of a scientific discovery gone berserk.[47]

A final observation on the subject concerns the prevalence during the early decades of the twentieth century of radiomedical fraud and pseudo-science: why were the concepts of radiation and radioactivity so rapidly and completely exploited by the makers of nostrums and patent medicines? Certainly, part of the answer has to do with the sheer excitement and enthusiasm produced by the technical glamour and medical triumphs of the new radiation science. It was at first difficult to define the reasonable limits of this novel set of medical processes and effects. Most claims of "revolutionary" technological breakthroughs initially inhabit an uneasy no-man's-land between acknowledged brilliance and refuted quackery, and radiation medicine was no exception.

However, the immediate and sustained appeal of radiation as the basis of putative cures for hundreds of diseases suggests that this concept touched a deep chord in popular culture of the time. One possible explanation involves the observation that successful promotion of medical quackery often uses one of two major classes of advertising strategies: (1) Romantic "secret-of-the-ages" promotional strategies seeking to label a product or process as a long-lost "natural" or "native" cure for various intractable diseases; or (2) "Scientific breakthrough" promotional strategies identifying a particular nostrum as the final triumphant result of years of intensive work by a heroic group of dedicated scientists.

The discovery of radioactivity and its translation into radiation medicine were readily adaptable to both promotional philosophies. Radiation science encompassed both the study of age-old natural forces and the development of certifiable high-tech marvels. In appealing successfully to both types of promotional images, radio-medical nostrums, inhalers, and other apparatus quickly captured the public's interest and succeeded in sustaining that interest through the 1920s. Radioactive forces seemed the perfect physical embodiment of the high-energy, fast-moving spirit of those years. And, at least for a while, the radioactive future seemed limitless. As William Bailey noted in one of his promotional pamphlets for Radithor, "Radioactivity is one of the most remarkable agents in medical science. The discoveries relating to its action in the body have been so far-reaching that it is impossible to prophesy future developments. It is perpetual sunshine."[48] And this sunshine would appeal to several generations of Americans, only to be obscured by the mushroom clouds of mid-century concern—shadows which linger today in the public perception of medical radiation.

▶ ▶ ▶ ▶ ▶ ▶

REFERENCES

1. Osler, Sir William.
2. Weart, Spencer. *Nuclear Fear.* Cambridge, Mass.: Harvard University Press, 1988.
3. Ibid.
4. "Hidden solids revealed: Prof. Routgen experiments with Crooke's vacuum tube," *New York Times* (16 Jan. 1896):9, col 5.
5. Caufield, C. *Multiple Exposures: Chronicles of the Radiation Age.* 1st ed. New York: Harper & Row, 1989.
6. Ibid.
7. Ibid.
8. *Phototimes* 28 (1896):150.
9. Knight, Nancy, "The New Light: X-Rays and Medical Futurism," in Joseph J. Corn, *Imagining Tomorrow: History, Technology, and the American Future.* Cambridge: MIT Press, 1987.
10. Mould, Richard F. *A History of X-Rays and Radium: 1895–1937.* London: Westminster, 1990.
11. Caufield, *Multiple Exposures.*
12. *Anthony's Photo Bulletin*, April 1896.
13. Martin, T.C.; Wood, R.W.; Thomson, E.; et al., "Photographing the Unseen," *Century Magazine* (1896):120-130.
14. Caufield, *Multiple Exposures.*
15. Knight, "The New Light."
16. Weart, *Nuclear Fear.*
17. Meadows, J. *The Great Scientists* (New York: Oxford Press, 1986):89-208.
18. Weart, *Nuclear Fear.*
19. Ibid.
20. Meadows, *Great Scientists.*
21. Weart, *Nuclear Fear.*
22. Caufield, *Multiple Exposures.*
23. Weart, *Nuclear Fear.*
24. Soddy, F. *The Interpretation of Radium* (London: Murray Publishing, 1912):250-260.
25. Macklis, R.M., and Beresford, B., "Radiation Hormesis," *J. Nuclear Medicine* 32 (1991):350-359.
26. Weart, *Nuclear Fear.*
27. Caufield, *Multiple Exposures.*
28. Pfahler, George F., "Fifty Years of Trials and Tribulations in Radiology," in *American Roentgen Ray Society 1900–1950.* (Springfield, Illinois: Charles Thomas, Pub., 1950):15-29.
29. Macklis, R.M., "Radithor and the era of mild radium therapy," *J.A.M.A.* 264 (1990):614-618.
30. Borland, V., "Mild Radium Therapy," *Brit. J. Phys. Med.* 6 (1932):226-228.
31. Goodman, H., "The Romance of Radium," *Med. J. Res.* 131(1930):190-192.
32. Macklis, R.M., "The great radium scandal," *Scientific American* 269 (1993):94-99.
33. Macklis, "Radithor."
34. Bailey, W.J.A. *Modern Treatment of the Endocrine Glands with Radium Water: Radithor, the New Weapon of Medical Science.* East Orange, N.J.: Bailey Radium Laboratories, 1926.
35. Caufield, *Multiple Exposures.*
36. Ibid.
37. Ibid.
38. Stannard, J. *Radioactivity and Health: a History.* Springfield, Va.: National Information Service, 1988.
39. Macklis, "Radithor."
40. Macklis, "Great Radium Scandal."
41. Bailey, *Modern Treatment.*
42. Weart, *Nuclear Fear.*
43. Caufield, *Multiple Exposures.*
44. Macklis, R.M., "Magnetic healing, quackery, and the debate about the health effects of electromagnetic fields," *Ann. Int. Med.* 118 (1993):376-383.
45. Badash, Lawrence. *Radioactivity in America: Growth and Decay of a Science.* Baltimore:Johns Hopkins Press, 1979.
46. Young, J. Harvey. *American Health Quackery.* Princeton, N.J.: Princeton Univ. Press, 1992.
47. Weart, *Nuclear Fear.*
48. Bailey, Modern Treatment.

CHAPTER TWELVE

THE FUTURE OF RADIATION ONCOLOGY

Herman Suit, M.D., D. Phil.

▶ ▶ ▶ ▶ ▶ ▶ ▶ ▶ ▶ ▶ ▶ ▶ ▶ ▶ ▶

From virtually every perspective the future of radiation oncology for the next three decades must be reckoned of high promise. One of the most positive facts underlying this statement is the exceptional quality of young physicians entering this discipline. Certainly, in terms of academic credentials, we have never had such talent to choose from in the resident selection process. The contrast between our present situation and that of three decades ago could hardly be sharper. Then, there were small numbers of resident candidates, few quality teaching programs, only occasional research programs and a modest literature. Today, the scene is qualitatively and quantitatively better, and the prospects are for this improvement in professional staffing at all levels to continue.

Radiation oncology is in the midst of a period of rapid and profound change affecting nearly every facet of our discipline. In the near future there will be (1) important advances in technical proficiency resulting in the ability to use substantially smaller treatment volumes with proof that the target is within that treatment volume; (2) subspecialization within radiation oncology that will become the standard in large centers and practice groups; (3) radiation oncology subspecialists integrated with their subspecialty counterparts in surgery, medical/pediatric oncology, radiology, and pathology to form site-specific teams for patient management; (4) increasing combinations of radiation with chemotherapy and/or surgery; (5) a rationalized integration of agents to modify responses of tumor and/or normal tissues, e.g., chemotherapeutic agents, biological response modifiers, cytokines; (6) the use of new predictors of response of tumor and normal tissues so as to permit individualized, scientifically-based treatment strategies; (7) sustained efforts to reduce costs of care, i.e., manage our practices much more efficiently and reduce costs; and (8) at more distant times, inclusion of molecular-genetic-based therapy into management protocols.

National manpower requirements for all medical specialists are being revised downward. As part of this process, reductions in the number of residents in radiation oncology are currently being implemented. This may be appropriate and have long-term beneficial consequences for oncology.

However, serious consideration will have to be given to the increase in physician time in planning and implementing definitive treatments as the technical developments described below are introduced into clinical practice.

There are several significant negatives for our future, and they are all economic. The changing health care system in this country will effect reductions in the available funds for patient care, research and development, and education. This result will be to slow but not stop the gains projected above. A significant decrement in the economic status of the radiation oncologist and to a lesser degree radiation physicists, therapists, and nurses is almost certain.

Still, my long-term view of radiation treatment quality is clearly optimistic. This may not obtain over the short term in the United States as we modify our current health care system. The near certainty that radiation treatments can be made more effective will force development and implementation of new methods. Medicine and science are international; even though we in the United States may experience a slowing of the pace of development, we can confidently expect to gain significant amounts of knowledge from laboratory and clinical studies throughout the world. For the pessimists, an examination of the advances which occurred in the 1930s, a decade of severe economic depression, may bring some encouragement. These include the introduction of supervoltage X-ray machines, van de Graaff accelerators, and betatrons to limited clinical use. Fast neutron therapy trials were begun. There were major gains in diagnostic imaging during this same period.

Several of the anticipated changes will be considered here. For this paper, the plan is to discuss advances which seem certain to be realized and then those judged likely. Finally, comments will be offered on the probable consequences for radiation oncology of the exuberant growth of the field of molecular biology and the potential for its successful applications to oncology.

TECHNOLOGICAL ADVANCES AND THEIR IMPACT

An astounding series of technological advances are appearing on the radiation oncology scene and will enhance the quality of treatment. There will, of course, be some increase in cost and especially in time required of the physician in treatment planning. Most of these technical developments stem from (1) astonishing advances in computer power at steeply declining costs, which will yield vastly more effective software systems for treatment planning; (2) the appreciation of the clinical importance of smaller treatment volumes to patient outcomes; and (3) dramatic improvements in diagnostic imaging techniques. Several of the most important are here discussed.

Three-dimensional optimized treatment planning software

The concept for three-dimensional (3-D) conformal optimized treatment planning came from and has been clinically tested in academic departments. These software systems are under intensive development by major industrial firms, often in collaboration with the academic centers.

Treatment planning using mature 3-D software systems and state-of-the-art computers will demand significantly more of radiation oncologists' time. Although it may appear counterintuitive, the time required of the clinical physicist for the individual patient's plan will decrease, a reversal of the present planning efforts. The actual planning process for a definitive or radical treatment is expected to proceed according to the following steps. First, the treatment position will be selected and an immobilization device prepared which will permit a highly-reproducible setup with severely restricted motion of the patient during each treatment session. Then a treatment planning imaging study (e.g., computed tomography [CT] or magnetic resonance [MR] imaging) will be performed with the patient immobilized in the treatment position. Following this, the physician will define the boundaries of the targets (initial, intermediate, and final) and of each of the sensitive normal tissues of concern on each section of the CT/MR images. In addition, the physician will indicate the allowance to be made for patient/target motion. The dose constraints for each of the designated normal tissues and dose aims for the target will next be specified. This information would be transferred by a physicist/treatment planner into the treatment planning software program on a powerful computer able to construct a series of beams—gamma and/or electron—to cover the target with minimal margins. Next, the planning system will generate many hundreds of plans in the process of optimizing the dose distribution. For the planning, the system will be able to consider static and dynamic beams, noncoplanar, dynamic wedges, intensity modulation, inverse planning, etc. The yield being the best feasible dose distribution, that is, adequate dose constraints on normal tissues with achievement of an effective dose to target. Further, the uncertainty around each isodose contour will be displayed. The computer will submit several of the supposedly best plans for the clinician and physicist to review. These plans would be produced in a mere few minutes, as computer capabilities will continue to advance. By the end of this century every large treatment center is expected to have a computer the equivalent of the so-called "supercomputer" of just a few years ago. Dose distributions will be shown as the dose volume histogram for any desired organ or tissue or portions thereof. These systems will provide the clinician and clinical physicist with means of interactively viewing and comparing the computer's first offering of plans.

This treatment planning procedure will be enhanced by the inclusion of subroutines to display not only the physical dose distributions but also the predicted biological effect distribution.[1] Several groups are developing this capability. The clinician and clinical radiation biologist select values for the parameters of alpha/beta ratio, cell proliferation kinetics, etc., for each specific tissue of concern. There will be, of course, much greater uncertainty regarding the values for these biological effect distributions in the individual patient and the particular tissue than will obtain for the statement of the physical dose. Nonetheless, the display of the estimated biological effect distribution should be an aid in the evaluation of the several treatment plans. At a minimum, this additional biological information regarding the plan options could provide warning signals when a particular plan might be predicted to yield an unanticipated probability of morbidity despite an attractive physical dose distribution.

The selected optimized plan will be employed in the simulation, which will primarily serve to document that the clinical placement of the beams covers the volume intended in the plan. If the plan is found to be practical, this will be indicated to the computer, which in turn will issue instructions to the treatment unit for gantry position, collimator angle, setting(s) of the multileaf collimator system, intensity modulation, patient support assembly positions, and preparation of patient records. Thus, treatment planning in the year 2000 will be quite different from and superior to that now considered to be standard medical center quality.

Confirmation that the target is in the beam

Employment of the most elegant treatment plan—a projected treatment volume-to-target-volume ratio of approximately 1.05—is of little merit unless there are means for confirming that all of the target is in the beam throughout each individual treatment session. Developments in this area fall into four classes: (1) portal visualization systems, either "beam's-eye" or off-beam line techniques; (2) physician-introduced fiducial markers; (3) increased security of patient immobilization systems; and (4) methods for gating the irradiation with patient/anatomic part movement during the individual treatment.

On-line portal viewing system

Portal viewing systems (PVSs) are currently being introduced to the market by virtually all of the linear accelerator suppliers. The goal of a PVS is to provide the ready capability of confirming that the beam is covering the structures intended and, if not, to permit an easy—or, even better, an automatic—correction. However, available PVSs are still developmental and possess too little contrast to be of great practical merit for sites other than the head, neck, and, in certain cases, the thorax. The technical characteristics will undoubtedly evolve rapidly toward providing superior contrast and result in a clinically essential tool. Some researchers are considering developing systems that include a diagnostic radiographic unit in the treatment head to produce diagnostic quality in portal imaging. High-quality PVS could be a major advantage for treatment of lesions, which move appreciably between treatments and even during treatment. An important feature to be developed is that of digitization of the fiducial markers on the initial image, with a subsequent control system to turn the beam off if the fiducial markers move beyond a permitted range. Thus, continuous visual monitoring, which is labor intensive and error prone, would no longer be required.

Physician-introduced fiducial markers

Discrete anatomical features—edges of high contrast parts—are the fiducial markers used in current clinical practice for assessment of portal films. This is often quite difficult, as the markers are not well visualized, and only rarely is an anatomic marker properly seen on two beam paths. This means that determination of the position of the target in relation to the beam is often quite difficult or unsatisfactory. The future should bring about a much expanded use of a variety of physician-introduced fiducial markers of a high atomic number (Z) material, and hence, comparatively easy for visualization on portal films and or in the PVS.

Our experience in this area has been quite positive in the treatment of patients with intracranial, skull base, and paranasal sinus tumors. This has been based upon a technique which places two gold screws in the outer table on one side of the cranium and a single screw on the opposite side.[2] This is done prior to the treatment-planning CT scan. The position of the screws relative to the target and any sensitive structures of interest is defined and included in the data used in the 3-D treatment planning procedure. At treatment, with the patient in the immobilization device, a pair of orthogonal films are taken. The centers of the screws on each of the films are digitized, and, thus, the relative positions of the screws one to the others are defined. By a software routine, the patient support assembly is instructed to move to the correct position, and a repeat set of films is then evaluated. With few exceptions, the target has been placed correctly in relation to the beam (± 1 millimeter [mm]). Markers are frequently employed to aid in simulation of patients with tumors of selected sites in the head and neck region, pelvis, and other areas. Our judgment is that there can be substantial reductions in treatment volumes by more frequent utilization of such a simple technical maneuver as insertion of several markers.

Patient immobilization systems

The developments in this field have been significant. Of special note have been devices for immobilizing the head for the treatment of intracranial and other head and neck sites. The repositioning of the head for fractionated treatment is now accurate to 1 mm.[3] Thus, highly accurate beam localization of intracranial targets does not require the use of head frames, which are screwed into the cranium, or treatment by a single large dose. Target position definition in a space at most 1 mm is not feasible by use of even the highest resolution CT imaging techniques due to scan thickness, scan intervals, errors in marking target margins on the

CT/MR image, and variation between observers as to the concept of the target. The dose fractionation protocol will increasingly be based upon the radiation biological features of the tumor/normal tissue situation and not on the technical requirements for accuracy of immobilization.

Systems are now being developed and marketed for greater security and comfort in patient immobilization for most anatomic regions. The future is certain to see extended utilization of such systems and their improved derivatives regularly in definitive radiation treatments.

Gating of radiation treatment to patient/anatomic part movement

Structures in the thoracic and abdominal cavities move regularly with respiration and the heart beat. For radiation treatment in those sites based upon close margins, oncologists will have to employ techniques for gating treatment to such movements. There is no fundamental difficulty in devising the means for achieving progress in this area; only time, effort, and resources need be applied. Recent reviews of these technical developments include the November 1993 issue of *Radiotherapy and Oncology* and Meyer and Purdy's *Frontiers of Radiation Oncology* (1995).[4]

CHANGES IN CLINICAL PRACTICE PATTERNS

Multi-modality treatment strategies

Over the recent three decades, utilization of radiation as a sole modality has decreased sharply. This trend is likely to continue as (1) chemotherapeutic agents and multi-drug protocols are shown to yield clinical gains in terms of higher distant metastasis-free survival and local control probability; (2) conservative surgery is replaced with radiation and or chemotherapy; and (3) new strategies involving biological response modifier and gene therapy are shown to yield clinical benefits. At present chemotherapy is combined with radiation alone or with surgery in the management of a high proportion of patients with breast, esophageal, colorectal, anus, pancreas, bladder, high-grade glioma, pediatric tumors (Ewing's rhabdomyosarcoma, CNS, and the like), small cell lung cancer, lymphomas, and others. For many of these tumor types, the impact of the combined modality approach has been to augment both the rate of local control and metastasis-free survival. These results constitute one of the most important gains for patients whose treatment includes radiation. Further, quality of life has been greatly improved by replacing radical surgery with conservative surgery and radiation in cases of breast, head and neck, and rectal cancers and of sarcoma of soft tissues. Clinical studies of combined modality management will expand and provide further gains for patients.

Widespread subspecialization within radiation oncology

Specialization has been a characteristic of the development of the practice of medicine. There appears to be not only a continuation but an acceleration of this process in virtually all medical specialties. The present evidence is that there will be some shift toward greater numbers of young doctors entering primary care rather than a specialty practice. However, the individual physician to whom responsibility is given for management of the patient with a major medical problem is likely to be ever more specialized. This is a simple reflection of the growth of knowledge and of a growing literature in each facet of medicine. Changes in the general field of radiology serve as striking evidence for this trend over the past forty years of extensive subspecialization. Consider that, in major academic centers, diagnostic radiologists concentrate in one specific anatomic region or imaging technique—thoracic, breast, neuroradiology, gastrointestinal, genitourinary, skeletal, CT, MR, etc. General diagnostic radiologists are rarities or have become extinct in large medical centers. Radiation oncology has moved from being a subspecialty to a separate specialty.

The continuation of this process to subspecialization within radiation oncology is the natural consequence of the extraordinarily rapid expansion of our knowledge and experience bases. This includes for each type of tumor: the natural history, responses to various therapies administered alone or in combination, frequencies and severity of morbidities, tumor and normal tissue biology, radiobiology, epidemiology, efficacy of rehabilitation procedures, psychological components of patient care, basic and clinical genetics, cost of management, and many other factors. These advances have meant a virtual torrent of medical and scientific journal articles, monographs, books, symposia proceedings, and various governmental and nongovernmental reports. Patients expect their clinicians to be knowledgeable regarding this rapidly accumulating information base as it pertains to his or her disease. These papers and publications are emanating not only from this country but increasingly from countries across the globe. For example, in 1992 alone there were 7,200 papers published on cancers of the lung, 5,200 on cancers of the breast, 1,800 on cancers of the rectum, and 1,800 on cancers of the bladder.[5] Clearly, to be even moderately well informed about the activities in oncology related to a single tumor type requires a most serious effort. To be a generalist and also well informed is no longer feasible. The new "generalist" will have to limit practice to a relatively small number of tumor types.

If this seems a strong statement, consider how the practice of the general surgeon has changed during the past seven to eight decades; orthopedics, urologic, gynecologic, thoracic, plastic, and ear, nose, and throat procedures are infrequently done by a general surgeon today. Our patients will expect that we be more than technicians implementing treatments decided upon by surgical or medical oncologists. This obviously requires an intensive effort today and will be even more of a challenge with time.

Oncologic practice in large centers is rapidly evolving toward the use of different site/tumor specific teams, each of which is comprised of subspecialist representatives from radiation oncology, medical oncology, surgical oncology, pathology, diagnostic radiology, and nursing. In addition, there is often participation by basic scientists with interests in the specific tumor of concern. This means that for the major cancer treatment centers, there are now radiation oncologists whose practice is limited to patients with tumors at specific sites (such as head and neck, breast, central nervous system, gastrointestinal, genitourinary, gynecologic, thoracic), or to patients with specific histologies (lymphoma, sarcoma), or to patients of a specific age (pediatric oncology). Each of these have its counterpart in medical and surgical oncology as well as in pathology and radiology. Further, in radiation oncology, the physics group is also subspecializing, often working principally in stereotactic radiation techniques, intra-operative electron beam radiation therapy, or stereotactic brachytherapy. Additionally, some physicists work exclusively with one or a few of the site-specific teams. This trend is being seen also with radiation therapists, as patients with tumors in a given site tend to be treated on a single machine. The operation of such site/tumor specific centers in the larger hospitals clearly brings a higher level of experience and expertise into actual clinical practice for the diverse tumors and anatomic sites than is feasible for an individual attempting to cover the broad spectrum of patient problems.

This view implies that the present status of medical and radiation oncology as a single specialty in many countries will not persist. This is in part due to the fact that present chemotherapy drug and dose schedules are quite toxic and will probably become more so. Intensive use of present protocols requires a much more serious time commitment than is feasible for a clinician who is also involved in performing complex, high-technology radiation treatments.

Patients and their referring physicians will be increasingly likely to insist that their care be planned and administered by a multidisciplinary team which makes a special concentration of study and practice on their particular tumor. This trend is likely to be accelerated by the broad coverage of medical matters in the popular press and television programs, where the activities of the multidisciplinary centers are presented in favorable terms. That such teams could provide advantages in quality of care is likely to be readily accepted by the public.

There are, however, counter pressures to this trend toward subspecialization, particularly in the establishment of small community radiation facilities by entrepreneurial firms. These will attract many patients to the generalist who works either alone or with one or two associates. Further, if the health maintenance organizations (HMOs) and other health care organizations can negotiate lower prices with these smaller facilities, the tendency will be to encourage patients to be treated there. How the balance will be established between the large subspecialized centers and the small community-based unit will be under continuous discussion and review for the foreseeable future. With reference to potentially fatal diseases, this writer predicts that public pressures will ultimately result in substantial concentrations of experience and resources to maximize outcomes. For this to be realized, physicians should seek out opportunities to describe to the public the predicted benefits from patient management by subspecialty teams as well as from the clinical use of new technologies, predictors, and combined modality treatment methods.

Facility utilization

The staffing, technical equipment, and general facility requirements for a major treatment center providing multi-disciplinary care will become progressively more stringent and more costly. With reference to radiation oncology, at a minimum they will need: linear accelerators fitted with computer-controlled multileaf collimator (MLC) systems; portal imaging systems; 3-D treatment planning systems with optimization software; record and verification systems; simulators integrated with treatment planning systems and linear accelerators; and easily applied stereotactic techniques for external beam radiation therapy or brachytherapy. Requirements for cost-effective utilization of such facilities will likely mean operation for more than the current eight-hour day and five days a week. There should be no surprise to see these complex and costly facilities operating on a twelve- to sixteen-hour day and/or a six- or seven-day week. These extended operations are already commonplace for some MR imaging centers. This is not likely to be constrained by staffing shortages, as there appears to be rapidly increasing numbers of radiation oncologists, physicists, and other trained personnel.

Larger practice groupings

The move toward radiation oncology practice based upon tumor-type specific teams and supported by an associated spectrum of technical capabilities will mean larger and larger practice groups. Further, for purposes of cost efficiency, we may expect integration into multi-institutional groupings. Mergers and integration of medical services are commonplace and will almost certainly become more so. Easy referrals within such systems will make more likely the cost-effective utilization of specialized and expensive facilities.

A growing proportion of radiation oncologists will be working in managed care systems and to some extent on a capitation basis. This will mean that the choice of medical procedure will be determined not only by the individual physician but will be affected by the policies of the HMO.

SPECIAL TREATMENT FACILITIES

In addition to the growth of centers with highly focused teams for the management of patients with tumors at specific sites, a small number of cancer centers with highly specialized technical facilities will be established to serve as regional, national, and even international resources.

Fast neutrons

Fast neutron beams are under continuing clinical evaluation in the United States, Europe, Japan, and South Africa. The published results for clinical trials and practice since the late 1930s do not clearly demonstrate clinical gains over photon therapy, with the probable exception of treatment of locally advanced carcinomas of the parotid salivary glands.[6] Data from treatment of soft tissue sarcoma and carcinoma of the prostate are of interest and will be pursued but are not accepted as demonstrating superior results for neutrons. Therapeutic gain factors determined for fractionated (F=15) fast neutron irradiation of three spontaneous tumors systems of a C3H mouse (a mammary carcinoma, a fibrosarcoma, a squamous cell carcinoma) were not above 1, a disappointing result.[7]

The active fast neutron therapy centers in the United States are at the University of Washington, Seattle; Harper Hospital, Detroit; Fermi Laboratory, Chicago; and the Cleveland Clinic, Cleveland. Several fast neutron facilities in the United States have discontinued activities, including MANTA, Washington, D.C.; University of California–Los Angeles: and the M. D. Anderson Hospital in Houston. New fast neutron facilities are to be constructed in Germany and Sweden.[8] There is clearly needed in the relatively short term a definitive evaluation of the clinical efficacy (tumor control probability vs. treatment related morbidity) of fast neutrons relative to photons for those tumor and anatomic sites where the dose distribution for fast neutrons is approximately equivalent to that for photons. This would be valuable in assessing the potential of high linear energy transfer (LET) radiations in general as well as the emphasis to be placed on heavy ion beams.

There is one important difference in treatment protocols used in fast neutron treatments: essentially all treatments have employed overall times of three to four weeks. There might be a component of any observed gain due to the accelerated treatment rather than due to the radiobiological characteristics of the high-LET radiations. This needs to be considered in protocol designs used in the testing of the high-Z particle beam. A concern with the clinical application of high-LET beams is the apparently greater relative biological effectiveness for late damage, including radiation-induced tumors. Thus, the comparison of local control results at equivalent frequencies and severities of treatment related morbidity will demand observation over a prolonged time period.

Heavy Particle High-LET Radiations

The expectation for the high-Z particle beams—carbon, neon, argon—is that any gains relative to proton beams should be approximately comparable to gains of fast neutrons relative to photons, namely, similar dose distributions but quite different radiation biological characteristics.

The largest and most costly of these specialized radiation facilities is the HIMAC unit of the National Institute of Radiological Sciences in Chiba, Japan.[9] This facility is unique in that it has been designed primarily for medical purposes. Thus, the physicians should have optimal access to the unit and not be in competition with nuclear physics research programs for beam time. HIMAC commenced patient treatments in 1994. In addition, there are plans for access by clinicians to the high-Z beams at the nuclear research facilities located at Darmstadt and at Julich (near Düsseldorf). Small numbers of patients are being treated at the nuclear research facility at Dubna, near Moscow. Serious discussions are in progress regarding the feasibility of the establishment of a heavy particle therapy facility near Milan. These several programs constitute a most special testing of exotic particle beams in radiation oncology. This line of clinical research is not expected to have direct participation by an American facility in the near future. Earlier, Castro and colleagues had studied similar beams and acquired a major experience at the Bevalac at the University of California at Berkeley.[10] They judged that the results provided promise for the application of heavy ion beams.

Proton Treatment Facilities

Proton beams constitute an extremely attractive means for reducing treatment volumes. This is due to the physical characteristics of proton beams: (1) finite range, (2) energy-dependent range, (3) range dependent upon the density of the tissue along each particle path for a proton of a given energy, (4) dose deep to the end of range is zero, and (5) for a modulated energy beam the surface dose is in the range of 70 to 100 percent depending on the depth of the target from the surface, that is, less attractive than for the photon beams. However, regarding this last point, when multiple fields are utilized for deep lesions, this is not a factor of clinical significance.

The critical feature is that for each beam path there is no proton dose deep to the target. In comparing proton beams with 3-D conformal photon therapy, proton techniques can utilize as many beams (static or mobile), intensity modulations, etc. as a photon technique. Thus, the laws of physics demonstrate that integral dose will always be less for proton than for photon treatment.

To the extent that the smaller treatment volumes permit the employment of higher doses to the target, there would be anticipated increments in tumor control probability. The actual clinical experience in the use of proton beams in the treatment of cancer patients is based mainly on treatment of patients with uveal melanoma and skull base chondrosarcoma and chordoma. For those lesions there have been impressive five-year actuarial local control rates of 96 percent, 97 percent, and 55 percent, respectively.[11] These results are interpreted as constituting gains over those obtained by photon techniques. Other sites under current clinical trial include paranasal sinus, nasopharynx, pharyngeal wall, prostate, high-grade glioma, arteriovenous malformations, meningioma, acoustic neuromas of the spine and sacral sarcomas, retinoblastoma, and acoustic neuromas. By far the greatest number of patients treated with proton beams have been treated for benign intracranial lesions. The total world experience with proton therapy is some thirteen thousand patients.[12] Of this total, nearly 75 percent have been treated for benign intracranial lesions, uveal melanoma, and skull base/spine sarcomas.[13] Accordingly, data from proton treatment of the more common malignant neoplasms have accumulated.

Currently there are plans for about twelve new proton beam radiation therapy centers throughout the world.[14] Many of these are projected to be designed and built for radiation treatment and not as physics research facilities with a medical annex. At present the only facility built exclusively for medical purposes is that at Loma Linda University in California. The second one is the facility under construction at the Massachusetts General Hospital. It is to be hoped that there will be collaboration between centers as they are brought into clinical operation so that data may be rapidly generated for a number of important disease sites. A start in this direction has been the formation of the Proton Therapy Oncology Group (PROG) by the National Cancer Institute and the American College of Radiology. Their intent is to facilitate collaboration between various proton beam centers within the United States and abroad.

Intraoperative Electron Beam Therapy

Intraoperative electron beam therapy (IORT), combined with external beam radiation and resection, is clinically advantageous in the treatment of patients with locally advanced rectal carcinoma and retroperitoneal sarcoma, the advantages being an increment in survival and local control relative to that obtained by external beam radiation and surgery. There appears to have been a modest gain in survival time of patients with carcinoma of the pancreas but not in long-term survival rate. With respect to facility design and use, there is movement to have linear accelerators installed in operating room suites. This contrasts with the cumbersome and costly procedure of transferring the patient under full anesthesia from the operating room into a prepared treatment room in the radiation oncology department or dedicating the treatment room for the entire day of the IORT procedure. Such special facilities should increase the ability to test IORT against a greater spectrum of tumor types and sites.

High Technology Brachytherapy

The advantages of brachytherapy are well recognized and yield minimal treatment volumes in well-performed implantations. For complex anatomic situations, brachytherapy is being performed by stereostatic techniques (especially for intracranial lesions) and by CT or ultrasound for other sites. Preliminary studies are in progress at the Brigham and Women's Hospital using online MR imaging to facilitate high-precision implantation. The applications of these several sophisticated methods are allowing an expansion of brachytherapy and should result in some improved results.

Boron Neutron Capture Therapy

This strategy, initially proposed by Locher in 1932, has had long and serious appeal.[15] The principal interest among clinicians involved in the boron neutron capture therapy (BNCT) programs is high-grade glioma. This is based largely upon the fact that these tumors almost exclusively are local problems; nearly all patients succumb to the local growth and or complications of treatment. Secondly, there is the expectation of a greater concentration differential of the boron-containing compound between tumor and normal tissue. Early work on these tumors was not successful and has been interpreted as the result of poor differential concentrations of boron.[16] New compounds are now being investigated. There are active clinical therapy programs at Tokyo for gliomas and cutaneous lesions.[17,18] New programs are being developed at Tufts/Massachusetts Institute of Technology (MIT), Boston; Brookhaven National Laboratory (BNL)/Stony Brook, Long Island; Petten, the Netherlands; Sydney University, Australia; Ohio State University; and Idaho National Engineering Laboratory. Three facilities have the epithermal beams most desired for BNCT of deeply-sited lesions: Tufts/MIT, BNL, and Petten.

Radiolabeled Antibodies and Metabolites

This is a field of high potential and certainly sustained interest due to the very attractive rationale: incorporation of a radioactive atom into a cancer-specific compound, such as an antibody, metabolite, or antisense oligonucleotide, that is selectively concentrated in tumor tissue. Some worthwhile achievements have been realized in the use of this approach for diagnostic imaging. Research in this field for treatment of solid tumors has not yielded obvious clinical gains. Nonetheless, my expectation is that intensive research into tumor biology will yield molecules able to accumulate highly selectively in tumors. Then radiolabeling should not be beyond the capabilities of pharmacologists and will result in a new modality. This category of treatment may not require highly specialized facilities but instead would need a specialized staff.

INDIVIDUALIZED PREDICTION OF PATIENT RESPONSE TO RADIATION

Currently, clinicians utilize a substantial number of useful response indicators in regular practice. These are: tumor size, histological type, histological grade, anatomic site, hemoglobin, sex, age, clinical presentation (exophytic vs. infiltrative), presence of certain genetic diseases (for example, ataxia telangectasia), and presence of active autoimmune disease. In addition, there are physical factors; chief among these are dose per fraction, time between fractions, and total dose. LET is also a parameter of response. Thus, an experienced clinician can provide quite useful estimates of the likelihood of local control and of clinically important treatment-related morbidity following a specified treatment protocol in the individual patient. Even so, some uncertainty remains for the specific patient.

There are several additional potential predictors of response of tissues to radiation under active clinical and laboratory evaluation. The potential for success in the research for new predictors will be limited by: (1) the accuracy with which the value of the new test parameter may be determined, (2) the relationship between the value of the parameter and local outcome, and (3) heterogeneity with respect to this new parameter among tumors accessed into the study.

Were the predictive power to be high, there would be a potential of identifying tumors of a specified type, size, site, etc., for which the standard treatment would be predicted to fail in an unacceptably high proportion of patients. The value for the new predictors might indicate the use of an alternate treatment method for an expected gain.

Intertumoral heterogeneity is the principal determinant of the slope of the dose response curve.[20] Consider these simple calculations. Cell kill follows Poisson statistics. For a population of tumors each of which contains about 10^8 identical clonogens, the slope of the dose response curve for tumor inactivation would correspond to a γ_{50} of about 7.[21] However, the slope of an actual population of clinical tumors is not likely to be greater than 2; the difference being the consequence of intertumoral heterogeneity with respect to one or more of the determinants of response. This difference between γ_{50}s of 7 and 2 is the basis for anticipation of clinical gains from efforts to develop new physiological and radiation biological predictors. By critical assessment of the values obtained from measurements of one or more of the new predictors, there is a potential for devising an improved treatment strategy for the individual patient.

For optimal design of a clinical trial of a procedure which modifies a particular response, the eligibility requirement for accession into the trial must include evidence that the tumor (patient) is expected to benefit from that procedure. For example, a trial of a method which improves tissue pO_2 but accesses tumors that do and do not have hypoxic regions would have little prospect of demonstrating a gain even though a substantial gain would be obtained when applied only to tumors with hypoxic regions.

Current investigations of response predictors are directed principally toward three general lines of research: (1) determination of cellular radiation sensitivity of the tumor or normal tissue cell; (2) characterization of the tumor in physiological and biochemical terms, (for example, pO_2, [SH]); and (3) definition of the proliferation kinetics of the tumor and normal tissue cells. Additional predictors being investigated include genetic assessment of the factors that determine DNA damage repair, predispose the development of distant metastases, and indicate sensitivity to radiation and chemotherapeutic agents.

Inherent radiation sensitivity is being measured *in vitro* on freshly derived cell lines for determination of SF_2, MID, D_0, α, β, and α/β of cells in exponential or plateau growth phases, using colony formation as the endpoint for cell viability. The results of the various published studies have been mixed. There is one strongly positive result: the Courtney Mills assay on fresh cells from carcinoma of the uterine cervix studied at Manchester.[22] For SF_2 values <0.4 and >0.4 there was a wide separation between actuarial local control and local failure rates. These various SF_2 assay techniques are characterized by large coefficients of variation, around 40 percent.[23] This means that the absolute SF_2 as measured on an individual patient is of uncertain value. Further, there is no close correlation in SF_2 values determined on the same cell lines in different laboratories or for different assay techniques.[24] These factors mean real uncertainty, at this writing, as to the value of available assay methods in determining cellular radiation sensitivity *in vitro*.

Radiation sensitivity of tumor cells as they live in tumor tissue is, of course, the property of interest rather than the sensitivity of cells *in vitro* under near optimal metabolic conditions. The principal end-points for estimation of cell sensitivity *in vivo* under current investigations include micronuclei formation, comet assay, and the frequency of specific chromosome breakage, etc. There are at this writing no data that establish any of these techniques as a valid predictor. Experimental error inherent in these methods is not likely to be less than that for cell survival *in vitro*. Quite encouraging results have been published which indicate that the severe response of normal tissue to radiation correlates with the sensitivity of the cells of that normal tissue when assayed *in vitro*.[25] Bentzen and colleagues have discussed the choice of end-points of normal tissue response and the *in vitro* assay.[26]

As pO_2 is a proven powerful determinant of cell lethal response to radiation and hypoxic regions have been demonstrated in many human tumors, the measurements of pO_2 in the individual tumor are a subject of much clinical research. Each of three reports of measured pO_2 before radiation treatment has shown a strong predictive power for local control.[27] Further, tissue pO_2 can be manipulated for potential therapeutic advantage. Hence, the intense interest in the role of tumor pO_2 in determination of treatment outcome.

The electrode technique is the only one that measures pO_2 directly. Limitations of this method include applicability only to accessible tumors, its invasive nature, and uncertainties of measured pO_2 values, especially at ≤5 mm mercury. Electrode measurements of pO_2 in human tumors are being performed for carcinomas of the oral cavity, uterine cervix, skin, anus, and metastatic nodes in the cervical regions. There are indirect methodologies which could be utilized for deep-sited tumors.[28] The accuracy of such techniques is not well defined at this time. Among the indirect procedures expected to provide evidence in patients of the presence of hypoxia are tissue blood flow, binding of certain molecules (for example, misonidazole), comet assay, and MR spectroscopy.[29]

Presently, there is a need to define the ability of *measured* pO_2 values to predict outcomes on a much larger database than those of published results to date. When and if this is achieved, the efforts should be pushed to devise other means for assessing tissue pO_2 so that deep-sited tumors can also be investigated.

Proliferation kinetics of tumor clonogens during a course of fractionated dose irradiation increase the number of clonogens to be inactivated and, hence, reduce tumor control probability. Thus, for rapidly dividing tumor cell populations, treatment might be more effective when administered on an accelerated schedule, for example, two or three fractions per day. Currently RTOG is sponsoring a four-arm trial of accelerated fractionation irradiation against head and neck cancer: a control treatment of 2 gray (Gy) once per day to 70 Gy and three different accelerated methods (RTOG protocol 90-03). The most extreme test of accelerated dose fractionation is the CHART trial at Mt. Vernon Hospital, which features radiation administered in three fractions per day for twelve days; this is compared with conventional fractionation.[30] For this category of trials, the measurement of the proliferative activity of each tumor is made. There are means available, such as potential doubling time (T_{pot}), for these measurements.[31]

Results of a European Organization for Research and Treatment of Cancer (EORTC) trial (protocol 22851) indicate that tumors with short T_{pot} values (rapid proliferation) treated by conventional dose fractionation did less well than those with long T_{pot} values.[32] For tumors with long T_{pot} values, those greater than 4.5 days, results were the same for treatment by the accelerated or conventional dose fractionation. Additional studies are in progress on this question in several centers and interinstitutional groups. These should provide improved estimates of the merit of shortening overall times of treatment. Further methods are needed to estimate clonogen division rate during treatment.

Withers and group have analyzed a very large body of clinical local control data and concluded that the dose to achieve a specified local control probability was independent of time up to three to four weeks; after that point, the dose for that response increased at some 0.6 Gy/day.[33] This gives additional emphasis to the attention which needs to be paid to overall treatment times.

An alternate approach is to employ biological techniques which suppress cell proliferation. Studies are in progress in the laboratory on biologicals which might achieve a blockage of cell division over the relatively brief period of treatment.

For trials of both pO_2 and cell proliferation kinetics modification, there is uncertainty as to the adequacy of measurements made only prior to the commencement of treatment. Should similar determinations be performed at one or even more points during the course of fractionated dose irradiation?

LONG STANDING QUESTIONS—FUTURE ANSWERS?

There are a number of questions which have vexed radiation oncologists for many years that are likely to be answered definitively within the next two to three decades. This effort will be aided greatly by the design of trials with eligibility requirements limiting access to tumors with appropriate biological characteristics (see above).

Since the early 1950s radiation oncologists and radiation biologists have been studying the role of hypoxia in tumors as an important causative factor of local failure.[34] Strategies that have been investigated to minimize the importance of hypoxia include respiration of oxygen or carbogen at normal pressure, oxygen at increased pressure, intraarterial H_2O_2; administration of hypoxic cell sensitizers (such as misonidazole), agents specifically toxic to hypoxic cells; erythropoietin prior to the commencement of radiation to bring hemaglobin up to normal levels, and high-LET radiations and radiation therapy under conditions of local tissue hypoxia (tourniquet technique for extremity sarcomas). Despite the lack of clear successes, extensive laboratory research has continued unabated. The availability of reasonably reliable methods for measuring pO_2 in accessible human tumors is stimulating renewed interest in attempts to modify tumor tissue pO_2 distributions. Results of well-designed trials are expected to provide an evaluation of the importance of hypoxic tumor cells to outcome in radiation treatments. A current series of trials is based upon the work of the Gray Laboratory.[35] These feature respiration of carbogen at one atmosphere pressure to reduce diffusion limited hypoxia and administration of nicotinamide to reduce perfusion limited hypoxia, and these are integrated with an accelerated treatment protocol to decrease the importance of cell proliferation.

Due to the very active clinical trial programs in many parts of the world of accelerated dose fractionation, the clinical gain for each of several tumor types and for specific cell proliferation kinetics profiles should be firmly defined within one to two decades. In parallel with these endeavors, the benefit of administration of radiation in quite small doses per fraction—at less than 1.4 Gy—is also expected to become known.

The clinical efficacy of high-LET radiations in clinical radiation oncology should be well established in the future. This will come about from analyses of outcome data from closed Phase III trials and ongoing trials of fast neutron therapy. To a lesser extent there will be data from the testing of high-Z particle beam therapy. In addition, the predictive power of measured values of cellular radiation sensitivity *in vitro* and even *in vivo* will be defined.

Included here should be the characterization of the immunological status of the patient and the immunogenetic relationship between the patient and the tumor. There will surely be developed new and more effective methods for altering specific components of the immunological status of the patient, so as to augment the efficacy of the antitumor reaction.

This work on the immune reaction has important possibilities for enhancing the diagnostic accuracy of radioisotope labeled antibodies. This may well extend to therapeutic success for a few selected tumors, for which there is great specificity of the antibody to the tumor associated antigen, a large number of receptors/tumor cell, and good access of the antibody to the tumor cell and, hence, the antigen. This is an area which has been the subject of substantial attention in experimental animal tumor systems but has seen modest success when studying spontaneous autochthonous tumors. The level of research in tumor immunology gives good prospects for some gains in either diagnosis or treatment of the cancer patient.

INCREASED AWARENESS OF THE LATE SEQUELAE OF RADIATION TREATMENT

There will be a much heightened sensitivity to and awareness of the late changes following radiation treatments as large numbers of patients will be surviving twenty to forty years postradiation. This concern will result in many studies that will contribute to the understanding of the role of dose, anatomic part irradiated, treatment volume, patient age, use of chemotherapy concomitantly or at separate times, and observation time. With reference to radiation-induced tumors, there will almost certainly be an increase (probably of a nontrivial magnitude) over the currently recognized frequency of 0.5 percent at twenty years after treatment with radiation alone in adult patients and significantly higher for patients treated by radiation and chemotherapy. This will place much greater pressure to use treatment techniques which involve the irradiation of smaller volumes of nontarget tissues. Radiation-induced morbidity does not develop in unirradiated tissues, and society is likely to be decreasingly tolerant of morbidity developing in nontarget tissues. This is significant, as most of the major treatment related morbidities arise in nontarget tissues.

STANDARDIZATION OF RADIATION ONCOLOGY PRACTICE

The trend toward the use of accepted or standard treatment techniques is expected to be strengthened with the continued publication of "good" results from large series. By this, I mean that for the more common tumors there can be expected to be a generally accepted radiation treatment technique and dose (dose per fraction, total dose, and overall time). Treatment by other than an approved method will be less easy and surely more risky than at present, given the exigencies of both litigation and managed care. The exception will be limited to evaluation of new strategies on institutional- and or governmental-authorized protocols.

RADIATION ONCOLOGY EDUCATION AND MANPOWER

The available data indicate a substantial surplus of radiation oncologists within the next decade. D. Flynn estimates that there is a net increase in radiation oncologists in the United States of 120 per year (160 graduates less 35 retirements and 5 deaths among active practitioners).[36] This would mean some 1,200 additional radiation oncologists by 2003. The impact would be a reduction in the number of new patients per year per full time equivalent to only 172 by 2003. This would be down from 194 in 1993 and 212 for the period 1974 to 1990. These estimates make no allowance for a probable decrease in the number of patients being referred to radiation oncology. Currently, some 48 percent of cancer patients receive radiation. Further, the health care system can be expected to specify that each clinician be responsible for more rather than fewer patients. However, definitive treatments will require more physician time as the more sophisticated techniques come into clinical use. These figures, then, stress the urgency to assess critically the large number of young doctors being trained in this specialty.

There is an obligation to enhance the quality of education of young doctors accepted into our residency programs. Education of residents in radiation oncology will be modified and, it is hoped, enhanced by the introduction of many computer-aided teaching tools. For example, teaching of human anatomy should be much improved by the 3-D visualization of any anatomic part or region from all angles, external and internal. Perhaps virtual reality, holograms, and other methods will become available and, if so, should be real boons. Interactive teaching programs should aid in all aspects of didactic teaching. Further, one can expect there to be simulations of all of the procedures in radiation oncology. These same capabilities will also become available in all facets of medical practice, facilitating and accelerating the learning and retention processes. Surely these projected tools will make for better informed and more continuously up-to-date physicians.

Through regulation of payment for radiation oncology services, the number and the distribution of radiation oncologists and radiation oncology facilities will be controlled to meet the expected needs of the American population. There will probably be pressure in the United States for the number of patients treated per radiation oncologist per year to go up toward the levels current in other countries.

RADIATION ONCOLOGY IN TWENTY TO THIRTY YEARS

Despite the virtual certainty of large gains in the effectiveness of radiation oncology, a serious question is whether or not advances in basic genetics and the applications of genetics to clinical oncology will be so great that we have to consider whether radiation will even be needed as a major therapeutic modality within thirty to forty years. There are serious and highly respected scientists, in addition to the popular press, who are so optimistic about the potentials of the clinical applications of molecular biology and gene therapy to oncology that they have predicted that surgery, radiation, and chemotherapy as currently employed would be unnecessary. However, the number of years required for even partial realization of this remarkable goal is clearly uncertain. Lewis Thomas, one of the most highly regarded thinkers and spokespersons in medicine, stated in 1983 that "cancer would no longer be a problem by the year 2000."[37] Achievement of this nirvanic state in oncology will require the resolution of a number of quite important problems within a very short time period. Regrettably, at this writing such a happy turn of events does not appear to be a serious prospect for the near or even for the not-so-near future. Popular weekly magazines such as *Time* and *Newsweek* detail glowing accounts of the changes to be wrought in oncology as a consequence of the research in molecular biology.[38]

There can, however, be little doubt that there will be: (1) dramatic increments in the knowledge of the oncogenic process in molecular terms, (2) important gains from the utilization of that knowledge in diagnosis and prognostication, and (3) major advantages accrued from the capacity to identify individuals with high risk of development of a neoplasm, in many instances of a specific type. The practical clinical utilization of these genetic insights in the eradication of solid tumors is less clear.

Genetically defined diagnostic categories

The genetic analysis of the biopsy specimen from an established tumor should bring an extensive revision of diagnostic categories; the new pathological types will be based not only on morphological appearance and immunohistochemical profile but, importantly, upon a genetic characterization with special reference to genes which determine, among other things, the metastatic potential, capacity to repair radiation damage, ability to cope with cytotoxic chemicals, and rate of cell proliferation.

Already, this area of research has led to quite accurate predictions of the clinical course of neuroblastoma. Namely, the extent of amplification of the N-*myc* gene is inversely related to survival probability.[39] The number of copies of N-*myc* is now, by a substantial margin, the most important diagnostic feature of neuroblastoma. There is every confidence that there will be developed highly reliable genetic indicators of the degree of malignancy for many tumors. Preliminary experience with genetic markers as prognostic markers for metastatic probability appears positive for carcinomas of the breast, ovary, and lung. This information will mean a sophisticated individualization of the management strategy.

Identification of individuals at high risk for development of specified tumors

A parallel series of developments will result in the capacity to identify individuals with the genetic alteration which will at some point be associated with the appearance of specific neoplasms. The consequence will be that these particular persons can be monitored carefully for the earliest evidence of that specific tumor and treatment implemented at a minimal stage. Alternatively, there may be "preemptive" therapy in identified patients. For example, at a specified age, bilateral mastectomy in women carrying defined genetic abnormalities associated with exceptionally high probabilities of developing breast cancer is being performed in a few clinics.

As the skill in defining the genetic characteristics that are regularly associated with development of particular neoplasms is expanded, the frequency with which patients are seen with more than minimal disease should plummet or at least decrease appreciably. This will, for some categories of tumors, mean that those patients will be managed by rather simple procedures and experience a higher cure rate. We need only consider the impact of the extensive use of the Pap smear on the frequency distribution of stages of carcinoma of the uterine cervix. In communities where this test is commonly employed, very large proportions of patients with carcinoma of the uterine cervix are treated by conization or hysterectomy for carcinoma *in situ* or early stage disease. Genetic study of the general population or of defined high risk individuals is likely to have a similar impact on at least some diseases. The result will be fewer patients requiring radiation therapy.

For many tumor categories, knowledge of a virtual certainty of a later development of tumor in an individual patient is not likely to provide an opportunity to diagnose the tumor at a sufficiently early stage to augment sharply the therapeutic outcome, at least from the use of currently available treatment methods. This limit might apply to, for example, carcinoma of the pancreas or high-grade glioma. As there is virtually no clinical experience with these tumors at very small size, up to 1 centimeter, we do not know the efficacy of even current treatment methods against such small lesions. There might be a pleasant surprise.

Gene therapy

As recent reviews demonstrate, the potential for use of genetic approaches in therapy has clearly attracted intense interest and excitement in biomedicine.[40] The basic strategy is to define the genetic abnormality and to determine the gene construct which would reverse and eradicate the pathological process and then proceed to correct the genetic abnormality. This requires (1) the availability of needed gene(s), (2) an appropriate vector for the delivery of the new genetic material to some or all of the affected cells, (3) integration of the new gene(s) into the genome of those cells and commencement of proper function in both quantitative and qualitative terms, (4) the new gene function be stable, and (5) little toxicity of the procedure and risk of secondary morbidity from the use of a viral or other vector relative to the clinical effect on the disease process being treated.

To date partial success in gene therapy has been realized in the treatment of a small number of patients whose diseases are due to defects in a single gene. This has apparently been accomplished with ADA (adenosine deaminase deficiency) and LDL receptor deficiency resulting in familial hypercholesterolemia. For these treatments the patients' target cells were transduced *ex vivo* and then reimplanted. These brilliant achievements bode well for advances to come. The major question is the effort and time required for these to be realized. The current issue of the influential London *Economist* gives the year 2015 as the time for success against the recessive single gene mutation diseases.[41]

For genetic treatment of this class of disease, there is no need to affect all cells of the particular category in the patient. The requirement is to have altered sufficient cells to produce a minimal quantity of the deficient compound, be it an enzyme, hormone, metabolite, etc. However, the transfected cells must function in a stable manner for very long times. This contrasts with the situation for anticancer treatment where all cells of the tumor must be altered genetically unless there is a significant bystander effect of the gene product. On the positive side is the fact that the action of the gene need persist only for the period required to inactivate the cell.

That success of gene therapy of even the single gene diseases is far from straightforward is evident in the fact that results to date are limited to only two such diseases, small numbers of patients, relatively short follow-up observation periods and a palliative result. This remains true despite the cloning of the altered gene some years ago. For example, the genetic defect for hemophilia was defined some years ago with no major clinical benefit to date. This indicates the difficulties in carrying detailed understanding of the gene alteration forward to make an effective clinical treatment strategy.

The clinical testing of these methodologies has just commenced. At present, much of the effort is and will continue for some undefined period toward overcoming the technical problems of transfecting appropriate numbers of target cells of the patient either using *in vivo* or *ex vivo* techniques.[42] Extension of these strategies to treatment of the cancer patient is sure to prove more difficult. This is due to several factors: the genetic abnormality in the cancer cell is more than one gene alteration; there is extensive genetic heterogeneity of tumor cells; the delivery of the genetic material is complicated by the pathophysiology of tumor tissue, that is, poor to nearly absent perfusion of regions of tumor; and the new genetic material must be delivered to all or nearly all of the tumor cells. For some proposed mechanisms there would be an important bystander effect so that the cell kill would be greater than the actual number of transfected cells.

The inherent pathophysiological characteristics of solid tumors impose such severe constraints on the selective delivery of gene therapy agents to a useful proportion of tumor cells that the eradication of established tumors by such strategies does not appear to be a high likelihood in the near term. These factors are almost certain to prove a serious constraint on the efficacy of gene therapy against epithelial and mesenchymal tumors. However, were the gene therapy strategy to achieve a cell kill by methodologies which were to be relatively nontoxic of only 90 percent, this would not be insignificant. This is especially so if gene therapy could be repeated one or more times. Such a treatment could be combined with other modalities, say, radiation, to achieve eradication of all tumor clonogens at a cost of lower treatment-related morbidity than that following the use of radiation alone.

An alternate genetic approach is to introduce genes designed to make the transfected cell strongly antigenic and trigger an effective response by the host. As a complimentary approach, the cells of the host immune system could be modified genetically in such a manner as to enhance their cytotoxicity.

That some clinical benefit will accrue from these diverse efforts in gene therapy is judged to be of an extremely high probability. The real question is the time frame and the magnitude of the gain and the determination of which specific tumors. The writers for the *Economist* have given 2040 as the year for elimination of cancer as a threat to life. One is not necessarily skeptical to observe that if the estimated great day has been moved from 2000 to 2040 there might be some additional upward revisions to come.

A concern in radiation oncology research is the present and future impact that the optimism reflected by the Thomas statement and articles of related short-term "pay-off" of research in molecular biology, such as that in the *Economist*, are having and will have on the allocation of research funds in oncology. From the perspective of a clinician with only a limited appreciation of the prospects for the modern biology in this area, I expect that there will remain major difficulties in the eradication of gross neoplastic disease despite creative and aggressive attempts to apply the techniques of molecular biology. According to this opinion, we should proceed to develop radiation oncology with the expectation that radiation will continue to be an essential component of the management strategy for a large fraction of cancer patients well into the future.

Gene activation by radiation

There is now convincing evidence that radiation doses of the magnitude employed clinically activate a variety of human genes. There is potential for the use of this category of response to therapeutic advantage. This effect is for many genes of small degree and the duration of the activation is rather short. Nonetheless, this phenomenon is of interest and is being pursued in the laboratory.

SUMMARY

From the perspective of this radiation oncologist, there is an impressive array of important advances in radiation oncology to be brought into clinical practice in the coming several decades. This is to be accompanied by a startling increase in the understanding of the biology of tumor initiation, progression, metastasis, and response to the diverse therapies employed and being developed. Within our specialty, the elegance of actual treatment will be advanced markedly. Further, we almost certainly will have the capability to predict the tumor and normal tissue response in the individual patient to a greater degree than now obtains. This will provide the basis for the design of treatment strategies which maximize the therapeutic efficacy of the treatment modalities available. Moreover, the questions which have occupied so large a portion of our intellect, time, energy, and resources will have been answered, and these capabilities will be directed to new problems. An important basis for a high and general level of optimism is the exceptional quality of new entrants into the century-old field of radiation oncology.

▶ ▶ ▶ ▶ ▶ ▶

REFERENCES

The author is most appreciative of the excellent comments on this paper by Drs. Edward R. Epp, Simon N. Powell, Ira J. Spiro, and C. C. Wang. Ms. Claire Hunt provided valuable help in preparing this manuscript.

This work was supported in part by USPHS/NCI Grants CA13311 and CA21239.

1 Niemierko, A., and Goitein, M., "Implementation of a Model for Estimating Tumor Control Probability for an Inhomogeneously Irradiated Tumor," *Radiother. Oncol.* 29 (1993):140-147; Ling, C.C.; Burman, C.; Chui, C.S.; et al., "Perspectives of Multidimensional Conformal Radiation Treatment," *Radiother. Oncol.* 29 (1993):129-139; and Lyman, J.T., "Complication Probabilities as Assessed from Dose Volume Histograms," *Radiat. Res.* 104 (1985):513-519.

2. Gall, K.P., and Verhey, L.J., "Computer-Assisted Positioning of Radiotherapy Patients Using Implanted Radiopaque Fiducials," *Med. Phys.* 20 (1993):1153-1159.
3. Gill, S.S.; Thomas, D.G.T; Warrington, A.P.; and Brada, M., "Relocatable Frame for Stereotactic External Beam Radiotherapy," *Int. J. Radiat. Oncol. Biol. Phys.* 20 (1991):599-603.
4. *Radiotherapy and Oncology* 29 (1993):81-284; Meyer, J., and Purdy, J., eds. *Frontiers of Radiation Therapy and Oncology.* vol. 29. Basel, Switzerland: Karger, 1994.
5. BRS Colleague.
6. Laramore, G., and Griffin, T., "Fast Neutron Radiotherapy: Where Have We Been and Where Are We Going? The Jury Is Still Out—Regarding Maor et al.," *Int. J. Radiat. Oncol. Biol. Phys.* 32 (1995): 599-604, 879-882.
7. Suit, H.D.; Silver, G.; Sedlacek, R.S.; and Walker, A., "Experimental Condition and the Acute Reaction of Mouse Skin to Ionizing Radiation," *Radiat. Research* 95 (1983):427-433.
8. Personal communication, T. Griffin, 1994.
9. Kawachi, K.; Kanai, T.; Endo, M.; et al., "Radiation Oncological Facilities of the HIMAC," *J. Jpn. Soc. Ther. Radio. Oncol.* 1 (1989):19-29.
10. Castro, J.R., "Heavy Ion Therapy: Bevalac Epoch," Proceedings of the International Symposium on Hadron Therapy, Como, Italy, 18-21 Oct. 1993, in *Nuclear Instruments and Methods in Physical Research*, 1994.
11. Munzenrider, J.E.; Verhey, L.J.; Gragoudas, E.S.; et al., "Conservative Treatment of Uveal Melanoma: Local Recurrence after Proton Beam Therapy," *Int. J. Radiat. Oncol. Biol. Phys.* 17 (1989):493-498; and Liebsch, N.J.; Ojemann, R.; and Munzerider, J.E., unpublished data, 1994.
12. *Particles. A Newsletter for those interested in proton, light ion and heavy charged particle radiotherapy* 13 (1994). Ed. by J. Sisterson.
13. Suit, H.D., and Urie, M., "Clinical Gains to be Realized from Proton Beams in Radiation Therapy," *J. Natl. Cancer Inst.* 84 (1992):155-164.
14. *Particles.*
15. Wazer, D.E.; Zamenhof, R.G.; Harling, O.K.; and Madoc-Jones, H., "Boron Neutron Capture Therapy," Chapter 7 in Mauch, P., and Loeffler, J., eds. *Radiation Oncology: Technology and Molecular Biology*. Philadelphia: W.B. Saunders, 1994.
16. Sweet, W.H.; Soloway, A.H.; and Brownell, G.L., "Boron-Slow Neutron Capture Therapy of Gliomas," *Acta Radiol. Ther.* (Stockholm) 1 (1963):114.
17. Hatanaka, H.; Kamano, S.; Amano, K.; et al., "Clinical Experience of Boron Neutron Capture Therapy For Gliomas—A Comparison to Conventional Chemo-Immuno-Radiotherapy," in Hatanka, H., ed. *Boron Neutron Capture Therapy for Tumors* (Niigata, Japan: Nishimura Co., 1986):349.
18. Mishima, Y.; Honda, C.; Ichihashi, M.; et al., "Treatment of Malignant Melanoma by Single Thermal Neutron Capture Therapy with Melanoma-Seeking ^{10}B Compound," *Lancet* 31 (1989):388.
19. Personal communication, D. Wazer, 1994.
20. Suit, H.D.; Skates, S.; Taghian, A.; et al., "Clinical Implications of Heterogeneity of Tumor Response to Radiation Therapy," *Radiother. Oncol.* 25 (1992):251-260.
21. γ_{50} is the descriptor of slope proposed by Brahme. This is the percent point increase in response probability for a 1 percent increase in dose for treatment at or near the 0.5 response probability. See Brahme, A., "Dosimetric Precision Requirements in Radiation Therapy," *Acta Radiol. Oncol.* 23 (1984):379-391.
22. West, C.M.L.; Davidson, S.E.; Roberts, S.A.; and Hunter, R.D., "Intrinsic Radiosensitivity and Prediction of Patient Response to Radiotherapy for Carcinoma of the Cervix," *Brit. J. Cancer* 68 (1993):819-823.
23. Suit, Skates, and Taghian, "Clinical Implications."
24. Taghian, A.; Geara, F.; Dahlberg, W.; et al., "Inter-Researcher, Inter-Laboratory and Inter-Technique Variations in the Determination of SF2 as Predictive Assay," (Abstract) *Radiat. Res.* (1994).
25. Geara, F.B.; Peters, L.J.; Ang, K.K.; et al., "Radiosensitivity Measurement of Keratinocytes and Fibroblasts from Radiotherapy Patients," *Int. J. Radiat. Oncol. Biol. Phys.* 24 (1992):287-293.
26. Bentzen, S.M.; Overgaard, M.; and Overgaard, J., "Clinical Correlations Between Late Normal Tissue Endpoints after Radiotherapy: Implications for Predictive Assays of Radiosensitivity," *Euro. J. Cancer* 29A (1993):1373-1376.
27. Kolstad, P., "Intercapillary Distance, Oxygen Tension and Local Recurrence in Cervix Cancer," *Scand. J. Clin. Lab. Invest. Suppl.* 106 (1968):145-157; Gatenby, R.A.; Kessler, H.B.; Rosenblum, J.S.; et al., "Oxygen Distribution in Squamous Cell Carcinoma Metstases and Its Relationship to Outcome of Radiation Therapy," *Int. J. Radiat. Oncol. Biol. Phys.* 14 (1988):831-838; and Hockel, M.; Knoop, C.; Schlenger, K.; et al., "Intratumoral pO2 Predicts Survival in Advanced Cancer of the Uterine Cervix," *Radiother. Oncol.* 26 (1993):45-50.
28. Chapman, J.D., "Measurement of Tumor Hypoxia by Invasive and Non-Invasive Procedures: A Review of Recent Clinical Studies," *Radiother. Oncol. Suppl.* 20 (1991):13-19.
29. Olive, P.L.; Banath, J.P; Durand, R.E., "Heterogeneity in Radiation-Induced DNA Damage and Repair in Tumor and Normal Cells Measured Using the 'Comet' Assay," *Radiat. Res.* 122 (1990):86-94; Parliament, M.B.; Wiebe, L.I.; and Franko, A.J., "Nitroimidazole Adducts as Markers for Tissue Hypoxia: Mechanistic Studies in Aerobic Normal Tissues and Tumour Cells," *Brit. J.*

Cancer 66 (1992):1103-1108; and Raleigh, J.A.; Zeman, E.M.; Rathman, M.V.; et al., "Development of an ELISA for the Detection of 2-Nitroimidazole Hypoxia Markers Bound to Tumor Tissue," *Int. J. Radiat. Oncol. Biol. Phys.* 22 (1992):403-405.

30   Saunders, M.I.; Dische, S.; Grosch, E.J.; et al. "Experience with CHART," *Int. J. Radiat. Oncol. Biol. Phys.* 21 (1991):871-78.

31   Begg, A.C.; McNally, N.J.; Shrieve, D.C.; and Karcher, H., "A Method to Measure the Duration of DNA Synthesis and the Potential Doubling Time from a Single Sample," *Cytometry* 6 (1985):620-626;Wilson, G.D., "Assessment of Human Tumour Proliferation Using Bromodeoxyuridine: Current Status," *Acta Oncol.* 30 (1991):903-910.

32   Horiot, J.C., "Present Status and Updated Results of the Trials on Hyperfractionation (FH) and/or Accelerated Fractionation (AF)," (Abstract) *Euro. J. Cancer* Suppl. 6. 29A (1993):S43.

33   Withers, H.R.; Taylor, J.M.G.; and Maciejewski, B., "The Hazard of Accelerated Tumor Clonogen Repopulation during Radiotherapy," *Acta Oncol.* 27 (1988):131-146.

34   Gray, L.H.; Conger, A.D.; Ebert, M.; et al., "The Concentration of Oxygen Dissolved in Tissues at the Time of Irradiation as a Factor in Radiotherapy," *Brit. J. Rad* 26 (1953):638-648.

35   Kjellen, E.; Joiner, M.C.; Collier, J.M.; et al., "A Therapeutic Benefit from Combining Normobatic Carbogen or Oxygen with Nicotinamide in Fractionated X-ray Treatments," *Radiother. Oncol.* 22 (1991):81-91; and European Organization for Research and Treatment of Cancer. Protocol no. 22.923. ARCON: Accelerated Radiotherapy with Carbogen and Nicotinamide. A phase I/II study. Study coordinator: Bernier J, Dept. of Radiotherapy, Ospedale San Giovanni, 6500 Bellinozona, Switzerland.

36   Personal communication, D. Flynn, 1995.

37   Thomas, L. *The Youngest Science* (New York: Viking Press, 1983):205. Quote from *New York Times,* 20 February 1983.

38   *Time*, 17 January 1994; and Cowley, G., "Family Matters: The Hunt for Breast Cancer Gene," *Newsweek* 6 December 1993:46-55.

39   Seeger, R.C.; Brodeur, G.M.; Sather, H.; et al., "Association Multiple Copies of the N-MYC Oncogene with Rapid Progression of Neuroblastoma," *New Eng. J. Med.* 313 (1985):1111-1116.

40   Blaese, R.M.; Mullen, C.A.; and Ramsey, W.J., "Strategies for Gene Therapy," *Pathol-Biol* (Paris) 41 (1993):673-676; Friedman, T., "A Brief History of Gene Therapy," *Nature Genetics* 2 (1992):93-98; Miller, A.D., "Human Gene Therapy Comes of Age," *Nature* 357 (1992):455-460; Tepper, R.I., and Mule, J.J., "Experimental and Clinical Studies of Cytokine Gene-Modified Tumor Cells," *Human Gene Therapy* 5 (1994):153-164; Gutierrez, A.A.; Lemoine, N.R.; and Sikora, K., "Gene Therapy for Cancer," *Lancet* 339 (1992):715-721; and Kozarsky, K.; Grossman, M.; and Wilson, J.M., "Adenovirus-Mediated Correction of the Genetic Defect in Hepatocytes from Patients with Familial Hypercholesterolemia," *Somat. Cell. Mol. Genet.* 19 (1993):449-458.

41   "The Future of Medicine," *The Economist* 330 (19-25 March 1994): 17.

42   Mulligan, R.C., "The Basic Science of Gene Therapy," *Science* 260 (1993):926-932.

43   Fornace, A.J., Jr., "Mammalian Genes Induced by Radiation: Activation of Genes Associated with Growth Control," *Annu. Rev. Genet.* 26 (1992):507-526; Hallahan, D.E.; Virudachalam, S.; Schwartz, J.L.; et al., "Inhibition of Protein Kinases Sensitizes Human Tumor Cells to Ionizing Radiation," *Radiat. Res.* 129 (1992):345-350; Fukunaga, N.; Burrows, H.L.; Meyers, M.; et al., "Enhanced Induction of Tissue-Type Plasminogen Activator in Normal Human Cells Compared to Cancer-Prone Cells Following Ionizing Radiation," *Int. J. Radiat. Oncol. Biol. Phys.* 24 (1992):949-957; and Markiewicz, D.A.; McKenna, W.G.; Flick, M.B.; et al., "The Effects of Radiation on the Expression of a Newly Cloned and Characterized Rat Cyclin B mRNA," *Int. J. Radiat. Oncol. Biol. Phys.* 28 (1994):135-144.

Index

AAPM. *See* American Association of Physicists in Medicine
AB Scanditronix
    microtron development, 71
Abbé, R., 193
    afterloading, 23, 118, 196
    studies with Marie Curie, 173
    vaginal application of radium for cervical cancer, 187
The ABC of the X Rays, 166
ABR. *See* American Board of Radiology
Absolute dosimetry, 74
Absorbed dose, 63, 74, 108
Ackerman, L., 30
    cancer textbook, 28
Acme Machine Shop and Electric Company
    cobalt-60 unit, 114
ACS. *See* American Cancer Society
ADA. *See* Adenosine deaminase deficiency
Adair, F.
    Memorial Hospital staff member, 194
Adams, G.E.
    betatron research, 105
    bioreductive drug research, 151
    Gray Laboratory and, 152
    hypoxia research, 143, 144
Adenosine deaminase deficiency, 309
Advisory Committee on the Biological Effects of Ionizing Radiations, 121
Advisory Committee on X-ray and Radium Protection
    radiation protection recommendations, 110
AERE. *See* Atomic Energy Research Establishment
AES. *See* American Endocurietherapy Society
Aetna Insurance Company
    health insurance plan, 205
African American radiation oncologists. *See also specific persons by name*
    history of African American physicians, 264-66
    table, 271
    timeline history of, 266-73
Afterloading, 23, 118, 153, 195-97
AHA. *See* American Hospital Association
Aikins, W.
    American Radium Society president, 192
    ARS president, 219
"Air wall" ionization chambers, 55, 108-9
Albers-Schönberg, L.
    Americans going to study with, 173
    spermicidal properties of X rays, 17, 130
Alexander, L.L.
    career and accomplishments, 270
Allen, S.J.
    dosimetric method research, 54
Allen, S.W.
    pain relief experiments, 12
Allen, W.E., 272
    accomplishments of African American radiologists, 263-64
    career and accomplishments, 267, 268
Allis-Chalmers
    betatron development, 52, 58, 104, 105
Almond, P.
    electron research, 115
Alper, T.
    neutron research, 142, 143
AMA. *See* American Medical Association
American Association of Physicists in Medicine
    dosimetric protocols, 63-64, 75
    stereotactic radiosurgery report, 75
    Task Group 21, 117
American Board of Ophthalmology, 174, 224
American Board of Radiology
    Committee on Graduate Radiologic Training
        internship recommendations, 175
    founding of, 192
    incorporation of, 175, 176
    radiation oncology and, 224-25
American Cancer Society
    cigarette smoking statement, 223
    linear accelerator grant, 61
    prostate cancer survival rates, 40
American Club of Therapeutic Radiologists, 180
American College of Radiology
    Annual Report of 1965
        radiation oncology items, 223
    billing for radiological services, 203, 204, 207-8
    budget, 222
    Commission on Cancer
        standards for radiotherapy in approved cancer centers, 223
    Council of Affiliated Regional Radiation Oncology Societies
        formation of, 222
    founding of, 175
    history of, 221-23
    in-training examination, 181-82
    income range survey, 206
    Intersociety Committee for Radiology, 221
    joint venture policy, 226
    local and state chapter development, 222
    Medicare and, 215-16
    membership, 221
    original purpose of, 221
    Proton Therapy Oncology Group, 301
    radiation oncology and, 221-23
    RVS development and implementation, 214
    "A Standard Nomenclature for Radiation Therapy, ACT Supplement #2, 1975," 211, 212
    UCR reimbursement and, 210-11
American Electrotherapeutic Association, 4
American Endocrine Laboratory
    radioactive medical rejuvenation devices, 288
American Endocurietherapy Society
    founding of, 197
American Hospital Association
    billing for radiology services, 203
The American Journal of Electrotherapeutics and Radiology 239
American Journal of Roentgenology and Radium Therapy, 192
    founding of, 220
    Janeway's article on radon seeds, 193-94
American Medical Association
    admission of African Americans, 265
    Council of Medical Education
        list of hospitals with approved internships, 173, 180
        specialty conference committee sponsor, 180
    hospital billing of radiology services and, 203
    interaction with radiation oncology, 220-21
    overutilization of hospitals, 203
    radium rental issue, 218
    self-referral issue, 227
American Medicine, 169
The American Quarterly of Roentgenology, 170
American Radium Society
    ABR formation and, 220

INDEX

accreditation efforts, 192
founding of, 192, 219
history of, 16, 219-20
inappropriate uses of radium and, 217-18
Janeway lectures, 192
membership, 201
proposal to become section of AMA, 220
representation on ACR Board of Chancellors, 221
American Roentgen Ray Society
ABR formation and, 220
radiation burns reports, 285
Transactions, 170
American Society for Therapeutic Radiology and Oncology, 178
history of, 223-24
name changes, 224
representation on ACR Board of Chancellors, 221
American Women's Hospitals, 242
The American X-Ray Journal, 170
Americium-241, 197
Analgesic applications of radium, 194
Analgesic applications of X rays, 11-13
Antineoplastic drugs
early research, 37
Archives of Clinical Skiagraphy, 166
Army Institute of Pathology
Registry of Radiologic Pathology, 176
Arnesen, A.N.
isoeffect curve research, 131
Arneth, G.
Hodgkin's disease treatment, 11
ARRS. *See* American Roentgen Ray Society
ARS. *See* American Radium Society
Ashayeri, E.
work with and for African American radiologists, 273
Associated Radium Chemists
patent medicines, 288
ASTRO. *See* American Society for Therapeutic Radiology and Oncology
Atomic Energy Commission
establishment, 136
iodine-125 and, 195
licensing use of radioelements, 219
radiation protection, 117
restriction on use of iridium-192, 195
Atomic Energy of Canada Ltd.
remote afterloader, 197
Therac-25 machines, 68
Atomic Energy Research Establishment
linear accelerator research, 53
Atomic medicine development, 195
Attix, F.H.
exposure rate measurement, 97
Autoradiography, 143-44

Baetjer, F., 171, 172, 173
effects of X rays on hen's eggs, 17
European study, 173
Bagg, H.
Memorial Hospital staff member, 194
Bagshaw, M.
effectiveness of ^{60}Co therapy, 33
prostate cancer research, 39
Bailey, H., 190
cervical cancer treatment, 192
Bailey, W.J.A.
radioactive patent medicines, 287-89
Baily, N.A.
real-time imaging research, 70

Bainbridge, W.S.
career and accomplishments, 266
Banting, F.
insulin research, 284
Barendsen, E., 141
linear-quadratic model, 145-46
Barkla, C. G.
polarization of X rays, 91
Barringer, B.
cystoscopic introduction of radon, 189
Memorial Hospital staff member, 194
textbook, 190, 191
work with Failla, 94
Bartlett
"shockproof and rayproof" system, 46
Baskerville, C., 187
Baylor University Hospital
health insurance plan, 202
Beam parameter measurements, 75-76
Beck, C.
osteoscope, 53
textbook, 170
Béclère, A.
Americans going to study with, 173, 187-88
Becquerel, A.H.
radium properties discovery, 185, 279
uranium discoveries, 185
Begg, A.
flow cytometer research, 153
Gray Laboratory and, 152
Bell, A.G.
interest in radioactivity, 280
interstitial radium use, 97
sealed radium treatment suggestion, 185-86
Bellevue Hospital, New York
African American radiologist training and, 267-68
Benign conditions. *See also specific conditions by name*
early X ray treatments, 4, 6-9
Benoist, L.
penetrometer, 53
Bentel, G.C.
universal wedge, 66
Bentzen, S.
fractionation research, 146
Bergonié, J.
spermicidal properties of X rays, 17, 130, 133
Berson, S.
radioimmunoassay research, 105
Best, C. H.
insulin research, 284
Betatrons
availability of units, 37
early development, 51-52, 58, 104-5, 107, 111
introduction of, 28, 30, 31
small field limitation, 58
Billing for radiological services, 202-3, 206-7
Binkley, G.E.
Memorial Hospital staff member, 194
Biological Laboratory of Radium, 186
Bladder cancer
cystoscopic introduction of radon treatment, 189
Blanc, J.
cell response to radiation, 17
Bleehen, N.
high pressure oxygen research, 140
Blue Cross-Blue Shield plan, 205
Blue Cross Hospital Insurance Plan, 202
BNCT. *See* Boron neutron capture therapy

314

INDEX

Boag, J.
- Gray Laboratory and, 152

Boggs, R.H.
- ARS vice president, 219

Bone marrow
- early effects of X rays on, 17

Bone portrait studios, 2

Boot, H.
- magnetron development, 53

Boron neutron capture therapy, 302

Boston Female Medical College
- as first women's medical school, 234

Boston Medical and Surgical Journal, 169

Bothe, 94-95

Bousfield, M.O.
- history of African American physicians, 264-65

Bowing, H., 24, 25
- Mayo Clinic radium therapy chief, 193

Braasch, A.
- electron beam therapy research, 114

Brachytherapy, 23-24, 38
- early history of radium in the United States, 187-92
  - founding of the American Radium Society, 192
  - Memorial Hospital and, 189-91
  - uterine cancer treatment, 191-92
- 1910-1950 era
  - brachytherapy systems, 98-99
  - radium and radon availability, 97
  - radium source strength and, 96-97
  - radon plants, 97-98
  - standardization of radium sources, 96
- 1920-1940: golden era of radium, 192-94
- 1950-1970 era
  - computer calculations and, 111, 113
  - dark age of brachytherapy, 194-97
  - afterloading of radioactive sources, 195-96
  - atomic medicine advances, 195
  - problems with radium implants, 194-95
  - remote afterloading, 196-97
- megavoltage era: 1970 to present, 118
- 1980 to the present, 197-98
- applicator types, 190
- definition, 187
- dose-rate effect and, 153
- early radium discoveries, 185-87
- high dose rate, 118, 153, 197
- low dose rate, 197
- online MR imaging and, 302
- pulsed-dose-rate, 197

Brachytron, 197

Brady, L.
- mentoring of African American radiologists, 273

Bragg, W.H.
- cavity ionization chamber research, 109
- X-ray spectrometer, 93

Bragg-Gray theory, 109, 115, 117

Brahme, H.
- microtron research, 71

Brauer, C.H.
- tuberculosis treatment with X rays, 6

Breast cancer
- brachytherapy, 23
- breast conservation, 40
- treatments, 9-10, 27, 186
- Grubbé's early experiments, 5, 6
- mortality rates, 40
- oncogenes and, 155-56
- surgical adjuvant radiation therapy, 24, 33

Brecher, E., 90
Brecher, R., 90
Brenner, D.J.
- high dose-rate brachytherapy research, 118

BRH. *See* Bureau of Radiological Health

Brigham and Women's Hospital
- brachytherapy using online MR imaging, 302

British Empire Cancer Campaign, 138, 142

British Hospital Physicists Association
- codes of practice, 115

British Ministry of Health
- linear accelerator research, 53, 60

British X-ray and Radium Protection Committee
- radiation protection recommendations, 110

Brock, W.
- predictive assay research, 152

Brookhaven National Laboratory
- boron neutron capture therapy, 302

Brooklyn Post-Graduate School of Clinical Electrotherapeutics, 168

Brown, I.G.
- career and accomplishments, 266

Brown, J.M.
- etanidazole research, 150, 151
- radiosensitivity of cells research, 138

Brown-Boveri
- betatron development, 52, 58

Brown-Roberts-Wells head ring, 69

Burdick, G.G.
- textbook, 171

Bureau of Health Insurance
- Medicare management, 208

Bureau of Radiological Health
- linear accelerator study, 57, 65

Burnam, C.
- source calibration, 95-96

Burry, James
- tuberculosis treatment with X rays, 6

Buschke, F.J.
- supervoltage radiation research, 28, 29

Byers, E.M.
- use of radioactive patent medicine, 286-90

Caesar (runaway slave)
- article on treatment for poisoning, 264

Cahal, Mac, 214-15

Caldwell, E., 23, 171
- lupus vulgaris treatment with X rays, 6
- textbook, 170, 171

Calibration protocols, 74-75

Califano, Joseph A., Jr., 208

California Institute of Technology
- supervoltage research, 50

California Relative Value Schedule, 208-10, 209, 210, 213

Calorimetry, 115

Cameron, J.
- thermoluminescence research, 116

Cancer, 30

"Cancer: A Brief Study in, with Special Reference to Its Surgical Treatment," 266

Cancer: Diagnosis, Treatment, Prognosis, 28

Cancer Research, 28

Cantril, S.T.
- supervoltage irradiation research, 29

"Carcinoma of the Breast: Preoperative and Postoperative X-Ray Treatment," 266

"Carcinoma of the Cervix: Diagnosis and Treatment," 269

"Carcinoma of the Lip," 269

INDEX

▶ ▶ ▶ ▶ ▶ ▶

Cargill, W.H.
    career and accomplishments, 269
Carman, R.
    ACR presidential address, 222
Carr, Mr., 5, 6
Carter, Dr., 50
Cascaded tubes, 49
Case, J.
    radiologist training and education, 173, 174
    therapy transformer units, 47
Cassisi, N.
    collaboration with Million, 37
Catcheside, D.C.
    linear-quadratic model, 145
Catterall, M.
    neutron research, 142
Cavanagh, J.
    radiosensitivity of cells research, 138
Cavendish Laboratory
    neutron research, 133-34
Cavity ionization chambers, 55, 97, 108-9
Century Magazine
    "Symposium on the Roentgen Rays," 279
Certification of radiologists, 174-76, 224-25. *See also* American Board of Radiology
Cervical cancer
    afterloading technique, 197
    brachytherapy, 23, 97
    ^{60}Co teletherapy, 31
    "curietherapy," 23
    early techniques, 27, 186, 192
    Fletcher's treatment of, 37-38
    radium treatment, 187
    radium tube treatment, 23, 131
    vaginal applicator for radiation, 23
Cesium-137, 195
Chadwick, J.
    neutron research, 134, 135
Chamberlain, W.E.
    ABR examining board representative, 224
    ACR chairman of Board of Chancellors, 221
Chaoul contact therapy tubes, 49
Chapman, J.D.
    predictive assay research, 152, 153
Charlton, E.E.
    resonant transformer generator development, 50
    supervoltage research, 103
Check point genes, 155
Chemotherapy
    prior to radiation therapy for Hodgkin's disease, 39
Chism, S.E.
    independent collimator research, 66
Chodorow, M.
    klystron research, 53, 59
Christen, T.
    HVL to quantitate beam quality, 54
Christie, A.C.
    ABR examining board representative, 224
Christie Hospital, Manchester, England
    gantry-mounted linear accelerator, 61
Chu, F.
    autonomy issue, 247
    as department head, 247
    discrimination issue, 256
    family issues, 245, 254
    honors, 251
Churchill-Davidson, I.
    high pressure oxygen research, 140
    hypoxia research, 143

Clapp, J.
    Hodgkin's disease treatment, 11
Clark, R.L., 34
    radium treatment, 192
Cleaves, M.
    biographical data, 236-37
    brachytherapy, 23
    cervical cancer treatment, 187
    "curietherapy" for cervical cancer, 23
    family issues, 254
    uterine cancer treatment, 191
Cleveland Clinic
    fast neutron facility, 300
    radiation physics department, 193
Clifton
    clone research, 139
Clinac 4 linear accelerator, 62
Clinac 6 linear accelerator, 62
Clinac 35 linear accelerator, 62-63
Clinical and radiobiological research. *See also* Clinical practice
    X-ray era: 1910-1950
        early leaders, 129-31
        isoeffect curves and overall treatment time, 131-32
        neutrons and heavy ions, 133-36
        oxygen effect, 132-33
    supervoltage era: 1950-1970
        cell, tissue, and tumor kinetics, 143-45
        mouse tumors hypoxic cells: reoxygenation and high pressure oxygen, 139-41
        neutrons, 141-42
        neutrons and charged particles: the effect of radiation quality, 141
        perceived problem of hypoxia, 143
        radiosensitivity through the cell cycle and potential lethal damage, 137-38
        split dose experiments, 137
        in vitro survival curves, 136-37
        in vivo survival curves, 138-39
    megavoltage era: 1970 to present
        dose-rate effect and brachytherapy, 153
        fractionation concepts, 145-47
        Gray Laboratories, 152-53
        hyperthermia, 153-55
        hypoxic cell radiosensitizers and bioreductive drugs, 150-51
        molecular techniques, 155-57
        particle therapy, 147-50
        predictive assays, 151-52
Clinical Applications of the Electron Beam, 250
Clinical practice. *See also* Clinical and radiobiological research 1910-1950, 21-29
    supervoltage era: 1950-1970, 29-36
    radiation oncology: 1970-1995, 36-40
    brachytherapists, 23-24, 28
    generalists, 22-23
    Hodgkin's disease, 38-39
    prostate cancer, 39-40
    radiation oncologists, 24-29
    schools and practice policies, 37-38
Clinical Roentgen Therapy, 26
Cloud, S.M.
    electron beam therapy research, 114
^{60}Co therapy. *See* Cobalt-60 teletherapy
Cobalt-60 teletherapy, 195
    availability of units, 37
    development of, 113-14, 204, 213
    effectiveness of, 33

▶ ▶ ▶ ▶ ▶ ▶

316

INDEX

Fletcher's design for unit, 29, 31
original uses for, 31
Cockroft, J. D.
  neutron research, 134
Cockroft-Walton generators, 50
Codman, E.A.
  X-ray burns report, 169
Cohen, L., 148
  nominal standard dose research, 131
Cold-cathode tubes, 44-45
Coley, W.
  Memorial Hospital staff member, 194
Collaborative Working Group on the Evaluation of Treatment Planning for External Photon Beam Radiotherapy, 119
Columbia University, New York
  African American radiologist training and, 267-68
Committee for Radiation Oncology Studies, 181
Committee for Radiation Therapy Studies, 181
  endorsement of cooperative study of radiation oncology, 36
  formation of, 35-36
  fractionation survey, 35
  Radiation Therapy Oncology Group, 36, 181
Compton, A.H., 93-94
Compton scattering, 93-95
Computed tomography
  availability of, 119
  stereotactic radiosurgery/radiotherapy and, 69
Computer calculations. See also Three-dimensional optimized treatment planning software
  brachytherapy and, 111, 113, 197
  for X-ray beams, 116, 119-20
Computer control of X-ray treatment machines
  accident timeline, 68
  accidental overdoses, 68-69
  advantages and disadvantages, 67-68
Cook County Hospital, Chicago
  African American radiologist training and, 267-68
Coolidge, W.D.
  ARS honorary member, 192
  biographical data, 98
  ductile tungsten production, 101
  hot-cathode tube development, 45-46, 100, 101-2, 282
  supervoltage machine construction, 103
Cooper, G.
  Medicare comments, 215
Correspondence schools for radiology, 168
Corrigan, H.
  radiation dosage measurement, 253
Cosman
  GTC frame development, 70
Council of Affiliated Regional Radiation Oncology Societies formation of, 222
Courtney Mills assay, 303
Coutard, H.
  protracted-fractional method, 25-26, 27, 130-31
  supervoltage radiation research, 29
CPT. See Physicians' Current Procedural Terminology
Crabtree, H.G.
  oxygen effect research, 132-33
Cramer, W.
  oxygen effect research, 132-33
Craver, L.
  Memorial Hospital staff member, 194
Cromelin, P.
  ABR incorporation and, 225

Crookes tube, 45, 279
CRTS. See Committee for Radiation Therapy Studies
CRVS. See California Relative Value Schedule
Cuff, J.R.
  career and accomplishments, 267, 269
Cunningham, J.
  computer calculation of X-ray beams, 119
  computer use in radiotherapy, 116
Curie, M.
  Americans traveling to study with, 173
  ARS honorary member, 192
  atomic weight of radium discovery, 187
  biographical data, 232-33
  Duane and, 92
  Failla and, 94
  Harding's presentation of radium to, 193
  international radium standard, 96
  interview with Meloney, 192-93
  mobile X-ray service for French army, 282-83
  Nobel Prize for chemistry, 21
  radium discovery, 185, 186, 277, 279
  United States tour, 284
Curie, P.
  Americans traveling to study with, 173
  biographical data, 232-33
  calorimetry research, 115
  first planned experiment in radiobiology, 129, 130
  radium discovery, 185, 186, 277, 279
Curry, J.
  mentoring of African American radiologists, 273
Cyclotrons, 134, 142, 147
Cytotoxic drugs
  early research, 37

Dailey, U.G.
  career and accomplishments, 267, 268
Dally, C.
  X ray burns and, 14-15, 285
Daniel, J.
  hair loss and X ray treatments, 13
Daniels, F.
  thermoluminescence research, 115-16
Danlos, H.
  lupus treatment, 185
Danne, J.
  Biological Laboratory of Radium physics director, 186
Davenport, A.
  career of, 238
  family issues, 254
de Broglie, L.
  wave-like properties for particles, 95
de Vries, H.
  X ray effect on plants, 17
Dean, A.
  Memorial Hospital staff member, 194
Debye, P., 93
Deep/orthovoltage therapy X-ray systems, 47
Degrais, P.
  radium institute founding, 186
Delclos, L., 38
del Regato, J., 90
  American Club of Therapeutic Radiologists founding, 224
  areas to master for successful education, 180
  biographical data, 30
  cancer textbook, 28
  Committee for Radiation Therapy Studies member, 181

317

INDEX

▶ ▶ ▶ ▶ ▶ ▶

effectiveness of ^{60}Co therapy, 33
mentoring of African American radiologists, 273
prostate cancer research, 39
separation of radiation therapy from diagnosis, 177
Soiland biography, 24
Delany, M.R.
    education, 264
Denekamp, J.
    contributions, 146
    fractionation research, 146
    Gray Laboratory and, 152
    neutron research, 142
Dental oncology, 37
Department of Veterans Affairs
    lung cancer study, 33
Depilatory uses of X rays, 8
Deprex interrupter, 45
Derham, J.
    medical practice of, 264
DES. *See* Diethylstilbestrol
Desjardins, A.
    acute and late effects of irradiation of normal tissues, 24-25
Despeignes, 129
Dewey, W.C.
    hyperthermia research, 154
Diagnosis related groups, 216
Diethylstilbestrol
    prostate cancer and, 39
Discrimination issue
    for women, 234, 248, 255-56
Domen, S.R.
    absorbed dose research, 74
Dominici, H.
    Biological Laboratory of Radium director, 186
Donovan, M.
    family issues, 254
    internship of, 241
Dose-rate effect, 153
Douglas, B.G.
    linear-quadratic model, 145
Douglas, J.
    radium mining and donation, 23, 188
    radium research, 188-89
Dresser
    supervoltage research, 51
DRGs. *See* Diagnosis related groups
Duane, W.
    accomplishments, 91
    ARS honorary member, 192
    biographical data, 92
    debate with Compton, 93-94
    free-air ionization chamber system, 54, 107, 108
    interstitial use of radon gas, 97
    radium emanation extraction and purification plant, 189
    work with Hunt, 92
    X-ray measurement, 107
Duffy, J.
    isoeffect curve research, 131
DuSault, L.A.
    time/dosage research, 253
Dutreix, A.
    Paris system for brachytherapy, 113
Dynamic wedges, 66-67, 75

Early years of radiation therapy
    analgesic and palliative applications, 11-13, 27
    benign conditions, 4, 6-9
    cancer therapy, 9-11
    electrotherapists, 3-4
    excitement over discovery of X rays, 1-2
    medical practice in the United States, 3
    plant and protozoa experiments, 15, 17
    "regular" and "irregular" practitioners, 2-4
    Röntgen's discovery, 1
    technological developments, 43-56
    X-ray burns, 13-15, 169, 196
Eastern Cooperative Group in Solid Tumor Chemotherapy
    clinical research in oncology, 35
Eberhart, M.
    textbook, 171-72
E.C.H.O. *See* Endocurietherapy/Hyperthermia Oncology
Economic relations. *See* Intersociety, government, and economic relations
Economist
    2040 as year for elimination of cancer as a threat to life, 310
Edgerton, Dr. C., 24
Edison, T.
    "Edison effect," 100
    hope for an X-ray unit in every home, 2
    mass-produced home X-ray kits, 278
    tuberculosis treatment with X rays, 4, 6
Education. *See* Teaching hospitals and medical schools; Training and education
Edward, L.V.
    isoeffect curve research, 131
"Effects of Fractionated Total Body Irradiation on Sarcoma Cells in the Peritoneal Exudate of the Mouse," 269
8 MeV linear accelerators, 60, 61
Eine Neue Arte von Strahlen, 166
Eldorado Mining and Refining Company
    cobalt-60 unit, 114
Electrical Engineer, 167
Electrical Institute of Correspondence Instruction, 168
Electron beam therapy, 114-15
Electronic portal imaging, 70, 71
Electrotherapy, 3-4, 12, 239
Elements of General Radio-therapy for Practitioners, 171
Elkind, M.
    split dose experiments, 137, 153
    in vivo survival curve research, 139
Ellis, E.
    needle aspiration biopsy, 194
Ellis, F.
    fewer treatments with larger doses, 33
    isoeffect curve research, 153
    nominal standard dose concept, 131-32
Endicott, K., 181
    CRTS formation, 35-36
Endocurietherapy/Hyperthermia Oncology, 197
Endometrial cancer
    afterloading technique, 197
Eppendorf probe, 152
Equipment. *See also specific items by name;* Technological developments
    early inadequacies, 10, 25
Equitable Insurance Company
    health insurance plan, 205
Ernst, E.C.
    ABR examining board representative, 224
Errington, R.F.
    cobalt-60 research, 114
Etanidazole, 150-51

INDEX

European Organization for Research and Treatment of Cancer, 153, 304-5
Eve, A. S. 93
Ewing, J., 188, 239
    acceptance of Guttmann as intern, 245
    ARS honorary member, 192
    Memorial Hospital staff member, 194
    pioneering work in radiation and radium therapy, 172
Extended-field radiation, 33, 39
External beams
    1910-1950 era
        cavity ionization chambers for X-ray and gamma ray measurement, 108-9
        high-energy X-ray machines, 102-7
        radium cannon, 99
        radium irradiators, 99-100
        X-ray measurement standardization, 107-8
        X-ray tubes, 100-102
    supervoltage era: 1950-1970
        calorimetry, 115
        cobalt irradiators, 113-14
        computer calculation, 116
        electron beams, 114-15
        thermoluminescent dosimeters, 115-16
    megavoltage era: 1970 to present
        computer calculation, 119-20
        optimum energy for, 118-19
        stereotactic radiation treatment, 120-21
External wedge filters, 66
Eye diseases
    radium treatment, 188

Facilities Master List
    ^{60}Co units, 31
    linear accelerators, 31
Failla, G., 133, 193
    beta particle problem and, 97-98
    biographical data, 94
    brachytherapy accessory invention, 190
    cavity ionization chamber research, 108
    expression of dose in terms of energy deposited in tissue, 56
    glass radon seed plant, 193
    Memorial Hospital staff member, 194
    necrotic effect of radon seeds, 194
    as radioactivity authority, 189
    radium packs, 100
    radon plant design, 194
    supervoltage research, 103
    textbook, 190, 191
    work with Quimby, 112
Familial hypercholesterolemia, 309
Family issues
    for women, 253-55
Farmer-chamber dosimetry systems, 75
Farrow, J.
    Memorial Hospital staff member, 194
Fast neutron beams, 300
FDA. See Food and Drug Administration
Federal government relationship with radiation oncology, 214-16. See also specific agencies and national laboratories by name
Federal Trade Commission
    Byers case and, 289
    CRVS ban, 213
    RVSs and, 211
Feldman, A.
    in vivo survival curve research, 139
Fermilab

    neutron facility, 148, 300
Ferraux, R.
    fractionation proponent, 130-31
Fiducial markers, 296
Field, C.E., 220
Field, S.
    neutron research, 142
Filters
    early research, 47-48, 90
    linear accelerators and, 60
    universal wedges, 66
Fisher, J.
    Gray Laboratory and, 153
Flammarion, C.
    report on Mars, 280
Flanders, P.H.
    high pressure oxygen research, 140
Fletcher, G., 177
    biographical data, 34
    cervical cancer treatment, 37-38
    cobalt-60 research, 113, 114
    Committee for Radiation Therapy Studies chair, 181
    as CRTS chair, 35
    design of ^{60}Co teletherapy unit, 29, 31
    fractionation research, 147
    isodose research, 113
    as mentor to women radiologists, 249-50
    radiation training school, 37
Fletcher-Suit applicator, 38
Flexner Report on medical education, 173, 238
Flinn, F.B.
    Byers case and, 289
Florance, D.C.H., 93
Florida Medical Association
    self-referral issue, 227
Flow cytometry, 153
Fluorescent X ray measurement, 95
Flynn, D.
    estimate of increase in radiation oncologists, 306
Focht, E.
    radiation effects on the eye, 253
Food and Drug Administration
    campaign for broader powers, 289
    linear accelerator approval, 219
    nonregulation of radium, 284
    report on iodine-125, 195
Forssell, G.
    Radiumhemmet founding, 187
    separation of radiation therapy from diagnosis, 176
4 MeV gantry-mounted linear accelerators, 61
Fowler, J., 141
    Gray Laboratory director, 152
    linear-quadratic model, 145, 147
    neutron research, 142
Fox, E.
    family issues, 254
    internship of, 241
    nontraditional training, 242
    setting up of her own practice, 243
Fractionation, 25-26, 27, 35, 130-31, 145-47
Franklin, M.
    dosimetric method research, 54
Fraud. See Radiomedical fraud and popular perceptions of radiation
Frazell, E.
    Memorial Hospital staff member, 194
Free-air ionization chambers, 54, 96, 107, 108
Freedmen's Hospital
    establishment of, 265

319

INDEX

▶ ▶ ▶ ▶ ▶ ▶

Freund, L.
    Americans going to study with, 173
    hairy nevus treatment, 129
    lupus vulgaris treatment, 6
    textbook, 171, 172
Fricke, H., 95, 193
    cavity ionization chamber research, 55, 108-9
Friedman, M.
    Committee for Radiation Therapy Studies member, 181
Friedrich, W., 96
    roentgen measurement, 196
Frontiers of Radiation Oncology, 297
Fry, D.W.
    linear accelerator research, 31, 53, 59, 60
    waveguide development research, 59
Fuchs, W., 167
Fullagar, P.
    mentoring of African American radiologists, 273
Fuller, L.
    discrimination issue, 256
    family issues, 255
    family support for, 249
    honors, 251
    Kligerman as mentor, 249
    promotions, 250
    work with Fletcher, 249
Furey, W.
    Residency Review Committee member, 180
Future of radiation oncology
    clinical practice pattern changes
        facility utilization, 299
        larger practice groupings, 299
        multi-modality treatment strategies, 297
        subspecialization, 297-99
    gene therapy, 308-10
    genetically defined diagnostic categories, 308-10
    hypoxia question, 305
    immune reaction question, 306
    individualized prediction of patient response to radiation, 302-5
    late sequelae of radiation treatment, 306
    long-standing questions, 305-6
    negatives, 294
    optimistic long-term view, 294
    radiation oncology education and manpower, 293-94, 306-7
    special treatment facilities
        boron neutron capture therapy, 302
        fast neutron therapy, 300
        heavy particle high-LET radiations, 300-301, 305-6
        high technology brachytherapy, 302
        intraoperative electron beam therapy, 301-2
        proton treatment facilities, 301
        radiolabeled antibodies and metabolites, 302
    standardization of radiation oncology practice, 306
    technological advances and their impacts
        three-dimensional optimized treatment planning software, 294-97
    twenty to thirty years from now, 307-10

"Gamma knife," 120
Garcia, M.
    Committee for Radiation Therapy Studies member, 181
Gardner lymphosarcoma
    high pressure oxygen treatment, 140

Gas-filled X-ray tubes
    unpredictability of, 90, 100
Gating of radiation treatment to patient/anatomic part movement, 297
Geiger, H. W., 94-95
Gene therapy, 308-10
General Electric Company, 49-50, 102
    cobalt-60 unit, 114
    supervoltage research, 103
General Electric X-Ray Corporation
    supervoltage radiation research, 29
Genetic approaches to treatment
    gene activation by radiation, 310
    gene therapy, 308-10
    identification of individuals at high risk for specified tumors, 308
    popular view of, 307
George, F.
    effectiveness of ^{60}Co therapy, 33
    prostate cancer research, 39
German relative value scale, 207
Gilbert, R.
    Hodgkin's disease research, 38
    irradiation of lymphatic regions, 33
Gilchrist, T.
    X-ray burn report, 13
Gill, S.S.
    GTC frame development, 70
Gillin, M.T.
    comparison of Paris and Manchester systems, 113
Gilman, J.E., 5
Gilman, P.
    effects of X rays on hen's eggs, 17
Ginzton, E.
    klystron research, 53, 59
    linear accelerator research, 32, 61, 106
    microwave device development, 31
Girdwood, G.P., 171
Glasser, O., 95, 193
    biographical data, 106
    cavity ionization chamber research, 55, 108-9
    glass window tube, 47
    work with Quimby, 112
Gocht, H.
    breast cancer treatment, 9-10
Goitein, M.
    precise dose delivery, 119
Gold-198, 195
Goldfeder, A.
    cancer research, 253
Goldson, A.L.
    career and accomplishments, 272
Goodspeed, W., 172
Gore, J.C.
    beam parameter measurement, 75
Gould
    clone research, 139
Graduate Hospital of the University of Pennsylvania
    African American radiologist training and, 267
Gray, A.
    ARRS address, 17
Gray, H., 141
    British Empire Cancer Campaign, 138
    cyclotron research, 142
    high pressure oxygen research, 140, 141
    linear accelerator development, 60
    neutron research, 134-36
    oxygen effect research, 133

▶ ▶ ▶ ▶ ▶ ▶

INDEX

Gray, J.A., 93
Gray, L.
    cavity ionization chamber research, 109
Gray Laboratory, 142, 152-53
Green, D.T.
    cobalt-60 research, 114
Green, H.M.
    career and accomplishments, 266
Grenz-ray tubes, 43, 46-47
Griffin, T., 148
Grigg, E.R.N., 90
    book section "Negroes in American radiology," 264
    Coolidge's hot cathode tube and, 101
    description of the Snows, 239
Grimmett, L.
    cobalt-60 research, 29, 31, 113-14
Groover, T.
    elimination of radiation therapy from private practice, 176
Grubbé, E.
    cancer experiments with X rays, 5-6, 9, 129
    Illinois School of Electro-Therapeutics and, 235
    life prolongation effect of X-ray treatments, 13
    overexposure to X rays and, 13
    radiographic education efforts, 167-68
GTC frame, 70
Guttmann, R.
    autonomy issue, 247
    as department head, 247
    discrimination issue, 256
    family issues, 255
    internship experience, 245
    reason for choice of radiology, 245
    World War II and, 245

Haagensen, 27
Hahn, G.
    potentially lethal damage of radiation, 138
Haimson, J.
    linear accelerator design, 31
Hair loss
    early X-ray experiments and, 13
Hall, E., 146
    high dose-rate brachytherapy research, 118
    misonidazole research, 151
    radiosensitivity of cells research, 138
Halphide, A.C., 5
Hamann, A.
    biographical data, 246
    as department head, 247
    family issues, 255
    medical school challenges, 244, 245
    retirement, 250
    World War II and, 245
Hammer, W.
    interest in radioactivity, 280
Hammersmith Hospital, London
    linear accelerator installation, 60, 61
    neutron research, 142, 147
Handbook 59, 116-17
Hanks, M.
    comments on neglect of X-ray therapy, 239
    family issues, 254
Hansen, W.W., 31
    linear accelerator research, 106
    rhumbatron development, 53
Harding, Pres. Warren, 193

Hare, H.
    electron beam therapy research, 114
Harper Hospital, Detroit, MI
    fast neutron facility, 300
    supervoltage system, 50, 103
Harris, J.R.
    breast conservation research, 40
Harris, W., 26, 27
Harrison, C.
    career and accomplishments, 267
Hart, A.L.
    name change, 243, 244
    setting up of her own practice, 243
Harvard Medical School
    coeducational status, 245
Harvard University
    radium donation, 188
HCFA. *See* Health Care and Financing Administration
Head rings
    stereotactic radiosurgery and, 69, 70
Healing waters, 283-84
Health Care and Financing Administration
    billing of services, 203
    Medicare and, 208
    UCR reimbursement and, 210-11
Health maintenance organizations, 202, 299
Healy, W., 190
    cervical cancer treatment, 192
Heavy ion research, 133-36, 148-50
Heineke, H.
    irradiated bone marrow effects, 17
Henschke, U.
    afterloading, 196
    iridium-192 as substitute for radium needles and/or seeds, 195
    work with and for African American radiologists, 273
Hereditary nonpolyposis colorectal cancer, 157
Herendeen, R.
    Memorial Hospital staff member, 194
Hertwig, W. A. O., 17
Hewitt, H.
    collaboration with Gray, 138, 139
    Gray Laboratory and, 152
    high pressure oxygen research, 140
Hickey, P.M., 173
    "cones" terminology, 23
High-energy X-ray machines. *See* Supervoltage radiation
High linear energy transfer radiations, 300-301, 305-6
High pressure oxygen, 139-41
High Voltage Engineering Corporation
    supervoltage generators, 51, 57-58
"The Histologic and Clinical Response of Human Cancer to Irradiation," 269
The History of Medicine, 207
HNPCC. *See* Hereditary nonpolyposis colorectal cancer
Hockel, M.
    predictive assay research, 152
Hocker, A.
    Memorial Hospital staff member, 194
Hodes, P.
    mentoring of African American radiologists, 273
Hodges, P., 6
Hodgkin's disease
    changes in treatment of in the 1970s, 38-39
    ^{60}Co therapy, 33
    CRTS trials, 36
    early research, 2, 10-11

INDEX

"mantle" field treatment, 38
MOPP therapy, 38-39
staging classifications, 38
Hollander, L.
    ABR examining board representative, 225
Holmes, G.
    ABR examining board representative, 225
    Memorial Hospital staff member, 194
    simplification of ABR certification, 176
    supervoltage research, 50-51
Holthusen, H.
    oxygen effect research, 132
Hornsey, S.
    neutron research, 142
Hospital Physicists Association
    dosimetric protocols, 63-64
Hospitals. *See also specific institutions by name;
    Teaching hospitals and medical schools*
    cost of hospitalization in 1940, 204
    first X-ray units in, 202
    flat rate plans, 202
    nonphysicians practicing radiology, 203-4
    percentage versus salary payment of radiologists, 204-5, 207-8
    radiology service billing, 202-3
Hot-cathode tubes, 45-46, 100-102, 173, 282
Houdek, P.V.
    stereotactic radiation research, 120
Howard, A.
    cell cycle research, 144
Howard University Medical School
    admission of women, 265
    African American radiologist training, 263, 264-65, 268
HPA. *See* Hospital Physicists Association
HPO. *See* High pressure oxygen
Hunt, B.
    establishment of field hospital, 242
    family issues, 254
    partnership with father, 242
    setting up of her own practice, 243
Hunt, F.
    work with Duane, 92
Hurst, B.P.
    career and accomplishments, 266
Hyams Trust
    supervoltage research grants, 51
Hyman, S.
    autonomy issue, 247-48
    discrimination issue, 256
    family issues, 245, 254
    male colleagues and, 245, 247
Hyperthermia, 153-55
Hypoxia research, 143, 305
Hyslop, G.
    Memorial Hospital staff member, 194

IAEA. *See* International Atomic Energy Agency
ICR. *See* International Congress of Radiology and Electricity
ICRP. *See* International Commission on Radiological Protection
ICRU. *See* International Committee on Radiological Units
Idaho National Engineering Laboratory
    boron neutron capture therapy, 302
Illinois School of Electro-Therapeutics
    founding of, 168
    Rice as faculty member, 235
"*In Vitro* Sensitivity to X-rays of Free Tumor Cells in the Exudate of the Mouse," 269
In vitro survival curves, 136-37
In vivo survival curves, 138-39
Income for radiologic practice and medical practice, 205-6
Independent collimators, 66-67
Induction coils, 45
Institute for Radiation Research, 196
International Atomic Energy Agency
    calibration protocol, 75
    computers in radiotherapy meeting, 116
    Directory of High-Energy Radiotherapy Centres, 57, 65
International Club of Radiotherapists, 223
International Commission on Radiological Protection
    radiation protection recommendations, 110
International Committee on Radiological Units
    dosimetric protocols, 63-64
    HVL method for characterizing roentgen-ray beams, 54
    maximum permissible dose, 116
    rad study, 56
International Conference on Computers in Radiotherapy, 116
International Conference on Hyperthermia Oncology, 155
International Congress of Radiology and Electricity
    Committee on Units and Measures, 175
    consensus definition for measuring radiation exposure, 54-55
    fractionation presentation, 131
    international radium standard, 96
    radiation protection recommendations, 110
    X-ray measurement, 107-8
International Journal of Radiation Oncology, Biology, Physics, 76, 119, 179
International Workshop on Electronic Portal Imaging, 71
The Interpretation of Radium, 281
Intersociety, government, and economic relations. *See also specific societies by name*
    development of third party payers
        billing for radiological services, 206-7
        radiologic practice income, 205-6
        relative value scale, 207-13
        RVS and CPT and, 213-14
    government and HCFA relationship, 214-16
    hospital relationship, 202-5
    joint ventures and, 225-27
    malpractice issues, 216-17
    radioactive mineral regulation, 217-19
Intersociety Committee for Radiology *Bulletin*
    economic problems of medicine, 204
    hospital report, 202
Intersociety Council for Radiation Oncology, 181
Intraoperative electron beam therapy, 301-2
Iodine-125, 195
Ionization chambers, 54, 55, 64, 107, 108-9
IORT. *See* Intraoperative electron beam therapy
Iridium-192, 195
Ising, G.
    linear accelerator design, 52-53
Isoeffect curves, 131-32

Jacobi, J.P.
    belief that women were the equal of men in medical practice and abilities, 240
Janeway, H.
    glass radon seeds, 99
    necrotic effect of radon seeds, 194

INDEX

radium therapy developments, 189, 193-94
textbook, 189-90, 191
uterine cancer treatment, 192
work with Failla, 94
JCRT. *See* Joint Center for Radiation Therapy
Jefferson Hospital, Philadelphia
African American radiologist training and, 267
Jenkinson, E.
ABR examining board representative, 225
Jesse, R.H., 37
Jirtle
clone research, 139
Johns, H.
betatron research, 104
calorimetry research, 115
Johns Hopkins Hospital
admission of women medical students, 234, 235, 238
formal training site, 173
radium donation, 188
Johnson, W.
skin cancer treatment with X rays, 9
Johnston, G.
radiology as auxiliary service, 172
X-ray burn experiments, 14
Johnston, Z.A.
family issues, 254
organization leadership, 242
partnership with father and Boggs, 242
Joint Center for Radiation Therapy
breast conservation research, 40
Joint ventures, 225-27
Jolly, J.
interstitial radiation research, 190
Jones, P.M.
lupus vulgaris treatment with X rays, 6
Journal of the American Medical Association, 169
cautionary article on X rays, 2
radium rental, 218
Journals. *See also specific journals by title*
popular magazines and radiography development, 279-80
as source of radiology education, 169
specializing in radiology, 170

Kaempffert, W.
atomic energy and societal transformation, 281
Kallman, R.
high pressure oxygen research, 140
reoxygenation research, 138
Kang, Y.S.
beam parameter measurement, 75
Kansas Medical Society
professional radiologist administration of X-ray therapy and, 203
Kaplan, H., 177
biographical data, 32
Committee for Radiation Therapy Studies member, 181
effectiveness of high dose extensive radiation, 33, 39
Hodgkin's disease research, 38, 39
linear accelerator research, 31, 61, 105, 106
Karzmark, C.J.
accelerator operation description, 58
linear accelerator design, 31
Kassabian, M.
breast cancer treatment, 10
dosimetric method summary, 54

management of radiation dermatitis, 22
skin cancer treatment with X rays, 9
textbook, 170, 171
X-ray burn report and experience, 13-14, 15
X-ray suite recommendation, 15
X-ray treatment of "epileptic" patients, 8
Kellerer, A.M.
linear-quadratic model, 145
Kelley, C.H., Jr.
career and accomplishments, 268
Kellogg, W.W., 50
Kellogg Laboratories
supervoltage research, 50
Kelly, H., 172
ARS honorary member, 192
cystoscope invention, 191
National Radium Institute founding, 188
radium irradiator construction, 100
radium mining and donation, 23, 97
source calibration, 95-96
uterine cancer treatment, 191
Kelly, I.R., 7
Kelly-Koett Company
supervoltage system, 50
Kerr, H.D.
Residency Review Committee member, 180
Kerst, D.
betatron research, 51-52, 104, 105
Kienbock, R.
radiosensitivity research, 130
Kinetics of cells, tissues and tumors, 143-45
Kinney, L.
ABR examining board representative, 225
Kirklin, B.
ABR examining board representative, 225
residency program recommendation, 175
Residency Review Committee member, 180
Kligerman, M.
Committee for Radiation Therapy Studies member, 181
medical pion facility, 148
as mentor to women radiologists, 249
Klystrons, 53, 59
Knox, J.T.
lupus vulgaris treatment, 169
Knudsen, A.G.
suppressor gene research, 156-57
Koh, W.J.
predictive assay research, 153
Kolle, F.S.
X-ray meter, 53
Kramer, S., 250
intravenous methotrexate therapy, 36
mentoring of African American radiologists, 273
prostate cancer research, 39
Kusserow, R.
acceptable business practice comments, 226
Kutcher, G.J.
high-energy X-ray beam research, 118-19

Laborde, A.
calorimetry research, 115
Lamar, E.
supervoltage research, 50-51
Lamperti, P.J.
absorbed dose research, 74
Lange, F.
electron beam therapy research, 114

323

INDEX

Langmuir, I.
    hot cathode tube research, 101
Lanzl, L.
    betatron research, 105
Laughlin, J., 90
    betatron research, 105
    electron therapy, 105
    high-energy X-ray beam research, 118-19
Lauritsen, C.C.
    radiotherapy clinical trial, 24
    supervoltage research, 29, 50, 102-3
Lawlah, J.W.
    career and accomplishments, 268
Lawrence, E.
    cyclotron invention, 134, 135
Lawrence, J.
    neutron research, 134, 135
Lawrence Berkeley Laboratory
    negative pi meson research, 148, 149, 150
    neutron research, 133-34
Laws, B.
    ABR incorporation and, 225
Lea, D., 146
Leavitt, D.D.
    dynamic wedge research, 67
    multidetector array research, 75
Lee, B.
    ARS president, 192
    etanidazole research, 150
Lee, Mrs. Rose, 5, 6
Leksell, L.
    "gamma knife," 120
Lenz, M., 26, 27
LET radiation. *See* High linear energy transfer radiations
Leukemia
    early treatments, 10
    radiation and, 109
Leveroos, E.
    Residency Review Committee member, 180
Leveson, N.G.
    Therac-25 accidents, 68
Liaison Committee for Graduate Medical Education, 180
Lichter, A.S.
    editorial, 120
Lilienfeld, J. E.
    hot cathode tube research, 101
Linderman window, 47
Linear accelerators
    advancements in, 58-63
    availability of units, 37
    block diagram, 58
    buncher design improvements, 59
    compared with microtrons, 71
    early designs, 31, 52-53, 105-7, 204, 213
    electron gun improvements, 59
    filters, 60
    first clinical accelerator features, 60-61
    gantry mounts for, 61
    moving field, 63
    problems with, 58
    stereotactic radiosurgery/radiotherapy, 69-70
Linear-quadratic model, 145-46, 147
Linton, O.
    comments on debate regarding the definition of radiology, 208
Little, J.
    potentially lethal damage of radiation, 138

Lodge, Sen. Henry Cabot
    federal health insurance program, 215
Loeb, J., 284
Loma Linda University
    proton facility, 301
Lopriore, G.
    plant experiments, 15
Los Alamos National Laboratory
    medical pion facility, 148
    negative pi meson research, 148
Loshek, D.D.
    dynamic wedge research, 67
Loucks, R.E.
    ARS recording secretary, 219
Ludlam, R., 5
Luminescence dosimetry systems, 64, 115-16
Lung cancer
    interoperative radiation therapy, 196
    supervoltage therapy, 33
Lupus erythematosus
    early X-ray treatment, 7, 8
Lupus vulgaris
    early X-ray treatment, 6-7, 8, 169
    Grubbé's early experiments, 5, 6
    radium treatment, 185
Lysholm, E.
    radium irradiator construction, 100

MacComb, W.
    isoeffect curve research, 131
    Memorial Hospital staff member, 194
Macleod, J. J. R.
    Insulin research, 284
MacKay, J.
    cobalt-60 research, 114
Mackie, T.R.
    tomotherapy research, 73, 74
Macomb, W.S., 37
Magnetic resonance imaging
    stereotactic radiosurgery/radiotherapy and, 69
Magnetrons, 53, 59
Major, R., 207
Malaise, E., 145
    inherent radiosensitivity assay research, 152
Malignant disease of the hypopharynx, 26
Malpractice issues, 216-17
"The Mammary Gland in 702 Autopsy and 9220 Surgical Specimens," 269
Manchester (Holt Radium Institute) variation of the Paris technique, 27, 37-38, 99, 113
Manges, W.F.
    ABR examining board representative, 224
Manhattan District Project, 136, 144
Mansfield, C.M.
    career and accomplishments, 271-72
    method for combining open and wedged fields, 66
Mantel, J.
    multileaf collimator research, 67
Marcial, V.
    cervical cancer treatment, 38
    Committee for Radiation Therapy Studies member, 181
    fractionation survey, 35
Marcus, P.I.
    in vitro survival curve research, 136-37, 145
Marcuse, W.
    X-ray burn report, 13
Marie Curie Radium Fund, 193

INDEX

Martin, C.
    brachytherapy, 23-24
    low intensity needle use for brachytherapy, 98
Martin, H.
    Memorial Hospital staff member, 194
    "nerve injection" of radium, 194
Martin, J.L.
    career and accomplishments, 266
Marting, E.
    discrimination issue, 256
    home-based practice, 254
    reason for choice of radiology, 245
Massachusetts General Hospital
    African American radiologist training and, 267-68
    formal training, 173
    proton facility, 301
    Van de Graaff generators, 51
Mauch, H., 196
Maximar 250, 102
Mayneord, W.V.
    exposure rate measurement, 96-97
    "gram-roentgen" as unit for dose measurement, 56
Mayo Clinic
    formal training, 173
    organization into radiodiagnosis and radio-therapy sections, 24
    radiology department separation, 193
McCullock, E.A.
    clone research, 139
McGinley, P.H.
    stereotactic radiosurgery/radiotherapy research, 69
McKee, G.
    textbook, 170
McKenna, W.G.
    oncogene research, 156
McNeer, G.
    Memorial Hospital staff member, 194
M.D. Anderson Hospital
    AAPM oversight of Radiological Physics Center, 117
    breast conservation research, 40
    clinical training site, 177
    fractionation research, 147
    isodose research, 113
    neutron facility, 148
    number of new cases treated, 181
    radiation training school, 37
Meadowcroft, W., 166
The Mechanistic Conception of Life, 284
Medical News, 169
Medical physics
    X-ray era: 1910-1950
        brachytherapy, 96-99
        cavity ionization chambers for X-ray and gamma ray measurement, 108-9
        Compton scattering, 93-95
        exposure rate relation to radium source strength, 96-97
        external beams, 99-109
        as golden age of radiology, 89, 244
        high-energy X-ray machines, 102-7
        histories of radiation therapy, 90
        major advances timeline, 91
        pair production, 95
        photoelectric effect and X-ray cutoff, 91-93
        radiation interactions, 90-96
        radiation protection, 109-10
        radium and radon availability, 97
        radium irradiators, 99-100
        radon plants, 97-98
        relative measurement, 95-96
        standardization of radium sources, 96
        status in 1910, 90
        X-ray measurement standardization, 107-8
        X-ray tubes, 100-102
    supervoltage era: 1950-1970
        brachytherapy, 111-13
        calorimetry, 115
        cobalt irradiators, 113-14
        computer calculations and brachytherapy systems, 111, 113
        computer calculations for X-ray beams, 116
        electron beams, 114-15
        extension of calibrations to high energy X rays and electrons, 115
        external beams, 113-16
        major advances timeline, 111
        radiation protection, 116-17
        status in 1950, 111
        thermoluminescent dosimeters, 115-16
    megavoltage era: 1970 to present
        brachytherapy, 118
        computer calculations for external-beam treatments, 119-20
        external beams, 118-21
        major advances timeline, 118
        optimum energy for X-ray beams, 118-19
        radiation protection, 121-22
        status in 1970, 117-18
        stereotactic radiation treatment, 120-21
        physicists in radiotherapy: future activities, 122-23
Medical practice in the United States
    early years, 3, 166-67
Medicare
    advent of, 205, 208
    division of, 213-14, 215-16
    guidelines on acceptable business practices, 226
    need for standardization of nomenclature and billing of radiology services, 211
    OBRA '88 and, 214
    RVS and, 208, 211
    usual and customary charges, 208, 210-11
Meertens, H.
    real-time imaging research, 70
Megavoltage radiation. *See also* Medical physics; *specific machines by name;* Supervoltage radiation
    brachytherapy, 118
    external beams, 118-21
    fractionation, 145-47
    radiation protection, 121-22
    status in 1970, 117-18
Meharry University
    African American radiologist training, 263, 268
Meloney, Mrs.
    interview with Marie Curie, 192-93
Memorial Hospital, New York
    brachytherapy research, 189-91
    clinical training site, 177
    formal training, 173, 174
    General Radium Service, 194
    number of new cases treated, 181
    pioneering work in radiation and radium therapy, 172
    prostate cancer treatment, 119

325

INDEX

▶ ▶ ▶ ▶ ▶ ▶

    radiation physics department, 193
    radium donation, 188
    radon seed research, 113
    remote afterloading technique, 196-97
    supervoltage machine, 103
Mendelsohn, M.L.
    "growth fraction" research, 144
Merrill, W.
    skin cancer treatment with X rays, 9
Methotrexate
    use prior to radiation therapy, 36
Metronidazole, 151
Metropolitan Life
    health insurance plan, 205
Metropolitan Vickers
    linear accelerator development, 60
    supervoltage system, 50
Meurk, M. L.
    computer calculation of isodose patterns, 111, 113
Mevatron 8 linear accelerator, 62
Meyer, J., 297
Michaels, B.
    Gray Laboratory and, 152
Microtrons
    compared with linacs, 71
    development of, 71-72
Miller, H.
    linear accelerator development, 60
Millikan, R.
    radiotherapy clinical trial, 24
    supervoltage research, 29, 50
Million, R.
    University of Florida radiation oncology department establishment, 37
Minton, R.
    career and accomplishments, 268
Misonidazole, 150
Mitchell, J.S.
    cobalt-60 research, 113
MLCs. See Multileaf collimators
MM 22 microtrons, 71
Mohan, R.
    dynamic wedge research, 67
    high-energy X-ray beam research, 118-19
Molecular biology
    check point genes, 155
    mutator genes, 157
    oncogenes, 155-56
    repair genes, 155
    suppressor genes, 156-57
Monell, S.
    teaching of radiology, 168
    textbook, 170
Montague, E.
    colleagues and, 258
    family issues, 249, 255
    family support for, 249
    honors, 250
    Kligerman as mentor, 249
    work with Fletcher, 249
Moore, E.B.
    cobalt-60 unit, 114
Moore, S., 29
MOPP therapy for Hodgkin's disease, 38-39
Morkovin, D.
    in vivo survival curve research, 139
Morphine
    X-ray treatments and overuse of, 12-13

Morton, W., 171, 172
    beneficial properties of X rays, 12
    European study, 173
    X-ray prints, 166
Mossell, F.
    education, 264
Mottram, J.C.
    oxygen effect research, 133
Mt. Sinai Hospital, New York
    African American radiologist training and, 267
Muller, H. J.
    genetic mutations from radiation, 109
Multi-modality treatment strategies, 297
Multi-sectioned tubes, 49-50
Multidetector arrays, 75
Multileaf collimators, 67
"Multiplex Pathology and the Cancer Problem," 266
Murphy, W.
    Committee for Radiation Therapy Studies member, 181
Mutator genes, 157

NACP. See Nordic Association of Clinical Physicists
National Academy of Sciences
    radiation protection reports, 121
National Advisory Cancer Subcommittee for Diagnosis and Treatment, 181
National Cancer Institute
    Collaborative Working Group on the Evaluation of Treatment Planning for External Photon Beam Radiotherapy funding, 119
    cyclotron construction grant, 148
    founding of, 215
    Proton Therapy Oncology Group, 301
    radiation oncology training, 35
    RTOG funding, 181
National Council on Radiation Protection and Measurement
    radiation protection definitions, 109, 110
    stochastic effect risks, 121-22
National Institute of Radiological Sciences, Chiba, Japan
    HIMAC high LET facility, 300
National Institutes of Health
    linear accelerator grant, 61
National Medical Association
    founding of, 265
National Radium Institute
    founding of, 188
National Research Council
    radiation protection report, 121
NCI. See National Cancer Institute
NCRP. See National Council on Radiation Protection and Measurement
Nedzi, L.A.
    dynamic field shaping research, 70
Nelson, R. F.
    computer calculation of isodose patterns, 111, 113
Neutron research, 133-36, 141-42. See also Fast neutron beams
New York State Institute for the Study of Malignant Disease
    radiation physics department, 193
Newbery
    ionization chamber construction, 64
Newcastle Upon Tyne, England
    linear accelerator, 61
Newcomer, E.
    family issues, 254

INDEX

Newell, R.R.
    filtration necessity, 47
Nickson, J.
    Committee for Radiation Therapy Studies member, 181
NIH. See National Institutes of Health
Nitroimidazole, 151-52
NMA. See National Medical Association
Nominal standard dose concept, 131-32, 147
Nordic Association of Clinical Physicists
    calibration protocol, 75
    dosimetric protocols, 63-64
Norris, C.
    Byers case and, 287
Notter, G.
    fractionation research, 146
NRC. See Nuclear Regulatory Commission
NSD. See Nominal standard dose concept
Nuclear Fear, 277
Nuclear reactors
    development of, 195, 219
Nuclear Regulatory Commission
    licensing of radioisotopes, 219

"Observations on the Results of Combined Fever and X-ray Therapy in the Treatment of Malignancy," 269
OER. See Oxygen enhancement ratio
Oestreich, A.
    work with and for African American radiologists, 273
Office of the Inspector General
    guidelines for acceptable health-care provider business practices, 226
Office of Veterans Affairs
    clinical research in oncology, 35
Ogura, J., 37
Ohio State University
    boron neutron capture therapy, 302
OIG. See Office of the Inspector General
Omnibus Budget Reconciliation Act of 1988
    Medicare charges, 214
On-line imagers, 70
Oncogenes, 155
1 MV X-ray machine, 50
Operative afterloading, 196
Orbital accelerators. See Betatrons
Orthotron linear accelerator, 61
Orton, C.G.
    high dose-rate brachytherapy research, 118
Osler, W.
    radium as metaphor for human condition, 277
Osteoscopes, 53
"Outline of the Present Status of the Surgical Treatment of Hyperthyroidism," 268
Ovadia, J.
    betatron research, 105
Ovarian cancer
    early treatments, 10
Oxygen effect, 132-33
Oxygen enhancement ratio, 141

Pack, G.
    Memorial Hospital staff member, 194
Packard, C., 17
Pain relief. See Analgesic applications of X rays
Pair production, 95
Palladium-103, 197

Palliative applications of X rays
    early research, 11-13, 27
Pancoast, H., 172
    ARS corresponding secretary, 219
Parberry, M.E.
    teaching of radiology in his office, 167
Paris (Radium Institute) technique, 27, 113, 153
Parker, R., 148
    dosage table, 194
    training assessment, 181
Parsons, C.
    National Radium Institute founding, 188
    radium mining and donation, 23
Particle therapy, 147-50
Pasteur Pavilion of the Institut du Radium, 21
Paterson, R.
    dosage table, 194
    isoeffect curve research, 153
    work with Bowing, 193
Patient immobilization systems, 69, 70, 296-97
Patient response to radiation
    characterization of the tumor in physiological and biochemical terms, 304
    indicators, 302-3
    inherent radiation sensitivity, 303-4
    proliferation kinetics, 304-5
    research for new predictors, 303
Patterns of Care Studies
    early equipment, 31
    prostate cancer, 39-40
PCS. See Patterns of Care Studies
Peer Review Organizations, 216
Pelc, S.R.
    cell cycle research, 144
Pendergrass, E.
    Residency Review Committee member, 180
Penetrometers, 53
Penrose Cancer Hospital, Colorado Springs
    clinical training site, 177
    residents' program, 180
Perry, H.
    career and accomplishments, 270-71
Perthes, G.
    cell response to radiation, 17
    depth dose measurements, 47, 54
Peters, L., 145, 148
    fractionation research, 147
    Gray Laboratory and, 153
    predictive assay research, 152
Peters, V.
    autonomy issue, 248
    biographical data, 252
    discrimination issue, 256
    family issues, 254-55
    Hodgkin's disease research, 38
    irradiation of lymphatic regions, 33
    medical school challenges, 245
Petry, E.
    oxygen effect research, 132
Petten, the Netherlands
    boron neutron capture therapy, 302
Petterson, C.
    electron research, 115
Pfahler, G., 172
    European study, 173
    filtration necessity, 47
    mentoring of African American radiologists, 266, 273
    "saturation" method of radiotherapy, 22-23

327

INDEX

▶ ▶ ▶ ▶ ▶ ▶

Phelps-Dodge Corporation, 188
Philadelphia General Hospital
    radiation physics department, 193
Philadelphia Post-Graduate School of Roentgenology, 173-74
Philadelphia Roentgen Ray Society, 173
Philips Medical Systems
    universal wedge, 66
Philips Metalix tube, 46
Phillips, C.E.S.
    dosimetric method research, 54
Phosphorus-32, 195
Photoelectric effect, 91-93
Photography
    radiographic technique and, 278
Physicians' Current Procedural Terminology, 211-14
Picker X-Ray Corporation
    cobalt-60 unit, 114
Pierquin, B.
    breast conservation research, 40
    Paris system for brachytherapy, 113, 153
Pierson, J.W.
    ABR examining board representative, 224
Plants
    early experiments on, 15, 17
Pohle, E.A., 26, 27
Pollard, E., 141
Popular perceptions of radiation. *See* Radiomedical fraud and popular perceptions of radiation
Portal viewing systems, 296
Porter, A.T.
    career and accomplishments, 272
Portmann, U.V.
    cavity ionization chamber research, 109
    survey of current state of radiation therapy instruction, 176-77
"The Possibilities of X-Ray Therapy in the Treatment of Cancer," 268
Postoperative afterloading, 196
Potentially lethal damage of radiation, 138
Powers, W.
    high pressure oxygen research, 140
    radiobiology research, 137-38
Practical X-Ray Therapy, 171-72
Pratt, H.P.
    tuberculosis treatment, 167
Predictive assays, 151-52
"The Present Status of the Diagnosis and Treatment of Carcinoma of the Uterus," 268
Princess Margaret Hospital, Toronto
    breast conservation research, 40
    fractionation, 38
Princeton particle accelerator, 148
Professional Standards Review Organizations, 216
PROG. *See* Proton Therapy Oncology Group
Prostate cancer
    brachytherapy, 190
    changes in treatment in the 1970s, 39-40
    CRTS trials, 36
Prostate-specific antigen, 40
Proton Therapy Oncology Group, 301
Proton treatment facilities, 301
Protozoa
    early experiments on, 15, 17
Prudential Insurance
    health insurance plan, 205
PSA. *See* Prostate-specific antigen

Puck, T.T.
    in vitro survival curve research, 136-37, 139, 140, 145
Puletti, J.K.
    honors, 251
Purdy, J., 297
Pusey, W.
    breast cancer treatment, 10
    episodic reports of treatment, 169
    Hodgkin's disease treatment, 10, 11
    lupus vulgaris treatment with X rays, 6
    palliative effect of X rays, 13
    skin cancer treatment with X rays, 9, 23
    textbook, 170, 171
PVSs. *See* Portal viewing systems

Quastler, H., 52
    betatron research, 105
    cell cycle research, 144
Quick, D.
    Carman Lecture to American Radium Society, 176
    Residency Review Committee member, 180
Quimby, E., 193, 253
    biographical data, 112
    brachytherapy system, 98
    Glasser and, 106
    intensity distribution tables, 194
    isoeffect curve research, 131
    supervoltage research, 103
    work with Failla, 94, 190-91, 193
Quimby System, 98, 113

Racetrack microtrons, 71-72
Radiation interactions
    Compton scattering, 93-95
    pair production, 95
    photoelectric effect and X-ray cutoff, 91-93
    relative measurement, 95-96
Radiation oncology. *See also* Future of radiation oncology; *specific topics and researchers by name*
    American Board of Radiology and, 224-25
    clinical practice period, 1970-1995, 36-40
    consideration as sideline, 28
    early development of, 24-29
    federal government and HCFA relationship with, 214-16
    interrelationship with American Medical Association, 220-21
    joint ventures and, 225-27
    relationship with American College of Radiology, 221-23
    RVS and CPT relation to, 213-14
    training program expansion, 35
Radiation protection
    early efforts, 15, 46, 166, 285
    1910-1950 era, 109-10
    1950-1970 era, 116-17
    1970 to present, 121-22
    maximum permissible dose, 116-17
    potentially lethal damage of radiation, 138
    stochastic effect risks, 121-22
"Radiation Therapy in Carcinoma of the Cervix," 267
Radiation Therapy Oncology Group, 36, 181
Radiation therapy treatment planning, 76
Radioactive medical rejuvenation devices, 288
Radiobiological research. *See* Clinical and radiobiological research
Radiobiological Research Institute, 141
Radiolabeled antibodies and metabolites, 302

▶ ▶ ▶ ▶ ▶ ▶

328

INDEX

Radiological Society for North America
    ABR formation and, 220
Radiologists
    billing for services, 202-3, 206-13
    hospital payment of, 204-5, 207-8
    income for radiologic practice and medical practice, 205-6
    nonphysicians practicing radiology, 203-4
    subspecialization, 297-99
"A Radiologist's View on the Treatment of Carcinoma of the Breast," 268
Radiomedical fraud and popular perceptions of radiation
    advertising strategies for medical quackery and, 291-92
    after World War I, 283
    Byers affair: a landmark tragedy in radio-entrepreneurship, 286-90
    discovery of radium, 279-80
    Edison's home X-ray kit, 278
    emergence of professional science commentators, 280-81
    mild radium therapy movement, 283-85
    occult power of radium, 281-82
    over-the-counter availability of radium, 284
    photography and, 278-79
    popular descriptions of Röntgen's experiments, 277-78
    popular magazines and, 279-80
    post-World War II disenchantment, 290
    proliferation of radium-containing nostrums and home-use products, 285
    radiation as life force, 281-82, 290-91
    World War I and, 282-83
Radionuclides, 195, 197, 219. *See also specific radionuclides by name*
Radiosensitivity of cells research, 137-38
"Radiosensitivity of Free Sarcoma: 37 Cells Grown in Peritoneal Exudate of the Mouse *In Vivo*," 269
Radiotherapeutic Research Unit, 60
Radiotherapy and Oncology, 297
Radithor, 287, 288-89
Radium
    early history of, 187-92
    1920-1940: golden era of, 192-94
    Byers affair and, 286-90
    concerns about exposure to, 195
    donations of, 188
    early discoveries, 185-87, 277, 279
    expense of, 191-92, 282
    exposure rate relation to source strength, 96-97
    half-life effect, 280
    industrial workers and, 285, 286, 289
    occult power of, 281-82
    over-the-counter availability, 284
    problems with, 194-95
    radium-containing nostrums and home-use products, 285
    regulation of, 217-19, 285-86
    source standardization, 96
Radium institutes, 186, 187
Radium needles, 98, 131, 190
Radium Therapy in Cancer, 94
Radiumhemmet, 187
Radon plants, 97-98, 194
Radon seeds, 99, 111, 113, 193, 240
Randall, J.
    magnetron development, 53
Rasey, J.
    Gray Laboratory and, 153

Rayle, A.
    billing for radiological services, 206
    socialized medicine comments, 215
RBE. *See* Relative biological effectiveness
Read, J.
    cyclotron research, 142
    oxygen effect research, 133
    split dose experiments, 137
Reagan, Pres. Ronald
    health policy, 225-26
Real-time accelerator control systems, 64
Real-time imaging
    development of, 70-71
Regaud, C.
    ARS honorary member, 192
    cell response to radiation, 17
    characterization of task of radiation oncologists, 21, 22
    fractionation proponent, 130-31
    Pasteur Pavilion director, 187
Reisner, A.
    isoeffect curve research, 131
Reistad, D.
    microtron research, 71
Relative biological effectiveness, 141, 142
Relative measurements for X rays, 95-96
Relative value scale
    congressional mandate for, 212
    CRVS, 208-10
    FTC ban on publication of, 211
    German system, 207
    Medicare and, 208, 211
    national RVS for selected radiation oncology procedures (1995), 212
    relation to radiation oncology, 213-14
    resource-based, 213
    summary of procedures and values, 211
    usual and customary payment system, 208, 210-11
Remote afterloading, 196-97
Reoxygenation, 140-41
Repair genes, 155
Residency Review Committees, 180
Resonant transformer generators, 50
Retinoblastoma
    suppressor genes and, 156-57
Rhumbatrons, 53, 59
Rice, H.A.
    honors, 265
Rice, M.C.
    Illinois School of Electro-Therapeutics faculty member, 235
Rice, R.K.
    beam parameter measurement, 75
Richards, G.E., 26
Richardson, O. W.
    emission of electrons and temperature relation, 100
Ritz, V.H.
    exposure rate measurement, 97
Robarts, H.
    founding of *The American X-Ray Journal,* 170
    teaching of radiology in his office, 167
Roberts, J.E.
    exposure rate measurement, 96-97
Robinson, E.
    hyperthermia research, 154
Robotic therapy, 72-73
Roentgen Rays and Electro-Therapeutics, 22, 170, 171

INDEX

Roentgen Rays in Medicine and Surgery, 22, 170, 187
Roentgen Rays in Therapeutics and Diagnosis, 23, 170, 171
Roizin-Towle, L.
    misonidazole research, 151
Rollins, W.
    internal shielding, 46
    textbook, 170
    X-ray burn report, 13
Röntgen, W.C.
    beam hardening by filtration, 47
    high-voltage source, 45
    popular descriptions of experiments, 277-78
    X ray discovery, 1, 45, 165, 186
Rose, C.B.
    family issues, 254
    internship of, 241-42
Rosenberg, S.
    effectiveness of high dose extensive radiation, 33, 39
Rosenthal, J.
    hot cathode tube research, 101
Rosh, R.
    career of, 243, 244
Rossi, H.H.
    linear-quadratic model, 145
Royal Institute of Technology, Stockholm
    racetrack microtron research, 71
RRCs. *See* Residency Review Committees
RSNA. *See* Radiological Society for North America
RTOG. *See* Radiation Therapy Oncology Group
RTP. *See* Radiation therapy treatment planning
Ruhmkorff coil, 45
Rush, B.
    comments on Derham, 264
Rush Medical School
    African American radiologist training and, 267-68
Rutherford, E.
    calorimetry research, 115
    interest in radioactivity, 280, 281
    neutron research, 133
RVS. *See* Relative value scale

Sagittaire model of linear accelerator, 31, 62-63
San Francisco Hospital
    supervoltage system, 50
Sante, L.
    ABR examining board representative, 225
"Saturation" method of radiotherapy, 22-23
Sauerbruch, F., 196
SCAROP. *See* Society of Chairmen of Academic Radiation Oncology Programs
Schaudinn, F.
    protozoa experiments, 15, 17
Schiff, E.
    lupus vulgaris treatment with X rays, 6
Schinz, H., 29
Schmincke, 26-27
Schmitz, H.
    ABR examining board representative, 225
    American Radium Society founding, 192
Schulz, M.D., 90
Schulz, R.
    AAPM chair, 117
    beam parameter measurement, 75
Schwartz, G.,17
Schwarz, W.
    oxygen effect research, 132

Schwinger, J.
    racetrack microtron research, 71
The Science of Radiology, 106
Scott, L.D.
    career and accomplishments, 269
Scott, O.
    Gray Laboratory director, 152
Seitz, V.B.
    cavity ionization chamber research, 109
Self-rectifying Coolidge radiator tube, 46
Self-referral issue, 226-27
Senn, N.
    Hodgkin's disease treatment, 10-11
Sequeira, A.
    skin cancer treatment with X rays, 9
Sherman, F.G.
    cell cycle research, 144
Siemens
    betatron development, 52, 58
    three phase generators, 48
Silen, W.
    breast conservation research, 40
Simon (runaway slave)
    medical practice of, 264
Simpson, J.R.
    career and accomplishments, 272
Sinclair, W.
    radiation effects research, 109-10
    radiosensitivity of cells research, 138
6 MeV linear accelerators, 62
Sjögren, T.
    skin cancer treatment with X rays, 9
Skaggs, L.
    betatron research, 105
    electron beam therapy research, 115
Skin as dose-limiting normal, 25, 31, 55-56
Skin cancer
    early treatments, 9, 186
    radiation and, 109
    radium treatment, 188
Skin lesions. *See also specific types of lesions*
    early X-ray treatment, 6-8, 21-23
Skinner, C.E.
    ovarian cancer treatment, 10
Skinner, G.C.
    X-ray burn report, 13
Sklar, M.D.
    oncogene research, 156
SL-75 linear accelerator, 62
Slepian, J.
    betatron research, 51
Sloan, D.
    supervoltage research, 103
Smith, A.R.
    editorial, 120
Smith, F.
    Memorial Hospital staff member, 194
Smith, I.
    cobalt-60 research, 114
Smith, J.
    Hodgkin's disease treatment, 11
Smithsonian Institution
    linear accelerator exhibit, 62
Snook, H. C.
    autotransformer with push-button control, 46
    interrupterless coils, 45
Snow, M.A.
    electrotherapy and, 239
    family issues, 254

INDEX

Snow, W.B.
    electrotherapy and, 239
    Hodgkin's disease treatment, 11
    textbook, 172
Social Security Administration
    Medicare and, 208
Social Security Amendments of 1965
    Medicare and, 216
Society of Chairmen of Academic Radiation Oncology Programs, 206
Soddy, F.
    radioactivity research, 281
Sofia, J.W.
    multileaf collimator research, 67
Soiland, A.
    ABR examining board representative, 225
    ARS annual dues proposal, 219
    efforts to have Section on Radiology authorized by AMA, 175
    founding of American College of Radiology, 175
    as one of the original radiation oncologists, 24
    publication of ARS proceedings, 219-20
    skin cancer treatment, 188
    supervoltage radiation research, 29, 50, 103
    surgical adjuvant radiation therapy for breast cancer, 24
Something About X Rays for Everybody, 165
Southard, M.
    honors, 251
    promotions, 250
    specialization, 251
Spear, F.G.
    in vivo survival curve research, 139
Spermicidal properties of X rays
    early experiments, 17
Spiers, F.W.
    radiation protection recommendations, 110
Squamous cell carcinoma
    isoeffect curve, 131
St. Bartholomew's Hospital, London
    supervoltage system, 50
Stacy, L.
    career of, 238
    "curietherapy" for cervical cancer, 23
    discrimination issue, 256
    family issues, 254
Stanbridge, E.J.
    suppressor gene research, 156
Standard Chemical Co.
    carnotite ore mining, 97
"A Standard Nomenclature for Radiation Therapy, ACT Supplement #2, 1975," 211, 212
Standing-wave accelerating, 59
Stanford University Medical Center
    clinical training site, 177
    first linear accelerator, 61-62
    robotic therapy, 72
Steel, G.
    "cell loss factor" research, 144
    histological categories for tumors, 152
Steinbeck
    betatron research, 51
Steiner, J.
    Byers case and, 289
Stenbeck, T.
    skin cancer treatment with X rays, 9
Stenstrom, K., 32, 193
Stephens, W.B.
    career and accomplishments, 268

Stereotactic radiosurgery/radiotherapy, 69-70, 120-21
Sterility
    early experiments, 17
Stevens, R.
    ABR examining board representative, 225
Stevenson, W.C.
    interstitial radiation research, 190
Stewart, F.
    Memorial Hospital staff member, 194
Stockholm (Radiumhemmet) technique, 27
Stomach cancer
    interoperative radiation therapy, 196
Stone, R.
    neutron research, 134
    supervoltage research, 103
Stovall, M.
    isodose research, 113
Strandqvist, M.
    isoeffect curve research, 131
Stroebel, H.
    afterloading, 195-96
"Subperiosteal Osteogenic Sarcoma," 267
Subspecialization, 297-99
Suit, H.
    fractionation research, 147
    heavy ion research, 150
    high-energy X-ray beam research, 119
    high pressure oxygen research, 140, 141
    precise dose delivery, 119
    precision radiation therapy, 122
"Summary of the Results of Combined Fever and X-ray Therapy in the Treatment of Hopeless Malignancies," 269
Supervoltage radiation. *See also* Medical physics; Megavoltage radiation; *specific machines by name*
    early research, 29-36, 50-56, 102-7
    1950-1970 era
        cell, tissue, and tumor kinetics, 143-45
        mouse tumors hypoxic cells: reoxygenation and high pressure oxygen, 139-41
        neutrons, 141-42
        neutrons and charged particles: the effect of radiation quality, 141
        perceived problem of hypoxia, 143
        radiosensitivity through the cell cycle and potential lethal damage, 137-38
        split dose experiments, 137
        in vitro survival curves, 136-37
        in vivo survival curves, 138-39
        X-ray dosimetry, 63-64
        X-ray production and delivery, 57-63
    1968 distribution of high-energy machines, 57
    overdosage and underdosage effects, 31, 33
Supervoltage Roentgentherapy, 29
Suppressor genes, 156-57
Surgery Adjuvant Breast Group
    clinical research in oncology, 35
Sutherland, R.M.
    metronidazole research, 151
Svensson, H.
    electron research, 115
Sydney University, Australia
    boron neutron capture therapy, 302
Synchotrons, 58
System of Instruction in X-ray Methods, 170
Szilard, B.
    dosimetric method research, 54
    ionization chamber, 108

INDEX

▶ ▶ ▶ ▶ ▶ ▶

Takahashi, S.
    multileaf collimator research, 67
Tapley, N.
    education of, 248
    family issues, 255
    honors, 250
    Kligerman as mentor, 249
    promotions, 250
    text authorship, 250
    work with Fletcher, 249
Tarleton, G.J.
    career and accomplishments, 269
Tatcher, M.
    method for combining open and wedged fields, 66
Tax Equity and Fiscal Responsibility Act, 216
Taylor, L.
    constant potential equivalent method for obtaining the simple number, 54
    work with Quimby, 112
    X-ray quality study, 49
Teaching hospitals and medical schools. *See also* Hospitals; *specific institutions by name*
    African American radiologists, 263-73
    formal training, 173-74
    number of institutions training postgraduates in radiation therapy, 180-81
    reform of medical education and, 173
    women and, 234-35, 238-42, 265
Technological developments
    early development: 1910-1950
        timeline, 44
        X-ray dosimetry, 53-56
        X-ray production and delivery, 44-53
    supervoltage, cobalt, and technological advancement: 1950–1970
        distribution of high-energy machines, 1968, 57
        technology introductions, 57
        timeline, 56
        X-ray dosimetry, 63-64
        X-ray production and delivery, 57-63
    1970 to the present
        absolute dosimetry, 74
        beam parameter measurements, 75-76
        calibration protocols, 74-75
        computer control systems, 67-69
        distribution of high-energy therapy machines, 65
        independent collimators and dynamic wedges, 66-67, 75
        linac-based stereotactic radiosurgery/radiotherapy, 69-70
        microtrons, 71-72
        multileaf collimators, 67
        real-time imaging, 70-71
        robotic therapy, 72-73
        technology benefits, 65
        tomotherapy, 73-74
        treatment planning systems, 76
        universal wedges, 66, 75
        X-ray dosimetry, 74
        X-ray production and delivery, 66
    characteristics of the X-ray categories, 44
TEFRA. *See* Tax Equity and Fiscal Responsibility Act
Telecommunication Research Establishment
    linear accelerator research, 106
10 MeV microtron, 71

Tesla, Nikola
    cancer treatment with X rays, 9
Textbook of Radiotherapy, 34
Thames, H.
    fractionation research, 146, 147
    linear-quadratic model, 145
Therac-25 accidents, 68-69
Thermoluminescence dosimetry systems, 64, 115-16
Thimble ionization chambers, 55
Third party payers
    billing for radiological services, 206-7
    income from radiological practice, 205-6
    relative value scale and, 207-13
    RVS and CPT relation to radiation oncology, 213-14
Thomas, G.T.
    GTC frame development, 70
Thomlinson, H.
    high pressure oxygen research, 140, 141
Thompson, C.W., 269
Thomson, E.
    X-ray burn report and experiments, 13
Thornell, H.E.
    career and accomplishments, 268
Thorone Company
    patent medicines, 288
3-D Radiation Treatment Planning and Conformal Therapy, 119-20
Three-dimensional optimized treatment planning software. *See also* Computer calculations
    confirmation that target is in the beam
        gating of radiation treatment, 297
        on-line portal viewing system, 296
        patient immobilization systems, 296-97
        physician-introduced fiducial markers, 296
    simulations, 295
    steps, 294-95
    time required for, 294
Three-phase generators, 48
Thyroid cancer
    "afterloading" method, 23
Till, J.E.
    clone research, 139
Tobias, C.A.
    heavy ion research, 148
Tolmach, L.
    high pressure oxygen research, 140
    radiobiology research, 137-38
Tomotherapy, 73-74
Tonsils
    radium treatment of hypertrophic, 188
Training and education
    early years: 1896-1915, 165-73
    becoming a profession: 1915-1950, 173-77
    1951 to present: toward radiation oncology training, 177-82
    African American radiologists, 263-73
    future needs, 306-7
    radiation oncologists, 35, 37-38
    separation of radiation therapy from diagnosis, 176-77, 180
    subspecialization, 297-99
    teaching hospitals and medical schools and, 172
    textbooks, 170-72
    women and, 234-35, 238-42, 248, 265
Traveler's Insurance
    health insurance plan, 205
Traveling-wave accelerating structures, 59

INDEX

Travis, L., 143
    Gray Laboratory and, 153
"Treatment of Cancer of the Breast," 267
"Treatment of Fibroid Tumors of the Uterus with Radium," 266
"Treatment of Uterine Carcinoma," 192
Treatment planning systems, 76
Trenkle
    hot cathode tube research, 100-101
Tretter, P.
    education of, 248
    family issues, 255
    family support for, 249
Trevert, E.
    layman/professional definitions, 165-66
Treves, N.
    Memorial Hospital staff member, 194
Tribondeau, L.
    spermicidal properties of X rays, 17, 130, 133
Trump, J.
    electron beam therapy research, 114
    supervoltage research, 50-51, 104
Tsien, K.C.
    computer calculation for external beams, 116
Tuberculosis
    early X ray treatment, 4, 6, 7, 167
Tubiana, M., 145
Tufts/Massachusetts Institute of Technology
    boron neutron capture therapy, 302
Tungsten
    ductile tungsten uses, 101
    high-energy cutoff for, 92
Turesson, I.
    fractionation research, 146
Turner, C.S.
    Therac-25 accidents, 68
Twombly, G.
    Memorial Hospital staff member, 194

United Nations of Scientific Committee on the Effects of Atomic Radiation, 121
Universal wedges, 66, 75
University of California at Los Angeles
    neutron facility, 148
University of Florida
    oncology department establishment, 37
University of Michigan
    first coeducational medical school, 234
University of Pennsylvania Hospital
    formal training, 173
University of Washington in Seattle
    neutron facility, 148, 300
University of Western Ontario
    racetrack microtron research, 71
U.S. Public Health Service
    Bureau of Radiological Health, 117
U.S. Radium Company
    luminous watch dials, 285
User's Guide for Radiation Oncology, 211, 211-12, 214
Uterine cancer
    Pap smear and, 308
    radium treatment, 191-92

Vacuum tube amplifiers, 55
Vacuum-tube rectifiers, 48
"The Value of the Roentgen Ray in the Treatment of Uterine Fibroids," 239
Van de Graaff, R.
    electron beam therapy research, 114
    supervoltage research, 50-51, 104
Van de Graaff generators, 31, 51, 57-58, 104, 111
Van de Kogel, Bert
    fractionation research, 146
van den Brenk, H.A.S.
    high pressure oxygen research, 140
    hypoxia research, 143
van Putten, L.M.
    high pressure oxygen research, 140
Varian brothers
    linear accelerator design, 31, 53
Veksler, V.J.
    microtron research, 71
Verhey, L.J.
    precision radiation therapy, 122
Veronesi, U.
    breast conservation research, 40
Villard, P.
    dosimetric method research, 54
    X-ray measurement, 107, 108
Vincent Bill
    prohibition of practice of radiology by nonphysicians, 203-4
Voigt, 129
Von Kolliker, A.
    hand radiograph, 166
von Liebig, J.
    analysis of healing waters, 283
Von Poswick, G.
    family issues, 254
    internship of, 241
    setting up of her own practice, 243
von Schocky, S.
    luminous paint invention, 285
von Schubert, E.
    supervoltage therapy, 50

Waite, H. F.
    "shockproof and rayproof" system, 46
Walker, J.J.
    use of radioactive rejuvenation device, 288
Walton, H.
    hospital practice of radiology comments, 205
    neutron research, 134
Wasson, W.
    ABR examining board representative, 225
Watson, T.
    betatron research, 52
    cobalt-60 research, 114
Weart, S.
    early perceptions about radioactivity, 277, 280
Weatherwax, J. L., 193
    work with Quimby, 112
Wehnelt, A.
    hot cathode tube research, 100-101
Wendt, G.
    radiation research, 284
West, C.M.L.
    predictive assay research, 152
Westra, A.
    hyperthermia research, 154
Wheatland, M.
    career and accomplishments, 266
Wickham, L.
    radium institute founding, 186
Wideröe, R.
    betatron research, 51, 52, 134
Williams, C., 188

INDEX

Williams, F., 129, 172
    ARS honorary member, 192
    brachytherapy, 23
    chart of relative distances and intensities for X rays, 170
    European study, 173, 187-88
    fluoroscopy proponent, 22
    Hodgkin's disease research, 22
    lupus vulgaris treatment with X rays, 6
    textbook, 170
    X-ray burn report, 13
Willis, F.W.
    career and accomplishments, 266
Wilson, C.T.R.
    Compton's work and, 94
Wilson, C.W.
    collaboration with Gray, 138, 139
Wilson, I.M.
    medical school experience, 234-35
Wintz, H.
    caustic radiation proponent, 130
Wiot, J.
    joint venture policy, 226
Wiprud, T.
    billing for radiological services comments, 206-7
Withers, H.R.
    fractionation research, 146
    Gray Laboratory and, 153
    linear-quadratic model, 145
    radiosensitivity of cells research, 138, 305
    in vivo survival curve research, 139
Women in radiation oncology and radiation physics.
    See also specific women by name
    radiation experimenters: 1895-1905, 234-35
    lean years: 1905-1920, 235-40
    opportunities open: 1920-1938, 240-44
    search for autonomy: 1938-1950, 244-48
    recognition at last: 1950-1960, 248-53
    African American women, 265
    appeal of radiation therapy to women, 256-58
    biological arguments against, 231
    "defeminization" and, 231
    defining issues
        discrimination, 234, 248, 255-56
        family decisions, 253-55
    male physician opposition to, 234
    postgraduate training, 240-41, 245
    qualities of successful women, 258-59
    therapy certificates, men and women, 257
    therapy vs. medicine, men and women, 257
Workman, P.
    etanidazole research, 150
Wright, E.A.
    high pressure oxygen research, 140

"X-Ray and Electrotherapeutic Laboratory," 167
X-Ray and High-Frequency in Medicine, 171
X-ray burns
    early experiments and, 13-15, 169, 196, 285
X-ray cutoff, 91-93
X-ray measurement standardization, 107-8
X-ray meters, 53
"X-ray Treatment of Hyperthyroidism," 266

Yalow, R.
    betatron research, 105
Ytterbium-169, 197
Yttrium-90, 195

DATE DUE

7-30-02			

DEMCO 38-297